Mathematical Modelling of Chemical Processes

Advances
in
Science
and
Technology
in
the USSR

Chemistry Series

ADVANCES IN SCIENCE AND TECHNOLOGY IN THE USSR

Mathematical Modelling of Chemical Processes

LEO M. RABINOVICH

Translated from the Russian by B. Rassadin

Mir Publishers
Moscow

CRC Press
Boca Raton   Ann Arbor   London

Library of Congress Cataloging-in-Publication Data

Rabinovich, Leo M.
  Mathematical modelling of chemical processes / Leo M. Rabinovich; translated from the Russian by B. Rassadin.
       p. cm. — (Advances in science and technology in the USSR)
    Includes bibliographical references and index.
    ISBN 0-8493-7132-5
    1. Chemical processes—Mathematical models. I. Title. II. Series.
TP155.7.R33   1991
660′.281′015118—dc20   91-41101
                                                                                        CIP

This book represents information obtained from authentic and highly regarded sources. Reprinted material is quoted with permission, and sources are indicated. A wide variety of references are listed. Every reasonable effort has been made to give reliable data and information, but the author and the publisher cannot assume responsibility for the validity of all materials or for the consequences of their use.

All rights reserved. This book, or any parts thereof, may not be reproduced in any form without written consent from the publisher.

Direct all inquiries to CRC Press, Inc., 2000 Corporate Blvd., N.W., Boca Raton, Florida, 33431.

© 1992 by CRC Press, Inc.

International Standard Book Number 0-8493-7132-5

Library of Congress Card Number 91-41101

Printed in the United States   1 2 3 4 5 6 7 8 9 0

CONTENTS

Preface		9
1	**Design and Analysis of Kinetic Models for Heterogeneous Catalytic Reactions**	13
	G.M. Ostrovsky, A.G. Zyskin, and Yu.S. Snagovsky	
1.1	Introduction	13
1.2	Function and Structure of SAKC	15
1.3	Programs for Computing the Derivatives of a System of Explicit or Implicit Functions	20
1.4	Program Complex for Deriving Rate Equations and Steady-State Equation Systems	24
1.5	Nonlocal Method for Solution of Systems of Nonlinear Equations	31
1.6	Dynamics of Heterogeneous Catalytic Reactions in Closed Nongradient Reactors	37
1.7	Behaviour of Closed Chemical Systems Near Equilibrium	43
1.8	Conditions for Stability of Chemical Equilibria in Closed Systems of Various Type	48
1.9	An Example of Nonuniqueness of Equilibrium Computation in Heterogeneous Catalytic Reactions	52
1.10	Conclusion	56
	References	57
2	**Modelling of Polymerization Processes**	60
	E.B. Brun and V.A. Kaminsky	
2.1	Introduction	60
2.2	Free-Radical Polymerization	61
2.2.1	Physical Aspects of Chain Termination Reaction Model	63
2.2.2	Polymerization Kinetics and Polymer Molecular Weight Distribution	68
2.2.3	Free-Radical Polymerization in Heterogeneous Systems	80
2.2.3.1	Heterophase Polymerization of Vinyl Chloride in Batch and Continuous Reactors	81
2.2.3.2	Emulsion Polymerization	91
2.2.4	Nonisothermal Processes of Free-Radical Polymerization	101
2.3	Modelling of Copolymerization Processes	111
2.3.1	Statistical Characteristics of Linear Copolymers	112
2.3.2	Molecular Structure and Properties of Copolymers	121

2.3.3	Free-Radical Copolymerization	129
2.3.3.1	Analysis of Compositional Heterogeneity of Multicomponent High-Conversion Polymerizates	132
2.3.3.2	Universal Classification of Terpolymerization Processes	139
2.3.3.3	Experimental Verification of the Theory and Its Use in Predicting the Copolymer Properties	146
2.4	Conclusion	152
	References	152
3	**Modelling and Macrokinetics of Electrochemical Processes**	**157**
	A.B. Goldberg and *L.I. Kheifets*	
3.1	Introduction	157
3.2	Mass Transport and Charge Transfer in Ionic Solutions	159
3.2.1	Phenomenological Transport Equations	159
3.2.2	Determination of Transport Characteristics of Solutions	163
3.2.3	Potential-Theory Problems	167
3.2.4	Convective-Diffusion Problems	170
3.2.5	Electric-Field Effects in Mass Transport	174
3.2.6	Concentrational Overvoltage	177
3.2.7	Interdependence of Various Factors in a Cell Model	178
3.3	Magneto- and Electrohydrodynamic Approximations in the Theory of Transport Phenomena in Electrochemical Systems	181
3.4	Dynamics and Stability of Electrochemical Reactors	186
3.5	Origination of Dissipative Structures	190
3.6	Percolation Models as Applied to Electrochemical Systems	195
3.7	Nonequilibrium Membrane Thermodynamics	197
3.7.1	Discontinuous Systems	197
3.7.2	"Dusty Gas" Model	199
3.7.3	System Analysis Under Strongly Nonequilibrium Conditions	202
3.8	Macrokinetics of Chlorine Solid-Cathode Cells	209
3.8.1	Thermodynamic and Transport Properties of NaCl-NaOH-H_2O System	209
3.8.2	Macrokinetics of Chlorine Dissolution in Brine	216
3.8.3	Current Efficiency Theory for Diaphragm-Type Cell	221
3.8.4	Electrolytic Cell as Electro- and Hydrodynamic System	228
3.8.4.1	General Method for Calculating the Volt-Ampere Characteristic and Current Density Distribution Along the Cell Height	229
3.8.4.2	Gas Content in Circulation Electrolyte Flow	235
3.8.4.3	Similarity Criteria for Electric Fields	238
3.8.4.4	Ion Transport as Effected on Perforated Electrodes	240

3.8.5	Features of an Electrochemical Process Using Solid Polymer Electrolyte	248
3.9	Diaphragm-Type Chlorine Cells Studied by Computer Experiment Method	263
3.9.1	Requirements for Design Parameters	264
3.9.2	Performance Characteristics as a Function of Current Load	268
3.10	Certain Technologic Aspects in Modelling the Performance of Membrane-Type Recycle Cells	272
3.11	Conclusion	282
	References	283
4	**Problems in Modelling and Intensification of Mass Transfer with Interfacial Instability and Self-Organization**	**289**
	L.M. Rabinovich	
4.1	Introduction	289
4.2	Interfacial Instability Phenomena and Their Physical Nature	293
4.3	Theoretical Studies of Interfacial Instability and Self-Organization	311
4.3.1	Basic Equations in Mathematical Models	312
4.3.2	Conditions and Criteria for Interfacial Instability	318
4.3.3	Dynamics of Interfacial Structures and Computation of Convection Pattern Parameters	333
4.3.4	Computer Simulation of Interfacial Convection	347
4.4	Mass Transfer Models Involving Spontaneous Interfacial Convection	351
4.4.1	Empirical Correlations	353
4.4.2	Semiempirical Models	359
4.4.3	Analytical Relations	370
4.4.4	Computer Mass Transfer Models	386
4.5	Applied Aspects of Interfacial Self-Organization	389
4.6	Conclusion	405
	References	410
5	**Hierarchical System Studies of Energy-Saving Schemes for Mixture Separation**	**418**
	V.M. Platonov, A.G. Kolokol'nikov, L.V. Baburina, and G.A. Meskhi	
5.1	Introduction	418
5.2	Analysis of Exergy Losses in Distillation	418

5.3	Molecular Level	420
5.3.1	Binary Systems	421
5.4	Supramolecular Level	437
5.4.1	Three- and Four-Component Systems	437
5.4.2	Five-Component and n-Component Systems	447
5.4.3	Universality of the Azeotropy Rule	455
5.4.4	Regime of Infinite Separation Efficiency	459
5.4.5	Minimum Reflux Theory	464
	List of Symbols	478
	References	479
6	**Optimum Design of Chemical Plants**	**480**
	G.M. Ostrovsky, Yu.M. Volin, and *T.A. Berezhinsky*	
6.1	Introduction	480
6.2	Traditional Problem of PFS Optimization	480
6.3	Optimization of Chemical Processes with Uncertain Parameters	491
6.3.1	One-Stage Problem	492
6.3.2	Two-Stage Problem	495
6.4	Discrete Optimization	504
6.4.1	Design Optimization Problem	511
6.4.2	Synthesis of Optimum Process Flow-Sheet Structure	516
6.5	Multiobjective Optimization	548
	References	552

PREFACE

This collected volume is concerned with problems arising in mathematical modelling of chemical processes, in particular, catalytic and electrochemical processes as well as polymerization, separation, heat and mass transfer.

Contributors to this book are competent Soviet specialists professionally engaged in this field which, during the last decades, has taken a firm stand in the foreground of scientific and technological research. The issues considered reflect the conceptual ideas of mathematical modelling at different hierarchical levels, starting from studies of chemical kinetic and macrokinetic relationships, including elementary steps of hydrodynamics and transport processes, via model construction of separate reactors or apparatuses, and ultimately ending in modelling and optimization of complex chemical engineering schemes.

Despite seemingly different nature of the objects of study, a common feature of the adopted approaches has been the use of experimental physicochemical data as the basis for constructing a model, as well as the application of advanced analytical and numerical methods to process modelling.

An exceptionally important aspect of modelling, along with the study of major features of a chemical process, is the upgrading of operative commercial technologies, including the development of process intensification methods and the design of innovative high-efficient technologies, especially those aimed at energy- and material saving.

The opening paper by G.M. Ostrovsky, A.G. Zyskin, and Yu.S. Snagovsky considers a number of problems concerned with the software and mathematical support of comprehensive studies on kinetics of complex heterogeneous catalytic reactions. Dedicated program complexes are described that have been developed to make computer-aided kinetic computations more effective. To this effect, the System for Automatization of Kinetic Computations (SAKC), including a kinetic rate computation program for multi-step mechanisms, has been developed. A rigorous analysis of the dynamics of complex chemical reactions with nonideal kinetics for closed systems has been carried out; also, conditions for uniqueness and stability of a chemical equilibrium in systems of different type have been examined.

The paper by E.B. Brun and V.A. Kaminsky is concerned with problems specific of polymerization modelling. Major issues dealt with in this paper are the physical fundamentals of the theory of free-radical homo- and copolymerization and the current aspects of the theory of diffusion-controlled reactions in polymer solutions as applied to free-radical polymerization. A number of new problems related to mathematical modelling of polymerization in heterogeneous systems, in particular, emulsion and suspension systems, have been considered. Specific features of free-radical polymerization regimes have been analyzed from the standpoint of the theory of combustion and thermal explosion. Within the framework of an ultimate model for multicomponent copolymerization, a quantitative theory has been developed enabling prognostication of the properties of copolymers produced at high monomer conversions.

The paper of A.B. Goldberg and L.I. Kheifets deals with major steps in mathematical modelling of electrolytic cells within the framework of an hierarchical approach. A special emphasis has been put on the charge transfer kinetics and on the problems arising in situations that have no analogues in nonelectrolytic systems. Referring to such situations, one should mention the electric potential distribution in solutions of varied composition; the relation between hydrodynamic factors and current density distribution, magneto- and hydrodynamic effects. The data thus obtained have been used for calculation of the volt-ampere reactor characteristics, for optimal design of electrolyzer bus arrangement, also for estimating the current efficiency for formation of target products, and for handling other problems arising in electrolysis technology. A special merit of this paper is an unconventional approach to constructing the mathematical model of a chlorine solid-cathode electrolyzer.

The paper by L.M. Rabinovich focuses on the theory of self-organization phenomena as applied to the surface tension instability at the liquid-liquid or gas-liquid interface in heterophase systems with mass transfer and chemical reactions. This paper summarizes original results on hydrodynamics of interfaces and mathematical modelling of mass transfer attended by Marangoni convection. Potential utility of the developed methods for intensification of solvent extraction and gas absorption using novel technologies based on interfacial instability is also discussed.

The main issue of the paper by V. M. Platonov and coauthors is the problem of synthesis of energy-saving systems for multicomponent distillation and its solution within a hierarchical-system approach. A structural-analysis method, based on the concept of conjugate singular points of the initial vapour-liquid simplex has been developed. Of special interest is the minimum-reflux theory that has been worked out for a separate column section in a manner enabling its extension to a distillation scheme of arbitrary configuration. This approach allows synthesizing energy-saving schemes for distillation of pseudoideal mixtures in levelling such an essential parameter as the number of theoretical plates at separation stages.

Finally, the concluding paper by G.M. Ostrovsky, Yu.M. Volin, and T.A. Berezhinsky considers various formulations and approaches to the problem of optimization of chemical engineering systems, including issues concerned with synthesis, global optimization and multiobjective optimization under the conditions of partial uncertainty of the input data and model parameters. The authors have also considered a continuous optimization approach and situations where certain variables can take only discrete values.

The limited space of the book has precluded the presentation of other papers on the mathematical modelling of chemical processes that might with certainty find their interested readers. We entertain a hope that the materials collected in this volume will be useful for both scientific researchers and practical engineers and can give a clearer idea of the current achievements of Soviet specialists in certain areas of chemical engineering science.

This book is primarily intended for chemical engineering scientists; still, we feel that the presented material may prove to be useful also for practical chemical engineers, mathematicians, physicists and post-graduate students interested in applied problems of chemistry.

L. M. Rabinovich

1 Design and Analysis of Kinetic Models for Heterogeneous Catalytic Reactions

G. M. Ostrovsky, A. G. Zyskin, and *Yu. S. Snagovsky*

Karpov Institute of Physical Chemistry, Obukha 10, Moscow 103064, USSR

1.1 Introduction

The construction of kinetic models for heterogeneous catalytic reactions (HCR) is an essential stage in computer-aided design and optimization of reactors and processes of chemical engineering.

The kinetic model [1, 2] is defined as a family of equations that describe the rates of sequential chemical and physical steps through which the initial reactants and intermediates are converted to end products. The kinetic model is also presumed to take into account effects arising from the eventual change in the catalyst state at different reaction steps.

The kinetic model, derived from an analysis of process mechanism, experimental reaction rates, and the fundamental laws of chemistry, constitutes the basis for designing reactors and specifying critical conditions, stability regions, and other performance characteristics of the process.

Two kinds of problems emerge in the mathematical and computer-assisted treatment of kinetic models: (i) development of software facilities for kinetic computations; (ii) theoretical analysis of mathematical models chosen for a given kinetic model. These problems may arise in the analysis of HCR carried out under both steady-state and dynamic conditions. The authors wish to present in this paper certain results relevant to the aforementioned problems.

Sections 1.2-1.4 are concerned with computational algorithms and automation of programming in a computerized study of steady-state regimes of complex HCR. Commonly, the final target is to obtain rate equations, that is, expressions for reaction rates as a function of partial pressure, temperature, and kinetic parameters. Rate equations can be derived in an explicit form only for linear reaction mechanisms, that is, those in which each elementary reaction involves a single intermediate particle. For nonlinear mechanisms, in the general case, only a computational algorithm for reaction rates can be developed. Next, a program for computing the reaction rates and their derivatives with respect to kinetic parameters is to be designed and debugged. The derivatives are used in the search for kinetic parameters, in statistical analysis and kinetic experiment planning. The design and debugging of such programs performed "manually" is a laborious and time-consuming procedure. An essential point is that, if a computer is made use of, the computations should be carried out rationally, that is, the computational algorithm and designed programs must provide for a minimal expenditure of computer time.

The problem of deriving the rate equations for linear mechanisms was dealt with in [3, 4]. In [5] a program was described for deriving the rate equations for HCR linear mechanisms on the basis of a structured form of rate equations expressed in terms of reaction mechanism graph.

We have succeeded in solving completely a problem of computer-assisted programming for reaction rates and their exact derivatives with no restrictions imposed on the reaction mechanism. To accomplish this objective, a number of theoretical postulates for complex steady-state reactions have been formalized enabling one to develop the appropriate mathematical apparatus and program complexes realizable within the System for Automation of Kinetic Computations (SAKC). A version of SAKC, designed on the basis of the operating system (OS) for IBM- or EC-type computers, has been described in detail elsewhere [6-9] et al. Therefore, in Section 2, we have confined ourselves to a brief description of the function and structure of SAKC. In Sections 1.3 and 1.4, a description is given of newly-designed SAKC software facilities: a syntactic analysis (SYA) program package and programs for deriving the rate equations for linear mechanisms and steady-state equation system for nonlinear mechanisms in character form on the basis of REDUCE-2 computer language. We are unaware of reports concerned with automated programming methods for kinetic computations at a sufficiently high level of sophistication.

Section 1.5 is concerned with the development of a nonlocal method for solution of nonlinear equation systems. The program designed on the basis of this method permits exploration of the multiplicity of steady-state HCRs, which is essential for the analysis of their regimes—both stationary and dynamic.

The dynamic properties of complex reactions carried out in closed nongradient systems are discussed in Sections 1.6-1.9. The dynamics of complex chemical reactions in closed and open nongradient systems has been under intense study for over two last decades. Major results, achieved in this field, have been outlined in monographs [10-13]. The most spectacular achievements bearing upon the uniqueness and types of steady state, stability and existence of Lyapunov functions in HCR mechanisms have been signalled for closed systems. The authors of this paper were the first to explore the dynamics of complex HCR within the framework of generalized, correctly defined kinetic laws, with allowance made for nonideality of the adsorption layer [14-17]. The relevant results are given in Sec. 1.6.

Presented in Secs. 7 and 8 are the results of studies on the mode of approach of a system to equilibrium [18] and on the conditions providing for the stability of equilibrium in a closed system [19]. A specific feature of the adopted approach is that the reported results have been obtained for arbitrary chemical systems without specifying the kinetic law, but through applying strict mathematical methods to general thermodynamic relations. Finally, in Sec. 1.9, a nonideal HCR exhibiting a complex dynamic behaviour in a closed system has been exemplified.

As has already been mentioned, the dynamics of closed chemical systems is at present well understood, and major properties of such systems have been characterized and formulated. However, there remain a number of problems that await further consideration. These will be dealt with below, each in the context of appropriate sections of the book.

1.2 Function and Structure of SAKC

In what follows, we understand by the study of steady-state HCR regimes the choice of a reaction mechanism that would provide for the best agreement with the available experimental data and would enable determination of respective kinetic parameters. The adopted approach may be divided into the following stages.

Stage 1. A hypothetical mechanism is advanced for the process in question. This is understood as the devising of a set of reaction steps constitutive of the mechanism, and the specification of both types for catalyst surface and for adsorption of intermediate species (single-site adsorption or multiple-site adsorption) on the catalyst surface. The mechanism, thus accepted, enables deriving expressions for the rates of elementary reactions.

Stage 2. Expressions for reaction rates of key species (rate equations) are derived, preferably within the framework of a steady-state reaction theory [20-24]. The reaction rates and the steady-state system for reaction steps may be written as

$$\mathbf{r}(\mathbf{x}, \theta, \mathbf{z}) = B_1^T \mathbf{u}(\mathbf{x}, \theta, \mathbf{z}) \tag{1}$$

$$B_2^T \mathbf{u}(\mathbf{x}, \theta, \mathbf{z}) = 0 \tag{2}$$

where $\mathbf{r}$ is the reactant rate vector; $\mathbf{u}$ is the reaction step rate vector; $\mathbf{x}$ is the experimental condition vector; $\mathbf{z}$ is the concentration vector for a reaction intermediate expressed in fractions of occupied surface area (surface coverage); θ is the kinetic parameter vector; B_1 is the stoichiometric submatrix for the reactants (initial and end products); and B_2 is the stoichiometric submatrix for the reaction intermediates. The rates **r** are obtained—either in an explicit form, or as computed values—from Eq. (1) by substituting the expression for **z** defined by Eq. (2).

Stage 3. At this stage, the kinetic parameters are derived and the agreement between computational and experimental rates is estimated. This procedure is carried out by minimizing a function S chosen as optimization criterion with respect to its parameter θ, for example:

$$S = \sum_{n=1}^{N} \sum_{p=1}^{N_1} W_{np} [\omega_i r_{np} - r_p(\mathbf{x}_n, \theta)]^2 \tag{3}$$

where r_{np} is the experimental pth reaction rate in the nth experiment; r_p is the computed reaction rate; W_{np} is the weight factor; ω_i is the catalyst activity correction factor (to be determined); N is the number of experiments performed; N_1 is the number of elementary reactions (respectively, reaction rates); and $\mathbf{x}_n$ is the experimental condition vector for the nth experiment.

The S function having been minimized, the subsequent procedure comprises a comparison of experimental and computational results (reaction orders, selectivity, inhibition by end products, etc.) and a statistical analysis of the derived kinetic parameters.

Also, compared among themselves are the data computed within the framework of feasible hypotheses. The results thus obtained enable one either to suggest the most probable reaction mechanism, or to outline further research program, especially if more than one of the hypotheses tested have been found to compare well with the experimental evidence.

The origination of data for computer-aided solution of a problem includes: (i) the construction of a system of steady-state equations (2); (ii) the development, if possible, of rate equations; (iii) the development, in an explicit form, of the derivatives of **r** and S with respect to θ and their programming, or the programming of computational algorithms for the derivatives.

This procedure is to be done for each of the reaction mechanisms. The development and programming of rate equations and their derivatives, if performed manually, are laborious and consume much time to obtain an acceptably representative set of mechanisms; in addition, human errors cannot be excluded. Our experience has shown that the preliminary work prior to minimization, if carried out manually for a 7-10 reaction step mechanism, requires weeks or even months of strenuous efforts.

To circumvent these inconveniences, we have designed, and continue to improve, the System for Automation of Kinetic Computations (SAKC). It is based on a mathematically correct method of deriving the steady-state system (2) with allowance made for fast reaction steps [17] (with the aim to reduce the number of fitting parameters and to avoid the appearance of excessively large rate constants for fast reaction steps) and on a method of computer-aided programming for the exact derivatives, starting from the concept of "basic" and "adjoint" computing procedures [25].

The major idea of the latter consists in representing the sequence of operations for computing the reaction rates in the form of a computational graph. The blocks of the graph are brought in correspondence to definite operations, and its edges, to the variables involved therein. In the general case, the equation for the Nth graph block of the basic process (BP) has the following form:

$$y_N = f_N(\mathbf{x}, \mathbf{a}) \quad (\dim \mathbf{x} = n, \quad \dim \mathbf{a} = l) \tag{4}$$

where $\mathbf{x}$ is the input variable vector, and $\mathbf{a}$ is the input block constant vector. It is presumed that the functions y_N are further fed as input variables to m blocks. The BP graph computation having been performed (to this effect, a structural analysis for the graph may be needed to carry out including the separation of graph cycles and determination of an optimal set of edge "cut" variables), the reaction rates r and function S are obtained.

To compute the derivatives, the graph of the adjoint process (AP) is made use of. The topologic structure of AP graph is inverted to that of the BP graph. The equation of an AP block, correspondent to the Nth BP block, is

$$\mu_i = \frac{\partial f_N}{\partial x_i} \sum_{j=1}^{m} \lambda_j, \quad i = 1, \ldots, n \tag{5}$$

where λ is the Nth block input variable, and μ is the output variable. As distinct from Eq. (4), the AP block equations are linear with respect to the input variables, which facilitates the AP computation.

The AP graph computation allows obtaining the exact derivatives of reaction rates with respect to kinetic parameters. The sequential computation of BP and AP blocks having been performed, one can develop a computing program for the reaction rates and their exact first derivatives.

A flow chart for SAKC is shown in Fig. 1.1. The SAKC includes five interactive program complexes, or PPRED complex, for deriving the steady-state equation system and rate equations in character form. The output information is a stoichiometric reaction mechanism matrix and a fast reaction step indica-

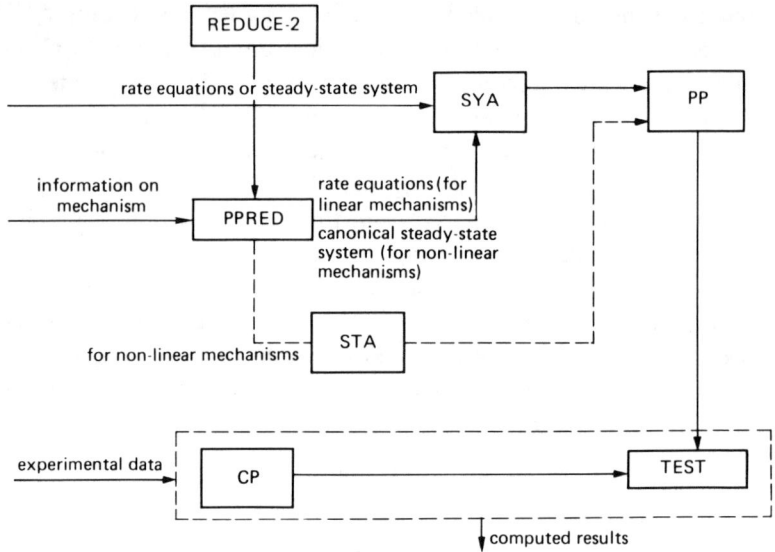

Fig. 1.1 Flow chart of the System for Automation of Kinetic Computations (SAKC)

tor. The PPRED complex yields either the so-called canonical system of steady-state equations (with allowance made for the fast reaction steps in non-linear mechanisms), or analytical expressions for reaction rates. At need, expressions for surface coverage can be obtained.

The syntactic analysis program complex (SYA) performs a syntactic analysis for a given system of explicit or implicit functions to effect the division of sequential computing procedure into unary and binary operations. The result of SYA performance is a series of tables representative of the computational BP graph. The PPRED complex output information, viz., rate equations and steady-state system, can be used as an input information for SYA.

In practice, empirical rate equations are occasionally used in kinetic computations. These (or a steady-state system), in some instances, can be derived manually. Be such the case, the derived expressions in character form can be fed directly into SYA, in bypass of PPRED.

The programming program complex (PP) receives, from the SYA complex, a program for computing the given system of functions and their exact first derivatives in PL/1 language (TEST program). An essential point is that the derived programs are optimal in the sense that if an expression is encountered more than once and/or in a number of formulas, it is, nonetheless, computed only once (as distinct from the known indiscriminate systems of analytical differentiation).

The derived programs are used by the computational program complex (CP). The CP performs minimization of the S function defined by Eq. (3) in both the isothermal case (search for elementary reaction rate constant) and the nonisothermal (search for logarithmic preexponential factor, activation energy, heat of adsorption, and other kinetic parameters). Any zero-order or first-order minimization method can be used. Also, the use of any relevant à priori information about the parameters (constraint conditions, linear relations, etc.) is admissible. The CP complex performs a statistical estimation of the derived parameters.

If the rate equations are intractable, plotting the mechanism's graph is possible with no preliminary development of the steady-state system. In such a situation, the SYA complex is functionally replaced by the program complex for structural analysis (STA), its major tasks being the construction of a closed graph for the steady-state system, graph cycle separation, determination of an optimal set of graph "cut" variables, development of a computing sequence for the open graph blocks).

The SYA and CP complexes can also be applied separately to any problems that require programs for computing a system of functions and their derivatives. Such problems, for example, can be concerned with the solution of a system of nonlinear equations, computation of reaction order, sensitivity analysis, solution of stiff systems of differential equations, stability analysis, and so forth.

The SAKC programs have been written in PL/1 for IBM- or EC-type computers in OC and VM operating systems. The PPRED complex has been realized only in VM and functions when assisted by the REDUCE-2 language of analytical computations and the text editor file facilities XEDIT. All procedures have been written in EXEC-2 procedural language.

The use of SAKC has shown a high efficiency of this system. The time needed for reaction mechanism analysis has been reduced several tenfolds. The SAKC system has been effectively applied to a variety of catalytic chemical engineering reactions in search for plausible mechanism.

1.3 Programs for Computing the Derivatives of a System of Explicit or Implicit Functions

Suppose, a system of functions

$$Y_1 = Y_1(\mathbf{x}, \mathbf{z}, \mathbf{b}),\ Y_2 = Y_2(Y_1, \mathbf{x}, \mathbf{z}, \mathbf{b}),\ \ldots,\ Y_n = Y_n(Y_1, \ldots, Y_{n-1}, \mathbf{x}, \mathbf{z}, \mathbf{b}) \quad (6)$$

$$R_1 = R_1(\mathbf{Y}, \mathbf{x}, \mathbf{z}, \mathbf{b}),\ \ldots,\ R_m = R_m(R_1, \ldots, R_{m-1}, \mathbf{Y}, \mathbf{x}, \mathbf{z}, \mathbf{b}) \quad (7)$$

and a system of equations

$$F_1(\mathbf{Y}, \mathbf{x}, \mathbf{z}, \mathbf{b}) = 0,\ \ldots,\ F_l(\mathbf{Y}, \mathbf{x}, \mathbf{z}, \mathbf{b}) = 0\ (\dim \mathbf{z} = l), \quad (8)$$

are given in an analytical form, where $\mathbf{Y} = (Y_1, \ldots, Y_n)$, $\mathbf{R} = (R_1, \ldots, R_m)$, $\mathbf{F} = (F_1, \ldots, F_l)$, and $\mathbf{b}$ is the constant vector.

Equations (8) define $\mathbf{z}$ as an implicit function of $\mathbf{x}$, $\mathbf{b}$. It is implied that Eqs. (8) are solvable for $\mathbf{z}$ which, once found, is then substituted into $\mathbf{Y}$ and $\mathbf{R}$. A problem is thus posed: to be developed are computer-assisted programs for computing the values of $\mathbf{Y}$, $\mathbf{R}$ and exact derivatives $\partial \mathbf{R}/\partial \mathbf{x}$ starting from the analytical function systems (6)-(8).

The Y_i's are intermediate quantities that must be computed and, at need, printed out. A special case is the absence of equation system (8) implying that R functions are explicit. In order to solve this problem a syntactic analysis program complex (SYA) has been developed, composed of FORMULA, FG, and POSL programs.

The FORMULA program effects the input of symbolic expressions for (6)-(8) and their conversion to a special digital code. The FG program is employed to derive digital codes for binary operations and algebraic functions corresponding to the sequential computations of (6)-(8). The source information for FG program is an MP data array, with its rows made up of the digital codes for respective expressions. The FG function may be exemplified as shown below. Let the expressions

$$Y_1 = (x_3 + x_1) \uparrow b_1 \quad (9)$$

$$R_1 = \exp(x_2 + b_2 * (x_3 + x_1)), \quad (10)$$

be computed, the arrow $\uparrow$ symbolizing the raising to power.

Three input variables x_1, x_2, and x_3, constants b_1 and b_2, and two output variables Y_1 and R_1 (denoted x_4 and x_5, respectively) are thus involved in Eqs.

(9) and (10). The digital notations for these variables are placed in the MP(1) and MP(2) elements: $x_4 = Y_1 \to$ MP(1), $x_5 = R_1 \to$ MP(2).

In the notation system we have adopted, the code MP(1) = (15,0) (1,1) (3,3) (6,1) (3,1) (2,1) (6,5) (4,1) (−1) is brought in correspondence to expression (6) (the commas and parentheses are used conventionally for separating the digital codes of operations and variables, and are virtually exempt from the MP data arrae elements); 15 is the number of characters in the line; −1 denotes the end of the line. Now, to exemplify certain currently adopted codes: (1, k) is the opening parenthesis of order k; (2, k) is the closing parenthesis of order k; (5,1) is exp; (6,1) is a plus sign (+); (6,3) is a multiplication sign (∗); (6,5) is ↑; (3,i) is x_i; (4, j) is b_j; (6,6) is an identity sign ≡.

Each expression and its respective code may be subdivided into elements and operations. The elements are: (1) parenthetic expressions; (2) b_j terms; (3) x_i terms; and (4) functions. The operations are further divided into groups, with their hierarchy increasing in the order:

$$\begin{pmatrix} \text{addition} \\ \text{subtraction} \end{pmatrix} \to \begin{pmatrix} \text{multiplication} \\ \text{division} \end{pmatrix} \to \begin{pmatrix} \text{algebraic} \\ \text{function} \end{pmatrix}$$

In analyzing the code of an expression, the program singles out its constituent elements, identifies a minor operation, and performs the division of the code expression by this operation into the codes of two operands. If the elements of a code are related to each other through more than one minor operations of the same level (for example, several additions and/or subtractions), then the division into operands is carried out by the code of the last right-side minor operation of an expression.

The code of each expression (a row in the MP array) is analyzed only once. In doing so, two operands and the operation between them are identified in the code. Code elements and combinations thereof can serve as operands. If an operand is structurally more complex than x_i or b_i, its code is compared to the codes of all preceding variables (these being already constituent elements of the MP array). If the operand code is not correspondent to any of the encoded variables, the operand is identified as a new variable x_k, and its code is assigned to the next row in the MP array. If the operand code corresponds to the code of a variable x_k, its operand is identified with the x_k variable, and no new entry is made into the MP array. Owing to this conditions, the same expressions encountered in different parts of formulas (6)-(8) are liable to a single computation only.

The two operands having been identified with their respective variables, the codes of these variables, with the operation code interposed between them, are entered into the next row of the MK array (the MK array has six columns, and its kth row is identified with x_k). In computing a function, the code of only one variable is entered into the MK array.

By way of example, let us consider the sequential build-up of the MP and MK arrays in computing Eqs. (9) and (10); by convention, the newly-derived array rows will be written in character form. Initially, three rows $x_i \equiv x_i$ ($i = 1$, 2, 3) are selected in the MK array for entering input elements. Then two new variables x_6 and x_7 are singled out and identified with the output variables x_4 and x_5.

$$\text{MK}(4): x_4 = x_6$$
$$\text{MK}(5): x_5 = x_7$$

The syntactic analysis of (9) reveals a new variable,

$$x_8 \to \text{MP}(3): x_3 + x_1 \tag{11}$$

which is entered into the sixth row of MK array,

$$x_6 = Y_1 \to \text{MK}(6): x_8 \uparrow b_1 \tag{12}$$

The analysis of (10) yields

$$x_9 \to \text{MP}(4): x_2 + b_2*(x_3 + x_1) \tag{13}$$
$$x_7 \to \text{MK}(7): \exp(x_9) \tag{14}$$

The analysis of (11) reveals no new variables, and the respective row is entered into MK:

$$x_8 \to \text{MK}(8): x_3 + x_1$$

The analysis of (13) gives

$$x_{10} \to \text{MP}(5): \ b_2*(x_3 + x_1) \tag{15}$$
$$x_9 \to \text{MK}(9): \ x_2 + x_{10} \tag{16}$$

Further, in analyzing (15), the program reveals that the $(x_3 + x_1)$ operand is equivalent to x_8; therefore,

$$x_{10} \to \text{MK}(10): b_2*x_8 \tag{17}$$

The MK array thus formulated is shown in Table 1.1. It specifies the division of the original formulas (6)-(8) into unary and binary operations and enables determination of the sequence in which these operations are performed. Next,

1 Design and Analysis of Kinetic Models

Table 1.1 Results for Exemplary Equations (9) and (10) Processed by FG Program

Row number	Row content in MK array	Operation	Note
1	3 1 6 6 3 1	$x_1 \equiv x_1$	Identical blocks
2	3 2 6 6 3 2	$x_2 \equiv x_2$	for input
3	3 3 6 6 3 3	$x_3 \equiv x_3$	graph variables
4	3 4 6 6 3 6	$x_4 \equiv x_6$	Identical blocks for output
5	3 5 6 6 3 7	$x_5 \equiv x_7$	graph variables
6	3 8 6 5 4 1	$x_8 \uparrow b_1$	Computational blocks
7	0 0 5 1 3 9	$\exp(x_9)$	
8	3 3 6 1 3 1	$x_3 + x_1$	
9	3 2 6 1 3 10	$x_2 + x_{10}$	
10	4 2 6 3 3 8	$b_2 \cdot x_8$	

the POSL program is called into play to generate decision tables in the MK array that determine the computational and topological structure of the corresponding basic process (in detail, this has been described elsewhere [6]).

The decision tables, if introduced into the programming program complex PP, generate computing programs for basic and adjoint processes in PL/1 language. These programs can be edited to the resident CP computing programs for performing minimization. They can also be edited to the resident programs of a complex for autonomous computation of functions and their derivatives. In this case, they can be utilized by any program devised for computing the derivatives of a system of functions (for example, computing programs for nonlinear equation systems or stiff systems of differential equations, programs for calculating the order of reactions and for studying problems related to the stability of reaction systems, and so forth).

It is noteworthy that there are known systems of analytical computations (for example, REDUCE-2 [26]) that either perform the analytical differentiation of a system of explicit functions, or give out programs for computing derivatives. Programs that can be obtained by our method have an essential advantage: an expression which is encountered in the formulas more than once is computed only once, as distinct from the aforementioned programs that do not discriminate this feature either. In addition, the PPRED complex can receive programs for computing derivatives of a system of implicit functions.

1.4 Program Complex for Deriving Rate Equations and Steady-State Equation Systems

The algorithm of the program complex in question (PPRED) is based on a method enabling one to derive correct (that is, canonical) systems of steady-state equations for intermediate reactants with allowance made for fast irreversible and quasi-equilibrium steps involved in the reaction mechanism [17]. Briefly, the method is in concept as follows. A system of steady-state equations is written in the form

$$B_2^T \mathbf{u}(\mathbf{x}, \mathbf{z}) = 0 \tag{18}$$

where B_2 is the stoichiometric matrix for intermediates, $\mathbf{u}$ is the reaction step vector; $\mathbf{x}$ is the vector accounting for both the partial pressures of reactants and kinetic parameters; and $\mathbf{z}$ is the vector of surface coverages.

For each type of catalytic site, the balance equation should be considered:

$$\sum_i z_i = 1 \tag{19}$$

Further, it is presumed that each fast irreversible step involves only one intermediate which is consumed only at this particular step and remains unaltered at other reaction steps. The canonical system of equations is constructed by the following algorithm. From (18), linearly independent equations are selected. They are reduced, via nonsingular transformation, to a form in which the rate of a fast reaction step is entered only in one equation, and no equation includes more than one fast reaction step rate. This having been done, the equations for fast irreversible reaction steps are dismissed from further consideration. The concentrations of intermediates, consumed only in these reactions, are presumed to be equal to zero and are excluded from the balance equation (19). The equations for quasi-equilibrium (equilibrium) steps are thus translated into conditions for a quasi-equilibrium (equilibrium) state:

$$u_s = u_{+s} - u_{-s} = 0 \tag{20}$$

where u_{+s} and u_{-s} are, respectively, the rates of the sth quasi-equilibrium (equilibrium) reaction step in the forward and reverse directions. The remained equations which include linear combinations of slow reaction step rates are unchanged, but they are augmented with balance conditions for each type of catalytic site.

In [17], two statements have been proved that lend support to the algorithm:
— all canonical equation systems are equivalent;

— the fast reaction step rate constants having been defined as $k_i = k_i^0/\varepsilon$ (where ε is a small parameter), the solution of the original system tends to the solution of a canonical system as the parameter ε tends to zero, $\varepsilon \to 0$.

In our program, the following procedure for deriving the canonical system of steady-state equations has been adopted:

(a) linearly dependent rows in the B_2^T matrix are deleted;

(b) the B_2^T matrix columns are rearranged in such a manner as to make the columns for fast irreversible reactions, in number KR, lead the column numbering, followed by KB columns for quasi-equilibrium and equilibrium steps. Let us denote $R = \text{rank } B_2$. Now we add $(R-KR-KB)$ columns of matrix B_2^T to the first $(KR + KB)$ columns so that the first R columns of the new A_1 matrix formed a nonsingular matrix M. Then in the equation system

$$A_2 \mathbf{u} = M^{-1} A_1 \mathbf{u} = 0 \tag{21}$$

the rate for each fast reaction step appears in only one equation, which enables obtaining a canonical system by the above-described method. Let us denote the total number of reaction steps by KS. Since the first L columns in A_2 matrix represent a unit matrix, the first reaction step rates $\mathbf{u}_f$, in number R, can be expressed from (21) through the slow reaction step rates $\mathbf{u}_s$, in number $(KS - R)$:

$$\mathbf{u}_f = A_3 \mathbf{u}_s, \quad \mathbf{u} = \begin{pmatrix} \mathbf{u}_f \\ \mathbf{u}_s \end{pmatrix} = \begin{pmatrix} A_3 \\ I \end{pmatrix} \mathbf{u}_s \tag{22}$$

The rates for formation of reactants involved in the process are

$$\mathbf{r}_{\text{reac}} = B_1^T \mathbf{u} = B_1^T \begin{pmatrix} A_3 \\ I \end{pmatrix} \mathbf{u}_s = A_4 \mathbf{u}_s \tag{23}$$

where B_1 is the stoichiometric matrix for reactants (assuming that in the $\mathbf{u}$ vector, the reaction step rates have been grouped as described above). By substituting the surface coverages defined by the canonical system into the right-hand side of Eq. (23), the expressions for reaction rates (rate equations) are obtained.

The canonical system of rate equations is written in the general form as

$$\begin{aligned} z_i &= 0 & (i &= 1, \ldots, KR) \\ u_l(\mathbf{x}, \mathbf{z}') &= 0 & (l &= KR + 1, \ldots, KR + KB) \\ A_2' \mathbf{u}' &= 0 \\ \sum_i z_i &= 1 \end{aligned} \tag{24}$$

In (24), A_2' is a matrix made up of $(L\text{-}KR\text{-}KB)$ lower rows of matrix A_2; $\mathbf{u}'$ is the slow reaction step vector composed of the last $(KS\text{-}KR\text{-}KB)$ elements of $\mathbf{u}$; $\mathbf{z}'$ is the surface coverage vector, subtracted the vector for surface coverage consumed in fast irreversible reactions.

The PPRED complex is realized in PL/1 language within the VM operating system; as has been noted above, it makes use of the facilities of XEDIT text editor and those of REDUCE-2 computer language. A block diagram for this complex is shown in Fig. 1.2. The source information for the PPRED complex is the stoichiometric matrix and the fast reaction step indicator. The STEX program gives out a canonical system, it performs the algorithmic transpositions of the B_2^T columns, identifies the A_1, M, and A_2 matrices in (21), and computes both the A_4 matrix in (23) and the A_2' matrix in (24). In addition, this program specifies the type—linear or nonlinear—of the rection mechanism. In the former case, the PSRED program starts operating to generate the character coefficients for a linear canonical steady-state system, the linear system solution operators, and the operators for formula (23) (file F1). Then the program is processed by the XEDIT text editor and is thus ready for operation in REDUCE-2 language. As the program is run, the derived formulas for r_{reac} and, if required, for the surface coverage are directed, in character form, to the file F_2. This file is processed by the text editor to be adapted to the SYA input file to be further processed by SYA as has been shown in Sec. 1.3.

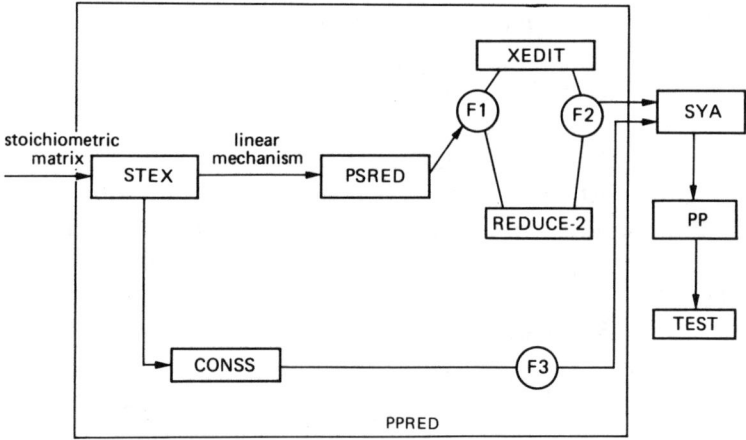

Fig. 1.2 Flow chart of PPRED program complex for deriving rate equations and systems of steady-state equations

In the case of a nonlinear canonical system, the CONSS program comes into play. This program writes, in character form, the input file F3 for SYA, following the scheme outlined for formulas (6)-(8), with the slow reaction step rates and surface coverages (those defined by balance relation (19)) included in the **Y** vector. In file F3, the partial pressures and surface coverages assigned to the second and third rows of formula (24) are generated. Since the zero surface coverages are excluded and the other coverages are, in part, accounted for by (19), this operation reduces the order of the canonical system.

The interaction of PPRED programs, the modification of the files and their transfer are accomplished by a procedure written in EXEC-2 language.

To exemplify the use of PPRED program, let us consider a mechanism for oxidative chlorination of ethylene over a $CuCl_2/\gamma\text{-}Al_2O_3$ catalyst that has been studied by us in collaboration with A. I. Gelbstein and Yu. M. Bakshi:

$$
\begin{array}{ll}
1.\ Z_2Cl_2 + 4HCl \equiv Z_2Cl_2 4HCl & 1 \\
2.\ Z_2Cl_2 4HCl + C_2H_4 \equiv Z_2Cl_2 3HClC_2H_4 + HCl & 2 \\
3.\ Z_2Cl_2 3HClC_2H_4 \rightarrow Z_2 3HCl + C_2H_4Cl_2 & 2 \\
4.\ Z_2 3HCl + HCl \equiv Z_2 4HCl & 2 \quad (25)\\
5.\ Z_2 4HCl + O_2 \rightarrow Z_2 O_2 4HCl & 1 \\
6.\ Z_2 O_2 4HCl + Z_2 4HCl \overset{f}{\rightarrow} Z_2Cl_2 + Z_2Cl_2 4HCl + 2H_2O & 1
\end{array}
$$

$$2C_2H_4 + 4HCl + O_2 = 2C_2H_4Cl_2 + 2H_2O$$

(shown in the column on the right side are the stoichiometric numbers).

We now arrange the reactants involved in the following order: reactants C_2H_4, O_2, HCl; end products $C_2H_4Cl_2$, H_2O; intermediates Z_2Cl_2, $Z_2Cl_2 4HCl$, $Z_2Cl_2 3HClC_2H_4$, $Z_2 3HCl$, $Z_2 4HCl$, $Z_2O_2 4HCl$; rate constants K_1, K_2, k_3, K_4, k_5.

The source information includes: (i) an extended ST matrix of stoichiometric coefficients for both left- and right-hand sides of reactions (25) (Table 1.2); (ii) an index array for the reaction steps, YK = 'q', 'q', 'i', 'q', 'i', 'f', where 'q' stands for quasi-equilibrium 'i' for irreversible, and 'f' for fast irreversible reaction steps. For mechanism (25), one has $KB = 1$, $KR = 3$, $KS = 6$, and $R = 5$.

The functional commitment of STEX program is:
(a) Transposition of the reaction step numbers: 6, 2, 4, 1, 5, 3.
(b) Construction of A_1 matrix via deletion of the last row from the B_2^T

Table 1.2 Extended ST Matrix of Stoichiometric Coefficients for Mechanism (25)

Reactants and intermediates	Reaction step numbering (columns numbered k and $-k$ refer, respectively, to stoichiometric coefficients in the lhs and rhs of a reaction step)											
	1	2	3	4	5	6	-1	-2	-3	-4	-5	-6
C_2H_4	0	1	0	0	0	0	0	0	0	0	0	0
O_2	0	0	0	0	1	0	0	0	0	0	0	0
HCl	4	0	0	1	0	0	0	1	0	0	0	0
$C_2H_4Cl_2$	0	0	0	0	0	0	0	0	1	0	0	0
H_2O	0	0	0	0	0	0	0	0	0	0	0	2
Z_2Cl_2	1	0	0	0	0	0	0	0	0	0	0	1
Z_2Cl_24HCl	0	1	0	0	0	0	1	0	0	0	0	1
$Z_2Cl_23HClC_2H_4$	0	0	1	0	0	0	0	1	0	0	0	0
Z_23HCl	0	0	0	1	0	0	0	0	1	0	0	0
Z_24HCl	0	0	0	0	1	1	0	0	0	1	0	0
Z_2O_24HCl	0	0	0	0	0	1	0	0	0	0	1	0

matrix:

$$A_1 = \begin{pmatrix} 1 & 0 & 0 & -1 & 0 & 0 \\ 1 & -1 & 0 & 1 & 0 & 0 \\ 0 & 1 & 0 & 0 & 0 & -1 \\ 0 & 0 & -1 & 0 & 0 & 1 \\ -1 & 0 & 0 & 0 & -1 & 0 \end{pmatrix}$$

(c) Construction of M matrix:

$$M = \begin{pmatrix} 1 & 0 & 0 & -1 & 0 \\ 1 & -1 & 0 & 1 & 0 \\ 0 & 1 & 0 & 0 & 0 \\ 0 & 0 & -1 & 0 & 0 \\ -1 & 0 & 0 & 0 & -1 \end{pmatrix}$$

(d) Construction of A_2 matrix:

$$A_2 = \begin{pmatrix} 1 & 0 & 0 & 0 & 0 & -0.5 \\ 0 & 1 & 0 & 0 & 0 & -1 \\ 0 & 0 & 1 & 0 & 0 & -1 \\ 0 & 0 & 0 & 1 & 0 & -0.5 \\ 0 & 0 & 0 & 0 & 1 & -0.5 \end{pmatrix}$$

(e) Construction of A_3 matrix:

$$A_3 = \begin{pmatrix} 0.5 \\ 1 \\ 1 \\ 0.5 \\ 0.5 \end{pmatrix}$$

and the relation (22) is expanded as: $u_6 = 0.5u_3$; $u_2 = u_3$; $u_4 = u_3$; $u_1 = 0.5u_3$; $u_5 = 0.5u_3$.

(f) Definition of the rates for reactants or products. For example, for C_2H_4, according to (23),
$B_1^T = (0 \; -1 \; 0 \; 0 \; 0 \; 0)$, and $r_{C_2H_4} = -u_2$. The matrix A_4 is $A_4 = (-1)$ and, consequently, $r_{C_2H_4} = -u_2 = -k_3z_3$.

In accordance with the adopted convention in our program, the rate for reactants consumed in the reaction is assigned a negative sign $(-)$. Therefore, $r_{C_2H_4} = k_3z_3$. The steady-state system (24), processed for obtaining the rate equations, is expanded as

$$z_6 = 0, \; u_2 = 0, \; u_4 = 0, \; u_1 = 0, \; u_5 - 0.5u_3 = 0$$

The character field of file F2, processed by PPRED complex looks like shown below:

'Y1 = $-$P3 ↑ 4 * (K4 * ((2 * P2 * K5 + K3) * P3 * P1 * K2
+ 2 * P3 ↑ 2 * P2 * K5 + 2 * P2 * K5) + P1 * K3 * K2 * K1);'
'Y2 = ($-$2 * K4 * K5 * P2 * P3 ↑ 2)/Y1;'
'Y3 = ($-$2 * K1 * K4 * K5 * P2 * P3 ↑ 6)/Y1;'
'Y4 = ($-$2 * K1 * K2 * K4 * K5 * P1 * P2 * P3 ↑ 5)/Y1;'
'Y5 = ($-$K1 * K2 * K3 * P1 * P3 ↑ 4)/Y1;'
'Y6 = ($-$K1 * K2 * K3 * K4 * P1 * P3 ↑ 5)/Y1;'
'Y7 = ∅;'
'R1 = ($-$2 * K1 * K2 * K3 * K4 * K5 * P1 * P2 * P3 ↑ 5)/Y1;'

here $Y_2, \ldots, Y_7$ are the surface coverages $Z_1, \ldots, Z_6$, and $R_1 = r_{C_2H_4}$.

We now consider, as an example, the development of a steady-state system as effected by CONSS program. Let us take a nonlinear model mechanism involving two types of reactive sites,

$$
\begin{array}{l|c|c}
1.\ Z_0 + A_1 \rightleftarrows Z_1 & 1 & 0 \\
2.\ Z_1 + A_2 \rightarrow Z_2 & 1 & 0 \\
3.\ Z_2 \stackrel{f}{\rightarrow} Z + A_3 & 1 & 0 \\
4.\ Z^{(0)} + A_4 \equiv Z^{(1)} & 0 & 2 \\
5.\ Z^{(1)} \rightarrow Z^{(2)} & 0 & 2 \\
6.\ 2Z^{(2)} \rightleftarrows 2Z^{(0)} + A_5 & 0 & 1
\end{array}
\qquad (26)
$$

$$A_1 + A_2 = A_3$$
$$2A_4 = 2A_5$$

where the Z's with a superscript denote species adsorbed on the 2nd type sites. Let the reactants be arranged in the order: A_1, A_2, A_3, A_4, A_5, and the intermediate species, in the order: Z_0, Z_1, Z_2, $Z^{(0)}$, $Z^{(1)}$, $Z^{(2)}$; the reaction step rate constants are k_1, k_{-1}, k_2, K_4, k_5, k_6, k_{-6}. The reaction rates are determined by the reactants A_3 and A_4. The character field of file F3 is obtained as:

'Y1 = 1 − Z1;'
'Y2 = 1 − Z2 − Z3;'
'Y3 = K4 * P4 * Z2 − Z3;'
'Y4 = K1 * P1 * Z1 − K2 * Y1;'
'Y5 = K5 * Z3;'
'Y6 = K3 * P2 * Y1;'
'Y7 = K6 * Y2 * Y2 − K7 * P5 * Z2 * Z2;'
'F1 = Y3;'
'F2 = −Y4 + Y6;'
'F3 = −Y5 + 2 * Y7;'
'R1 = Y6;'
'R2 = 2 * Y7;'

/* correspondence of the true number of a surface coverage to the values of Z and Y */:
/* surface coverage No. 1 is equal to Z1 */
/* surface coverage No. 2 is equal to Y1 */
/* surface coverage No. 3 is equal to 0 */
/* surface coverage No. 4 is equal to Z2 */

1 Design and Analysis of Kinetic Models   31

/* surface coverage No. 5 is equal to Z3 */
/* surface coverage No. 6 is equal to Y2 */
/* correspondence of the number of Y to the number of a step */
/* Y3 is equal to the rate of reaction step 4 */
/* Y4 is equal to the rate of reaction step 1 */
/* Y5 is equal to the rate of reaction step 5 */
/* Y6 is equal to the rate of reaction step 2 */
/* Y7 is equal to the rate of reaction step 6 */

The entry enclosed within the /* ... */ brackets is a supplementary reference information given out by the computer. In fact, written in the file F3 is a system of equations with respect to Z1, Z2, Z3:

$$u_4 = 0, \quad -u_1 + u_2 = 0, \quad -u_5 + 2u_6 = 0$$

We wish to emphasize that our approach is algorithmically more simple than those suggested in [3-5], since it makes use of the potentialities of analytical computation REDUCE-2 language. In addition, as distinct from the cited sources, our method provides for a correct account of fast reaction steps in the reaction mechanism.

The use of REDUCE-2 language permits deriving analytically the rate equations for certain mechanisms involving second-order reaction steps and, in certain cases, allows one to reduce the dimensionality of the steady-state system of interest. For example, in a mechanism involving only one nonlinearly expressed intermediate, an analytical computer-aided procedure can reduce the steady-state system to a single algebraic equation.

1.5 Nonlocal Method for Solution of Systems of Nonlinear Equations

Let us attend to the problem of finding all of the roots of a system of nonlinear equations,

$$f_1(\mathbf{x}) = 0, \ \ldots, \ f_N(\mathbf{x}) = 0 \tag{27}$$

within a region G defined by inequalities $\{c_i \leqslant x_i \leqslant d_i; i = 1, \ldots, N\}$. The functions f_i are presumed to be continuously differentiable.

Most of the known methods for solution of nonlinear equation systems are local in character and provide, in the case of a good initial approximation, for the finding of only one root of system (27). Simultaneously, the extant methods of nonlocal type are not exempt from shortcomings, and none of them necessarily ensures the finding of all the roots [63].

For example, an interesting class of nonlocal methods is represented by the methods featuring the movement along the trajectories of a system of differential equations in which the roots defined by (27) are singular points. In [27], a nonlocal version of the continuous Newton method was suggested. Chao et al. [28] proposed a modified approach in which a system of equations f_1, ..., $f_{N-1} = 0$ was initially solved for one of the system roots, and then the movement was effected along the integral curves making the above equalities remain valid in the direction of decreasing $|f_N|$ and then, the next root having been found, in the direction of increasing $|f_N|$. However, these methods exhibited inherent inconveniences arising from a complicated approach of the trajectories to a singular point of a differential equation system and also from a complicated departure from it. In addition, a multiple computation of the derivatives of functions f_i and the inversion of Jacobian matrices were required. The numerical solution of the respective systems of differential equations was performed through small integration steps, with the ensuing substantial increase in the computer time.

The nonlocal method that has been proposed in [7, 29] et al. is free from these shortcomings.

We now consider a subsystem of $(N - 1)$ equations from system (27):

$$f_1(\mathbf{x}) = 0, \ldots, f_{i-1}(\mathbf{x}) = 0, f_{i+1}(\mathbf{x}) = 0, \ldots, f_N(\mathbf{x}) = 0 \qquad (28)$$

Within the R^N space, subsystem (28) defines a curve L_i. We introduce quantities

$$M_{ij} = \det\left\{\frac{\partial f_l}{\partial x_k}\right\} \quad l \neq i,\ k \neq j$$

The proposed method consists in the following. Initially, system (28) is solved for one of its roots $\mathbf{x}_0$ by making use of any local method. Then a curve L_i, passing through the root $\mathbf{x}_0$, is selected (starting from $i = 1$) and analyzed, by moving along the curve trajectory, for the values of $\mathbf{x}$ at which the function $f_i(\mathbf{x})$ becomes zero. In such a manner, new roots for (27) are identified. In each instance, the number i of a curve in which the given root has been found is stored. The movement along the L_i curve is continued until one of the two situations occurs: (1) the curve L_i has reached the G region boundary. In this case, either the direction of movement along the L_i curve is reversed, or the next curve is picked out for analysis; (2) the curve L_i starts moving in cycles.

Here, to proceed with the movement, the next curve, passing through x_0, is selected.

The newly found roots are compared with the roots found previously, and those of the former not identified with the latter are stored. In any case, all numbers of the curves in whose scanning the given root has been identified are stored. The whole family of curves that pass through x_0 having been explored, the same procedure is applied to the next root over all the curves except those instrumental in its initial identification.

The movement along a curve L_i can be effected in a number of ways. In the work [29], the movement along L_i was suggested through solving a system of differential equations

$$\frac{d\mathbf{x}}{dt} = \frac{\mathbf{h}_i}{\|\mathbf{h}_i\|} \tag{29}$$

where $\mathbf{h}_i = (M_{i1}, -M_{i2}, \ldots, (-1)^{j-1}M_{ij}, \ldots, (-1)^{N-1}M_{iN})$ is the direction vector for the curve L_i. The integration of (29) was carried out by the fourth-order Runge-Kutta method with a sufficiently small integration step to provide for the reliable identification of the running point in L_i. Such an approach required a multiple computation of both the Jacobian matrix and the determinants, which rendered the method less efficient. Snagovsky and Ostrovsky [30] reported on the application of this method to finding the parameters of a kinetic model for deuterium exchange reaction between hydrogen and water vapour over a Ni catalyst.

In [6, 7], an alternative variant for movement along L_i, the parameter prolongation method, has been considered. The major idea of the method is as follows. In moving from the starting point x_0, the jth coordinate x_j is chosen as the movement parameter. On passage from the nth to the $(n + 1)$th step, the incremental change in x_j is

$$x_{0j}^{n+1} = x_{0j}^n + \delta_j^n \tag{30}$$

where n is the step number. Then system (28) is solved by a quasi-Newtonian method—for example, that of Broyden [31], with respect to $(x_{01}^{n+1}, \ldots, x_{0(j-1)}^{n+1}, x_{0(j+1)}^{n+1}, \ldots, x_{0N}^{n+1})$, with a simultaneous analysis of the behavior of f_i functions. The Broyden method is known to make use of the inverse Jacobian matrix approximation, with recursive computation

$$Q_{l+1}^n = F(Q_l^n) \tag{31}$$

where the subscript l is the iteration number at the nth step. To obtain Q_{l+1}^n, one must only compute $f_1, \ldots, f_{i-1}, f_{i+1}, \ldots, f_N$. As initial approximations for **x** and **H**, we use the roots $(x_{k1}^n, x_{k(j-1)}^n, x_{k(j+1)}^n, \ldots, x_{kN}^n)$ and the Q_k^n (K being the number of the last iteration at the nth step):

$$Q_0^{n+1} = Q_k^n$$
$$x_{01}^{n+1} = x_{k1}^n, \ldots, x_{0(j-1)}^{n+1} = x_{k(j-1)}^n, x_{0(j+1)}^{n+1} = x_{k(j+1)}^n,$$
$$\ldots, x_{0N}^{n+1} = x_{kN}^n, \qquad (32)$$

which makes the Broyden method converge faster.

This method requires neither the computation of Jacobian matrices, nor their inversion. The iteration steps δ_j^n may be chosen much larger than those used in numerical methods for solving the system of differential equations (29).

In moving along a curve L_i with the variable x_j parameter, major difficulties arise for small values of $|M_{ij}|$, that is, when the coordinates of a point in the curve become strongly dependent on x_j, which may result in their non-uniqueness dependence with respect to x_j. Be such the case, the following strategy of movement is adopted: If, with x_j incremented by δ_j^n, no solution for system (28) has been found with the number of attempted iterations greater by a definite factor over the average iteration number applied to the given curve, then the movement parameter is varied $x_j \to x_{j+1}$ (at $j = N$, the next j is taken equal to 1). In addition, the ratio $|\Delta x_i/\Delta x_j|$ ($i \neq j$) is analyzed at each step, and if it becomes greater over a certain predetermined value, the iteration switches to a parameter x_i with the same number i at which the overshoot has been reached.

If all of the parameters x_i ($i = 1, \ldots, N$) have been tried, and there is no way of moving along the curve (an eventual cause may be the degeneration of the curve), our program interrupts the movement along the curve and allows choosing either the reversal of movement, or the passage to the next curve. In either instance, two variants are envisaged: (i) the increment δ_j^n is chosen smaller, and an attempt is made to run the critical stretch of the curve at a smaller step; (ii) the critical stretch is run by integrating the differential equations (29).

The proposed algorithm by no means warrants the identification of all of the roots for system (27). The eventual failure can be due either to the curves (28) being disconnected, or to the occurrence of a boundary in the G region preventing the run through the whole stretch of curves (28) [8, 9]. Nonetheless, this method provides a determinate search for the roots of system (27). Our experience has shown that computations carried out with a variety of starting points enable identification of all of the roots of the chosen test systems of nonlinear equations.

The algorithm in question has been realized as a package of programs written in PL/1 language. In debugging the program, a test system of three equations with nine sets of roots from [28] has successfully been solved.

To demonstrate the performance of our program, we apply it to the computational experiment with a model borrowed from [32]. We consider a reaction

$$A \to B \to C \tag{33}$$

carried out in a stirred tank reactor. The reactant A is the only one supplied at the input into the reactor; both consecutive reactions are presumed to be of first order.

The system of nonlinear equations for nondimensional steady-state concentrations u and v, respectively, of species A and B and nondimensional temperature ω is that borrowed from [32]. The meaning of the variables employed will become clear in the sequel.

$$\begin{aligned} 1 - u\{1 + \alpha E(\omega)\} &= 0 \\ \alpha u E(\omega) - v\{1 + \alpha \sigma E^v(\omega)\} &= 0 \\ (1 + \varkappa)\omega + \alpha \beta u E(\omega) + \alpha \beta \varrho \sigma v E^v(\omega) &= 0 \end{aligned} \tag{34}$$

where $E(\omega) = \exp\{\gamma\omega/(\gamma + \omega)\}$.

The method for steady-state multiplicity analysis suggested by Aris relies on the reduction of system (34) to a single nonlinear equation. To be noted, it will not take much effort to suggest a more complex kinetic model for reaction (33) in which the system of steady-state equations is not reducible in concentration and temperature to a single equation.

We have carried out a study of the stirred tank reactor model for reaction (33) and its modifications. Initially, system (34) has been solved making use of the original parameters from [32]: $\beta = 9$; $\varkappa = \nu = \varrho = 1$; $\sigma = 0.01$; $\gamma = \infty$. For example, with $\alpha = 0.1$, three solutions have been obtained:

(1) $u_1 = 0.15445$; $v_1 = 0.80156$; $\omega_1 = 4.0015$
(2) $u_2 = 0.45395$; $v_2 = 0.53987$; $\omega_2 = 2.4879$
(3) $u_3 = 0.80779$; $v_3 = 0.19195$; $\omega_3 = 0.86787$

Since system (34) was derived presuming that both reactions were of first order and the B species was not supplied at the input to the reactor, we have further used a system with natural variables borrowed from [33].

$$q(c_{Af} - c_A) - Vr_1 = 0,$$
$$q(c_{Bf} - c_B) + V(r_1 - r_2) = 0 \tag{35}$$
$$q\varrho_f c_p(T_f - T) + (-\Delta H_1)Vr_1$$
$$+ (-\Delta H_2)Vr_2 - UA_c(T - T_c) = 0$$

where T_c and T_f are the temperatures of the cooler and the inlet flow; U is the heat transfer coefficient; V is the reactor volume; q is the feed flow rate; c_A and c_B are the concentrations for A and B species, and c_{Af} and c_{Bf} are their respective feed concentrations; ΔH_1 and ΔH_2 are the heats of reaction; A_c is the cooling surface area; r_i is the elementary reaction rate; c_p is the heat capacity; ϱ_f is the feed density.

At $c_{Bf} = 0$, system (35) is converted to (34) by the formulas shown below [33]:

$$u = c_A/c_{Af}, \quad v = c_B/c_{Af}; \quad \varkappa = UA_c/q\varrho c_p; \quad \overline{T} = (T_f + \varkappa T_c)/(1 + \varkappa);$$
$$\omega = (E_1/R\overline{T}^2)(T - \overline{T}); \quad \alpha = (VA_1/q)\exp(-E_1/R\overline{T}); \quad \gamma = E_1/R\overline{T};$$
$$\varkappa = UA_c/q\varrho C_p; \quad \beta = \{(-\Delta H_1)C_{Af}/\varrho C_p\}(E_1/R\overline{T});$$
$$\nu = E_2/E_1; \quad \varrho = \Delta H_2/\Delta H_1; \quad \sigma = (A_2/A_1)\exp\{(E_1 - E_2)/R\overline{T}\}$$

To obtain the previously reported values of α, β, γ, ν, ϱ, σ, and $\varkappa$, the following parameters have been chosen for (35):

$$T_f = T_c = 300 \text{ K}; \quad \Delta H_1 = \Delta H_2 = -3000; \quad V = 1; \quad q = 10;$$
$$q\varrho_f C_p = 1000; \quad UA_c = 1000; \quad c_{Af} = 1.125; \quad A_1 = \exp(80);$$
$$A_2 = 0.01 A_1$$

In this case, $\overline{T} = 300$. The value of γ has been taken as large as reasonably possible in order to retain the physical meaning of E_1 as activation energy. The adopted values are $\gamma = 40$ and $\gamma = 80$, which correspond to $E_1 = E_2 = 24\,000$ or $E_1 = E_2 = 48\,000$.

System (35) has been solved for c_A, c_B and $\theta = 500/T$ within the respective ranges of $0 \leqslant c_A \leqslant 4$; $0 \leqslant c_B \leqslant 4$; $1.2 \leqslant \theta \leqslant 2$ by making use of the nonlocal search program starting from a variety of initial points. In all cases, a single solution has been found. Interestingly, this situation is distinct from that reported in [32] at $\gamma = \infty$.

Next, a reaction $A \rightleftarrows B \rightarrow C$ has been considered under the following assumptions:
(a) The reaction $B \rightarrow A$ is a first-order reaction.
(b) The reaction $B \rightarrow A$ is a second-order reaction.
(c) The reactions $A \rightarrow B$ and $B \rightarrow A$ are second-order reactions.

(d) The reactions A → B and B → A are second-order reactions; both A and B are supplied at the inlet to the reactor.

In cases (a), (b), and (c), with the preexponential factor A_{-1} for the reaction B → A varied within a range of 10 A_1 to 90 A_1, a single solution for system (35) has been found to exist.

In case (d) at $A_{-1} = 90A_1$, $c_{Bf} = 1$, $E_{-1} = 51\,000$, three sets of roots have been obtained:

(1) $c_A^{(1)} = 1.0677$, $c_B^{(1)} = 1.0560$, $T^{(1)} = 300.88$
(2) $c_A^{(2)} = 0.27241$, $c_B^{(2)} = 0.26397$, $T^{(2)} = 336.61$
(3) $c_A^{(3)} = 0.049401$, $c_B^{(3)} = 0.043150$, $T^{(3)} = 346.62$

Another example of a profitable use of the designed program will be given in Sec. 1.9.

1.6 Dynamics of Heterogeneous Catalytic Reactions in Closed Nongradient Reactors

A number of works were concerned with the study of dynamic properties of closed chemical systems. In [34], axioms were formulated which, when fulfilled, enabled one to consider the free energy as a Lyapunov function for an appropriate system of differential equations. The kinetics of mass action law (MAL) satisfied that axiomatics, that is, the point of detailed equilibrium (PDE) for an ideal homogeneous system was asymptotically stable in the reaction simplex. In [35], a more complete and correct study of MAL systems was carried out; as distinct from [34], the approach used by those authors enabled the obtained results to be extended also to ideal heterogeneous systems. Akramov and Yablonsky [36] showed that for reactions with the MAL kinetics, the PDE was a stable node in the reaction simplex.

However, of interest are also nonideal systems intractable in terms of MAL. In [37] isothermal systems with a Marselen-De Donde (MDD) kinetics [38] were studied and shown to exhibit properties similar to those of systems with MAL kinetics. As shown in [39], the PDE is a stable node for such systems. However, a number of results [40, 41] have provided evidence that the MDD kinetics has little to offer in treating even an ideal HCR.

We now consider a closed isothermal system of constant volume in which reversible elementary reactions, in number of S, of type as shown below, occur (for simplicity, only one type of catalytic sites is assumed to be active):

$$\sum_{i=1}^{N} b_{\sigma i} B_i + \sum_{l=1}^{M} \beta_{\sigma l} ZI_l + (m_{ts} - \sum_{l=1}^{M} \beta_{\sigma l} m_l) Z$$
$$= TS \rightarrow \text{products} \tag{36}$$

where B_i are the reactants; ZI_l are the adsorbed particles; Z is the free site on the catalyst surface; TS is the activated complex; m_{ts} is the number of sites on the catalyst surface occupied by the activated complex; $\beta_{\sigma i}$ and $\beta_{\sigma l}$ are the stoichiometric coefficients; N is the number of species involved in the reaction; M is the number of adsorbed particles; m_l is the number of sites occupied by the lth absorbal particle; the subscript σ is either s (forward reaction) or $-s$ (reverse reaction).

Assuming that a definite chemical potential can be assigned to each sort of adsorbed particles (which implies thereby their fast surface migration) and that the gaseous phase is an adeal one, the rate of elementary reaction (36) is expressed [40, 41] as:

$$v_\sigma = \varkappa_s \frac{kT}{h} \exp\left(-\frac{\mu_s^{\neq}(\mathbf{z})}{kT}\right) \exp\left(\frac{\sum_{i=1}^{N} b_{\sigma i} \mu_{B_i}(\mathbf{c}_r) + \sum_{l=1}^{M} \beta_{\sigma l} \mu_{a_l}(\mathbf{z})}{kT}\right) \tag{37}$$

Here $\varkappa_s$ is the transmission coefficient; k and h are the Boltzmann and Planck constants; $\mu_s^{\neq}$ is the chemical potential of the reaction's activated complex minus $kT \ln c_\sigma^{\neq}$ ($c_\sigma^{\neq}$ denoting the activated complex concentration per unit reaction path; μ_{B_i} and μ_{a_l} are the chemical potentials for a gas-phase reactant and an adsorbed species; $\mathbf{c}_r$ is the reactant concentration vector; $\mathbf{z}$ is the surface coverage vector characterizing the portion of surface occupied by adsorbed species. A major distinction of (37) from the MDD formalism is that the vacant sites of catalyst surface are not included in the number of reactive species, and the rate "constant" is thus dependent on the surface coverage.

The temporal change of concentrations for HCR carried out in an isothermal constant-volume reactor is described by a system of differential equations:

$$\frac{d\mathbf{c}_r}{dt} = S_k B_1^T \mathbf{u}, \quad L\frac{d\mathbf{z}}{dt} = S_k B_2^T \mathbf{u} \tag{38}$$

where $\mathbf{u}$ is the reaction step rate vector ($u_s = v_s - v_{-s}$); L is the number of catalytic sites per unit reaction volume; S_k is the catalyst surface area per unit reactor volume; B_1 and B_2 are the submatrices of the stoichiometric matrix referred to the gas-phase reactants and adsorbed species, respectively.

Let us introduce definitions and notations as shown below:
R_{N+M}^+ is the nonnegative octant of the R_{N+M} reaction space;
R_{N+M}^+ is the positive octant of the R_{N+M} reaction space;
B is the matrix, $B = (B_1, B_2/L)$;
$$z_{oc} = \sum_{i=1}^{M} z_i \text{ is the fraction of the occupied catalyst surface;}$$
$$\mu = (\mu_{B_1}, \ldots, \mu_{B_N}, \mu_{a_1}, \ldots, \mu_{a_M});$$
Γ is the set $\{T \mid T > 0\}$;
c is the overall concentration vector; the c vector is called positive if each $c_i > 0$ and $z_{oc} < 1$, and nonnegative, if each $c_i \geq 0$ and $z_{oc} \leq 1$;
c_0 is the nonnegative concentration vector as specified by the initial conditions for (38).

The reaction subspace L_R is defined by us as a linear envelope for a row of the B matrix.

The set $L_S = \{L_R + c_0 \cap R_{N+M}^+ \mid z_{oc} < 1\}$ is named the positive reaction simplex.

In exploring system (38), the following conditions are presumed to hold true.

A1. (a) There exists a function $F(c, T)$, continuous within $\overline{R_{N+M}^+} \times \Gamma$ and twice differentiable within $R_{N+M}^+ \times \Gamma$ at $z_{oc} < 1$, so that $\mu = \dfrac{\partial F}{\partial c}$.

(b) The function F attains a minimum within each set L_S at its internal point.

A2. The matrix $\left\{\dfrac{\partial \mu}{\partial c}\right\} > 0$ within $R_{N+M}^+ \times \Gamma$ at $z_{oc} < 1$.

A3. (a) At $c > 0$, all the functions $v_\sigma > 0$ and are continuously differentiable.
(b) Functions

$$g_\sigma = \frac{u_\sigma}{\prod_i c_{ir}^{b_{\sigma i}} \prod_l z_l^{\beta_{\sigma l}}(1 - z_{oc})^{(m_{t_s} - \sum_l \beta_{\sigma l})}} \quad (39)$$

can be continuously extended onto $R_{N+M}^+ \times \Gamma$ to satisfy the Lipschitz condition within any limited region.

To satisfy the A1.b condition, it suffices that chemical potential of an adsorbed species is represented in the form [17]

$$\mu_i = \varphi_i(z, T) + \psi(z_{oc}, T) \quad (40)$$

providing that, for any fixed value of T, the function φ_i is limited on any compact set of $\overline{R_{N+M}^+}$ devoid of points at which z_i is zero. As z tends to z^i (where the ith component of z^i is equal to 0), φ_i tends to $-\infty$; ψ is limited at $z_{oc} \neq 1$, and $\psi \to \infty$ as $z_{oc} \to 1$.

One can readily see that the conditions A1-A3 are fulfilled for reactions carried out on the homogeneous surface of a catalyst. In this case [42],

$$\mu_s^{\neq}(\mathbf{z}) = \mu_s^0 - m_{ts}kT \ln(1 - z_{oc})$$

$$\mu_i(\mathbf{z}) = \mu_i^0 - kT \ln \frac{z_i}{1 - z_{oc}} \tag{41}$$

$$v_\sigma = k_\sigma \prod_l z_l^{\beta_{\sigma l}} \prod_i C_{iy}^{b_{\sigma i}}(1 - z_{oc})^{(m_{ts} - \sum_l \beta_{\sigma l})}$$

The conditions A1-A3 remaining valid for system (38), the following statements can be proved.

1. Given nonnegative initial parameters, a solution of the Cauchy problem for system (38) exists at $t \in [0, \infty)$, and it is nonnegative. The solution, with positive initial parameters given, is positive. If, for a certain l ($1 \leq l \leq N + M$), $c_l(t_0) = 0$, then $c_l(t) = 0$ at $t \in [0, t_0)$. The same holds true for $(1 - z_{oc})$.

2. Within each L_s, there exists a PDE, and it is unique. All nonnegative stationary points are the PDE.

3. A positive PDE is stable within R_{N+M}^+ and is asymptotically stable within L_s.

4. A set of ω-limiting points consists of PDEs. If there is no stationary point at the L_s boundary, then any trajectory that originates in L_s will converge to a positive PDE.

5. No nonnegative periodic solution exists within L_s.

6. At a positive PDE, the linear approximation matrix for system (38) has nonnegative eigenvalues. All the concentrations having been expressed in independent concentrations (from material balance equations), the linear approximation matrix for a shortened (in terms of independent concentrations) system (38) has negative eigenvalues at a positive PDE, implying that the PDE is a stable node in L_s.

In full, the proofs of statements 1-6 were given in [17] based in part on the ideas expounded in the works of other authors [35, 39, 43].

Of interest appears to be the HCR carried out in an adiabatic constant-volume reactor. Here, the equation system (38) is supplemented with a heat-balance equation

$$C_v(\mathbf{c}, T)\frac{dT}{dt} = S_k \sum_{j=1}^{s} h_j(\mathbf{c}, T)u_j(\mathbf{c}, T) \qquad (42)$$

where C_v is the heat capacity per unit system volume; and h_j is the thermal effect at the jth reaction step. In [44] and [45] nonisothermal systems were considered within the framework of MAL and MDD formalisms, respectively, presuming a constancy of both heat capacity and thermal effects of reactions and the validity of Arrhenius equation for the rate constants of elementary reactions. An analysis of MAL kinetics in [46] showed the eigenvalues of the linear approximation matrix at the PDE to be nonpositive.

In this paper, our objective is to study the kinetic law (37), with no necessary presumption of a constancy of either heat capacity or thermal effects.

The following statements are valid:

7. At any $T > 0$, there exists a stationary point for a system defined by (38) and (39). All the stationary points are PDEs.

8. If, at a positive PDE, $C_v \neq 0$, then the PDE is stable.

9. The eigenvalues of linear approximation matrix (38, 39) at a positive PDE are nonpositive.

Since the system in question is an adiabatic one, the internal energy per unit volume is a constant value,

$$U(\mathbf{c}, T) = \text{const.} \qquad (43)$$

At equilibrium, C_v is always greater or equal to zero [47]. If, at PDE, $C_v \neq 0$, then temperature may be expressed, according to (43), as $T = T(\mathbf{c})$, at least locally. By substituting $T(\mathbf{c})$ into (38), we obtain a holonomic system for concentrations, according to the terminology in [34]. If we express concentrations c_i and T from the balance relationships and (43) in terms of independent concentrations, the following statement can be proved:

10. The eigenvalues of the linear approximation matrix at the positive PDE of a holonomic system expressed in terms of independent concentrations are negative, implying that the PDE is a stable node in L_S.

The proof of statement 7-10 relies on the construction of a Lyapunov function $S(\mathbf{c})$ such that

$$\frac{\partial S}{\partial c_i} = \frac{\mu_i(\mathbf{c}, T(\mathbf{c}))}{T(\mathbf{c})}$$

where $T(\mathbf{c})$ is defined by (43). The full proof of this was given in [17]; see also [16].

To exemplify the nonideal HCR, let us consider a reaction which is allowed to proceed on the biographically inhomogeneous catalyst surface at arbitrary surface coverage and symbatically varied heat of adsorption for the intermediates, each of these presumed to take a single site on the catalyst surface. Then the chemical potential of an adsorbed l-type particle is [48]

$$\mu_{a_l} = -kT \ln j_{a_l} + kT \ln \frac{z_l}{z_{oc}} - kT \ln U \tag{44}$$

where j_{a_l} is the internal partition function for the adsorbed particle; the parameter U is simplicitly defined by the function z_{oc}

$$z_{oc} = \frac{\gamma}{\exp(\gamma f) - 1} \int_0^f \frac{\exp[-(1-\gamma)\lambda]d\lambda}{U + \exp(-\lambda)}$$

where γ is the parameter of a number-of-sites versus heat-of-adsorption distribution function; f is the maximum relative desorption index λ. The chemical potential μ_{B_i} of an ideal gas mixture is

$$\mu_{B_i} = \mu_{B_i}^0 + kT \ln c_{B_i} \tag{45}$$

An expression for $\mu_s^{\neq}$ has been derived in [49]:

$$\mu_s^{\neq} = \mu_s^0 - m_{t_s} kT \ln I(\alpha_s - \gamma, U, f) \tag{46}$$

where

$$I(\alpha_s - \gamma, U, f) = \int_0^f \frac{\exp[-(\alpha_s - \gamma)\lambda]}{1 + U^{-1}\exp(-\lambda)} d\lambda$$

$\mu_{B_i}^0$ and μ_s^0 being the standard parts of chemical potentials, and α_s, the transfer coefficient.

For the kinetic law (37) with chemical potentials as defined by (44), (45), (46), the conditions A1-A3 are fulfilled [14-16] and, consequently, the statements 1-10 hold true.

The case of the so-called mean surface coverage approximation for a biographically inhomogeneous catalyst surface is treated in a similar manner.

1.7 Behaviour of Closed Chemical Systems Near Equilibrium

As has been noted in the foregoing section, for chemical systems with MAL and MDD formalisms and for HCR with a nonideal kinetics, there has been provided proof that the positive PDE is a stable node, that is, the eigenvalues of a linear approximation matrix ("relaxation spectrum"; [50]) are nonpositive or, with allowance made for the material balance, negative. The reality and nonpositivity of eigenvalues are presumed to be associated with the inexistence of oscillatory regimes near equilibrium [46, 51].

The interest in the behaviour of chemical systems tending to equilibrium bears relevance to relaxation studies of closed systems [50].

As has been stated in [51] from general thermodynamic considerations, the relaxation spectrum is nonpositive irrespective of the form of a concrete kinetic law. However, the analysis by [36] revealed an error in the proof of that statement, and the issue has thus remained open to dispute.

In this Section we wish to show that the line of reasoning as suggested in [51] can be modified to attain the originally alledged result: the eigenvalues of a linear approximation matrix for an arbitrary nongradient chemical system are nonpositive at the stable point of detailed equilibrium. To be specific, we consider a homogeneous system at constant temperature T and volume V.

The system under consideration is an S-step reaction involving N reactants; the sth reaction step in the general case is written as

$$\sum_{i=1}^{N} b_{si}^{+} B_i \rightleftarrows \sum_{i=1}^{N} b_{si}^{-} B_i, \quad i = 1, \ldots, S \tag{47}$$

where b_{si}^{+} and b_{si}^{-} are the stoichiometric coefficients and the B_i symbolize the reactants. The stoichiometric matrix for mechanism (47) is

$$B = \{b_{si}^{-} - b_{si}^{+}\}$$

The system of differential equations for concentrations is written as:

$$\frac{d\mathbf{c}}{dt} = B^T \mathbf{u} \tag{48}$$

where $\mathbf{c}$ is the concentration vector and $\mathbf{u}$ is the reaction step rate vector.

Both the scheme for proving the statement and the notations are those from [51]. We introduce new state variables: ξ, the number of acts at each reaction step (47), that is, the difference of the numbers of elementary acts at the given step in the forward and reverse directions per unit reaction space during a period of time t (the so-called "chemical variables"). We choose the equilibrium state of the system for an initial state and then construct a linearized system of differential equations for ξ near equilibrium. If the chemical variables are incremented each by $\{\delta\xi_\varrho\}$, then near the state of equilibrium the relation

$$\frac{d(\delta\xi_\varrho)}{dt} = \sum_{\varrho'=1}^{S} \alpha_{\varrho\varrho'} \delta\xi_{\varrho'}, \quad \varrho = 1, \ldots, S \qquad (49)$$

must hold true, where $\alpha_{\varrho\varrho'}$ are coefficients estimated at the given equilibrium state. Since, at equilibrium, the chemical affinity reduces to zero, in the vicinity of equilibrium it can be represented by the following expansion:

$$A_\varrho = \sum_{\varrho'=1}^{S} \beta_{\varrho\varrho'} \delta\xi_{\varrho'}, \quad \varrho = 1, \ldots, S \qquad (50)$$

Let us calculate $\beta_{\varrho\varrho'}$. The affinity of a reaction numbered ϱ is

$$A_\varrho = -\sum_{\gamma=1}^{N} b_{\varrho\gamma'} \mu_{\gamma'}, \quad \varrho = 1, \ldots, S \qquad (51)$$

Here $b_{\varrho\gamma'}$ is the stoichiometric coefficient of the γ'th reactant at the ϱth reaction step, and $\mu_{\gamma'}$ is the chemical potential of the γ'th reactant. From (50), we obtain

$$\beta_{\varrho\varrho'} = \frac{\partial A_\varrho}{\partial \xi_{\varrho'}} = \sum_{\gamma=1}^{N} \frac{\partial A_\varrho}{\partial c_\gamma} \frac{\partial c_\gamma}{\partial \xi_{\varrho'}} \qquad (52)$$

where c_γ denotes the number of molecules of reactant γ per unit reaction space. But, since

$$dc_\gamma = \sum_{\varrho'=1}^{S} b_{\varrho'\gamma} d\xi_{\varrho'}$$

then

$$\frac{dn_\gamma}{d\xi_{\varrho'}} = b_{\varrho'\gamma} \qquad (53)$$

We obtain from (51):

$$\frac{\partial A_\varrho}{\partial c_\gamma} = -\sum_{\gamma'=1}^{N} b_{\varrho\gamma'} \frac{\partial \mu_{\gamma'}}{\partial c_\gamma} \qquad (54)$$

and, from (52) and (54),

$$\beta_{\varrho\varrho'} = \sum_{\gamma=1}^{N} b_{\varrho'\gamma} \left(\sum_{\gamma'=1}^{N} b_{\varrho\gamma'} \frac{\partial \mu_{\gamma'}}{\partial c_\gamma} \right) \tag{55}$$

Let us denote $M = \{\partial \mu_\gamma / \partial c_\gamma\}$. Then from (55), the matrix

$$\{\beta_{\varrho\varrho'}\} = -BMB^T \tag{56}$$

If the equilibrium in question is stable, then the inequality

$$M \geqslant 0 \tag{57}$$

holds true [51].

From now and on, the signs "$\geqslant, \leqslant$" as applied to a matrix will signify that the matrix is symmetric and its corresponding quadratic form is defined in sign. It follows from (56) and (57) that the matrix $\{\beta_{\varrho\varrho'}\} \leqslant 0$.

The reaction rate change (or the thermodynamic flow) is linked to a change in affinity (strength) near equilibrium through the Onsager relation:

$$\delta u_\varrho = \frac{d(\delta \xi_\varrho)}{dt} = \sum_{\varrho'=1}^{S} L_{\varrho\varrho'} A_{\varrho'} \tag{58}$$

where $\{L_{\varrho\varrho'}\}$ is a symmetric matrix [50]. By substituting (50) into (58), we obtain

$$\{\alpha_{\varrho\varrho'}\} = \{L_{\varrho\varrho'}\}\{\beta_{\varrho\varrho'}\} \tag{59}$$

It can easily be shown that the eigenvalues of a matrix product XQ (where the matrix X is symmetric, and the matrix $Q \leqslant 0$) are real [18]. Be this the case, the eigenvalues of matrix $\{\alpha_{\varrho\varrho'}\}$ are also real, and their nonpositivity follows from the presumption that the equilibrium is stable.

Let us consider an inverse problem. Suppose, as has been done in Sec. 1.6, that the matrix $M > 0$. Now, let us see if it is possible, without attaching to a concrete kinetic law, to draw a conclusion on the stability of a chemical equilibrium from the analysis of a linear approximation system.

We shall consider "independent" chemical reactions, by analogy with a procedure suggested in [52]. We select linearly independent columns in the B^T matrix (suppose, these are the first R columns) and express the remaining $(S - R)$ columns as linear combinations of the independent columns:

$$b_i = \sum_{j=1}^{R} f_{ij} b_j, \quad i = R+1, \ldots, S \tag{60}$$

Now the system (48) can be rewritten in the form

$$\frac{d\mathbf{c}}{dt} = \tilde{B}^{T}\mathbf{u} \tag{61}$$

where $\tilde{B}^{T}$ is the matrix made up of the first R columns of B^{T}, and the components of the $\tilde{\mathbf{u}}$ vector are

$$\tilde{u}_k = u_k + \sum_{i=R+1}^{S} f_{ik} u_i, \quad k = 1, \ldots, R \tag{62}$$

We consider R "independent reactions" proceeding at rates defined by (62); we introduce the appropriate chemical variables $\tilde{\xi}_i$ ($i = 1, \ldots, R$). The formulas (49)-(55) hold true on condition that the indices ϱ and ϱ' are allowed to run from unity to R. Denoting the chemical affinity vector for the whole set of reactions involved as $\mathbf{A}$ and that for the independent reactions as $\tilde{\mathbf{A}}$, we obtain, by virtue of (51) and (60).

$$\mathbf{A} = F\tilde{\mathbf{A}} \tag{63}$$

where the matrix F of size $S \times R$ has the form

$$F = \left\{ \begin{array}{c} I \\ \{f_{ij}\} \end{array} \right\} \tag{64}$$

Let us introduce a matrix $\tilde{L} = F^{T}LF$. From (58) and (62), we obtain

$$\delta\mathbf{u} = LF\tilde{\mathbf{A}}, \quad \delta\tilde{\mathbf{u}} = \tilde{L}\tilde{\mathbf{A}} \tag{65}$$

We find, after having performed the appropriate transformations, that for the independent reactions

$$\{\tilde{\alpha}_{\varrho\varrho'}\} = \tilde{L}\{\tilde{\beta}_{\varrho\varrho'}\} \tag{66}$$

where

$$\{\tilde{\beta}_{\varrho\varrho'}\} = -\tilde{B}M\tilde{B}^{T} \tag{67}$$

If $M > 0$, then $\{\tilde{\beta}_{\varrho\varrho'}\} < 0$ by virtue of the second principe of thermodynamics [51],

$$(L\mathbf{A}, \mathbf{A}) = \sum_{i,j=1}^{S} L_{ij} A_i A_j \geq 0 \tag{68}$$

the equality holding true only at equilibrium, that is, at $\mathbf{A} = 0$. Since the elements of $\mathbf{A}$ are related through (63) and, therefore, are not chosen arbitrarily, one cannot make a definite conclusion as to the positive determinance of L. Let us change, by making use of (63), to independent variables of chemical affinity; we obtain, from (68),

$$(LF\tilde{A}, F\tilde{A}) = (F^T LF\tilde{A}, \tilde{A}) \geq 0 \tag{69}$$

To be noted, the equality of (69) to zero occurs only at $\tilde{A} = 0$. Therefore, the matrix $\tilde{L}$ is positively definite,

$$\tilde{L} = F^T LF > 0 \tag{70}$$

It follows from (66) that the eigenvalues of matrix $\tilde{\alpha}$ are negative. The chemical equilibrium point is asymptotically stable for the system of differential equations that describe the temporal dependence of the chemical variables of independent reactions. Thus, allowance made for the linear dependence within the matrix B, in the general case one can exclude, when applying the Onsager relation, the zero eigenvalues from the relaxation spectrum, by analogy with the treatment of a MAL case in [36]. It can be shown that the $\tilde{\alpha}$ matrix and the linear approximation matrix in system (61), after exclusion of dependent concentrations from them, are similar, that is, their relaxation spectra are identical.

To illustrate the aforesaid, let us consider examples where the reaction kinetics is governed by one of the three formalisms: MAL, MDD, or kinetic law from [15]. The reaction step rate is written as

$$u_i = k_i \exp\left(\sum_j b_{ij}^+ \mu_j\right) - k_i \exp\left(\sum_j b_{ij}^- \mu_j\right) \tag{71}$$

where k_i is a function of temperature or, in the case of HCR, a function of both temperature and surface coverage. We obtain from (71), to a linear approximation near equilibrium,

$$u_i = k_i \exp\left(\sum_j b_{ij}^- \mu_j\right) A_i = u_i^* A_i$$

where u_i^* is the rate of a forward or reverse elementary reaction of the ith step at equilibrium. In this case, the L matrix is a diagonal one. We obtain from (64)

$$\tilde{L} = \begin{pmatrix} u_1^* & 0 \\ & \ddots & \\ 0 & & u_R^* \end{pmatrix} + \{f_{ij}\}^T \begin{pmatrix} u_{R+1}^* & 0 \\ & \ddots & \\ 0 & & u_s^* \end{pmatrix} \{f_{ij}\} \tag{72}$$

It follows from (72) that $\tilde{L} > 0$; this is in conformity with (70).

1.8 Conditions for Stability of Chemical Equilibria in Closed Systems of Various Type

In a work by Gorban et al. [53], a general form for writing the system of differential equations for closed chemical systems has been suggested and sufficient conditions for the equilibrium stability have been specified. We show in this Section that no analysis of a closed chemical system for equilibrium stability conditions is necessarily needed, and that these conditions are interrelated.

For simplicity's sake, we consider a homogeneous system in which chemical reactions of type (47), in total number S, occur. The adopted notations are: N, a vector whose ith component is numerically equal to the number of ith-type molecules within the system; V, volume; P, pressure; T, temperature; U, internal energy; H, enthalpy; G, Gibbs energy; F, Helmholtz energy; C_P and C_V, heat capacities at constant P and V, respectively. A system with constants T and V is called the (T, V)-system; by analogy, the (T, P)-, (V, U)-, and (P, H)-systems are defined.

In [53], the system of differential equations for chemical kinetics was written in the form

$$\dot{N} = VB^T u \tag{73}$$

where B is the stoichiometric matrix defined in terms of the reaction mechanism. The rate at the sth reaction step is written as

$$u_s = u_s^0(N, V, T) \left[\exp\left(\frac{\sum_i b_{si}^+ \mu_i}{kT}\right) - \exp\left(\frac{\sum_i b_{si}^- \mu_i}{kT}\right) \right] \tag{74}$$

We presume, for the (T, P)-system, the volume V to be defined in terms of T, N, P parameters by the equation of state; for the (P, H)-system, T is defined in terms of N, P by equation of state under the conditions of $H = \text{const}$ and $V = \text{const}$; for the (V, U)-system, T is defined in terms of (N, V) by the law of energy conservation. Otherwise stated, the (73) system is always a system that can be defined in terms of N vector and a set of invariable parameters specific of this system.

As has been shown in [53], the sufficient condition (A) for the equilibrium point stability is: the matrix $\{\partial(\mu_i/T)/\partial N_j\}$ is symmetric, nonnegatively definite within the concentration space and positively definite within any hyperplane orthogonal to a vector with positive components (partial derivatives are taken at constant values of the variable that define the type of the system). This condition warrants the existence of a local Lyapunov function and the asymptotic stability of equilibrium in each reaction simplex.

The chemical potentials are presumed to be functions of concentration and temperature, that is

$$\mu_i = \mu_i\left(\frac{\mathbf{N}}{V}, T\right) \tag{75}$$

and for the (T, V)-system, the matrix $\{\partial\mu_i/\partial N_j\}$ is positively definite over the entire concentration space. Let us show that the stability condition (A) remains valid for other systems as well. The systems considered are:

(a) (T, P)-system.

It follows from Eq. (75) that

$$\left.\frac{\partial\mu_i}{\partial N_j}\right|_{T,P} = \left.\frac{\partial\mu_i}{\partial N_j}\right|_{T,V} - \frac{1}{V}\left.\frac{\partial V}{\partial N_j}\right|_{T,P} \cdot \sum_k \left.\frac{\partial\mu_i}{\partial N_k}\right|_{T,V} \cdot N_k \tag{76}$$

Since $G = \sum_i \mu_i N_i$ [47], we have

$$\left.\frac{\partial G}{\partial N_j}\right|_{T,V} = \mu_j + \sum_i \left.\frac{\partial\mu_i}{\partial N_j}\right|_{T,V} \cdot N_i \tag{77}$$

$$\left.\frac{\partial G}{\partial V}\right|_{T,N} = \sum_i \frac{\partial\mu_i}{\partial V} N_i = -\sum_i \sum_j \left.\frac{\partial\mu_i}{\partial N_j}\right|_{T,V}\left(\frac{N_i}{V}\right) N_j = -\frac{1}{V}(\tilde{M}\mathbf{N}, \mathbf{N}) \tag{78}$$

where $\tilde{M} = \left\{\dfrac{\partial\mu_i}{\partial N_j}\right\}\bigg|_{T,V}$. Taking into account that $G - F = PV$ [47] and making use of (77) and (78), one obtains

$$\left.\frac{\partial V}{\partial N_j}\right|_{T,P} = V \frac{\sum \left.\dfrac{\partial\mu_i}{\partial N_j}\right|_{T,V} N_i}{(\tilde{M}\mathbf{N}, \mathbf{N})} \tag{79}$$

Substitution of (79) into (76) yields an expression for $\bar{M}$:

$$\bar{M} = \left\{\frac{\partial\mu_i}{\partial N_j}\right\}_{T,P} = \tilde{M} - \frac{\tilde{M}\mathbf{N}\cdot\mathbf{N}^T\tilde{M}}{(\tilde{M}\mathbf{N}, \mathbf{N})} \tag{80}$$

Let us demonstrate the application of (80) to an ideal gas mixture defined as

$$\mu_i = \varphi_i(T) + kT\ln\frac{N_i}{V} \tag{81}$$

$$\tilde{M} = kT\{\delta_{ij}/N_j\} \tag{82}$$

where $\varphi_i(T)$ is the standard part of chemical potential.

In this case, $(\tilde{M}N, N) = kT\Sigma N_i$, and we get from (80)

$$\overline{M} = kT\{\delta_{ij}/N_j - 1/\sum_i N_i\} \tag{83}$$

In (81)-(83) and henceforth, δ_{ij} is the Kronecker delta. Formula (83) can be derived from (81) by substituting V expressed in terms of the state equations.

It ensues from (80) that for any vector

$$(\overline{M}\mathbf{x}, \mathbf{x}) = \frac{\langle N, N\rangle\langle x, x\rangle - \langle x, N\rangle^2}{\langle N, N\rangle} \tag{84}$$

where $\langle x, y\rangle = (\tilde{M}x, y)$ is a scalar product defined in [39]. It follows from the Cauchy-Bunyakovsky inequality that

$$(\overline{M}\mathbf{x}, \mathbf{x}) \geq 0 \tag{85}$$

The sign of equality in (85) is appropriate only if $x = aN$, here a being any number; this means that $\overline{M}$ is positively definite in any hyperplane which is orthogonal to a positive vector, that is, condition (A) is thus fulfilled.

Apparently, the converse is also true: let condition (A) be fulfilled for matrix $\overline{M}$, and $(\tilde{M}N, N) > 0$. It follows then from (85) that the condition (A) is also fulfilled for matrix $\tilde{M}$.

The above reasoning shows that the convexity of F is always conducive to a convexity of G for chemical potentials of (81) type. Note that in the classical work of Zel'dovich [54] this relation was tested on an ideal mixture separately for both F and G.

(b) (V, U)-system.

The major condition for this system is

$$U(N, V, T) = \text{const} \tag{86}$$

As follows from (86), if, at an equilibrium point, $C_V \neq 0$, then

$$\frac{\partial T}{\partial N_j} = -\frac{U_j}{C_V} \tag{87}$$

where $U_j = \left(\dfrac{\partial U}{\partial N_j}\right)_{T,V}$. By making use of the relation $F = U - TS$, one can show that

$$\left\{\frac{\partial\left(\frac{\mu_i}{T}\right)}{\partial N_j}\right\}_{V,U} = \frac{1}{T^2}\left\{\frac{U_i U_j}{C_V}\right\} + \frac{1}{T}\left\{\frac{\partial \mu_i}{\partial N_j}\right\}_{T,V} \tag{88}$$

As implied by (88), condition (A) is fulfilled for the (V, U)-system. To be noted, the converse statement is not necessarily true, since the validity of the (A) condition for (V, U)-system does not warrant, as can be seen from (88), its validity for (T, V)-system.

(c) (P, H)-system.

In can be proved that

$$\left\{\frac{\partial\left(\frac{\mu_i}{T}\right)}{\partial N_j}\right\}_{P,H} = \frac{1}{T^2}\left\{\frac{H_i H_j}{C_P}\right\} + \frac{1}{T}\left\{\frac{\partial \mu_i}{\partial N_j}\right\}_{T,P} \tag{89}$$

where $H_i = \left(\frac{\partial H}{\partial N_i}\right)_{T,P}$. Relation (89) shows that condition (A) remains valid also for the (P, H)-system. Therefore, in all cases, it suffices to verify this condition only for (T, V)-system, which is more easy to accomplish than for other systems.

The results obtained in this Section are also valid for noninteractive chemical subsystems, for example, as represented by a gas phase and an adsorption layer. For the latter, the system size is featured by the number of catalytic sites, and the pressure, by surface tension.

It should be emphasized that the applicability of the derived statements to interactive subsystems needs further elucidation. The conditions under which the stability of an adiabatic system provides for the stability of an isothermal system also remain a controversial issue.

In accordance with [53], if condition (A) is fulfilled for each reaction rate vector $\mathbf{N}$, then for the thermodynamic potentials with logarithmic singularities at the reaction simplex boundary, there exists a unique positive PDE on each simplex. A prerequisite for adiabatic systems is that T could be expressed in terms of the laws of energy (or enthalpy) conservation. For instance, for any $\mathbf{N}$ of the (V, U)-system, there must be $T = T(\mathbf{N}, V)$. In strict terms, this may be defined as follows (condition (B)):

(B) Let T be a function of $\mathbf{N}$ continuously differentiable in a Ω region, and $0 < T_0 \leqslant T(\mathbf{N}, V) \leqslant T_1$. Then, if the reaction simplex belongs to Ω, there exists a unique positive PDE on this simplex.

Condition (B) is sufficient, if for any $\mathbf{N}$ and V,

$$C_V = \frac{\partial U}{\partial T} > 0 \tag{90}$$

and
$$\lim_{T\to\infty} U(\mathbf{N}, T, V) = \infty \tag{91}$$

For example, for an ideal gas mixture, the internal energy is [55]:

$$U = \sum N_i \left(kT^2 \frac{\partial \ln f_i}{\partial T} \right) \tag{92}$$

where f_i is the partition function of the ith-type particles. It can be shown that both (90) and (91) hold true for (92), that is, the ideal gas possesses a single equilibrium point under adiabatic conditions. We are unaware of systems for which a violation of conditions (90) and (91) was reported potentially conducive to a nonuniqueness of equilibrium for closed systems.

1.9 An Example of Nonuniqueness of Equilibrium, Computation in Heterogeneous Catalytic Reactions

As has already been mentioned in Sec. 1.6, an analysis of steady-state multiplicity makes part of the study of closed chemical system dynamics. The closed chemical systems feature a common behaviour; the new results and relevant references have been discussed in Sec. 1.6. In particular, the matrix of partial derivatives for chemical potentials with respect to concentration, $\{\partial \mu_i / \partial c_j\}$ (for example, for a system at constant pressure and volume) has been shown to be positively definite in concentration space. This, in turn, leads to the property of a closed chemical system to exhibit a single equilibrium point (and, simultaneously, a minimum of Helmholtz energy) in each reaction simplex.

Zel'dovich [54] was the first to prove the uniqueness of the equilibrium point within each material balance plane for an ideal mixture at constant temperature and volume, or constant temperature and pressure. There have been reported reactions in solution [56, 57] involving more than one equilibria. We are unaware of heterogeneous catalytic reactions, except an artificially designed system in [13], that would have exhibited the said behaviour.

We wish to show now that in a reaction for coadsorption of two species, treated within the so-called semiempirical model of induced inhomogeneity [58], a non-unique numerical solution of the equilibrium problem is possible (at constant temperature T and volume V).

The reaction in question is

$$\begin{aligned} D_1 + Z &\rightleftarrows ZI_1 \\ D_2 + Z &\rightleftarrows ZI_2 \end{aligned} \tag{93}$$

where D_1 and D_2 denote the molecules of two gas-phase species, and Z, a free catalytic site; ZI_1 and ZI_2 are the coadsorbed species. The chemical potentials for species D_1, D_2, ZI_1, and ZI_2 are, respectively,

$$\mu_{D_1} = \mu_{D_1}^0 + kT \ln P_{D_1} \tag{94}$$

$$\mu_{D_2} = \mu_{D_2}^0 + kT \ln P_{D_2} \tag{95}$$

$$\mu_{ZI_1} = \mu_1^0 + kT \ln \frac{\theta_1}{1 - \theta_1 - \theta_2} + \frac{kT}{\gamma} \ln \{\theta_1 [\exp(\gamma \eta_1^2 f)$$
$$- 1] + 1\} + \frac{kT}{\gamma} \left[\ln(D\theta_2 + 1) + \frac{D\theta_2}{1 + D\theta_1} \right] \tag{96}$$

$$\mu_{ZI_2} = \mu_2^0 + kT \ln \frac{\theta_2}{1 - \theta_1 - \theta_2} + \frac{kT}{\gamma} \ln \{\theta_2 [\exp(\gamma \eta_2^2 f)$$
$$- 1] + 1\} + \frac{kT}{\gamma} \left[\ln(D\theta_1 + 1) + \frac{D\theta_1}{1 + D\theta_2} \right] \tag{97}$$

where

$$D = \exp(\gamma \eta_1 \eta_2 f / 2) - 1 \tag{98}$$

In formulas (94)-(98), θ_1 and θ_2 are the surface coverages of species D_1 and D_2; $\mu_{D_1}^0$, $\mu_{D_2}^0$, μ_1^0, μ_2^0 are the temperature-dependent standard parts of chemical potential; η_i is the effective charge; γ is the inhomogeneity index; f is the interval of adsorption heat change per molecule (in kT units); P_{D_1} and P_{D_2} are the partial pressures for species D_1 and D_2.

It follows from the atom balance equations that

$$P_{D_1} = P_{D_1}^0 - \tilde{L} kT \theta_1 \tag{99}$$

$$P_{D_2} = P_{D_2}^0 - \tilde{L} kT \theta_2 \tag{100}$$

where $P_{D_1}^0$ and $P_{D_2}^0$ are constant partial pressures; $\tilde{L}$ is the number of catalytic sites per unit gas volume.

To obtain equilibrium values of θ_1 and θ_2, a system of isotherms must be solved derived from the adsorption equilibrium conditions,

$$\mu_{D_1} = \mu_{ZI_1}, \quad \mu_{D_2} = \mu_{ZI_2} \tag{101}$$

taking into account the balance equations (99) and (100).

This problem has been solved using a program of nonlocal search for nonlinear equation system roots (see Sec. 5) with the following parameters chosen:

$$P_{D_1}^0 = P_{D_2}^0 = 0.5; \quad \tilde{L} kT = 0.1; \quad \gamma = -0.5; \quad \eta_1 = 1;$$
$$\eta_2 = -1; \quad f = 20;$$

$$a_1^0 = \exp[(\mu_A^0 - \mu_1^0)/kT] = 9.26 \times 10^{-3};$$
$$a_2^0 = \exp[(\mu_B^0 - \mu_2^0)/kT] = 2 \times 10^{-3}$$

Three sets of roots have been obtained

(1) $\theta_1^1 = 0.183$, $\theta_2^1 = 0.362$
(2) $\theta_1^2 = 0.0519$, $\theta_2^2 = 0.144$
(3) $\theta_1^3 = 8.18 \times 10^{-6}$, $\theta_2^3 = 1.00 \times 10^{-3}$

As is seen, for the model of interest (93-96), the matrix $\{\partial \mu_i / \partial c_j\}$ is not, generally speaking, a positively definite one. This signifies a violation of the necessary condition for the system's stability [59] as far as the phase separation is concerned. Simultaneously, the physical feasibility of phase separation in reactive systems of the specified type was considered in the literature from opposing standpoints [57, 60, 61]. In this work, we stay away from discussing this controversial issue and rather turn our attention to the study of dynamical behaviour of system (93) presuming the nonoccurrence of phase separation (which may, for example, arise from the reasons suggested in [57] or [61]).

The system of differential equations for surface coverages takes the form:

$$L \frac{d\theta_1}{dt} = v_{+1} - v_{-1} = g_1(\theta_1, \theta_2)$$
$$L \frac{d\theta_2}{dt} = v_{+2} - v_{-2} = g_2(\theta_1, \theta_2) \tag{102}$$

where the elementary reaction rates, according to [40] are

$$v_{+1} = \varkappa_1 \frac{kT}{h} \exp\left(\frac{-\mu_1^{\neq} + \mu_A}{kT}\right),$$
$$v_{-1} = \varkappa_1 \frac{kT}{h} \exp\left(\frac{-\mu_1^{\neq} + \mu_{ZI_1}}{kT}\right) \tag{103}$$

The rates v_{+2} and v_{-2} are expressed in a similar manner. In Eq. (103), L is the number of catalytic sites per unit surface area; $\varkappa_1$ is the transmission coefficient; h is the Planck constant; $\mu_1^{\neq}$ is the chemical potential for the activated complex minus $kT \ln N_1^{\neq}$, $N_1^{\neq}$ being the activated complex concentration at the first reaction step per unit reaction path. According to [58], the chemical potentials for activated one-centre complexes are:

$$\mu_{t_i} = \mu_{t_i}^0 + kT \ln N_i^{\neq} - kT \ln(1 - \theta_1 - \theta_2)$$
$$+ \frac{kT}{\gamma} \ln[\theta_1(\exp(\gamma \eta_{t_i} \eta_1 f/2) - 1) + 1]$$
$$+ \frac{kT}{\gamma} \ln[\theta_2(\exp(\gamma \eta_{t_i} \eta_2 f/2) - 1) + 1] + \frac{kT}{\gamma} \theta_1 \times$$

1 Design and Analysis of Kinetic Models

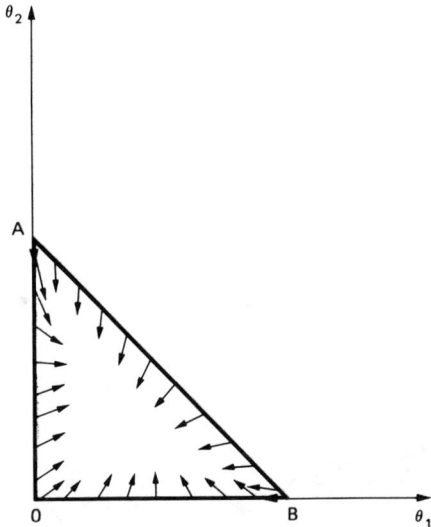

Fig. 1.3 Vector field as defined by differential equation system (102) at the Ω region boundary

$$\times (\exp(\gamma\eta_{t_i}\eta_1 f/2) - 1) + \frac{kT}{\gamma}\theta_2(\exp(\gamma\eta_{t_i}\eta_2 f/2) - 1),$$
$$i = 1, 2 \tag{104}$$

where η_{t_i} is the effective charge of the activated complex. It is presumed that $\eta_{t_i} = \eta_i$.

The vector field defined by Eq. (102) at the boundary of region $\Omega = \{\theta_1 \geqslant 0, \theta_2 \geqslant 0, \theta_1 + \theta_2 \leqslant 1\}$ is shown in Fig. 1.3. As has been reported in [62], the index of a closed curve with respect to a vector field pointed either to the inside, or to the outside of the closed curve is equal to unity. In our case, the vector field at two points A and B is tangent to the simplex Ω boundary; however, the validity of the above statement is easily proved in this particular case too. We consider a situation where all of the singular points of system (102) are simple points (in the paper cited, the singular points were also simple points). It can be shown that $\partial g_1/\partial \theta_1 < 0$ and $\partial g_2/\partial \theta_2 < 0$ at a singular point. Therefore, the singular points can be either stable nodes, or stable foci, or saddle points. However, as has been shown in Sec. 1.7, no stable foci occur in a closed chemical system. Since the Ω region boundary index is equal to unity, the only possible situation is envisaged: the two singular points of sys-

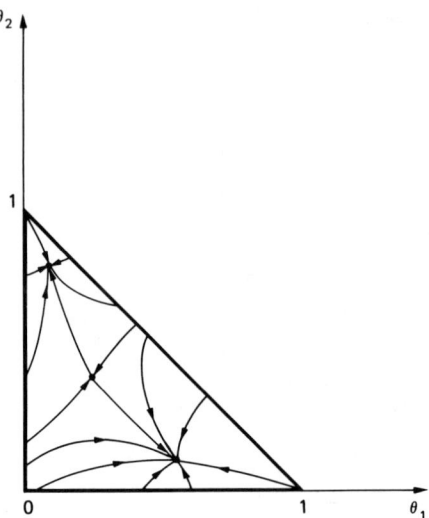

Fig. 1.4 Qualitative representation of a phase picture for differential equation system (102)

tem (102) are stable nodes, one of these being a saddle point. A qualitative representation of a phase picture for system (102) is shown in Fig. 1.4.

We also remark here that the semiempirical induced-inhomogeneity model has been applied in [57] to study two species coadsorbed in an open system. The reported sigmoidal "surface coverage versus partial pressure" relationship and the synergism of two-component coadsorption agree with experimental evidence available in the literature.

1.10　Conclusion

The present paper outlines the major results in the authors' pursuit to develop mathematical and software facilities for kinetic studies of heterogeneous catalytic reactions. Works in the field have been carried out at the L. Ya. Karpov Physico-Chemical Institute for about two decades.

The designed System for Automation of Kinetic Computations (SAKC) has been widely applied in a study of heterogeneous catalytic reactions. Currently, in active use is the PPRED program package for deriving rate equations. Further improvements of this system are oriented at (i) the employment of higher-order languages for analytical computations; (ii) development of methods for reducing the order of steady-state systems and the design of programs for deriving the rate equations in a variety of nonlinear mechanisms; (iii) the eventual adaptation of the system to personal computers; and (iv) design of facilities for interactive dialogue operation with the system.

The application of SAKC to studies of dynamic regimes of complex reaction kinetics appears to be quite a promising field. SAKC can be used for an automated development of the right-hand side of a system of differential equations of chemical kinetics, for deriving Jacobian matrices and variational differential equation system. This is very important, since the differential equations of chemical kinetics belong to a class of stiff equations. In addition, of frequent concern is also the differential-algebraic system emerging in the quasi-steady-state concentration method applied to certain reactants involved in the reaction. Currently, all of the aforementioned trends are under study.

The mathematical support of kinetic studies, that is, analysis of emerging mathematical models, is chiefly based, as is apparent from the presented material, on the methods of a qualitative theory of differential equations. As has already been mentioned, the major problems in the study of the dynamics of closed systems are at present solved. A number of problems awaiting elucidation have been touched upon in the course of the presentation of the material. A supplementary issue is also the problem of phase separation dynamics in studying the critical phenomena for closed systems.

As regards open systems, i.e. systems with distributed parameters [64], the sphere of action in this particular field appears to be vast indeed. Here may also be added specific problems of recent interest such as relaxation processes in HCR and the development of appropriate mathematical apparatus that may provide a deeper insight into the reaction mechanism. Novel experimental techniques and ever-increasing experimental evidence pose new challenging problems before mathematical chemistry. For more instructive details the interested reader is referred to the References, especially to the monographs [10-13].

References

1. M. G. Slin'ko, *Kinetika i Kataliz*, **13**: 566-580 (1972).
2. M. G. Slin'ko and A. K. Avetisov, *Kinetika i Kataliz*, **19**: 17-21 (1978).
3. A. J. Silvestry and J. S. Zahner, *Chem. Eng. Sci.*, **22**: 466-467 (1967).
4. V. V. Kafarov, V. N. Pisarenko, and A. V. Solokhin, *Zh. Fiz. Khimii*, **52**: 3143-3147 (1978).
5. G. S. Yablonsky, V. A. Evstigneev, and V. I. Bykov, in: *Primenenie teorii grafov v khimii* (Application of Graph Theory in Chemistry), Nauka, Novosibirsk, 1988, pp. 70-143 (in Russian).
6. G. M. Ostrovsky, A. G. Zyskin, and Yu. S. Snagovsky, in: *Fizicheskaya khimiya. Sovremennye problemy* (Physical Chemistry. Current Problems), Khimiya, Moscow, 1986, pp. 84-115 (in Russian).

7. G. M. Ostrovsky, A. G. Zyskin, Yu. S. Snagovsky, I. I. Malkin, and M. G. Slin'ko, *Dokl. Akad. Nauk SSSR*, **289**: 412-415 (1986).
8. G. M. Ostrovsky, A. G. Zyskin, Yu. S. Snagovsky, and M. G. Slin'ko, *Chem. Eng. Sci.*, **42**: 2579-2586 (1987).
9. G. M. Ostrovsky, A. G. Zyskin, and Yu. S. Snagovsky, *Comp. and Chem.*, **11**: 85-96 (1987).
10. G. S. Yablonsky, V. I. Bykov, and A. N. Gorban', *Kineticheskie modeli kataliticheskikh reaktsiĭ* (Kinetic Models of Catalytic Reactions), Nauka, Siberian Division, Novosibirsk, 1983 (in Russian).
11. G. S. Yablonsky, V. I. Bykov, and V. I. Elokhin, *Kinetika model'nykh reaktsiĭ geterogennogo kataliza* (Kinetics of Model Reactions of Heterogeneous Catalysis), Nauka, Siberian Division, Novosibirsk, 1984 (in Russian).
12. A. N. Gorban', V. I. Bykov, and G. S. Yablonsky, *Ocherki o khimicheskoi relaksatsii* (Notes on Chemical Relaxation), Nauka, Siberian Division, Novosibirsk, 1986 (in Russian).
13. V. I. Bykov, *Modelirovanie kriticheskikh yavleniĭ v khimicheskoi kinetike* (Simulation of Critical Phenomena in Chemical Kinetics), Nauka, Novosibirsk, 1988 (in Russian).
14. A. G. Zyskin, Yu. S. Snagovsky, and M. G. Slin'ko, *React. Kinet. Catal. Lett.*, **17**: 257-261 (1981).
15. A. G. Zyskin, Yu. S. Snagovsky, and M. G. Slin'ko, *React. Kinet. Catal. Lett.*, **17**: 263-267 (1981).
16. A. G. Zyskin, Yu. S. Snagovsky, and M. G. Slin'ko, *Kinetika i Kataliz*, **22**: 1031-1038 (1981).
17. A. G. Zyskin, *Teoreticheskiĭ i chislennyĭ analiz kineticheskikh zakonomernosteĭ slozhnykh geterogennykh kataliticheskikh reaktsiĭ* (Theoretical and Numerical Analysis of Complex Heterogeneous Catalytic Reaction Kinetics), Dissertation, Moscow, 1983 (in Russian).
18. A. G. Zyskin, Yu. S. Snagovsky, G. M. Ostrovsky, and M. G. Slin'ko, *Khim. Fizika*, **3**: 1730-1734 (1984).
19. A. G. Zyskin and Yu. S. Snagovsky, *React. Kinet. Catal. Lett.*, **27**: 213-217 (1985).
20. J. Horiuti, *J. Res. Inst. Catal., Hokkaido Univ.*, **5**: 1-26 (1957).
21. M. I. Temkin, *Dokl. Akad. Nauk SSSR*, **152**: 156-159 (1963).
22. M. I. Temkin, in: *Nauchnye osnovy podbora i proizvodstva katalizatorov* (Scientific Approach to Selection and Production of Catalysts), Izv. Akad. Nauk SSSR, Novosibirsk, 1964, pp. 46-47 (in Russian).
23. M. I. Temkin, *Dokl. Akad. Nauk SSSR*, **165**: 615-618 (1965).
24. M. I. Temkin, in: *Mekhanizm i kinetika slozhnykh kataliticheskikh reaktsiĭ* (Mechanism and Kinetics of Complex Catalytic Reactions), Nauka, Moscow, 1970, pp. 57-72 (in Russian).
25. G. M. Ostrovsky and Yu. M. Volin, *Metody optimizatsii slozhnykh khimiko-tekhnologicheskikh skhem* (Methods for Optimization of Complex Process Flow Sheets), Khimiya, Moscow, 1970 (in Russian).
26. A. G. Hearn, *REDUCE-2 USER'S MANUAL*, Computing Physics Group, Univ. of Utah, VCP-19, 1973.
27. F. H. Branin and S. K. Hoo, in: *Proc. of the Conf. of Numerical Methods for Nonlinear Optimization*, Univ. of Dundee, Academic Press, London, 1971, pp. 231-237.
28. K. S. Chao, D. K. Liu, and C. T. Pan, in: *Proc. IEEE Int. Symp. Circuits and Systems*, San-Francisco, New York, 1974, pp. 27-31.
29. I. I. Malkin, A. G. Zyskin, Yu. S. Snagovsky, and G. M. Ostrovsky, in: *Matematicheskie problemy khimii* (Mathematical Problems in Chemistry), Izv. Akad. Nauk SSSR, Novosibirsk, 1975, Part 2, pp. 165-171 (in Russian).
30. Yu. S. Snagovsky and G. M. Ostrovsky, *Modelirovanie kinetiki geterogennykh kataliticheskikh reaktsiĭ* (Simulation of Heterogeneous Catalytic Reactions), Khimiya, Moscow, 1976 (in Russian).
31. C. G. Broyden, *Math. Comput.*, **19**: 577-593 (1965).
32. R. Aris, in: *Proc. of the Intern. Chem. React. Eng. Conf.*, Poona, Wiley, New Delhi, 1983, Vol. 1, pp. 3-11.

33. R. Aris, *Chem. Eng. Sci.*, **38**: 45-53 (1983).
34. F. Krambeck, *Arch. Rat. Mech. Anal.*, **38**: 317-347 (1970).
35. V. M. Vasil'ev, A. I. Vol'pert, and S. I. Khudyaev, *Zh. Vychislit. Matematiki i Matem. Fiziki*, **13**: 683-697 (1973).
36. T. A. Akramov and G. S. Yablonsky, *Zh. Fiz. Khimii* **49**: 1818-1820 (1973).
37. T. A. Akramov, V. I. Bykov, and G. S. Yablonsky, in: *Matematicheskie problemy khimii* (Mathematical Problems in Chemistry), Izv. Vych. Tsentra SO AN SSSR, Novosibirsk 1975, Part 1, pp. 206-212 (in Russian).
38. M. Feinberg, *Arch. Rat. Mech. Anal.*, **46**: 1-41 (1972).
39. V. I. Bykov, A. N. Gorban', and V. I. Dimitrov, *React. Kinet. Catal. Lett.*, **12**: 19-23 (1979).
40. Yu. S. Snagovsky, *Kinetika i Kataliz*, **21**: 189-196 (1980).
41. F. S. Shub, A. G. Zyskin, M. G. Slin'ko, Yu. S. Snagovsky, and M. I. Temkin, *Kinetika i Kataliz*, **21**: 396-401 (1980).
42. M. I. Temkin, *Zh. Fiz. Khimii*, **11**: 169-189 (1938).
43. F. Horn and R. Jackson, *Arch. Rat. Mech. Anal.*, **47**: 81-116 (1972).
44. M. I. Lebedeva, in: *Matematicheskie problemy khimii* (Mathematical Problems in Chemistry), Izv. Akad. Nauk SSSR, Novosibirsk, 1975, Part 1, pp. 173-179 (in Russian).
45. N. N. Orlov, *React. Kinet. Catal. Lett.*, **14**: 149-154 (1980).
46. B. F. Gray, *Trans. Faraday Soc.*, **66**: Part 2: 363-371 (1970).
47. R. Kubo, *Thermodynamics*, North-Holland Publishing Company, Amsterdam, 1968.
48. Yu. S. Snagovsky, *Kinetika i Kataliz*, **19**: 932-941 (1978).
49. Yu. S. Snagovsky, *Kinetika i Kataliz*, **17**: 92-101 (1976).
50. M. Aigen and L. Mayer, in: *Investigation of Rates and Mechanisms of Reactions*, John Wiley and Sons, New York, London, Sydney, Toronto, 1974, pp. 79-172.
51. P. Glensdorf and I. Prigogine, *Thermodynamic Theory of Structure, Stability and Fluctuations*, Wiley-Interscience, New York, 1973.
52. G. R. Gavalas, *Nonlinear Differential Equations of Chemically Reacting Systems*, Springer-Verlag, Berlin, Heidelberg, New York, 1968.
53. A. N. Gorban', V. I. Bykov, and G. S. Yablonsky, *Kinetika i Kataliz*, **26**: 1239-1247 (1983).
54. Ya. B. Zel'dovich, *Zh. Fiz. Khimii*, **11**: 685-689 (1938).
55. Yu. V. Rumer and M. Sh. Ryvkin, *Termodinamika, statisticheskaya fisika i kinetika* (Thermodynamics, Statistical Physics, and Kinetics), Nauka, Moscow, 1972 (in Russian).
56. H. G. Othmer, *Chem. Eng. Sci.*, **31**: 993-1003 (1976).
57. H. S. Caram and L. E. Scriven, *Chem. Eng. Sci.*, 31: 163-168 (1976).
58. Yu. S. Snagovsky, in: *Khimicheskaya kinetika v katalize. Teoreticheskie problemy kinetiki* (Chemical Kinetics in Catalysis. Theoretical Problems of Kinetics), Izv. OIKhF, Chernogolovka, 1985, pp. 32-33 (in Russian).
59. A. V. Storonkin, *Termodinamika geterogennykh sistem* (Thermodynamics of Heterogeneous Systems), LGU, Leningrad, Parts 1 and 2, 1967 (in Russian).
60. R. A. Heideman, *Chem. Eng. Sci.*, **33**: 1517-1528 (1978).
61. V. F. Baibuz, V. Yu. Zitserman, L. M. Golubushkin, and Yu. G. Chernov, *Khimicheskoe ravnovesie v neideal'nykh sistemakh* (Chemical Equilibrium in Nonideal Systems), Izv. IVTAN, Moscow, 1986 (in Russian).
62. A. A. Andronov and S. E. Khaikin, *Teoriya kolebanii* (Theory of Vibration), ONTI NKTP SSSR, Part 1, 1973 (in Russian).
63. J. D. Sender, M. Kuno, W. J. Lin, et al., *Computers Chem. Eng.*, **14**: 71-85 (1990).
64. M. I. Koltsov, V. Kh. Fedotov and B. V. Alekseev, *Dokl. Akad. Nauk SSSR*, **317**, No. 2, 401-406 (1991).

2 Modelling of Polymerization Processes

E. B. Brun

State Research Institute of Chlorine,
Ugrezhskaya 2, Moscow 109088, USSR

V. A. Kaminsky

Karpov Institute of Physical Chemistry,
Obukha 10, Moscow 103064, USSR

2.1 Introduction

Mathematical modelling is now recognized as an effective tool in studying complex processes of chemical engineering. The remarkable progress that has been achieved in recent decades in various areas of chemical technology may to a great extent be accredited to practical applications of mathematical modelling methods. In a particular case of polymer synthesis processes, the use of these methods calls attention to a number of specific features. These features are mainly linked to the task of obtaining a polymer with desired performance characteristics. The route to attaining success via straightforward experimental study of the process conditions that might influence the properties of a target product appears to be little promising. First, it is practically impossible to try "blindfold" the entire variety of factors and conditions potentially capable of modifying the characteristics of a polymer product; second, even if one concedes to the practicability of such an approach under laboratory conditions, there remains largely uncertain the possibility of a scale-up to a commercial-sized technology. A poor aid in this is any formal mathematical model that ignores the specific features of a polymerization process arising from the long-chain structure of reactants and products.

The necessary requirement for a mathematical model of any polymerization process is the quantitative description of the molecular structure of polymers involved in the process. The experience has shown that the notion of a polymer specimen as a mixture of individual species (a concept traditional in the chemistry of low-molecular compounds) is in fact erroneous. For this reason, it appears necessary to consider a probabilistic approach, that is, to use functions describing the distribution of macromolecules in length (or in molecular mass), and, in the case of copolymers, branched and network polymers, also in chemical composition and structure. The statistical characteristics of a molecular polymer structure proffer the opportunity of relating the conditions for synthesis of a polymer to its properties and thus to predict the quality of a polymer product.

The mathematical modelling has revealed a number of interesting results in polymer technology; the interested reader is referred to excellent monographs and review papers [1-5]. To cite but few examples, multiple stationary states have been predicted for nonisothermal vinyl acetate polymerization in solution in a continuous stirred reactor. The relationships between the degree of conversion (or temperature) and reactor residence time have been shown to exhibit isolated regions of such states; the possibility for development of sustained oscillatory regimes has been revealed [5]. Such regimes have been shown to occur also in methyl methacrylate water-emulsion polymerization in a flow reactor [5]. In the former case, the complex dynamic behaviour is due to a nonlinear Arrhenius polymerization rate vs. temperature dependence, and in the latter, to the effect the colloidal properties of a disperse system exert on macrokinetics.

In the present paper, we focus on certain aspects of mathematical modelling of free-radical homo- and copolymerizations intimately linked to the molecular structure of a polymer—the aspect commonly ignored by most researchers in the field. Thus, we discuss, in Section 2, effects arising from the free-macroradical length distribution and the influence of diffusion-controlled reaction of kinetic chain termination on polymerization rate. Section 3 is concerned with methods for calculating statistical characteristics of linear copolymer chains and with the effect of compositional heterogeneity on the properties of copolymers, also with the prognostication of these properties. The suggested approaches are quite general in character and can be used in modelling a variety of synthetic processes and chemical modifications of polymers.

2.2 Free-Radical Polymerization

The free-radical polymerization may be regarded as a most well-studied process in polymer synthesis. Nonetheless, a major difficulty that arises in simulating the free-radical polymerization reactions is associated with the construction of a proper model on the basis of a sequential analysis of physico-chemical properties of polymerization medium. Thus, the major objective in modelling the free-radical polymerization processes is calculation of the polymerization rate and molecular weight distribution (MWD) of the polymer formed.

The kinetic mechanism of polymerization involving reactions of initiation, chain growth and bimolecular termination, as well as inhibition reactions and reactions of chain transfer to monomer, to solvent, and to other agents is known and can be summarized as shown below:

| Kinetic mechanism | Reaction rate |

Initiation

$$I \xrightarrow{k_d} 2R_{pri} \qquad V_{in} = 2fk_d I$$

$$R_{pri} \xrightarrow{k_{pri}} R_1$$

Chain growth

$$R_i + M \xrightarrow{k_p} R_{i+1} \qquad V_p = k_p M$$

Chain termination

$$R_i + R_j \begin{cases} \xrightarrow{\lambda k_t(i,j)} P_{i+j} \\ \xrightarrow{(1-\lambda)k_t(i,j)} P_i + P_j \end{cases} \qquad V_t = \sum_{i,j} k_t(i,j) R_i R_j$$

Chain transfer

to monomer

$$R_i + M \xrightarrow{k_M} P_i + R_1 \qquad V_{tr}^{(M)} = k_M R M$$

to solvent or other agents

$$R_i + S \xrightarrow{k_S} P_i + S^\cdot \qquad V_{tr}^{(S)} = k_S R S$$

Inhibition

$$R_i + Z \rightarrow \text{inert product} \qquad V_{inh} = k_Z R Z$$

Here I is the initiator; R_{pri} are primary radicals generated by initiator decomposition; R_i and P_i are, respectively, free-radical and "dead" polymer chains with polymerization degree i; M is the monomer; S is the chain transfer agent; Z is the inhibitor (the same letters are used to designate the concentration of the respective agent in the expressions of reaction rates); $R = \sum_i R_i$ is the overall concentration of free radicals; f is the efficiency of initiation.

Both bulk polymerization and solution polymerization are accompanied by a drastic change in physical properties of the medium because of the increased concentration of polymer solution, which entails an alteration in the rate constants. The rate constants of diffusion-controlled chain termination reactions undergo the most sizeable alteration. It is the decrease in these constants that has been suggested as the main cause of the effect of polymerization autoacceleration with increasing monomer conversion [6], typical of many monomers. Although the rate constant of a diffusion-controlled chain termination reaction is most certainly dependent not only on the polymer solution concentration (which determines the medium viscosity), but also on the chain length of reacting radicals, a single rate constant $k_t(V_t = k_t R^2)$ [7-14] is commonly used to define the chain termination reaction in modelling polymerization processes. As a rule, this rate constant is assumed to be a function of temperature and polymer concentration. Occasionally, dependence of this rate constant on average molecular mass is reported by certain authors [14]. An essential point is that the instantaneous MWD of a polymer as estimated by such a model is in fact the known Flory-Schulz distribution [15], even conceding that the rate constant k_t is dependent on the average MWD characteristics of the dead polymers or radicals. In what follows, we name such an approach the effective rate constant method.

Prior to discussing the models that take into account effects of radical chain length and reaction medium characteristics on the rate constant of a chain termination reaction, we consider in some detail the specific behaviour of diffusion-controlled reactions involving macromolecules in solution.

2.2.1 Physical Aspects of Chain Termination Reaction Model

A major feature of diffusion-controlled reactions between chain molecules, as distinct from reactions involving low-molecular reactants, is the wide spectrum of movements inherent in macromolecules: from microscale segmental movements to the motion of the polymer coil as a whole. A number of phenomenological models [16-19] have been attempted to account for the alteration in the movement mode due to the overlap of polymer coils. The progress achieved in the studies of dynamic properties of polymer solutions [20-25, 142] provides reasons to believe that a physically meaningful model for termination reaction can be designed, applicable both to the initial stage of polymerization and to the gel effect region. In a sense, the concepts of polymer solution physics have been used in the choice of a model for the termination reaction rate constant [26-30].

Consider a reaction between the active terminal units of like chain molecules in a polymer solution [143]. Let each of the interacting chains contains i units. Assuming the volume effect to be negligeable, one can write for the polymer coil radius, $\varrho_i \sim b(ia/b)^{1/2}$ (here b is the Kuhn segment size and a is the chain unit size). At first, we confine ourselves to considering highly flexible chains ($b/a \sim 1$). The reaction between active chain-end units is presumed to occur as they move to each other to a distance a (the so-called reaction radius).

The description of a reaction between the active units of different molecules can approximately be reduced to a diffusion equation, with a preferential singling-out of the movements that play a decisive role in changing the relative distance between these units. The distance between the polymer coils being $r > 2\varrho_i$, their relative movement can be characterized in terms of a quantity $2D_i$ (D_i being the diffusion coefficient of a single macromolecule). In a macromolecular coil overlapped with the neighbouring coils, the scale of displacement of the active unit and its effective diffusion coefficient as a major characteristic of mutual approach of active units are dependent on the size of the chain segment involved in the movement. The chain segment of n units is a major contributor to random movement of the chain to a distance of $r \sim an^{1/2}$; this distance in fact determines the size of a region this chain segment occupies in space. It follows therefore that each distance r between the active chain units in an overlapped combined coil can be brought in correspondence with the diffusion coefficient $D(r)$ which is a major factor characterizing the mutual approach of the active units. The functional form for $D(r)$ can be determined, if a dependence of diffusivity D_n on n (the number of units simultaneously involved in movement) is known. In the Zimm model [25], this dependence is $(D_n = D_1 n^{-1/2})$ $D(r) = D_1(a/r) = D_i(\varrho_i/r)$; in the Rouse model [25], $(D_n = D_1 n^{-1})$ $D(r) = D_1(a/r)^2 = D_i(\varrho_i/r)^2$. To estimate the limit stage of the chain termination reaction (translational diffusion of coils or differently scaled intramolecular movements), let us consider the probability $U_{ii}(r)$ for a reaction between polymer coils, each containing i units, in an infinite time, given the distance r between the active units. This probability satisfies the equation

$$\frac{d}{dr}\left[r^2 D_{ii}(r) \frac{dU_{ii}(r)}{dr}\right] = 0 \qquad (1)$$

2 Modelling of Polymerization Processes

the boundary conditions being $U_{ii}(a) = 1$, $U_{ii}(\infty) = 0$. Here $D_{ii}(r)$ is the relative diffusion coefficient for active units. The quantity $U_{ii}(2\varrho_i)$ determines the probability of a reaction after the collision of polymer coils:

$$U_{ii}(2\varrho_i) = \int_{2\varrho_i}^{\infty} \frac{dr}{r^2 D_{ii}(r)} \bigg/ \int_{a}^{\infty} \frac{dr}{r^2 D_{ii}(r)} \qquad (2)$$

We make use of the Smoluchowski approximation for expressing the diffusion-controlled reaction constant $k_t(i, i)$ defined as a diffusion flux of noninteracting particles per unit concentration across the spherical surface on which the reaction occurs:

$$k_t(i, i) = 16\pi D_i \varrho_i U_{ii}(2\varrho_i) \qquad (3)$$

The probability $U_{ii}(2\varrho_i)$ being of the order of unity, the limiting step of the reaction is the translational diffusion of polymer coils; these, when overlapped, generate a "compact exploration" regime which was considered by de Gennes [31]. The reaction probability has been shown to be practically independent of the true reaction radius a, and the effective reaction radius as defined by the expression $2\varrho_i U_{ii}(2\varrho_i)$ has a size about that of the coil. The transition to a compact exploration regime may be visualized through considering a particle subject to discrete random walking within a space subdivided into cells, each of size A. If the frequency at which the particle jumps between two neighbouring cells is τ^{-1}, then the total number of jumps by the particle within a period of time t is t/τ. The random motion of the particle is called noncompact if the total number of cells within a space volume $v(t)$ which the particle can visit in the time t is much greater over the number of jumps t/τ within the same period of time $t(v(t)\tau/A^3 t \sim t^\alpha, \alpha > 0)$. By contrast, the random motion is called compact if the particle can visit more than once all the cells in the space volume $v(t)$ ($\alpha < 0$) within the time t. The conditions for a transition to compact exploration regime using Eq. (2) are obtained, as distinct from de Gennes approach, without resorting to a nonstationary diffusion equation. By substituting into Eq. (2) the power expression for $D_{ii}(r)$

$$D_{ii}(r) = \begin{cases} 2D_i(2\varrho_i/r)^m; & r < 2\varrho_i \\ 2D_i; & r > 2\varrho_i \end{cases}$$

we obtain

$$U_{ii}(2\varrho_i) = \begin{cases} [1 + \ln(2\varrho_i/a)]^{-1}, & m = 1 \\ \left\{1 + \frac{1}{m-1}\left[1 - \left(\frac{a}{2\varrho_i}\right)^{m-1}\right]\right\}^{-1} & m > 1 \end{cases} \qquad (4)$$

Expressions (4) have been derived assuming that the D_{ii} vs. r relationship is the same for each r ($a < r < 2\varrho_i$). In reality, several types of chain movement effected in different time-space scales can be distinguished. Each scale is characterized by a specific D_{ii} vs. r dependence which can be derived from the mean-square displacement of a particular chain unit as a function of time ($<r^2(t)> \sim t^\beta$, $D(r) \sim r^{2(\beta-1)/\beta}$).

To be able to quantitate the probability $U_{ii}(2\varrho_i)$ and, accordingly, the chain termination rate constant $k_t(i, i)$, data on intramolecular dynamics in all scales are needed, with allowance made for the residual chain rigidity. The chain rigidity manifests itself in that the segment length b increases, when the region in which the compact exploration regime breaks down becomes larger ($r \sim b$), with the resultant decrease in probability $U_{ii}(2\varrho_i)$.

Although the chain termination, starting from the lowest conversions [32, 33], is a diffusion-controlled reaction, a comparison of its rate constants at the initial polymerization stage with those estimated by formula (3) reveals that $U(2\varrho) \ll 1$ at the initial polymerization stage. In other words, the effective reaction radius defined by the product $2\varrho U(2\varrho)$ is substantially smaller than the coil size. As the monomer conversion P grows in the course of polymerization, changed are the conditions that determine both the intramolecular dynamics and the translational mobility of polymer chains. In this process, the macroscale movements decelerate faster than the microscale movements [25], with the resultant increase in effective reaction radius. This provides a clue to the fact that the product $D \cdot U$ remains essentially invariant, and the chain termination reaction at initial stages can, as a rule, be described with the aid of a single effective rate constant k_t independent of monomer conversion, despite the quite sizeable variation in polymer chain diffusivity within this concentration range. As the reptation movement of polymer chains sets in due to the buildup of a physical entanglement network [20-22, 25], the translational diffusion coefficient decreases drastically. It may be presumed therefore that the formation of an entanglement network is the major factor conducive to a compact exploration regime for which $k_t(i, i) \simeq 16\pi D_i \varrho_i$.

Thus, in the gel effect region at $U_{ii} \simeq 1$, the main features of polymerization, associated with the termination rate constant decrease, are defined by the translational diffusion of macroradicals which, in turn, is dependent on the macroradical length and polymer solution concentration.

Expression (3) refers to a reaction between like macroradicals. However, in the course of polymerization, the macroradicals exhibit a broad chain-length distribution; therefore, one must be able to determine the chain termination rate constant $k_t(i, j)$ as a function of chain lengths i and j at an arbitrary ratio of these two variables. First, we consider a reaction between macroradicals markedly distinct in size ($i \ll j$). Since the diffusion coefficient decreases fastly enough with chain length ($D_i \sim i^{-1}$ in the Rouse model, or $D_i \sim i^{-2}$ in the reptation model), one may accept, to a first approximation, that at $r > 2\varrho_i$, the relative approach of active chains is determined by the mobility of the shorter chain only, whereas at $r < 2\varrho_i$, the intramolecular movements of both chains are equally contributive to the diffusivity. Finally, for the compact exploration regime (at $U_{ij}(2\varrho_i) \simeq 1$), we obtain $k_t(i, j) = 8\pi D_i \varrho_i (i \ll j)$. At an arbitrary i-to-j ratio, an additive model can approximately be accepted for $k_t(i, j)$, which treats nicely both extreme cases ($i = j$ and $i \ll j$):

$$k_t(i, j) = 8\pi(D_i\varrho_i + D_j\varrho_j) = k_t(1, 1)(\omega_i + \omega_j)/2 \tag{5}$$

In Eq. (5), $k_t(1, 1)$ is the rate constant for a reaction between short-chain radicals, the dimensionless parameters ω_i characterize the rate constant $k_t(i, j)$ vs. chain length relationship (they are normalized to unity at small values of i). Taking into account the coil radius ϱ_i and diffusion coefficient D_i as a function of i, we obtain the ω_i vs. i dependence for the Rouse model, $\omega_i \sim i^{-1/2}$, and for the reptation model, $\omega_i \sim i^{-3/2}$. Thus, after the formation of an entanglement network with characteristic scale i_e dependent on the polymer solution concentration ($i_e = i_e(p)$), the movement of polymer chains with $i > i_e$ is effected in such a manner that $\omega_i \sim i^{-n}(n > 1)$.

An important feature of the additive model (5), which makes this model distinct from a model of fast-coagulating rigid particles, is that in a reaction of polymer coils markedly different in size, the rate constant $k_t(i, j)$ is virtually independent of the size of the larger coil. Viewed from this standpoint, the multiplicative models of type $k_t(i, j) \sim (ij)^{-m}$ [34-39, 144] do not reflect specific features of diffusion-controlled reactions in polymer solution.

An important information on intramolecular chain movements and, in particular, on the conditions of compact exploration regime for intermolecular reactions, including the chain termination reactions, can be derived from the kinetics of intramolecular diffusion-controlled reactions. The rate constant of an intramolecular reaction between active units (for example, chain-end units) of a macromolecule can be expressed by the reaction probability $W(r, t)$ in a time t, given the initial distance r between the active chain units. Within the framework of the above diffusion model for intermolecular reactions, the equation for $W(r, t)$ can be written in the form:

$$\frac{\partial W(r, t)}{\partial t} = \frac{1}{r^2}\frac{\partial}{\partial r}\left[D(r)r^2 \frac{\partial}{\partial r} W(r, t)\right] - \frac{D(r)}{k_B T} \frac{dU(r)}{dr} \frac{\partial W(r, t)}{\partial r} \quad (6)$$

with the boundary conditions $W(a, t) = 1$, $W(\infty, t) = 0$ and the initial condition $W(r, 0) = 0$ at $r > a$ (a is the reaction radius). Here $D(r)$ is the relative diffusion coefficient of active units; $U(r)$ is the harmonic potential for the initial active-chain equilibrium distribution; k_B is the Boltzmann constant; and T is temperature. The intramolecular diffusion-controlled reaction rate constant is defined as

$$k(t) = \frac{d\overline{W}(t)}{dt} \quad (7)$$

where $\overline{W}(t)$ is the probability $W(r, t)$ averaged over the initial active-unit distance distribution.

Let us consider now the major kinetic features arising from the rate constant k_t as a function of macroradical chain length (i, j).

2.2.2 Polymerization Kinetics and Polymer Molecular Weight Distribution

The polymerization rate V is given by an expression

$$V = -\frac{dM}{dt} = k_p MR \quad (8)$$

Making use of a quasistationary condition for radical concentration ($dR_i/dt = 0$) and of a long-chain approximation enabling one to pass from summation to intergration over the radical chain length, a R_i-distribution is obtained for an additive model (5) for $k_t(i, j)$ from [40, 27]:

$$R_i = (2\gamma F^{1/2} + c)R \exp(-ci - \gamma F^{1/2} i - \gamma F^{-1/2} Q_i) \quad (9)$$

$$\gamma = \frac{(k_t(1, 1) V_{in})^{1/2}}{2k_p M}; \quad c = \frac{k_M}{k_p} + \frac{k_s S}{k_p M},$$

$$F = \int_0^\infty di \omega_i R_i/R; \quad Q_i = \int_0^i dj \omega_j$$

The factor F is related to the average rate constant of termination $\overline{k_t} = \iint di dj k_t(i, j) R_i R_j / R^2$ through a ratio $F = \overline{k_t}/k_t(1, 1)$; using expression (9) for

R_i, an implicit equation for F is obtained

$$F = (2\gamma F^{1/2} + c) \int_0^\infty di\omega_i \exp(-ci - \gamma F^{1/2} i - \gamma F^{-1/2} Q_i) \qquad (10)$$

F is a quantity enabling one to express all essential characteristics of a polymerization process, viz., the polymerization rate, $V = k_p M (V_{in}/k_t(1,1)F)^{1/2}$; the radical chain-length distribution, R (9); the average kinetic chain length $\nu = V/V_{in} = (2\gamma F^{1/2})^{-1}$; the average radical chain length $\nu_R = \nu(1 + \nu c)^{-1} = (2\gamma F^{1/2} + c)^{-1}$; the MWD of the polymer product and the statistical moments of MWD. Equation (10) allows determination of F for an arbitrary ω_i vs. i dependence and different rate constants [145].

The ω_i vs. i relationship, along with being instrumental in concrete calculations, also permits getting a deeper insight into the polymerization process. For this purpose, we explore Eq. (10) to obtain its asymptotic solutions for the case of a strong gel effect ($F \ll 1$). As has already been mentioned, the physical entanglement network that forms in a polymer solution has an entanglement scale i_e such that at $i > i_e$, the parameter ω_i decreases with i faster than i^{-1}. Two possible cases may be envisaged. If ω_i is but slightly varied with $i < i_e$, then the integral Q_{i_e} is, by the order of magnitude, practically coincident with i_e, that is, $Q_{i_e}/i_e \sim 1$. If, at $i < i_e$, the parameter ω_i varies as the $-n$th power of i ($\omega_i \sim i^{-n}$, $n > 1$; for example, $n = 1/2$ in the Rouse model), then $Q_{i_e}/i_e \ll 1$. In the former case ($Q_{i_e}/i_e \sim 1$) at $F \ll 1$, the first two terms in Eq. (10) can be neglected, and a simple algebraic equation for F is obtained,

$$F^{1/2} \ln \frac{2\gamma F^{1/2} + c}{\gamma F^{1/2} + c} = \gamma Q_\infty \qquad (11)$$

In this equation, the dependence of the chain termination rate constant on the radical chain length is characterized by a single integral parameter Q_∞. For certain ω_i vs. i relationships, the expression of Q_∞ takes the following forms. For the simplest "stepwise" model ($\omega_i = 1, i < i_e; \omega_i = 0, i < i_e$), $Q_\infty = i_e$. For a model that takes into account the decrease of ω_i only at $i > i_e$ ($\omega_i = 1, i < i_e; \omega_i = (i_e/i)^n, i > i_e, n > 1$), $Q_\infty = ni_e/(n-1)$. The weak dependence of Q_∞ on n provides a satisfactory description of the experimental data on polymerization kinetics using various models.

Depending on the ratio of the parameter c to the inverse kinetic length ν^{-1}, two cases can be distinguished ($F \ll 1$):

(1) $cv \ll 1$, $F \simeq (\gamma Q_\infty/\ln 2)^2$, $v \simeq v_R \simeq \ln 2/2\gamma Q_\infty$;
(2) $cv \gg 1$, $F \simeq cQ_\infty$, $v_R \simeq c^{-1}$; $v \simeq (4\gamma^2 cQ_\infty)^{-1/2}$

In the former case, the chain transfer reaction can be neglected; in the latter case, the chain transfer reaction is a decisive factor in the formation of a polymer.

If $Q_{i_e}/i_e \ll 1$ (for example, in a model $\omega_i \sim i^{-1/2}$, $i < i_e$), then Eq. (11) holds only at $F \ll Q_{i_e}/i_e$.

The foregoing expressions show that, in the gel effect region, the factor F and, consequently, the average termination rate constant $\bar{k}_t = k_t(1, 1)F$ become dependent, via parameter γ, on the initiation rate and the monomer concentration. The chain-transfer reaction having been neglected, $\bar{k}_t \sim V_{in}$ is obtained in the region of strong gel effect ($F \ll 1$). This signifies that in such a regime, the polymerization rate becomes independent of the initiation rate. Let us dwell on the physical reason of such a situation. Since the quantity F has the meaning of a fraction accounted for by "mobile" radicals ($F = \int d i \omega_i R_i/R$), at $F \ll 1$ the chain termination reaction is chiefly effected through the interaction of short-chain mobile radicals ($i \leq i_e$) with long-chain radicals of low mobility ($i > i_e$).

At an arbitrary c-to-v^{-1} ratio, for the kinetic order n (defined as a function of V and V_{in} by the ratio $n = d(\ln V)/d(\ln V_{in})$), we obtain from Eq. (11)

$$n = \left[\frac{\gamma Q_\infty}{c}(1 + \varepsilon)(2 + \varepsilon) + 1\right]^{-1}, \quad \varepsilon = 2cv = \frac{c}{\gamma F^{1/2}} \quad (12)$$

We have $n \sim \varepsilon$ at $\varepsilon \ll 1$, that is, $n \ll 1/2$.

Assuming that the onset of the polymerization autoacceleration is concomitant with the formation of an entanglement network, the condition for transition to the autoacceleration regime due to the termination rate constant vs. chain length relationship, $k_t(i, j)$, can be written in the form

$$v_R(p^*) \simeq i_e(p^*) \quad (13)$$

The i_e vs. p dependence as determined by scaling theory based on an equal number of blobs between the entanglement sites takes the form [21]

$$i_e(p) = i_{e0}/p^\alpha$$

where i_{e0} is the respective value for a polymer melt ($p = 1$); $\alpha = 1.25$ for a good solvent, and $\alpha = 3$ for the θ-solvent. To be noted, the applicability of scaling relations has a constraint that the blob size must be not smaller than the segmental size. Hence, an estimate for maximum conversion is obtained, to which the scaling relation can be applied; $p_{max} \simeq n^{-2}$ (n being the number of units in a segment). At $p > p_{max}$, one must use experimental data on the entanglement length i_e vs. polymer solution concentration relationship [24].

As the degree of monomer conversion p is allowed to increase, the rate constant $k_t(1, 1)$ is seen to change, which is due to the decrease in the mobility of shortchain radicals with increasing concentration of polymer solution. The decrease in $k_t(1, 1)$ over the entire range of p values can be described within the framework of the free volume theory [41, 42]:

$$k_t(1, 1) = k_t^{(0)} \exp[B(v_M^{-1} - v^{-1})] \qquad (14)$$
$$v = [v_M(1 - p) + v_P p(1 - \varepsilon)]/(1 - \varepsilon p)$$

where $k_t^{(0)}$ is the value of $k_t(1, 1)$ at $p = 0$; v_M and v_P are the free specific volumes of monomer and polymer melt; ε is the contraction factor; and B is a constant.

The polymerization acceleration due to the $k_t(1, 1)$ vs. p dependence may also be observed at $F \simeq 1$; however, this effect is more pronounced in the case of $F \ll 1$, taking into account that F is dependent on $k_t(1, 1)$ through parameter γ. As F decreases, the reactions of chain transfer and inhibition start to play an increasingly greater role in the presence of a weak inhibitor. The effect of the chain transfer reaction is determined by parameter $cv = c/2\gamma F^{1/2}$. The influence of a weak inhibitory reaction can be estimated by Eq. (10) for F, in which, with allowance made for the inhibition, the parameters c and γ should be replaced by the respective parameters $\hat{c}$ and $\hat{\gamma}$,

$$\hat{c} = c + c_Z; \quad \hat{\gamma} = (\gamma^2 + c_Z^2/16F)^{1/2} - c_Z/4F^{1/2},$$
$$c_Z = k_Z Z/k_p M$$

The polymerization rate in the presence of a weak inhibitor is described by the expression

$$V = k_p M \left\{ \left[\frac{V_{in}}{\bar{k}_t} + \left(\frac{k_Z Z}{2\bar{k}_t} \right)^2 \right]^{1/2} - \frac{k_Z Z}{2\bar{k}_t} \right\} \qquad (15)$$

An analysis of the above expression shows that, as $\bar{k}_t$ decreases with monomer conversion, the kinetic order n (determining the V vs. V_{in} relationship) also decreases until the chain-transfer and inhibition reactions can be neglected.

The enhanced role of these reactions with monomer conversion leads to an increase in the value of n to 1/2, if the inactive polymer chains are formed by the chain-transfer reaction, and to 1, if the inhibition reaction becomes predominant.

We have considered above the effects which are observed as the chain termination rate constants are made to vary in the course of polymerization, for example, as due to their dependence on the radical chain length. An account of such a dependence leads to an alteration in the kinetic function, that is, to a changed dependence of polymerization rate on kinetic constants and on monomer and initiator concentrations. The change in other kinetic constants (chain-growth rate constant and initiation efficiency) can be allowed for by making use of the derived expressions for the polymerization rate.

The above results have been obtained taking into account a quasistationary condition for free radical concentrations ($\partial R_i/\partial t = 0$). If the polymerization conditions, in particular, the initiation rate, undergo a marked alteration during the lifetime of free-radical chains, then the chain-length distribution in this case should be sought for from a nonstationary equation for R_i. As is known, of particular importance for determining the kinetic constants of polymerization are the results obtained under nonstationary conditions (for example, by making use of the rotating sector method or post-polymerization method [43]). It has been shown above that the formulation of the polymerization rate as expressed through kinetic constants under the quasistationary conditions for R_i becomes substantially altered if one takes into account the termination rate constant vs. radical chain length relationship. Let us consider, as exemplified by post-polymerization ($V_{in} = 0$, $t > t_s$), in what a manner the allowance made for a dependence of the rate constant $k_t(i, j)$ on the radical chain length affects the kinetics formalism under nonstationary (with respect to radical concentration) conditions [44].

A time distribution of radical concentration R_i is described by the equation

$$\frac{\partial R_i(\tau)}{\partial \tau} + \frac{\partial R_i(\tau)}{\partial i} = c[R(\tau)\delta(i) - R_i(\tau)] - \alpha R_i(\tau)(\omega_i + F) \tag{16}$$

in which τ is a dimensionless time variable defined by the relation $d\tau/dt = k_p M(\tau(t_s) = 0)$, taking into account the monomer concentration change during polymerization; $\alpha = k_t(1, 1)R/2k_p M$ is a dimensionless parameter; the subscript s indicates the time at which the initiation stops. The parameter F, similar to that under initiated polymerization conditions, is expressed as $F = \int di \omega_i R_i/R$.

The R_i distribution can be represented, using Eq. (16), in the form:

$$R_i(\tau) = \begin{cases} cR(\tau) \exp\left\{-ci + \int_{\tau-i}^{\tau} d\tau' \alpha(\tau')[F(\tau') - \omega_{i-\tau+\tau'}]\right\}, & i < \tau \\ R_{i-\tau}^{(s)} \exp\left\{-c\tau - \int_0^{\tau} d\tau' \alpha(\tau')[F(\tau') + \omega_{i-\tau+\tau'}]\right\}, & i > \tau \end{cases} \quad (17)$$

where $R_i^{(s)}$ is the radical chain-length distribution at the instant when initiation is stopped ($\tau = 0$).

The temporal change of the total radical concentration R after the initiation has been stopped is described by the equation

$$\frac{dR}{dt} = -k_t(1, 1)FR^2 \tag{18}$$

which allows, together with expression (17), calculating the post-polymerization kinetics whatever are the kinetic parameter values. During post-polymerization, the radical chain-length distribution is subject to alteration. Since the factor F is dependent on this distributions, it may be presumed that the simultaneous change in both F and overall radical concentration R will cause a change in the kinetic order of Eq. (18). We now consider typical extreme cases.

If the initiation has been stopped at the initial polymerization stage ($\omega_i \simeq 1$, $F \simeq 1$), then the radical concentration change is described by an ordinary differential equation,

$$\frac{dR}{dt} = -k_t R^2, \quad R = R_s, \quad t = t_s \tag{19}$$

If the initiation has been stopped in a regime of strong autoacceleration ($F_s \ll 1$), the above asymptotic expressions can be used for estimating F_s. Two scales, i_1 and i_2, can be introduced, which characterize, respectively, the lengths of short-chain mobile radicals and those of all other radicals and defined through the relations

$$\int_0^{i_1} dj \omega_j R_j = \int_{i_1}^{\infty} dj \omega_j R_j; \quad \int_0^{i_2} dj R_j = \int_{i_2}^{\infty} dj R_j$$

and their corresponding time scales $\tau_1 = i_1$ and $\tau_2 = i_2$. The τ_2 is bound from above by the value c^{-1} ($\tau_2 \leq c^{-1}$) whatever are the parameters used; i_1 is defined by the function ω_i and by the parameter $c/\alpha_s F_s$. If $\omega_i = 1$ at $i < i_e$ and $\omega_i = (i_e/i)^n$ at $i > i_e$, then

$$\tau_1 \simeq i_e[1 + (c/\alpha_s F_s)^{-1/(n-1)}]$$

The course of polymerization at $\tau \ll \tau_2$ is dependent on the ratio $c/\alpha_s F_s$. First, we consider the case when, till the instant of initiation interruption, the chain-transfer reaction may be disregarded. During the time $\tau \simeq \tau_1$, the concentration of short-chain mobile radicals drops down to a quasistationary level defined by the chain-transfer reaction,

$$R_i = cR \exp(-\alpha Q_i), \quad i < \tau \tag{20}$$

Whence, at $\tau > \tau_1$, an expression for F is obtained:

$$F = \frac{c}{\alpha}[1 - \exp(-\alpha Q_\infty)] \tag{21}$$

Substituting Eq. (21) into Eq. (18), an equation for the overall concentration of radicals is derived

$$\frac{dR}{d\tau} = -2cR[1 - \exp(-\alpha Q_\infty)] \tag{22}$$

In this particular case, at $F_s \ll 1$ and $c/(\alpha_s F_s) \ll 1$, one has $\alpha_s Q_\infty \simeq \ln 2$ and, neglecting the change in monomer concentration M in the course of postpolymerization, the final expression derived is

$$\frac{dR}{dt} = -2k_M M(1 - 2^{-R/R_s}) \tag{23}$$

As is seen, this equation has a variable kinetic order in R, the termination reaction rate being determined by the chain-transfer reaction rate.

The average length of active chains grows until the time $\tau \simeq c^{-1}$, and at $\tau > c^{-1}$, when $R \ll R_s$ and $\alpha Q_\infty \ll 1$, the radical chain-length distribution is determined by the parameter c,

$$R_i = cRe^{-ci}$$

The termination rate constant in this case is described by a second-order equation with the average constant $\bar{k}_t = k_t(1, 1)cQ_\infty \ll k_t(1, 1)$. It should be emphasized that the above relations are not in fact dependent on the concrete form of the ω_i vs. i relationship at $i > i_e$, since they contain a single integral parameter Q_∞.

2 Modelling of Polymerization Processes

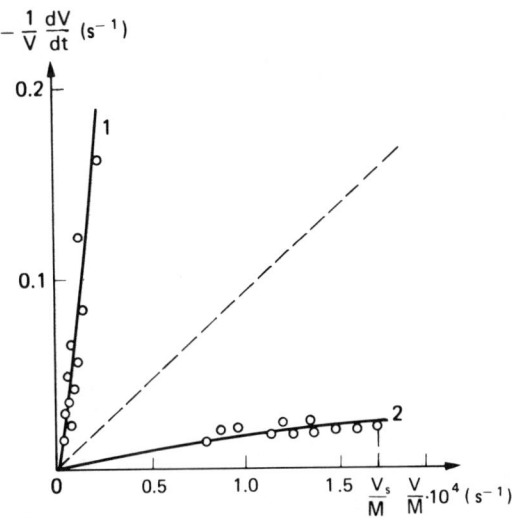

Fig. 2.1 Kinetics of methyl methacrylate post-polymerization: curve 1, $p_s = 0.07$; curve 2, $p_s = 0.35$ (curve 1 and the dashed line are calculated by Eq. (19) using k_t at $t = t_s$, while curve 2 is calculated by Eq. (23))

On the basis of the outlined concepts, we consider the published results on kinetic studies of methylmethacrylate post-polymerization at different values of p_s under the conditions of $c/\alpha_s F_s \ll 1$ [45, 46]. For the conditions specified, the onset of the gel effect corresponds to a conversion $p^* \simeq 0.15$. Figure 2.1 shows that at $p_s < p^*$ the experiment is nicely described by Eq. (19) with the value of k_t at the instant when the initiation has been stopped. However, a distinct situation emerges at $p_s > p^*$. Under continuous initiation to a conversion $p = 0.35$, a polymer has been obtained with the number-average degree of polymerization $P_N = 2 \times 10^3$. Since in the gel effect region, the value of i_e is markedly smaller than the average radical chain length and, consequently, than P_N, under the conditions specified, $\tau_1 \leqslant 10^3$, and, for example, at $k_p = 350$ litre mol^{-1} s^{-1} [26] and $M = 5.9$ mol/litre ($p = 0.35$), the corresponding time t_1 is $t_1 = \tau_1/k_p M \leqslant 1$ s. Having set $c = 1.2 \times 10^{-5}$ [47], we obtain $\tau_2 \simeq c^{-1} = 8.3 \times 10^4$ and $k_M = 4.2 \times 10^{-3}$ litre/mol^{-1} s^{-1}, whence $t_2 = \tau_2/k_p M = (k_M M)^{-1} = 40$ s. The experimental data in Fig. 2.1a for the case of $p_s = 0.35$ fall roughly into this time interval. As is seen, Eq. (23) with no fitting parameters included, describes fairly well the experiment, as distinct from Eq. (19) with a value of k_t corresponding to the effective bi-

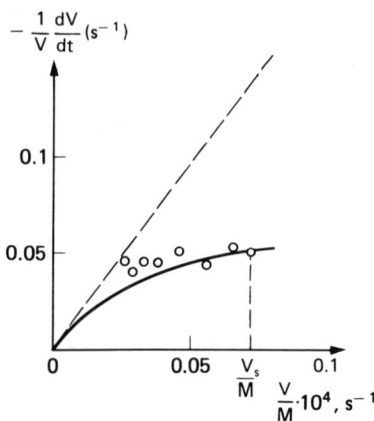

Fig. 2.2 Kinetics of styrene post-polymerization: solid line, $p_s = 0.56$ (calculated by Eq. (23); dashed line, calculated by Eq. (19) using $\bar{k}_t$ at $t = t_s$

molecular chain termination rate constant at instant t_s. After the initial sharp drop, the post-polymerization rate takes a practically constant value of V_{st} for a longer period of time until the limit conversion has been reached. According to Eq. (19) with the average rate constant $\bar{k}_t$, the characteristic time for change of concentration R and, consequently, of V is equal to $[k_t(1,1)cQ_\infty R]^{-1}$, and at concentration $R \sim 10^{-8}$ mol/litre corresponding to the value $V_{st}/M \sim 0.01\text{-}0.1\%/\text{min}$ and the value $k_t(1.1)Q_\infty \sim 10^9$ litre mol^{-1} s^{-1} (corresponding to $p = 0.35$) has been found to be of the order of 10^4 s. Considering a strong $k_t(1, 1)$ vs. Q_∞ dependence at $p > p_s$, this characteristic time can be even greater.

A similar analysis has been performed for the kinetics of styrene bulk post-polymerization at 80°C (Fig. 2.2; $p^* = 0.4$). The experimental values obtained in Ref. [48] are nicely described by Eq. (23) with $k_M = 8 \times 10^{-5} k_p$, in accordance with the data reported in the literature [49].

We now examine the effect of the chain termination rate constant vs. reacting radical chain length on the MWD of the polymer formed.

The rate for generation of inactive polymer molecules with the degree of polymerization i is given by the expression [27]

$$\frac{dP_i}{dt} = -\frac{V}{R}\frac{dR_i}{di} - \lambda R_i \int_0^\infty djk_t(i, j)R_j + \frac{\lambda}{2}\int_0^i djk_t(i-j, j)R_{i-j}R_j \quad (24)$$

where λ is the fraction of chains terminated by the combination mechanism.

Expressions for the normalized number and weight polymerization-degree distributions $f_N(i)$, $f_w(i)$ of inactive polymer molecules are easily derived by making use of an additive model for $k_t(i, j)$:

$$f_N(i) = \frac{\mu_p^{(1)}}{i\mu_p^{(0)}} f_w(i) = \frac{2}{2 - \lambda + \lambda c \nu_R} \left\{ [\theta_1 + \theta_2(\omega_i - \lambda \omega_i - \lambda F)] \right.$$

$$\times \exp(-i\theta_1 - Q_i\theta_2) + \frac{\lambda \theta_2}{\nu_R} \exp(-i\theta_1) \int_0^i dj \omega_j \exp(-Q_j\theta_2$$

$$\left. - Q_{i-j}\theta_2) \right\} \tag{25}$$

$$\theta_1 = c + \gamma F^{1/2} = \nu_R^{-1} - (2\nu)^{-1},$$
$$\theta_2 = \gamma F^{-1/2} = (2\nu F)^{-1}$$

Here $\mu_p^{(k)}$ is the kth order statistical moment for instantaneous distribution of dP_i/dt,

$$\mu_p^{(k)} = \int_0^\infty di \, i^k \frac{dP_i}{dt}$$

The number-average (P_N), weight-average (P_W), and Z-average (P_Z) degrees of polymerization are linked to $\mu_P^{(k)}$ through the known relations:

$$P_N = \frac{\mu_P^{(1)}}{\mu_P^{(0)}}; \quad P_W = \frac{\mu_P^{(2)}}{\mu_P^{(1)}}; \quad P_Z = \frac{\mu_P^{(3)}}{\mu_P^{(2)}}$$

The first two moments $\mu_P^{(0)}$ and $\mu_P^{(1)}$ are defined by expressions

$$\mu_P^{(0)} = V\left(\frac{2-\lambda}{2\nu_R} + \frac{\lambda c}{2}\right); \quad \mu_P^{(1)} = V$$

The general expression for $\mu_P^{(k)}$ ($k \geq 2$) takes the form

$$\mu_P^{(k)} = kV\mu_R^{(k-1)} + \lambda V \sum_{l=1}^{k-1} \binom{k}{l} \mu_R^{(k-l)} (l\mu_R^{(l-1)} - \theta_1 \mu_R^{(l)}) \tag{26}$$

where $\mu_R^{(k)}$ is the statistical moment of kth order for chain-length distribution of active chains R_i

$$\mu_R^{(k)} = \int_0^\infty dii^k R_i$$

Expressions for statistical polymer MWD moments in the case of a termination reaction effected only by recombination or disproportionation mechanisms have been obtained in [28]. The explicit formulas for P_N, P_W, and P_Z are given below:

$$P_N = \frac{2\nu_R}{2 - \lambda + \lambda c \nu_R}, \qquad P_W = 2(1 + \lambda)\mu_R^{(1)} - \lambda(c + \nu_R^{-1})(\mu_R^{(1)})^2,$$

$$P_Z = \frac{3}{2}\left(\frac{\mu_R^{(2)}}{\mu_R^{(1)}} + \lambda \frac{2\mu_R^{(1)} - \theta_1\mu_R^{(2)}}{1 + \lambda - \lambda\theta_1\mu_R^{(1)}}\right) \qquad (27)$$

These expressions enable calculating the statistical characteristics of a polymer whatever are the kinetic parameters and the ω_i vs. i dependence.

We now consider specific features of the MWD of a polymer in the region of strong autoacceleration ($F \ll 1$) by making use of the above asymptotic expressions for F. First, we write asymptotic expressions for the moments $\mu_R^{(k)}$ through which, according to (26), the moments $\mu_P^{(k)}$ are expressed. At $F \ll Q_{i_e}/i_e$, we obtain

$$\mu_R^{(k)} = k!(\mu_R^{(1)})^k = \frac{k!}{\theta_1^k} \qquad (28)$$

The respective asymptotic expressions for P_W and P_Z take the form

$$P_W = 2\theta_1^{-1}; \qquad P_Z = \frac{3}{2}P_W \qquad (29)$$

The moments $\mu_R^{(k)}$ at $k \geq 2$ become independent of λ for $F \ll 1$. Consequently, both P_W and P_Z are also independent of λ, that is, of the chemical mechanism for chain formation in the region of fast autoacceleration.

A point to be noted is that in the region of strong autoacceleration the chain termination occurs mainly through interaction of mobile short-chain radicals (which are a smaller part of all the radicals) with radicals of low mobility. This specific feature causes a substantial broadening of the instantaneous MWD. For example, for the polydispersity index of a polymer, $\varkappa = P_W/P_N$, we obtain, in neglecting the chain transfer reaction, $\varkappa = 2(2 - \lambda)$ rather than $\varkappa = (2 + \lambda)(2 - \lambda)/2$ at the initial stage of polymerization.

Another consequence of the specific behaviour of the termination reaction is the possibility for occurrence of a bimodal instantaneous distribution $f_W(i)$ as reported earlier by Boots [27]. For the simplest (stepwise) ω_i vs. i model ($\omega_i = 1$, $i < i_e$; $\omega_i = 0$, $i > i_e$), the condition for a bimodal instantaneous distribution $f_W(i)$ at $\lambda = 0$ (termination by disproportionation) is $i_e \theta_1 < 1$, which is with certainty valid in the region of fast autoacceleration.

The conclusions formulated above have been inferred with allowance made for a dependence of kinetic diffusion-controlled chain termination constants on the chain length of reactive macroradicals. In an approximation that neglects such a dependence (for example, the effective rate constant method), the shape of the instantaneous polymer MWD must be retained as the conversion rises, that is, the Flory-Schulz distribution must also describe the instantaneous molecular characteristics of a polymer in the region of gel effect. In particular, the polydispersity index for instantaneous MWD, $\chi = (2 + \lambda) \times (2 - \lambda)/2$, must retain its value.

The simple analytical expressions as given above for the factor F (through which the polymerization rate V is expressed) and for the statistical moments $\mu_R^{(k)}$ have been derived under the conditions of a strong gel effect ($F \ll 1$). The calculation of F can be made directly by equation (10) at arbitrary parametric values and over the entire range of monomer conversion. In analyzing the MWD of a forming polymer, it should be remembered that experimentally determined are the characteristics of a cumulative, rather than instantaneous, polymer at a given degree of conversion:

$$\overline{P_N}^{-1} = p^{-1} \int_0^p P_N^{-1} \mathrm{d}p; \quad \overline{P_W} = p^{-1} \int_0^p P_W \mathrm{d}p$$

The inclusion of k_t as a function of i and j produces a sizeable broadening of the instantaneous MWD and, consequently, leads to an increased value of polydispersity index of the cumulative polymer.

The above analysis shows that the inclusion of the chain termination rate constant vs. radical chain length relationship produces a marked change in the MWD of polymer product, especially at $F \ll 1$. As the conversion p is allowed to further increase and, accordingly, the entanglement scale $i_e(p)$ to diminish, the contribution of the chain-growth mechanism (in neglect, however, of the translational chain diffusion, $i > i_e$) to the termination reaction becomes increasingly pronounced (28, 50). Apparently, the respective termination rate constant k_t^* is independent of the radical length. The enhanced role of this chain termination mechanism leads to an increase in F despite the eventual rise of the polymerization rate. The kinetic characteristics of polymerization (kinetic order of polymerization rate in initiator and monomer concentration, polydispersity index for instantaneous polymer MWD) approach those typical of the initial polymerization stage. The condition for the predominance of chain termination by chain-growth mechanism can be expressed in the form $k_t(1, 1) \int_0^\infty d i \omega_i R_i / k_t^* R \simeq 1$. Finally, it should be pointed out that at a high polymerization degree, especially on approaching the glass-transition point of a reaction mixture, the polymerization rate drops sharply owing to the change of other kinetic characteristics, especially, initiation efficiency [51].

2.2.3 Free-Radical Polymerization in Heterogeneous Systems

In the foregoing, the free-radical polymerization in homogeneous systems has been characterized. In such systems the transfer processes exert influence on diffusion-controlled reactions, for example, chain termination reactions. In many known procedures, polymerization is carried out in heterogeneous systems (dispersion, emulsion, and suspension polymerizations). In these cases, the transfer of components is manifested on a macroscopic scale, and the involved processes determine in essence the macrokinetics of polymerization in disperse systems. Prior to starting a model analysis of such processes, let us consider the eventual occurrence of inhomogeneities in bulk polymerization providing that the monomer is a good solvent for its polymer. Basically, the possibility for the occurrence of polymer inhomogeneities is associated with the fact that in autoaccelerated polymerization ($dV/dp > 0$) the local-fluctuation amplitude of polymer concentration, Δp, can grow with time if its spatial dimension has reached a certain critical value. The temporal behaviour of concentrational fluctuation is determined by the ratio of the polymer formation rate to diffusion dissolution rate of this fluctuation region. By equating these rates, we obtain an estimate for the minimum size of a kinetically stable fluctuation: $L_{\min} \sim D_{\text{coop}}^{1/2}(p)(dV/dp)^{-1/2}$, where D_{coop} is the cooperative diffusion coefficient. By substituting the characteristic MMA polymerization values for $D_{\text{coop}} \simeq 10^{-6}$ cm^2/s [52] and $dV/dp \simeq 10^{-4}$ s^{-1} [53], we obtain $L_{\min} \simeq 5 \times 10^{-2}$ cm.

We now consider to what an extent can the fluctuation amplitude Δp grow within a period of autoaccelerated polymerization. To obtain a limiting estimate, we consider a fluctuation with $L > L_{\lim}$ assuming that the reverse process of diffusion dissolution is insignificant. During a period of autoaccelerated polymerization, the degree of monomer conversion in solution medium undergoes a change from p_1 to p_2, the respective changes within the fluctuation region being $p_1 + \Delta p_1$ and $p_2 + \Delta p_2$. Taking equal and synchronous periods of time for polymerization inside and outside the fluctuation region, we can write

$$\int_{p_1 + \Delta p_1}^{p_2 + \Delta p_2} \frac{dp}{V(p)} = \int_{p_1}^{p_2} \frac{dp}{V(p)}$$

and thus obtain an estimate for fluctuation amplitude increment Δp:

$$\frac{\Delta p_2}{\Delta p_1} \leqslant \frac{V_2}{V_1}$$

By substituting the thermodynamic fluctuation amplitude at $p = p_1$, calculated as a fluctuation ΔN of the number N of polymer molecules confined within the region of dimension $L_{\min}$ for Δp_1, we arrive at a conclusion that the density fluctuations of a polymer solution are incapable of exerting influence on the kinetics of autoaccelerated polymerization. Thus, the assumption that such systems may be regarded as homogeneous seems to be justified.

We now analyse free-radical polymerization models in heterogeneous systems.

2.2.3.1 Heterophase Polymerization of Vinyl Chloride in Batch and Continuous Reactors

Barret [54] suggested a classification for free-radical polymerization processes in heterogeneous systems and considered colloidal problems arising in the studies of dispersion polymerization in organic media. The dispersion polymerization may be exemplified by a polymerization of vinyl chloride (VC) in bulk. The polymer formed is sparingly soluble in its proper monomer and is obtained in particulate form, with no special measure provisioned for its colloid stabilization. The VC polymerization may be regarded as heterogeneous, since it proceeds in two phases: a monomer phase (the monomer containing a small amount of dissolved polymer), and a polymer phase (the polymer swollen in monomer). In constructing a mathematical model for polymerization, one must take into account phase-transfer processes involving all the components (monomer, initiator, growing chains). An essential factor is the influence of polymerization kinetics on the formation of the porous structure of the polymer. To describe the VC polymerization rate, a number of models have been proposed differing in hypothetical effects associated with chain termination reactions in monomer and polymer phases and with the influence of phase-transfer processes on polymerization rate in each phase [55-62].

It should be kept in mind that the heterophase character of polymerization presents difficulty for determining the kinetic constants of elementary reactions in each phase. The polymerization in polymer phase proceeds at high conversion, like in systems with a high polymer-in-monomer solubility. As has been shown above, it is essential, in describing a chain termination reaction under such conditions, to take into consideration the dependence of kinetic constants on the chain length. In this particular case, a welcome fact for reaction modelling is that the VC polymerization is characterized by a large rate constant for the reaction of chain transfer to monomer ($c_M \simeq 10^{-3}$), and the factor F (10) that provides a measure for the chain termination rate constant vs. chain length dependence can be represented by a simple expression $F \simeq c_M Q_\infty$, independent of the initiation rate. This indicates that the chain termination reaction in the polymer phase can be characterized in terms of a single average rate constant $\bar{k}_t$. Since the composition of the polymer phase (polymer-to-monomer ratio) remains constant until the monomer phase is present in the system, the average rate constant for a chain termination reaction in the polymer phase, as distinct from homogeneous polymerization, is independent of the degree of monomer conversion. After the monomer phase depletion, the kinetic constants are liable to alteration with a change in the polymer phase composition.

The distribution of growing chains in the bulk of polymer phase particles is determined by reactions of initiation, chain growth, chain transfer and chain termination in polymer phase with allowance made for the diffusion radical flow from the monomer phase. If the characteristic lifetime of radicals with respect to the termination reaction is shorter than the time of radical diffusion to a distance of the order of the size of a polymer phase particle, then the distribution of radicals in polymer phase becomes nonuniform, with their increased concentration in the layer adjacent to the interface. To assess the effect of various factors on the concentrational radical distribution in polymer phase and, consequently, on the kinetics of VC polymerization, let us consider a simplified model. In polymer phase, we define two regions differing in the radical concentration. We assume that, within either region, the radical distribution is uniform. The layer adjacent to the interface has a higher concentration owing to the supply of radicals from the monomer phase. We define the thickness δ of the adjacent layer subject to the condition that the average time of radical duffision into the δ-layer is equal to the average radical lifetime as determined by the chain termination reaction. The time $\bar{\tau}_\delta$ is better determined by solving the inverse problem for out-diffusion time profile in the δ-layer using an equation for spherical particles,

$$\frac{1}{r}\frac{d^2}{dr^2}(r\tau_\delta) = -\frac{1}{D_2}$$

with boundary conditions

$$\tau_\delta(\varrho) = 0; \quad \left.\frac{d\tau_\delta}{dr}\right|_{r-\delta} = 0$$

Here D_2 is the effective diffusion coefficient for macroradicals in polymer phase; ϱ is the radius of a polymer phase particle. The quantity $\tau_\delta(r)$ is the average time needed by a particle to travel a distance to the surface from an arbitrary interior point of coordinate r within the δ-layer. The averaging over the initial positions inside a spherical δ-layer yields

$$\bar{\tau}_\delta = \frac{3}{[\varrho^3 - (\varrho - \delta)^3]} \int_{\varrho-\delta}^{\varrho} \tau_\delta(r) r^2 dr = \frac{\delta^2}{3D_2} f(\delta/\varrho) \tag{30}$$

where

$$f(x) = \frac{1 - 2x + \frac{7}{5}x^2 - \frac{1}{3}x^3}{1 - x + \frac{1}{3}x^2}$$

In the extreme cases of $\delta/\varrho \ll 1$ and $\delta/\varrho = 1$, the derived average time for diffusion into a planar layer is $\bar{\tau}_\delta = \delta^2/3D_2$ and, respectively, into a spherical particle, $\bar{\tau}_\delta = \varrho^2/15D_2$.

The polymerization rate per unit volume, V, is obtained as the sum of polymerization rates in monomer phase, V_1, in the δ-layer, $V_{2\delta}$, and in the core of polymer phase, V_{2n}:

$$V = V_1 + V_{2\delta} + V_{2n}$$
$$V_1 = k_p M_1 R_1 \varphi_1 d^{-1}, \quad V_{2\delta} = k_p M_2 R_{2\delta} \varphi_{2\delta} d^{-1} \tag{31}$$
$$V_{2n} = k_p M_2 R_{2n} \varphi_{2n} d^{-1}$$

where k_p is the chain growth rate constant; M_1 and M_2 are the weight concentrations of monomer in monomer and polymer phases; R_1, $R_{2\delta}$, and R_{2n} are the radical concentrations in monomer phase, δ-layer, and the core of polymer phase; φ_1, $\varphi_{2\delta}$, and φ_{2n} are the volume fractions of the respective reactive zones.

The equations for the radical concentration in a quasistationary approximation take the form:

$$V_{\text{in}}^{(1)} - k_{t1}R_1^2 - j_{12}\varphi_1^{-1} = 0$$
$$V_{\text{in}}^{(2)} - k_{t2}R_{2\delta}^2 + j_{12}\varphi_{2\delta}^{-1} = 0 \qquad (32)$$
$$V_{\text{in}}^{(2)} - k_{t2}R_{2n}^2 = 0$$

where $V_{\text{in}}^{(i)} = 2f_i k_d I_i$ is the initiation rate in the ith phase (k_d being the rate constant for initiator decomposition; f_i and I_i, respectively, the initiation efficiency and the initiator concentration in the ith phase); k_{t_i}, the rate constant for chain termination reaction in the ith phase; j_{12} is the interfacial radical flux per unit volume of the polymerization system. It has been assumed that, at $p < 0.1$, the radicals are uniformly distributed over the entire space of monomer phase. On the other hand, considering that $k_{t_1}/k_{t_2} \geq 5 \times 10^2$, the termination reaction in monomer phase may be neglected at $p \geq 0.3$. The initiator concentrations are determined by the rate of initiator degradation and by its distribution (presumably close to thermodynamic equilibrium) between the reactive phases.

In Eqs. (32), the volume fraction $\varphi_{2\delta}$ is dependent on the depth of radical penetration into the polymer phase during the lifetime as determined by the chain termination reaction:

$$\frac{\delta^2}{3D_2} f(\delta/\varrho) = (k_{t_2}R_{2\delta})^{-1} \qquad (33)$$

In the case of a thin δ-layer ($\delta/\varrho \ll 1$), the second equation (32) can be rewritten, using the explicit expression for δ, in the form

$$V_{\text{in}}^{(2)} - k_{t_2}R_{2\delta}^2 + V_{\text{in}}^{(1)} \frac{1 - p/\alpha}{pS(p)d_{\text{VC}}} \left(\frac{k_{t_2}R_{2\delta}}{3D_2}\right)^{1/2} = 0 \qquad (34)$$

where $S(p)$ is the specific interfacial surface; α is the weight fraction of the polymer in the polymer phase; d_{VC} is the vinyl chloride density.

Equation (34) is further simplified, if the initiation in the δ-layer is neglected:

$$R_{2\delta} = \left[V_{\text{in}}^{(1)} \frac{1 - p/\alpha}{pS(p)d_{\text{VC}}}\right]^{2/3} (3D_2 k_{t_2})^{-1/3} \qquad (35)$$

In the model of interest, polymerization autoacceleration is associated not only with the accumulation of polymer phase, but also with the transition of polymer-phase radicals from an inhomogeneous to a homogeneous type of distribution as the monomer phase becomes gradually depleted. Obviously, if the overall initiation rate remains unchanged, the total amount of radicals and, consequently, the polymerization rate must increase on transition to a homogeneous radical distribution. If so, then the expected additional autoacceleration, strong enough to be manifest in the kinetic curve, will primarily be determined by the distribution of initiator between the two phases and by the rate the interfacial surface $S(p)$ decreases with monomer conversion. It is the latter dependence of $S(p)$, differing for various modes of VC polymerization (polymerization in bulk, microsuspension and suspension polymerizations), that may be related to different types of kinetic curves $V(p)$, other experimental conditions (initiation rate and temperature) remaining the same. It is seen, therefore, that the kinetics of heterogeneous VC polymerization is dependent on the characteristics of the pore structure of the polymer.

In batch reactors, as follows from the available evidence, the heterophase polymerization proceeds in a very similar manner as does the polymerization in bulk or in suspension. Minor distinctions may be attributed to different conditions for the pore structure formation: in suspension polymerization, an additional factor contributive to the polymer porosity is the interfacial tension in suspended particles.

Presented in Fig. 2.3 are the curves for the VC polymerization rate, V, and the specific interfacial surface, $S(p)$, calculated as functions of the degree of monomer conversion. The curve *1* for $S(p)$ corresponds to a system of isolated polymer particles. Presumably, this case can be envisioned in microsuspension polymerization, the size of emulsion droplets formed being of the order of 1 µm. Curve *2* is typical of bulk polymerization. Curve *3* can be assigned to suspension polymerization in which the specific PVC surface is commonly smaller than that in bulk polymerization.

Let us consider specific features of heterophase suspension polymerization as carried out in a continuous stirred reactor. Assuming the polymerization in separate particles to proceed independently (the so-called "microblock" ap-

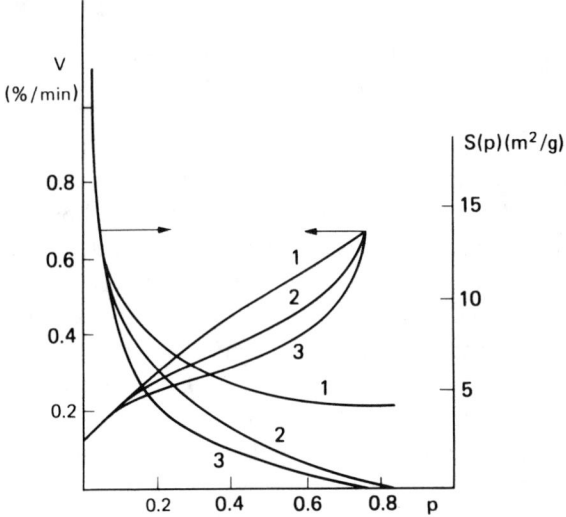

Fig. 2.3 Vinyl chloride (VC) polymerization rate as a function of degree of conversion p at different specific interface surface S

proximation), the degree of conversion in the reactor is easily estimated by averaging over the total set of particles, with allowance made for their reactor residence time distribution. Simultaneously, particles with different reactor residence time (of different age) differ in composition, that is, in polymer-to-monomer ratio, and such particles can exchange components (monomer and initiator) through aqueous phase. In this process, the monomer molecules are transported from the fresh particles to particles depleted in monomer phase. A comparison of the characteristic polymerization time and diffusion time (this latter including the times associated with the diffusion of monomer molecules from the aqueous phase into the suspended particle through the boundary diffusion layer, with the diffusion through water- or monomer-filled pores and, finally, with the diffusion in polymer-phase particles of characteristic size about 1 μm) shows that the monomer concentration in polymer phase is close to a thermodynamically equilibrated content. A change in the concentration of the initiator contained in the particles as a function of initiator decomposition rate or initiator solubility in aqueous phase occurs in a transient regime (intermediary between the kinetic and diffusion regimes).

To characterize the polymerization rate under the conditions of a continuous process, we introduce a function $f(x, y, z, t)$ for distribution of suspended particles according to their mass x, to polymer mass y and initiator mass z, at the instant t. The function $f(x, y, z, t)$ is normalized to total particle concentration at the time t by the relation

$$\iiint dx\, dy\, dz\, f(x, y, z, t) = N(t) \tag{36}$$

The equation for distribution function $f(x, y, z, t)$ takes the form:

$$\frac{\partial f}{\partial t} + \frac{\partial}{\partial x}(V_x f) + \frac{\partial}{\partial y}(V_y f) + \frac{\partial}{\partial z}(V_z f) = \frac{f^{(0)} - f}{\bar{\tau}} \tag{37}$$

Here V_x, V_y, and V_z denote, respectively, the rate of particle mass change due to a redistribution of monomer molecules between particles of different composition, the rate of polymer mass change in a particle due to polymerization, and the rate of initiator mass change; $\bar{\tau}$ is the average residence time of a particle in the reactor; and $f^{(0)}$ is the distribution function for particles in the reaction mixture fed into the reactor. All the particles can be divided into two fractions: particles of the first fractions contain both monomer and polymer phases, whereas in the particles of the second fraction, no monomer phase is present and the composition of polymer phase is determined by the limiting swelling of polymer in monomer ($y/x = \alpha$). Accordingly, the function $f(x, y, z, t)$ may be rewritten as

$$f = f_1 \eta(\alpha x - y) + f_2 \delta(\alpha x - y)$$

where $\eta(\xi)$ is the unit step function ($\eta(\xi) = 0$, $\xi < 0$; $\eta(\xi) = 1$, $\xi > 0$); $\delta(\xi)$ is the delta-function. Accordingly, for the rates V_ξ ($\xi = x, y, z$) we introduce a subscript $i(V_{\xi i}; i = 1, 2)$ to specify to which particular fraction the given quantity belongs. The polymerization rate V_{yi} can be calculated by making use of equations (31), (32).

The mass development for the particles of second fraction V_{x2} is determined by the rate of polymerization, with the subsequent swelling of the formed polymer in monomer.

$$V_{x2} = V_{y2}/\alpha$$

The mass development rate for the particles of the first fraction is determined by the following expression assuming that the mass-transfer rate is proportional to the surface area of a particle:

$$V_{x1} = -b(x/x_0)^{2/3}$$

where x_0 is the initial mass of the particle, and b is the mass-transfer parameter that can be found on condition that the total mass of all the particles remains constant:

$$\iiint dx\, dy\, dz\, V_x f(x, y, z, t) = 0$$

The polymerization rate $\overline{V}$ and the degree of monomer conversion $\bar{p}$ in the reactor are found by averaging over all the particles:

$$\overline{V}(t) = \frac{1}{N} \iiint dx\, dy\, dz\, V_y f(x, y, z, t)$$

$$\bar{p}(t) = \frac{1}{Nx_0} \iiint dx\, dy\, dz\, y\, f(x, y, z, t)$$

(38)

These expressions describe nonstationary regimes of suspension polymerization in a continuous stirred reactor.

Let us consider in some detail stationary regimes for VC suspension polymerization. We assume that the relative mixture fed into the reactor has a monodisperse distribution in monomer droplet size. In this case, under steady-state conditions, the polymerization processes within a single particle are determined by the age of the particle τ (its time of residence in the reactor). Since the variables x, y, z are functions of particle age, the particle-composition distribution function can be replaced by a particle-age distribution function $F(\tau)$. For a single continuous stirred-tank reactor, one has $F_1(\tau) = (\bar{\tau})^{-1} \exp(-\tau/\bar{\tau})$. The critical age τ_{12} can be defined as the period of time after which a particle is transported from the first into the second fraction. The value of τ_{12} is determined by the ratio $y(\tau_{12})/x(\tau_{12}) = \alpha$.

Accordingly, the polymerization rate $\overline{V}_{st}$ in the reactor and the degree of monomer conversion $\bar{p}_{st}$ under steady-state conditions are determined by the expressions below:

$$\overline{V}_{st} = \int_0^\infty d\tau V_y(\tau)\, F_1(\tau);$$

$$\bar{p}_{st} = \int_0^\infty d\tau y(\tau) F_1(\tau)$$

(39)

The variations in particle mass and in particle polymer content in the course of polymerization are convenient to represent as phase trajectories plotted on the x/x_0 vs. y/y_0 coordinates. The phase trajectories of polymer particles as

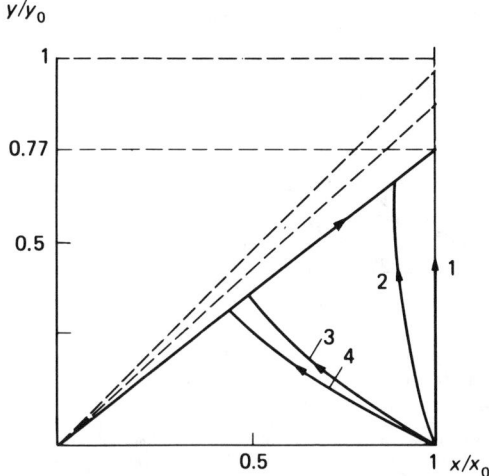

Fig. 2.4 Phase trajectories for composition variation of suspension particles in batch (1) and continuous (2-4) vinyl chloride polymerizations; initiator AIBN, I (mol/l): curve 2, $2 \cdot 10^{-2}$ ($p_{st} = 0.51$); curve 3, $3 \cdot 10^{-2}$ ($p_{st} = 0.64$); curve 4, $4 \cdot 10^{-2}$ ($p_{st} = 0.72$)

calculated for continuous VC suspension polymerization (initiator AIBN, $T = 50\,°C$) are shown in Fig. 2.4. Under the condition of a batch process in a stirred reactor, the phase trajectory is a straight line parallel to the y axis, since in the course of polymerization, there is no change of polymer content in a particle. Under the conditions of a continuous process in a stirred reactor, the fresh monomer droplets initially supply monomer molecules to particles with high polymer content ($V_{x1} < 0$); then, as the monomer phase becomes depleted, they are fed with monomer from incoming fresh droplets ($V_{x2} > 0$). In this process, the mass of a particle at the end of a certain time becomes greater over the initial mass ($x/x_0 > 1$). The intensity of monomer redistribution tends to increase with a higher initial concentration of initiator. If the rate of monomer redistribution is sufficiently high, a self-sustained oscillatory regime with a variable number of particles is possible.

The trajectories shown correspond to values of $\bar{p}_{st} < \alpha\,(1 - \alpha')$, where α' is the monomer fraction in aqueous phase. If the conversion in the reactor is $\bar{p}_{st} > \alpha(1 - \alpha')$, no monomer fraction is present in the polymer particles, while the polymer fraction in them is higher than that of the particles fully

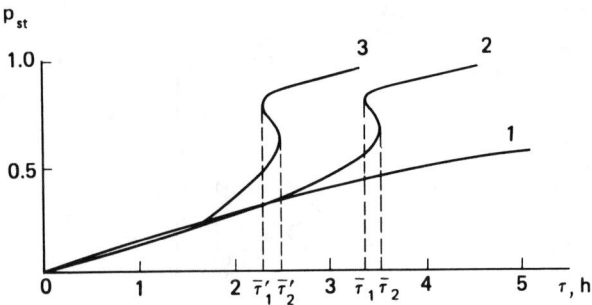

Fig. 2.5 Degree of conversion vs. average reactor residence time relationship for steady-state vinyl chloride polymerization in a continuously-stirred tank reactor:

(1) microbulk approximation (no redistribution of monomer and initiator between particles); (2) complete monomer redistribution, no initiator redistribution; (3) complete monomer-initiator redistribution

swollen in monomer ($y > \alpha x$). The composition of all the particles is the same, ($y/x = \bar{p}_{st}(1 - \alpha')$), which permits establishing a connection between the rates of continuous and batch polymerizations for this regime.

The steady-state conditions for continuous polymerization in a stirred reactor can be defined from the equation

$$\frac{\bar{p}_{st}}{\bar{\tau}} = \bar{V}_{st}(\bar{p}_{st}) \tag{40}$$

Curves depicting the $\bar{p}_{st}$ calculated as a function of average residence time $\bar{\tau}$ for continuous VC suspension polymerization in a stirred reactor (initiator AIBN, $I = 4.4 \times 10^{-2}$ mol/litre, $T = 50°C$) are shown in Fig. 2.5. The three curves correspond to different conditions for monomer and initiator redistributions. As is seen, the monomer redistribution leads to multiple stationary states within the interval $\bar{\tau}_1 < \bar{\tau} < \bar{\tau}_2$. The initiator redistribution, while affecting the shape of p_{st} vs. $\bar{\tau}$ relationship but little, leads to an increased polymerization rate within an interval of average values of $\bar{\tau}$ owing to the transfer of initiator molecules into the particles of the second fraction. In a microblock approximation, the steady-state solution is always unique.

Thus, the above results, as exemplified by VC polymerization, are illustrative, on the one hand, of specific features in modelling the heterogeneous polymerization and, on the other hand, provide evidence that the mode of distribution of monomer between polymer particles can produce a marked effect on the macrokinetic characteristics of a continuous polymerization process.

2.2.3.2 Emulsion Polymerization

The emulsion polymerization (also currently referred to as latex polymerization) is a widespread chemical engineering technique for manufacturing rubbers and latexes. The early ideas on the mechanism of emulsion polymerization were advanced by Harkins [63-65] and Yurzhenko [66]. According to these pioneering concepts, the initial reactive system contains monomer droplets dispersed in aqueous phase by mixing in the presence of an emulsifier. The major portion of the emulsifier persists in the form of micelles, whereas the initiator is dissolved in aqueous phase. The polymerization as a chemical act proceeds in polymer-monomer particles (henceforth abbreviated PMP) which are formed from monomer-containing micelles after they capture radicals from aqueous phase. The polymerization process is conventionally divided into three stages: (i) formation of PMP; this stage ends with depletion of the emulsifier which persists in micellar state and is consumed to provide a covering layer on the growing PMP surface; (ii) polymerization stage sequent to the PMP formation in the presence of monomer droplets which serve as a source for delivering the monomer molecules to PMP via aqueous-phase transport; (iii) final stage which sets in after the monomer droplets have been completely consumed. In the absence of aggregation, the number of particles is constant in stages (i) and (ii). This picture represents a conceptual basis for most works concerned with a quantitative treatment of emulsion polymerization. In this context, the pioneering works of Smith and Ewart [67] should be mentioned. In deriving the original equations for describing emulsion polymerization, commonly a number of simplifying conditions are introduced: at the first polymerization stage, the surface areas of both PMP and micelles are assumed to be invariant; at the second and third stages, the PMP concentration is assumed to be constant.

The major objectives of modelling a process are to predict, by making use of computational methods, the emulsion polymerization rate, the molecular weight distribution, and the PMP size distribution.

The problems of mathematical modelling of emulsion polymerization have been dealt with in a great number of works (see, for example, Refs. [68-70]); the main idea was to calculate the PMP concentration, N, as a function of emulsifier and initiator concentrations, and to determine the average number of radicals, $\bar{n}$, in polymer particles. In greater detail, the mathematical model for emulsion polymerization has been expounded in the work [71].

The studies in Ref. [72] show that, at the initial stage of emulsion polymerization, the formation of microemulsions is possible with microdroplets in size much superior to the monomer-solubilized micelles. The PMP generated from such microdroplets can produce a marked effect on the characteristics of emulsion polymerization. It was hypothesized in [72] that in certain instances the said effect might prove to be the decisive factor for the process. In view of this, in constructing a mathematical model for emulsion polymerization, one must take into account the eventual simultaneous formation of PMP from micelles and microdroplets.

Let us consider the structure of kinetic equations assuming that PMP can be formed not only from micelles, but also from monomer microdroplets produced by microemulsification of large drops. As distinct from thermodynamically equilibrated microemulsions [73], in emulsion polymerization thermodynamically unstable microdroplets can form whose size exceeds that of monomer-solubilized micelles and whose lifetime is long enough to enable the buildup of PMP from them [74].

The essential characteristics of emulsion polymerization (polymerization rate, PMP size distribution, and molecular weight distribution) can be calculated, given the PMP distribution functions $F_n(v, \varphi, i_1, \ldots, i_n; t)$ in volumes v, in volume fraction of monomer φ, in number n, and in lengths $i_1, \ldots, i_n$ of growing radicals at the instant t. In the particles produced from micelles, the monomer concentration, starting practically from zero-of-time, corresponds to the equilibrium swelling of a polymer particle in monomer phase with allowance made for interfacial tension ($\varphi = \varphi^*$) [75], since the initial concentration of monomer in the micelle is low. With PMP formed from moderately large microdroplets, the volume fraction of monomer φ during a time comparable to the duration of the first stage will be greater, $\varphi > \varphi^*$. Such particles cannot absorb monomer molecules.

The distribution function $F_n(v, \varphi, i_1, \ldots, i_n; t)$ satisfies a kinetic equation of the type

$$\frac{\partial F_n}{\partial t} + n \frac{\partial}{\partial v}(\theta_v F_n) + n \frac{\partial}{\partial \varphi}(\theta_\varphi F_n) = \sum_k V_k^{F_n} \qquad (41)$$

where θ_v and θ_φ are, respectively, the rates of volume increment of a particle and its monomer volume fraction increment, providing that the particle has a single growing chain; $V_k^{F_n}$ are the rates of various processes conducive to

2 Modelling of Polymerization Processes

an alteration of the distribution function (formation of PMP from micelles and microdroplets; change in the number of radicals in a particle as the radicals are captured from the aqueous phase or formed in the particle, if an oil-soluble initiator is used; change in the number of radicals in a particle by a bimolecular termination reaction; change in volume and in PMP concentration due to coalescence).

The mathematical model of emulsion polymerization includes, along with equations for the distribution function, material balance equations for each component of the system (initiator, emulsifier, and monomer).

We now attend in greater detail to certain issues that usually receive little attention in mathematical modelling of emulsion polymerization.

We consider a problem for calculating the average number $\bar{n}$ of radicals in polymer particles from the standpoint of the theory of free-radical bulk polymerization at a high degree of conversion. In fact, starting from the very onset of polymerization, the volume fraction of polymer in PMP is retained at a high level. Under such conditions, it appears reasonable to take into consideration the dependence of the termination rate constant not only on the polymer concentration in solution, but also on the chain length of interacting radicals, $k_t = k_t(i, j)$.

The system of equations for the distribution function F_n in a quasistationary approximation takes the form (variables v and φ are omitted for simplicity):

$$\frac{V_{in}}{N} \{nF_{n-1}(i_1, \ldots, i_{n-1})\delta(i_n) - F_n(i_1, \ldots, i_n)\}$$

$$+ \frac{1}{2v} \left\{ \iint dl\, dm\, k_t(l, m) F_{n+2}(i_1, \ldots, i_n, l, m) \right.$$

$$\left. - 2F_n(i_1, \ldots, i_n)(1 - \delta_{n1} - \delta_{n2}) \sum_{s=1}^{n} \sum_{r=s+1}^{n} k_t(i_s, i_r) \right\}$$

$$- k_p M \sum_{k=1}^{n} \partial F_n(i_1, \ldots, i_n)/\partial i_k = 0; \quad n = 1, 2, \ldots \tag{42}$$

Here V_{in} is the rate of formation of radicals in aqueous phase. Further, we use the additive model (5) for the termination reaction rate constant, $k_t(i, j) = k_t(1, 1)(\omega_i + \omega_j)/2$, and assume that the average time interval within which a radical is captured by the particle ($t_{in} = N/V_{in}$) is longer over the time required for the radical chain to grow to a length i_e equal to the entanglement scale; this having been surpassed, chain mobility is drastically reduced ($t_e = i_e/k_p M$). This indicates that the termination reaction proceeds primarily through interaction of the shortest-chain radical, the last one captured by the particle, with one of the longest-chain radicals resident in the particle. The termination rate constant is roughly expressed as $k_t(i, j) \simeq k_t(1, 1)\omega_i/2$ ($i < j$). In this case, the original system of equations can be simplified through replacing the distribution function $F_n(i_1, \ldots, i_n)$ by the concentration of PMP, $f_n(i)$, with the number n of growing radicals (included among these is the radical of length i, the last one to enter the particle), and by the concentration of PMP, g_n, with the number n of radicals presuming that the last-captured radical has already taken part in the termination reaction.

The functions f_n and g_n satisfy a system of quasistationary equations

$$\frac{V_{in}}{N}(g_{n-1}) + \int dj j f_{n-1}(j)\delta(i) - \frac{V_{in}}{N} f_n(i) \frac{k_t(1, 1)}{2}$$

$$\times \omega_i(n-1)f_n(i) - k_p M \frac{\partial f_n}{\partial i} = 0 \qquad (43)$$

$$- \frac{V_{in}}{N} g_n + \frac{n+1}{2v} k_t(1, 1) \int dj \omega_j f_{n+2}(j) = 0$$

The expression for f_n and g_n may be represented in the form

$$f_n(i) = \frac{N_{n-1}}{L} \exp\left[-\frac{i}{L} - (n-1)\frac{\alpha Q_i}{L}\right]$$

$$g_n = \frac{N_{n+1}}{L} \alpha(n+1) \int_0^\infty dj \exp\left[-\frac{j}{L} - Q_j\right] \qquad (44)$$

where $N_n = g_n + \int^i dj f_n(j)$ is the concentration of PMP with n growing radicals; $Q_i = \int_0^i dj \omega_j$; $\alpha = k_t(1, 1)N/2v V_{in}$; $L = k_p MN/V_{in}$.
N_n is defined by the relation

$$N_n = N_{n-1} Y_{n-1} - N_{n+1}(1 - Y_{n+1})$$

where

$$Y_n = \int_0^\infty dj \exp\left(-\frac{i}{L} - \frac{n\alpha}{L} Q_i\right)$$

As a result, we obtain expressions for distribution of particles over the number of radicals N_n and for the average number $\bar{n}$ of radicals per particle [76]:

$$N_n = \frac{N_0}{1 - Y_n} \prod_{k=1}^{n-1} \frac{Y_k}{1 - Y_k}, \quad n = 1, 2, \ldots$$

$$\bar{n} = \frac{1}{2} + \frac{\sum_{n=1}^{\infty} n \prod_{k=1}^{n} Y_k/(1 - Y_k)}{1 + \sum_{n=1}^{\infty} \prod_{k=1}^{n} Y_k/(1 - Y_k)} \tag{45}$$

Expression (45) allows one to calculate $\bar{n}$, if the ω_i vs. i relationship is known. An analysis of bulk polymerization has shown that the major features of this process can be treated within the framework of a simple step unit model for ω_i taking into account the sharp decrease of ω_i at $i > i_e$ ($\omega_i = 1$, $i < i_e$; $\omega_i = 0$, $i > i_e$). Applying this model to ω_i, we obtain from Eq. (45):

$$\bar{n} = \frac{1}{2} + \left(\sum_{n=1}^{\infty} \frac{1}{\alpha^n (n-1)!} \prod_{k=1}^{n} \frac{1 + \alpha k e^{-\gamma k}}{1 - e^{-\gamma k}} \right)$$

$$\times \left(1 + \sum_{n=1}^{\infty} \frac{1}{\alpha^n n!} \prod_{k=1}^{n} \frac{1 + \alpha n e^{-\gamma k}}{1 - e^{-\gamma k}} \right)^{-1} \tag{46}$$

where $\gamma = k_t(1, 1) i_e / 2 k_p M v$.

The condition $\gamma \gg 1$ corresponds to an approximation of fast termination ($\bar{n} \simeq 1/2$); at $\gamma \ll 1$, we obtain $\bar{n} = \gamma^{-1} \ln 2 \gg 1$. The calculation of $\bar{n}$ at the intermediate values of γ reveals that an increase in $\bar{n}$ occurs at $\gamma \geq 1$, that is, at $v \geq k_t(1, 1) i_e / 2 k_p M$. An increase in $\bar{n}$ at invariant number N of polymer particles must lead to the accelerated rate of emulsion polymerization. By substituting the parameters characteristic of MMA polymerization ($k_p = 500$ litre/mol s, $k_t(1, 1) = 10^5$ litre/mol s, $i_e = 600$), we obtain the size of particle starting from which the emulsion polymerization is expected to accelerate ($v^{1/3} \geq 10^2$ nm). The emulsion polymerization data for particles of various size have shown that the transition to autoacceleration is observed precisely in this range of particle size [77]. The above relations also define the conditions for applicability of fast termination approximation, with allowance made for the termination rate constant vs. radical chain length relationship.

Thus, the analysis outlined above allows characterizing, in a consistent and generalized manner, the kinetics of both high-conversion free-radical bulk polymerization and emulsion polymerization.

As has been noted in the foregoing, the monomer microemulsification occurs during emulsion polymerization. In this process, the formation of PMP is possible both from monomer-containing micelles and from microdroplets. In this context, it is thought of interest to focus on one of the mechanisms of PMP formation and to consider a system in which the formation of PMP from micelles would be the only possible way. To this end, we consider the emulsion polymerization at a monomer concentration not higher than the limiting concentration for micellar solubilization [78]. Under such conditions, PMP can form only from monomer-containing micelles. A specific feature of such a system is that micelles are not only the sites for formation of PMP, but also serve as reservoirs for supplying monomer to growing particles. Particles formation occurs during the entire polymerization process, and the cessation of the process is primarily due to complete depletion of monomer, rather than of emulsifier in the micellar state, as commonly is the case. Since the overall surface area of the micelles is much larger than the overall PMP surface area, the rate for PMP formation remains very little varied until the end of polymerization, and is equal to the initiation rate, $dN/dt \simeq V_{in}$. Here, an effect due to the eventual re-entry of a radical into PMP may be neglected.

The degree of conversion as a function of time $p(t)$, with allowance made for the incessant formation of new particles, takes the form:

$$p(t) = \frac{d_P}{d_M} \int_0^t d\tau \, \frac{dN}{dt} \, v(t, \tau)(1 - \varphi) \qquad (47)$$

where d_P and d_M are the polymer and monomer densities; $v(t, \tau)$ is the volume of a particle formed at the time τ; and φ is the volume fraction of monomer at the time t.

In order to be able to calculate the degree of conversion $p(t)$ and the polymerization rate dp/dt, one must know the volume fraction $\varphi(t)$ for monomer at the time t; the $\varphi(t)$ can be determined from the material balance equation assuming that the distribution of monomer between PMP, micelles, and aqueous phase is close to an equilibrium. In addition, if the monomer concentration is much lower than the solubilization limit, the linear proportionalities

$$\frac{\varphi(t)}{\varphi(0)} = \frac{\varphi_m(t)}{\varphi_m(0)} = \frac{\varphi_w(t)}{\varphi_w(0)}$$

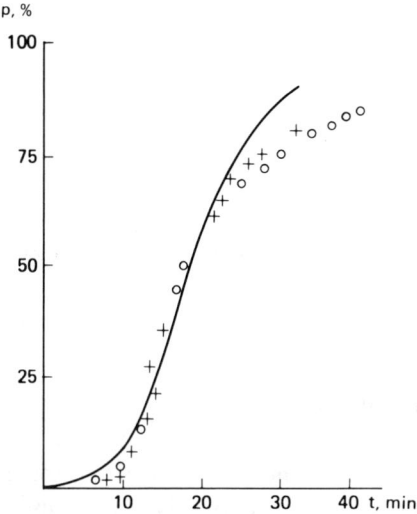

Fig. 2.6 Temporal dependence of degree of conversion for emulsion polymerization of styrene at monomer-to-water ratio 1:200 (+) and 1:100 (•); initiator $K_2S_2O_8$, $I = 3.75 \cdot 10^{-4}$ mol/l Calculated curve is shown in solid line

can be used in evaluating the distribution of monomer between PMP, $\varphi(t)$, micelles, $\varphi_m(t)$, and aqueous phase, $\varphi_w(t)$. Providing $\varphi(0) \ll 1$, simple approximate expressions can be obtained for $p(t)$ and dp/dt:

$$p(t) \simeq 1 - \exp(-\alpha t^2); \quad dp/dt \simeq 2\alpha t \exp(-\alpha t^2) \qquad (48)$$

where $\alpha = k_p \varphi(0) V_{in}/2M_\Sigma (1 - \varepsilon \varphi(0))$; $\varepsilon = (dp/d_M - 1)$ is the polymerization contraction; M_Σ is the net monomer concentration at zero-of-time.

Although the relation for $p(t)$ formally is not expected to exhibit a constant rate in the range of moderate conversions ($p \simeq 0.1$-0.6), the polymerization rate varies but little, which should be attributed to the simultaneous effect of two factors: everincreasing number of PMP in the system, and polymerization deceleration within a separate particle because of monomer depletion. Figure 2.6 shows a comparison of the results for $p(t)$, calculated by Eq. (48) with the available experimental data on emulsion polymerization of styrene at monomer/aqueous phase ratios of 1:200 and 1:100 [78] (initiator $K_2S_2O_8$, $I = 3.75 \times 10^{-4}$ mol/l).

The equation

$$\frac{\partial F}{\partial t} + \theta_v \frac{\partial F}{\partial v} = V_{in}\delta(v - v_0)$$

describes a volume distribution for PMP, $F(v, t)$, providing that each polymer particle contains only one radical (the probability for occurrence of two radicals in a single particle is presumed to be negligeably small). Here θ_v is the rate of particle volume growth at low monomer content and is expressed as

$$\theta_v \simeq k_p \frac{\mathrm{d}M}{\mathrm{d}P} \varphi(0) \exp(-\alpha t^2)$$

It should be emphasized that in calculating the polymer MWD in such a system (disregarding an eventual occurrence of two radicals in a particle and, therefore, the chain termination by a bimolecular reaction), the major factor to be accounted for is the chain transfer reaction. By making use of the growing chain-length PMP distribution function, $f(i, t) = \int \mathrm{d}v f(v, i, t)$, and the dead chain-length PMP distribution function (associated with the chain transfer to monomer), $g(i, t) = \int \mathrm{d}v g(v, i, t)$, the following equations are derived:

$$\frac{\partial f}{\partial t} = \left(V_{in} + k_M M \int \mathrm{d}j\, f\right) \delta(i) - k_p M \frac{\partial f}{\partial i} - k_M M f \quad (49)$$

$$\frac{\partial g}{\partial t} = k_M M f$$

where k_M is the reaction rate constant for chain transfer to monomer.

Figure 2.7 shows a comparison of calculated profiles for the weight fraction of the chains composed of l monomer units with the available experimental data [79] for emulsion polymerization of styrene of a low monomer concentration (monomer/aqueous phase ratio = 1:66).

To briefly summarize, the studies on emulsion polymerization in a system with a low monomer concentration have shown that the formation of PMP and the polymerization occurring therein proceed under kinetic conditions, that is, the monomer diffusion produces little (if any) effect on the macrokinetics of the process. In the real conditions of emulsion polymerization, when at the initial stage large monomer drops, monomer-containing micelles and microdroplets are present in the system, the PMP size distribution is primarily determined by the formation of polymer particles from both the micelles and the microdroplets.

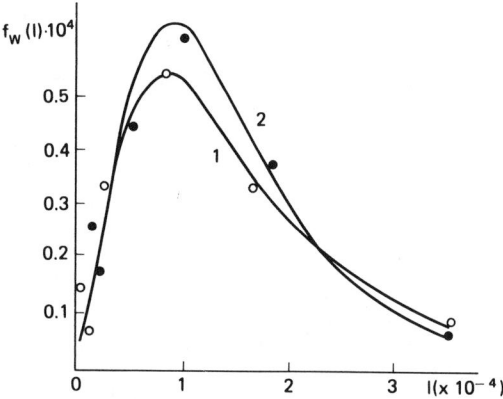

Fig. 2.7 Molecular weight distribution for a polymer obtained by emulsion polymerization of styrene at monomer-to-water ratio 1:66; degree of conversion $p = 0.3$ (curve 1) and 0.5 (curve 2)

In discussing microemulsions, one must distinguish between thermodynamically and kinetically stable types and apply, accordingly, a thermodynamic or a kinetic approach to their description. A consistent thermodynamic approach in treating the microemulsions implies a simultaneous analysis of phase diagrams, both in examining the phase separation and phase microstructure. Most commonly, in the analysis of microemulsions in a quasiternary system made up of water, "oil", and a surfactant, a variety of phenomenological parameters, such as the spontaneous interfacial curvature, the interface elastisity, and others, are made use of [80-85]. This allows expressing the size of stable microdroplets through these parameters, but does not permit constructing phase diagrams. The notion of interfacial tension, as applied within the framework of a lattice-cell model to subdivide the system into cells each accommodating in its interior one of the components (water, "oil") with the surfactant molecules concentrated at the interface, is used down to a molecular-scale level, with an arbitrary constraint imposed on the minimum cell size [86]. A consistent description of phase diagrams and the determination of the phase microstructure can be given within the framework of a lattice-cell model using the molecular interaction energies as parameters. Given the molecular interaction parameters, a phase diagram is determined assuming that the chemical potentials μ of all the components in different phases are the same. Thus, three different solutions (three roots) to a system of equations

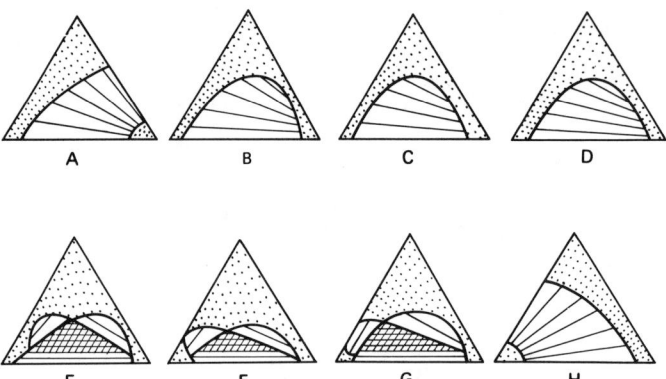

Fig. 2.8 Phase diagrams for three-component solutions containing an amphilic component. Parameters used in calculations are:

$\alpha_{21} = 0.17$; $\alpha_{13} = 9$; $\alpha_{24} = 3$; $\alpha_{34} = 1$; A: $\alpha_{14} = 1.3$; $\alpha_{23} = 0.1$; B: $\alpha_{14} = 1.4$; $\alpha_{23} = 0.18$; C: $\alpha_{14} = 1.3$; $\alpha_{23} = 0.3$; D: $\alpha_{14} = 1.1$; $\alpha_{23} = 0.3$; E: $\alpha_{14} = 0.4$; $\alpha_{23} = 0.3$; F: $\alpha_{14} = 0.19$; $\alpha_{23} = 0.3$; G: $\alpha_{14} = 0.1$; $\alpha_{23} = 0.3$; H: $\alpha_{14} = 0.1$; $\alpha_{23} = 0.9$.

$\mu_n^{(1)} = \mu_n^{(2)} = \mu_n^{(3)}$ characterize a three-phase region (the subscript n denotes the nth component of the system). The two-phase equilibrium region is determined through solving a system of equations $\mu_n^{(1)} = \mu_n^{(2)}$. The remaining region in the Gibbs triangle corresponds to a homogeneous solution. The computed results [87] for phase diagrams derived by the lattice-cell model [88] in a quasichemical approximation are shown in Fig. 2.8. It has been assumed that the three-component system contains two immiscible compounds whose molecules are located each at a single site in the lattice ($n = 1, 2$), and one amphiphilic component s whose molecule consists of two groups denoted by the respective indices 3 and 4 and occupies two sites in the lattice. Parameters α_{ij} are expressed through the molecular interaction energies ε_{ij} as shown below:

$$\alpha_{ij} = \exp\left(-\omega_{ij}/k_B T\right)$$
$$\omega_{ij} = \varepsilon_{ij} - (\varepsilon_{ii} + \varepsilon_{jj})/2, \quad i, j = 1, \ldots, 4$$

The values of unary N_i and binary N_{ij} correlation functions allow one to judge about the phase microstructure. The binary functions N_{ij} define the number of contacts between molecules (or groups) of type i and j, and allow drawing a conclusion on the eventual formation of aggregates from amphiphilic molecules and on the solubilization of one of the mixture components by these aggregates. Such an approach enables difining the type of the microemulsion phase formed depending on the values of molecular interaction parameters and the system composition.

As distinct from thermodynamically equilibrated microemulsions, the formation of microemulsions under nonequilibrium conditions with no interference due to a mechanical agitation is commonly referred to as the quasispontaneous microemulsification. A causative factor in such a microemulsification is the surfactant interphase transfer associated, at the initial stage, with a nonequilibrium distribution of the surfactant between the phases. The interphase transfer of surfactant can lead to an interfacial instability which, when further advanced, results in a breakdown of the interface boundary followed by the formation of microdroplets [89, 90]. Major parameters determinative of interfacial instability are the surface activity of the surfactant and the relation of diffusion coefficient to kinematic viscosity in different phases [89]. Simultaneously, the extent of interfacial instability is strongly dependent on the phase partition coefficient of a given surfactant. It may be presumed therefore that the intensity of microemulsification is also dependent on the direction of surfactant mass-transfer and is expected to increase as the surfactant is introduced into the phase of lower solubility. The available experimental data on the effect produced by a surfactant on the rate of microemulsion layer growth at planar interface lend support to this hypothesis [74].

Thus, the characteristics of emulsion polymerization are essentially dependent on the initial state of the emulsion system which, in turn, is dependent on the mode employed in preparing the emulsion.

2.2.4 Nonisothermal Processes of Free-Radical Polymerization

The free-radical polymerization belongs to a class of exothermal reactions proceeding with an intense evolution of heat. Therefore, nonisothermal conditions can occur practically in any mode the free-radical polymerization is implemented (in bulk, solution, or suspension). The thermal conditions for polymerization affect both the process rate and the molecular weight distribution of polymer. A rise in temperature produces a sharp acceleration of polymerization and, under slow heat removal, leads to conditions which are not much different from an adiabatic regime. In this connection, it is of interest, staying within the framework of the gel effect theory as outlined above, to consider in greater detail the specific features of free-radical polymerization under nonisothermal conditions; in particular, we shall make use of certain concepts developed in the theory of combustion for highly exothermal reactions in an attempt to define the conditions for a transition of free-radical polymerization to an adiabatic regime (thermal explosion conditions).

Commonly, in calculating the polymerization rate V and the statistical MWD moments for a polymer (number-average, P_N, and weight-average, P_W, degrees of polymerization) under nonisothermal conditions, the Arrhenius relations are used for the kinetic constants of initiation reactions (k_d), chain propagation (k_p), chain termination (k_t), also chain transfer to monomer (k_M) or to solvent (k_S). To calculate V, P_N, and P_W, the expressions are used which are, however, valid only at the initial stage of polymerization [7-13]:

$$V = k_p M (V_{in}/k_t)^{1/2}$$
$$P_N = (c + V_{in}/V) \qquad (50)$$
$$P_W = (2 + \lambda)(2 - \lambda) P_N / 2$$

The approach in which formulas (50) are assumed to hold also at high conversions, while admitting a variation in the kinetic constants themselves, may be termed the effective rate constant method. However, such an approach fails to provide an explanation to a number of features even for the polymerization under isothermal conditions: see, for example, the termination rate constant vs. initiation rate dependence [91], dependence of the kinetic order of polymerization on the initiation rate [53], and a marked broadening of MWD as a function of conversion [50]. In addition, the substantial rise in the activation energy of a chain termination reaction with conversion also remains poorly understood, which is at variance with the concept of a diffusion control of this reaction.

We shall apply the above concepts on high-conversion polymerization to an analysis of nonisothermal polymerization process taking into account the termination rate constant vs. radical chain-length relationship.

An expression for effective activation energy $E_t^{(ef)}$ of a chain termination reaction can be derived, taking into account the equality $\bar{k}_t = k_t(1, 1)F$, Eq. (10) for F, and the relation $E_t^{(ef)} = R_g \, d(\ln \bar{k}_t)/d(T^{-1})$

$$E_t^{(ef)} = E_t + m(E_{in} + E_t - 2E_p) + 2m'(E_{tr} - E_p)$$
$$m = \frac{1}{2} d(\ln F)/d(\ln \gamma)$$
$$m' = \frac{1}{2} d(\ln F)/d(\ln c) \qquad (51)$$
$$E_t = R_g d[\ln k_t(1, 1)]/d(T^{-1})$$

The parameters m and m' depend on the conditions for performance of a process and are liable to variation with a changing degree of monomer conversion and temperature. In the absence of gel effect ($F = 1$), one has $m = m' = 0$ and $E_t^{(\text{ef})} = E_t$. With a strong gel effect ($F \ll 1$), we obtain, using the asymptotic expression (11) for F,

$$m = (1 - 2m')$$
$$m' = [1 + \gamma^2 Q_\infty c^{-1}(1 + \varepsilon)(2 + \varepsilon)]^{-1} \qquad (52)$$

where $\varepsilon = c/(\gamma F^{1/2})$ is a parameter characterizing the polymer fraction formed by a chain-transfer reaction. If the chain-transfer reaction is neglected ($\varepsilon \ll 1$), one may write: $E_t^{(\text{ef})} \simeq E_{\text{in}} + 2(E_t - E_p)$. It is seen, therefore, that the gel effect leads to a significant increase in the observed activation energy of the chain termination reaction if estimated by kinetic expressions (50). A decrease in F and, consequently, in $\bar{k}_t$, in the gel-effect region enhances the role of the chain-transfer reaction. At $\varepsilon \gg 1$ ($m \simeq 0$, $m' \simeq 0.5$), we obtain $E_t^{(\text{ef})} = E_t + E_{tr} - E_p$. Thus, $E_t^{(\text{ef})}$ exhibits a nonmonotonic variation: the onset of the gel effect is concomitant with a rise in $E_t^{(\text{ef})}$ owing to the increasing parameter m, and the enhanced chain-transfer influence can subsequently lead to a drop in $E_t^{(\text{ef})}$. A similar effect on polymerization kinetics associated with a weak inhibitory reaction can be analyzed by making use of Eq. (15).

By way of example, given below are expressions for kinetic orders n and l (in terms of parameters m and m') that define the polymerization rate as a function of the initiation rate ($n = d(\ln V)/d(\ln V_{\text{in}})$) and monomer concentration ($l = d(\ln V)/d(\ln M)$); $n = (1 - m)/2$, $l = 1 + m + m'$.

Both the gel effect, associated with the influence of polymerization medium viscosity on the chain termination rate constant, and the warm-up due to the initiation reaction acceleration lead to an increase in the polymerization rate. It is to be noted, however, that a rise in the initiation rate weakens the gel effect. The simultaneous intervention of these two factors is nicely exemplified by suspension polymerization under adiabatic conditions. The kinetic conditions for polymerization are primarily defined by the bulk properties of polymerization medium, whereas the extent of warm-up is determined by phase composition, that is, by the fraction of disperse phase. The calculated F-factor curves for adiabatic suspension polymerization of MMA at different phase compositions are shown in Fig. 2.9 (initiator AIBN). The volume fraction of monomer being large, the warm-up in the system almost completely inhibits the gel effect ($F \simeq 1$); as distinct, at $\varphi < 0.2$, the polymerization proceeds under conditions close to isothermal. A time dependence for the degree of

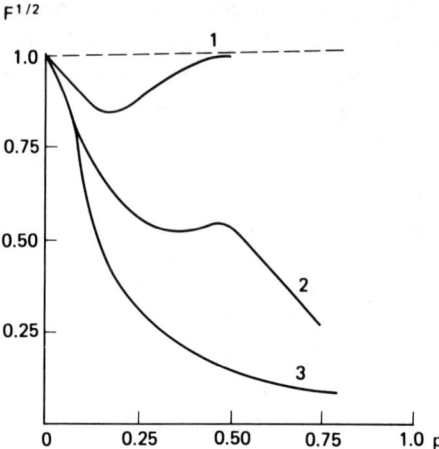

Fig. 2.9 $F^{1/2}$ vs. p relationship for adiabatic suspension polymerization of methyl methacrylate at monomer-to-water ratio φ_s 0.6 (curve 1), 0.4 (curve 2), 0.2 (curve 3); $T_0 = 293\,°C$; AIBN = 0.5 w % to monomer

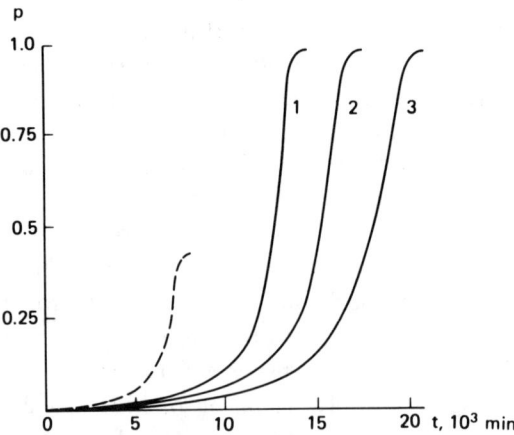

Fig. 2.10 Kinetics of adiabatic suspension polymerization for methyl methacrylate at different monomer-to-water ratio φ_s: 0.6 (curve 1); 0.4 (curve 2); 0.2 (curve 3); dashed curve refers to bulk polymerization

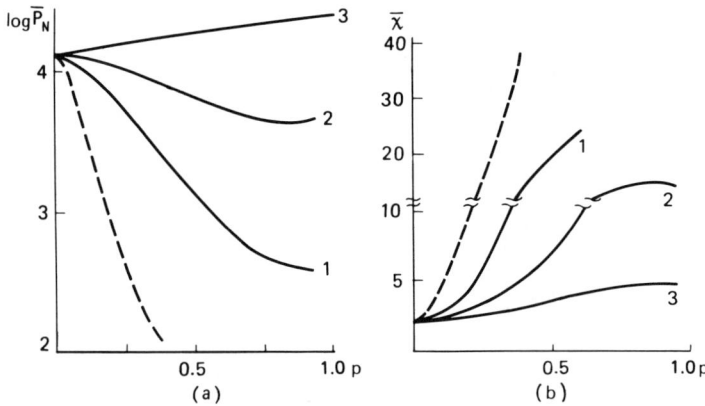

Fig. 2.11 Relationships $\overline{P_N}$ vs. p (curve a) and $\bar{x}$ vs. p (curve b) for adiabatic suspension polymerization of methyl methacrylate (for notations see Fig. 2.10)

monomer conversion under the same conditions is shown in Fig. 2.10. As is seen, in intense warm-up, the polymerization stops because of fast initiator decomposition, and no full conversion is reached. The respective relationships between the number-average degree of polymerization $\overline{P_N}$ and the polydispersity index $\bar{x} = \overline{P_W}/\overline{P_N}$ of cumulative polymer as defined by relations

$$\overline{P_N}^{-1} = p^{-1} \int_0^p P_N^{-1} \mathrm{d}p,$$

$$\overline{P_W} = p^{-1} \int_0^p P_W \mathrm{d}p$$

are represented in Fig. 2.11.

In distinction to suspension polymerization in which the effect of phase composition on polymerization kinetics is mediated only by the extent of warm-up, the concentrational change of an inert solvent in solution polymerization produces a straightforward effect on polymerization kinetics. The calculated results for the kinetics of adiabatic solution polymerization and the polymer MWD characteristics are shown in Figs. 2.12 and 2.13. An increase in solvent concentration leads to a sizeable (in comparison with suspension polymerization) increase in polymerization time. The incompleteness of the conversion degree is determined by depletion of the initiator. An increase in solvent concentration lowers the molecular mass of polymer formed at the

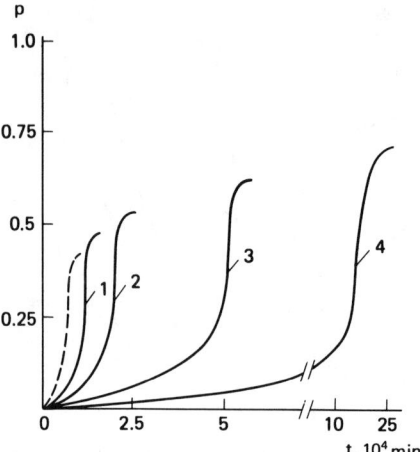

Fig. 2.12 Kinetics of adiabatic methyl methacrylate polymerization in solution at different monomer-to water ratio φ_s: 0.8 (curve 1); 0.6 (curve 2); 0.4 (curve 3); 0.2 (curve 4); dashed curve refers to bulk polymerization

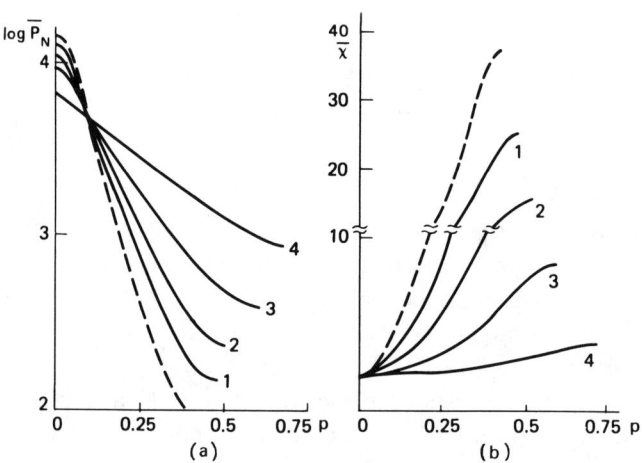

Fig. 2.13 The $\log \overline{P_N}$ vs. p and $\overline{x}$ vs. p relationships for adiabatic methyl methacrylate polymerization in solution (for notations, see Fig. 2.12)

initial stage; however, the attendant drop in warm-up leads, at the final polymerization stage, to values of $\overline{P_W}$ much the same for different solution concentrations. Simultaneously, an increase in solvent concentration is accompanied by a decrease in polydispersity index.

The presented results bear upon the adiabatic polymerization conditions. Under heat removal, the thermal regime is subject to variation within a wide range—from isothermal to adiabatic. To define a thermal regime, one must consider a system of equations for material and heat balances, taking into account the heat release by polymerization and the heat removal through a cooling surface. For a stirred reactor coupled to a simple cooling system at a given temperature T_0 of the cooling surface, provided the rate of heat removal is proportional to the difference $(T - T_0)$ (T being the reactor temperature), we obtain a system of equations in dimensionless variables [92]

$$\frac{dp}{d\tau} = e^\theta f(p, \theta)$$

$$\gamma_1 \frac{d\theta}{d\tau} = e^\theta f(p, \theta) - \frac{\theta}{\varkappa} = 0 \qquad (53)$$

$$\theta = \frac{E_{v0}(T - T_0)}{R_g T_0^2}; \quad \gamma_1 = \frac{dC_p R_g T_0^2}{QE_{v0}}$$

$$\varkappa = \frac{QE_{v0} V_0 W}{\chi S R_g T_0^2}$$

Here V_0 and E_{v0} are the initial rate and the activation energy of polymerization; R_g is the universal gas constant; d and C_p are the density and the heat capacity of reaction system; Q is the heat of polymerization; W and S are the reactor volume and the cooling surface area; and $\varkappa$ is the heat-transfer coefficient. Through equating the right-hand side of the second equation (53) to zero, we obtain an equation enabling one to arrive at a solution quasistationary in temperature. The applicability of the quasistationary temperature approximation is defined by a condition of the characteristic time for heat removal being much shorter than the time of polymerization; $\gamma_1 d\theta/dp \ll 1$. The system of equations (53) allows calculating, in a quasistationary regime, the temperature depending on the degree of conversion and the conditions for process implementation. The nonexistence of a solution for this system of equations signifies the transition of the reaction to a regime of thermal explosion. The extent of warm-up Q_{exp} at the time of transition to such a regime and the respective degree of conversion p_{exp} are defined by the value of parameter λ. There exists a critical value of this parameter, $\varkappa_{cr}$, such that at $\varkappa < \varkappa_{cr}$, the polymerization proceeds in a quasistationary regime up to limiting conversions, whereas for $\varkappa > \varkappa_{cr}$, a thermal explosion occurs at a finite degree of monomer conversion $p = p_{exp}$ and at a warm-up $\theta = \theta_{exp}$. At $\varkappa = \varkappa_{cr}$, both p_{exp} and θ_{exp} reach critical values, $p_{exp} = p_{cr}$, and $\theta_{exp} = \theta_{cr}$. The critical values $\varkappa_{cr}$, p_{cr}, and θ_{cr} can be obtained by solving the system of equations [93]:

$$e^{\theta} f(p, \theta) - \frac{\theta}{\varkappa} = 0 \tag{54.1}$$

$$\theta \left(1 + \frac{\partial}{\partial \theta} \ln f(p, \theta)\right) = 1 \tag{54.2}$$

$$\frac{\partial}{\partial p} f(p, \theta) = 0 \tag{54.3}$$

At $\varkappa > \varkappa_{cr}$, Eqs. (54.1) and (54.2) are used to calculate $\theta_{exp}(\varkappa)$ and $p_{exp}(\varkappa)$. At $\varkappa < \varkappa_{cr}$, the maximum warm-up $\theta_{max}(\varkappa)$ and the respective degree of conversion $p_{max}(\varkappa)$ are calculated by Eqs. (54.1) and (54.3).

Equations (53) and (54) hold for an arbitrary exothermal reaction. For free-radical polymerization, the kinetic function $f(p, \theta)$ in Eqs. (53) and (54) takes the form

$$f(p, \theta) = (1 - p)[k_{t0}/k_t(1, 1)F]^{1/2}$$

where k_{t0} is the chain termination rate constant at the initial stage of polymerization. An analysis of critical conditions for thermal explosion in free-radical polymerization without regard for the kinetic features in the region of gel effect has been given in Ref. [94]. At low warm-up ($\theta \ll 1$), the function $f(p, \theta)$ may be represented in the form

$$f(p, \theta) = f(p, 0) e^{-\beta \theta}$$

$$\beta = \frac{E_t^{(ef)} - E_{t0}}{2 E_{v0}} \tag{55}$$

where $E_{t0} = E_t|_{p=0}$, $E_{v0} = E_v|_{p=0}$; and E_v is the activation energy for polymerization rate. An increase in $f(p, 0)$ with rising p due to the gel effect can be offset by a decrease in the factor $e^{-\beta \theta}$ as the warm-up intensifies. It is clear that $\beta = 0$ at $p = 0$.

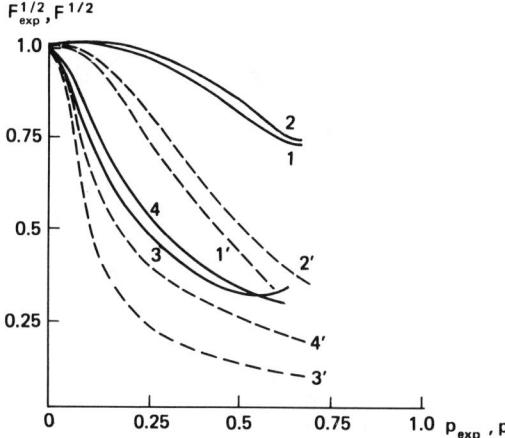

Fig. 2.14 The $F_{exp}^{1/2}$ vs. p_{exp} relationship for thermal explosion regimes of nonisothermal methyl methacrylate bulk polymerization at different rate parameters. The curves 1'—4' are respective F_{exp} vs. p relationships for isothermal MMA polymerization ($\theta = 0$) curves 1, 1', $\gamma = 5 \cdot 10^{-4}$, $c = 10^{-5}$; curves 2, 2', $\gamma = 5 \cdot 10^{-4}$, $c = 10^{-4}$; curves 3, 3', $\gamma = 5 \cdot 10^{-5}$, $c = 10^{-5}$; curves 4, 4', $\gamma = 5 \cdot 10^{-5}$, $c = 10^{-4}$

A consequence to the temperature dependence of function f in the gel-effect region is the rise of θ_{exp}, signifying an extension of the polymerization region under quasistationary conditions that are described by Eqs. (53). The calculation of critical conditions confined to the applicability of Eqs. (53) is performed, as has been noted above, on the basis of Eqs. (54). In assessing the critical conditions for thermal explosion, a measure allowing proper consideration of the dependence of the chain termination rate constant $k_t(i, j)$ on i and j is the deviation of factor F from unity. We wish to refer, by way of example, to the results of thermal regime calculations for MMA polymerization in bulk [94].

The $F^{1/2}$ vs. p_{exp} relationships, that is, those calculated at a thermal explosion limit, are shown in Fig. 2.14. The points in $\varkappa$ curves *1-4* have been calculated at different values of parameter $(\varkappa(p_{exp}) = \chi_{exp})$ and, therefore, correspond to various polymerization conditions. For comparison, shown in Fig. 2.14 are $F^{1/2}$ vs. p relationships for isothermal polymerization, $\theta = 0$ (curves *1-4*). As is seen, the values of $F^{1/2}$ at a thermal explosion limit ($\theta = \theta_{exp}$) are markedly

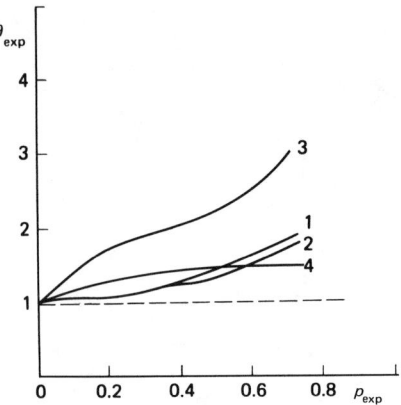

Fig. 2.15 The θ_{exp} vs. p_{exp} relationship for nonisothermal methyl methacrylate polymerization (for notations, see Fig. 2.14)

greater over their counterparts for isothermal polymerization. The limiting warm-up values corresponding to thermal explosion are shown as a function of p_{exp} in Fig. 2.15. The dependence of warm-up parameter on p under various conditions of heat removal is shown in Fig. 2.16; the intersection with the dotted line corresponds to thermal explosion. The calculations have been performed under a constraint of $p < 0.75$, that is, confined to the conversion region within which the polymerization rate is on the rise. An estimation of

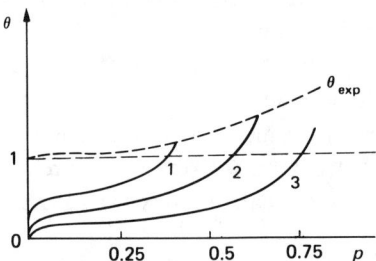

Fig. 2.16 The θ vs. p relationship for nonisothermal methyl methacrylate polymerization ($\gamma = 5 \cdot 10^{-4}$, $c = 10^{-5}$) at different values of heat-transfer parameter $\varkappa$: 0.1 (curve 1); 0.3 (curve 2); 0.3 (curve 3). Dashed line denotes the explosion limit

critical values of parameter $\varkappa_{cr}$ as a function of γ_0 has shown that, given $p_{cr} = 0.7$, the value of $\varkappa_{cr}$ varies from 0.126 at $\gamma_0 = 5 \times 10^{-4}$ to 0.034 at $\gamma_0 = 5 \times 10^{-5}$.

Thus, for nonisothermal conditions of free-radical polymerization, the polymerization kinetics and the thermal regime are interdependent—a feature characteristic of a large class of exothermal reactions. However, for free-radical polymerization, this dependence is kinetically more intricate owing to the intervenient gel effect.

2.3 Modelling of Copolymerization Processes

Synthesis of copolymers, that is, mixed polymers with the polymer chains made up of structural molecular units (monomeric units) of several types ($m \geqslant 2$) opens up wide frontiers for fabrication of novel polymer materials. Such an approach allows a marked diversification of the properties of conventional polymers of commercial large-scale production. However, the diversity of synthetic possibilities becomes a handicap in the product quality prediction and in the choice of optimal conditions for obtaining a material with desired properties. It is evident that the role of mathematical modelling which provides a clue to the solution of these problems can hardly be overestimated. An issue of prime interest is the establishment of quantitative correlations between operational parameters which are determined by the physico-chemical and mechanical properties of a copolymer and its molecular structure. Next step is to develop a formalism enabling one to relate computationally the molecular structure to the reactivity of reactants under different conditions and modes of implementation of a given process.

Following this line of reasoning, we wish to demonstrate in what a manner the molecular structure of a copolymer exerts a specific influence on the solution of the above problems. We concern ourselves with the processes of straightforward synthesis of a copolymer from two or more monomers by free-radical copolymerization. A point of special emphasis is that the methods that are expounded in this Section for describing the molecular structure of copolymers and the approaches to establishing the structure-property correlations are universal in character. Basically, they can be applied to other procedures of copolymer production, including polyaddition reactions and chemical modification of polymers on their interaction with low-molecular reactants. Certain aspects in modelling the processes of the latter type as exemplified by chlorination of polyethylene have been reported in Refs. [95, 96].

2.3.1 Statistical Characteristics of Linear Copolymers

The diversity of copolymer properties provides the possibility of varying the molecular structure of a copolymer, that is, the configuration of its macromolecules, in the course of the synthesis. For a linear homopolymer, the only "controllable" parameter is most commonly the degree of polymerization i; by contrast, in a copolymer, liable to variation are also the composition of a macromolecule (the relative fractions of units of different type) and the structure (the mode these units alternate in the chain). For this reason, the parameters that characterize the configurational structure of a macromolecule must include the fractions η of both separate units and the sequences U of two or more such units in the polymer chain; the fractions η must be defined in terms of i. The longer the chain, the greater is the number of possible sequences, and, consequently, the more diversified in detail becomes the description of a macromolecule. However, even if we content ourselves with considering the composition of a macromolecule and the distribution of various pairs of units within the macromolecular chain, we can arrive at important correlations between the molecular structure and properties of the copolymer.

Any parameter $\eta\{U\}$ characteristic of a polymeric species is a random quantity defined within an interval of $0 < \eta < 1$. The probability distribution $f(\eta)$ for this quantity determines the relative fraction of molecules (that is, of their configurations) characterized by a certain value of η. When the joint distribution for probabilities of several random quantities is to be known, they should be regarded as the components of a random vector $\vec{\eta}\{U, V, \ldots\}$ with a multidimensional probability distribution $f(\vec{\eta})$. For example, the relative fractions $\eta\{A_\alpha\} = \xi_\alpha$ of separate units A_α of different types $\alpha = 1, \ldots, m$ are the components of a compositional vector $\vec{\xi}$ of a given macromolecule. The function $f(\vec{\xi})$ determines the compositional distribution within a copolymer. In the general case of a polydisperse specimen

$$f(\vec{\xi}) = \int \Phi_i f^{(i)}(\vec{\xi}) \, di \qquad (56)$$

where Φ_i is the fraction of macromolecules with degree of polymerization i, and $f^{(i)}(\vec{\xi})$ is their composition distribution. The Φ_i being either a number distribution, or a weight distribution of the copolymer in degree of polymerization, one obtains, by formula (56), the respective number- or weight-composition distribution.

2 Modelling of Polymerization Processes

The components of the average composition vector of a copolymer are first-order moments of the distribution $f(\vec{\xi})$:

$$\bar{\xi}_\alpha = \int \xi_\alpha f(\vec{\xi}) d\vec{\xi}, \quad \alpha = 1, \ldots, m \tag{57}$$

The second- and third-order moments

$$\overline{\xi_\alpha \xi_\beta} = \int \xi_\alpha \xi_\beta f(\vec{\xi}) d\vec{\xi}$$
$$\overline{\xi_\alpha \xi_\beta \xi_\gamma} = \int \xi_\alpha \xi_\beta \xi_\gamma f(\vec{\xi}) d\vec{\xi} \tag{58}$$
$$\alpha, \beta, \gamma = 1, \ldots, m$$

permit calculating corresponding central moments

$$\lambda_{\alpha\beta} = \overline{\xi_\alpha \xi_\beta} - \bar{\xi}_\alpha \bar{\xi}_\beta, \quad \lambda_{\alpha\beta\gamma} = \overline{\xi_\alpha \xi_\beta \xi_\gamma} - \lambda_{\alpha\gamma} \bar{\xi}_\beta - \lambda_{\alpha\beta} \bar{\xi}_\gamma - \lambda_{\beta\gamma} \bar{\xi}_\alpha + 2\bar{\xi}_\alpha \bar{\xi}_\beta \bar{\xi}_\gamma \tag{59}$$

which characterize the width and the symmetry of distribution. In particular, the variance $\lambda_{\alpha\alpha}$ gives a quantitative measure of compositional heterogeneity in each of the copolymer components.

By virtue of the evident condition $\sum_{\alpha=1}^{m} \xi_\alpha = 1$, any single component out of the m components of the vector $\vec{\xi}$ is uniquely determined through the remaining components. Consequently, the statistical moments satisfy the relationships

$$\sum_{\alpha=1}^{m} \bar{\xi}_\alpha = 1, \quad \sum_{\alpha=1}^{m} \lambda_{\alpha\beta} = \sum_{\alpha=1}^{m} \lambda_{\alpha\beta\gamma} = 0, \quad \beta, \gamma = 1, \ldots, m \tag{60}$$

For example, the chemical composition of a macromolecule with $m = 2$ is given by a scalar quantity $\xi = \xi_1$, and the average composition and the compositional heterogeneity of a binary copolymer, by the respective moments,

$$\bar{\xi} = \int \xi f(\xi) d\xi, \quad \lambda = \bar{\xi}^2 - \bar{\xi}^2 = \int (\xi - \bar{\xi})^2 f(\xi) d\xi \tag{61}$$

The functions f and their statistical moments are the statistical characteristics of the molecular structure of a copolymer. They play a decisive role in establishing correlations between the conditions for production of a copolymer and its macroscopic properties.

These characteristics are conveniently calculated using a probabilistic (stochastic) approach by virtue of which each macromolecule is viewed as a particular realization of the random process of a conditional movement along the polymer chain. The summed-up probabilities for these realizations in which the parameter $\eta\{U\}$ takes fixed values determine the function $f(\eta)$ and are calculated by mathematical methods developed in the theory of appropriate random processes. The application of such methods, pioneered in the early 1950s [15] allows one to simplify significantly and to unify the procedure for obtaining final results. As an example, one may cite the use of random cascade processes for describing branched polymers [97] and also the use of Markovian chains for linear products of stepwise and chain-growth copolymerizations [98, 99]. The choice of the type of a random process for describing a given copolymer is determined by the conditions for copolymer production and can only be made by analyzing appropriate rate equations.

We now formulate, following the ideas exposed in the work [100], the general principles of a probabilistic approach to the description of linear copolymers and define essential universal rules for statistical characteristics of such copolymers.

For any copolymer specimen one may introduce a probability measure defined on the set of all sequences of polymer units (or states). For such a measure, the probability $P_l\{U\}$ of finding a chain-unit sequence U of length n at the sites $l, l+1, \ldots, l+n-1$ from a fixed end (origin) of a randomly selected macromolecule must be proportional to the concentration of molecules with such a sequence in the specimen.

The distribution of probabilities $P_l\{U\}$ defines completely the configurational structure of a copolymer. For this reason, any statistical characteristic of a copolymer is a function of these probabilities. We wish to show that an appropriately introduced random process provides a means for calculating this function and thus exploring its properties with reference to the structure of the probability measure for $P_l\{U\}$. Let us consider, by way of example, a problem for calculating the composition distribution $f^{(i)}(\xi)$ of a monodisperse fraction of length i of a binary copolymer made up of units A and B. We must identify the units of a definite type, for example, of A, as we step forward along the polymer chain, that is, we must know how to estimate the number of "right hits" into this particular unit type. It suffices to introduce a random process that takes a value of 1 at each right hit, and a value of 0, if the hit is missed: $I_l\{A\} = 1$ or 0, respectively, if a chain unit of type A is encountered or not at the lth step, that is, the $I_l\{A\}$ is a characteristic function of the unit type A. Then the composition of the given molecule is expressed by the formula

$$\xi^{(i)} = \frac{1}{i} \sum_{l=1}^{i} I_l\{A\} \tag{62}$$

2 Modelling of Polymerization Processes

Here the summation is carried out to a value of $l = i$, since in further progress ($l > i$) we fall into an absorptive state, where $I_l\{A\} = 0$.

The sequence of random quantities $I_l\{A\}$ ($l = 1, 2, \ldots$) forms a discrete random process which we refer to as "monadic". The relations below are easy to prove,

$$\overline{I_l\{A\}} = \overline{I_l^2\{A\}} = P_l\{A\}, \quad \overline{I_l\{A\}I_{l+n+1}\{A\}} = P_l\{AX^nA\} \quad (63)$$

where $\{AX^nA\}$ ($n \geq 0$) denotes an arbitrary sequence of length $n + 2$ flanked by A units, whereas the bar over a symbol signifies an averaging over all the realizations of the random process that delivers copolymer molecules. It will not take much effort to derive from formulas (62) and (63) the general expressions for statistical moments (61) of copolymer composition distribution that remain valid for any mode of the copolymer formation:

$$\overline{\xi^{(i)}} = \frac{1}{i}\sum_{l=1}^{i} P_l\{A\}, \quad \overline{(\xi^{(i)})^2} = \frac{1}{i^2}\left(i\overline{\xi^{(i)}} + 2\sum_{l=1}^{i-1}\sum_{n=0}^{i-1-l} P_l\{AX^nA\}\right) \quad (64)$$

For a multicomponent copolymer, the "monadic" random process is a vectorial process whose m steps $I_l\{A_\alpha\}$ are characteristic functions of respective chain units. The components of composition vector $\vec{\xi}^{(i)}$ are represented by the relation

$$\xi_\alpha^{(i)} = \frac{1}{i}\sum_{l=1}^{i} I_l\{A_\alpha\}, \quad \alpha = 1, \ldots, m \quad (65)$$

whereas the corresponding statistical moments of the first and second orders take the form

$$\overline{\xi_\alpha^{(i)}} = \frac{1}{i}\sum_{l=1}^{i} P_l\{A_\alpha\}, \quad \overline{(\xi_\alpha\xi_\beta)}^{(i)} = \frac{1}{i^2}\left[i\overline{\xi_\alpha^{(i)}}\delta_{\alpha\beta}\right.$$

$$\left. + \sum_{l=1}^{i}\sum_{n=0}^{i-1-l}(P_l\{A_\alpha X^n A_\beta\} + P_l\{A_\beta X^n A_\alpha\})\right],$$

$$\alpha, \beta = 1, \ldots, m \quad (66)$$

Henceforth, $\delta_{\alpha\beta}$ is the Kronecker delta equal to 1 at $\alpha = \beta$ and to 0 at $\alpha \neq \beta$.

Having summarized the right-hand sides of Eqs. (64) and (66) with the weight Φ_i according to formula (56), we obtain the statistical composition distribution moments for an arbitrary polydisperse specimen.

The above concepts can easily be extended to other statistical characteristics of a copolymer. For example, in calculating the distribution of molecules in different chain unit diads, one must consider a "diadic" random process as defined by the characteristic functions $I_l\{U\}$ of chain unit diads ($n = 2$). The corresponding random process enables deriving relations, similar to Eqs. (64) and (66), for statistical moments of any parameter of the configurational structure of a macromolecule.

Since, in real copolymers, there is $i \gg 1$, of primary interest are asymptotic results obtained at sufficiently large values of i. Commonly, the probability $P_l\{U\}$ with increasing l is rather fast to become independent of this index, and thus attains a certain limiting value $P\{U\}$. The disappearance of the l-dependence is associated with a decay of the correlation between the types of chain units spaced apart by a stretch of a polymer chain larger than a certain characteristic length l^*. On the l^*-scale, a sufficiently large number of transitions between different states of a random process are expected to occur. Therefore, for a statistical copolymer, the value of l^* is commonly not greater than 5-10, whereas in the case of a block copolymer it can be much larger. A reason for the rapid decay of correlations is the absence of long-range effects for the kinetic and thermodynamic reaction constants that determine the probability for formation of a unit in the copolymer chain. For example, in free-radical copolymerization, the chain growth constants responsible for the chain microstructure are, as a rule, independent of the chain length.

The probability distribution $P\{U\}$ forms a stationary (independent of l) measure on the set of all chain unit sequences, and the random process introduced onto this measure is also stationary [101]. We now wish to show which of the properties of this measure define the specificity of the configurational structure of a copolymer, in particular, the degree of its compositional heterogeneity. To this effect, we consider the relative frequency at which a certain sequence U is encountered in moving along the macromolecular chain. Two possibilities may be envisaged: (i) the limit to which this frequency tends at a sufficiently large number of steps is the same for all the macromolecules and is therefore equal to the respective stationary probability $P\{U\}$; (ii) there are macromolecules that exhibit different limits at this frequency. In the former case, the stationary random process possesses an ergodic property, and the probability measure is ergodic; in the latter case, none. Of prime importance is which of the two possibilities can be realizable. In the former case, the probability distribution $f^{(i)}(\eta)$ for any parameter $\eta\{U\}$ with growing i becomes increasingly narrower and, at the limit, degenerates into a Dirac delta-function $\delta(\eta - \bar{\eta})$. This statement ensues from the law of large numbers for the sum of random quantities of type (62) for ergodic stationary processes [102]. The corresponding copolymer which we name the ergodic copolymer can be fully described, at a sufficiently high degree of polymerization, by the averages $\bar{\eta}$ of configurational structure parameters η, since the composition and structure remain practically the same for all copolymer molecules. At the same time, for the processes not exhibiting an ergodic behaviour, the set of all realizations is divided into ergodic classes. The ergodicity is fulfilled only within each particular class, so that the probability P{U} for the same chain unit sequence for macromolecules that do not belong to a particular class is different. Therefore, the probability distribution $f^{(i)}(\eta)$ for a nonergodic copolymer is determined through summarizing the respective distributions for its separate classes with weights β equal to their relative fractions in the polymer. At sufficiently large i, when distinctions between macromolecules of the same class can be neglected, the $f^{(i)}(\eta)$ becomes altogether independent of i and is determined by the averages $\bar{\eta}$ for each class and by the fractions β. Therefore, the degree of inhomogeneity in both composition and structure of a nonergodic copolymer is independent of the copolymer polydispersity and can retain its sizeable value at any length of the macromolecule.

The ergodic copolymer may be exemplified by products of free-radical copolymerization at low monomer conversion p tractable in terms of a regular or a cyclic Marcovian chain [98]. On the other hand, the same process carried out at high conversion yields a nonergodic copolymer in which a separate ergodic class is represented by the molecules generated from a mixture of definite composition, that is, at a definite conversion p. A mixture of several homopolymers may be regarded as the limit case of a nonergodic copolymer, with each homopolymer constituting an ergodic class.

By making use of the kinetic model appropriate to a given technology for commercial production of a copolymer, one can relate the parameters of stationary probability measure P{U}, characteristic of the products, to the kinetic constants of the process. We wish to show now which of the characteristics of this measure determine the copolymer ergodicity; we now discuss in outline the methods for calculating the statistical characteristics of copolymers of both types.

At $i \gg l^*$, we can neglect the contribution due to the probabilities of terminal chain units with the number $l \leq l^*$ in the sums (64) and (66). Then, for example, relations (64) for the statistical moments of composition distribution of a binary copolymer, taking into account (61), can be rewritten in the form

$$\bar{\xi} = P\{A\}, \quad \lambda^{(i)} = \frac{1}{i}\left[P\{A\} - P^2\{A\} + 2\sum_{n=0}^{i-2}\left(1 - \frac{n+1}{i}\right)\omega_n[A, A]\right],$$

$$\omega_n[A, A] = P\{AX^n A\} - P^2\{A\} \tag{67}$$

For an ergodic copolymer, the variance $\lambda^{(i)}$ tends to zero with increasing i. It follows from (67) that the limit relation

$$\lim_{i \to \infty} \frac{1}{i} \sum_{n=0}^{i-2} \omega_n[A, A] = 0 \tag{68}$$

must be fulfilled. If so, the limit to which the frequency at which the A-type chain unit is encountered in moving along the polymer chain is the same for all the macromolecules and is equal to $P\{A\}$. The values of n at which, according to (68), this decay takes place are precisely those that determine the correlation scale l^* for the given ergodic copolymer. Condition (68) can be fulfilled even if ω_n do not tend to zero with n. Thus, for an alternating copolymer, one has $\omega_n[A, A] = (-1)^{n+1}/4$. Most commonly, however, the absolute value $|\omega_n|$ for an ergodic copolymer decreases rapidly with n and becomes negligeably small at $n \geq l^*$. For example, for Markovian-chain ergodic copolymers, this decay is exponential, with an exponent $\ln|1 - K_M| = (l^*)^{-1}$, where $K_M = P\{AB\}/P\{A\}P\{B\}$ is the microheterogeneity index [99].

It follows from formula (67) that, if the sequence $|\omega_n|$ tends to zero faster than n^{-1} (which is the case practically in any synthesis of an ergodic copolymer), then $\lambda^{(i)}$ can be represented in the form

$$\lambda^{(i)} = D/i, \quad D = P\{A\} - P^2\{A\} + 2\sum_{n=0}^{\infty}\omega_n[A, A] \tag{69}$$

where the parameter D, characteristic of the composition distribution width, may be named "the sequence homogeneity index" after the authors of Ref. [103] who suggested this term for a Markovian copolymer. In this case, the central limit theorem [102] holds stating that the composition distribution of an ergodic copolymer at sufficiently large i is asymptotically described by a normal law

$$f^{(i)}(\xi) = (2\pi\lambda^{(i)})^{-1/2} \exp\left[-(\xi - \bar{\xi})^2/2\lambda^{(i)}\right] \tag{70}$$

with the centre $\bar{\xi} = P\{A\}$ and the variance $\lambda^{(i)}$ (69). The ergodic copolymer distributions in other parameters of $\eta\{U\}$ are also described by a normal law.

The above results for a binary copolymer can readily be extended to a copolymer of any number of components. The ergodicity condition (68) thus reduces to a system of analogous limit equalities for the quantities $\omega_n[A_\alpha, A_\beta] = P\{A_\alpha X^n A_\beta\} - P\{A_\alpha\}P\{A_\beta\}$. The composition vector $\bar{\xi}^{(i)}$, providing that all the $\omega_n[A_\alpha, A_\beta]$ tend to zero faster than n^{-1}, is described, at $i \gg 1$, by a multidimensional normal law with the centre $\bar{\xi}^\alpha = P\{A_\alpha\}$ and the moment matrix

$$\lambda^{(i)}_{\alpha\beta} = D_{\alpha\beta}/i, \ D_{\alpha\beta} = \delta_{\alpha\beta}P\{A_\alpha\} - P\{A_\alpha\}P\{A_\beta\} +$$

$$\sum_{n=0}^{\infty}(\omega_n[A_\alpha, A_\beta] + \omega_n[A_\beta, A_\alpha]),$$

$$\alpha, \beta = 1, \ldots, m \tag{71}$$

To be noted, the third-order central moments $\lambda^{(i)}_{\alpha\beta\gamma}$ are identically equal to zero.

Let us now proceed to the statistical characteristics of nonergodic copolymer. We denote by $\beta(t)$ the relative number- or weight-fraction of molecules belonging to the ergodicity class t. The set of such classes can be finite, descrete, or continuous, in the last-mentioned case the quantity $\beta(t)$ is the density of distribution of molecules over different classes. By way of example, for high-degree copolymerizates, the monomer conversion p can serve as such a parameter.

For any random parameters η and sequences U, the following relations hold:

$$f^{(i)}(\eta) = \int \beta(t)f^{(i)}(t, \eta)dt,$$

$$P\{U\} = \int \beta(t)P\{t, U\}dt \tag{72}$$

Here the integrands are statistical characteristics for the molecules of ergodicity class t, and integration is carried out over feasible values of t.

For a high-molecular copolymer, the function $f^{(i)}(t, \eta)$ in the right-hand side of (72) can be replaced by $\delta(\eta - \bar{\eta}(t))$, where the average $\bar{\eta}(t)$ is independent of i for all values of t. The subsequent integration, for example, for the composition distribution of binary copolymer molecules and for the compositional statistical moments, yields i-independent expressions:

$$f^{(i)}(\xi) = f(\xi) = \sum_{t^*} \beta(t^*) \Big/ \left|\frac{d\bar{\xi}(t)}{dt}\right|_{t=t^*},$$

$$\langle \xi^k \rangle = \int \beta(t) |\bar{\xi}(t)|^k dt, \quad k = 1, 2, \ldots \tag{73}$$

where t^* is the root to equation $\bar{\xi}(t) = \xi$, and the summation sign is used here to mean that, with $\bar{\xi}(t)$ being a nonmonotonic function of t, the number of roots can be more than one. From now and on, we use the French quotes to denote the distribution averaging for nonergodic copolymer molecules to distinguish it from the respective averaging for ergodic copolymer molecules (or for those of a particular ergodic class). Expressions (73) hold also for the distribution of a multicomponent non-ergodic copolymer in any of its monomer. The corresponding multidimensional distribution and its statistical moments take the form:

$$f(\vec{\xi}) = \sum_{t^*} \beta(t^*) \prod_{\alpha=2}^{m} \delta(\xi_\alpha - \bar{\xi}_\alpha(t^*)) \Big/ \left|\frac{d\bar{\xi}_1(t)}{dt}\right|_{t=t^*}$$

$$\bar{\xi}_1(t^*) = \xi_1, \quad \langle \xi_\alpha \xi_\beta \rangle = \int \beta(t) \bar{\xi}_\alpha(t) \bar{\xi}_\beta(t) dt \tag{74}$$

$$\langle \xi_\alpha \xi_\beta \xi_\gamma \rangle = \int \beta(t) \bar{\xi}_\alpha(t) \bar{\xi}_\beta(t) \bar{\xi}_\gamma(t) dt, \quad \alpha, \beta, \gamma = 1, \ldots, m$$

Till now, we understood by $\eta\{U\}$ the number fraction of sequences U in a macromolecule. However, of frequent need is also to calculate the distribution of molecules in volume or in mass fraction of U. In the next Section, we will need to know a function $\Phi_N f^{(N)}(\vec{\xi})$ denoting the relative volume fraction of macromolecules of volume N which are characterized by the volume fraction ξ_α of structural units A_α ($\alpha = 1, \ldots, m$). Here the volume fraction for macromolecules of volume N is denoted by Φ_N, whereas $f^{(N)}(\vec{\xi})$ is meant to denote the probability distribution density for a volume-composition vector $\vec{\xi}$ within the set of macromolecules of fixed volume N. The above results can easily be extended to the case of $f^{(N)}(\xi)$ distributions. To this end, it will suffice to have the probability of any sequence U renormalized in such a manner that, when augmented by a new chain unit A_α, the corresponding probability is multiplied by v_α/v, here v_α being the structural unit volume for A_α, and $v = \sum_{\alpha=1}^{m} v_\alpha P\{A_\alpha\}$. For example, in formulas (71), one must have $P\{A_\alpha\}$ and

$P\{A_\alpha X^n A_\beta\}$ replaced by

$$P'\{A_\alpha\} = v_\alpha P\{A_\alpha\}/v, \quad P'\{A_\alpha X^n A_\beta\} = v_\alpha v_\beta P\{A_\alpha X^n A_\beta\}/v^2 \tag{75}$$

and the index i replaced by N.

2.3.2 Molecular Structure and Properties of Copolymers

At present, no quantitative correlations between the molecular structure and properties of a copolymer can be obtained on the basis of a rigorous molecular theory. Attempts have been made to advance in that direction, however, without much success [104]. For this reason, a more realistic approach is the use of simple and, simultaneously, generalizing empirical relations for modelling the copolymerization process. Correlations between the composition average and properties of a copolymer are established most easily. For example, an impressive body of such correlations has been established for chlorinated polymers [105]. In addition, more intricate and imaginative relationships linking the macroscopic properties to the microstructure and compositional heterogeneity have been found for copolymers.

For example, a formula has been proposed for determining the glass-transition temperature T_g in statistical copolymers [106, 107]:

$$T_g = \sum_{\alpha,\beta=1}^{m} T_{\alpha\beta} P\{A_\alpha A_\beta\} \tag{76}$$

where each summand in the right-hand side represents an additive contribution of chain-unit diads of different type to the value of T_g; $T_{\alpha\beta}$ denotes the glass-transition temperatures of corresponding homopolymers ($\alpha = \beta$) or alternating binary copolymers ($\alpha \neq \beta$) that appear in (76) as parameters. For a binary copolymer, expression (76) may be rewritten in the form

$$T_g = T_{11} P\{A\} + T_{22} P\{B\} + 2\Delta T P\{A\ B\},$$
$$\Delta T = T_{12} - (T_{11} + T_{22})/2 \tag{77}$$

A simple rule ensues from Eq. (77) relating T_g to the composition average of the copolymer [108]: at $\Delta T = 0$, the T_g vs. composition relationship is a straight line: at $\Delta T > 0$, it is a convex curve; at $\Delta T < 0$, it is a concave curve. To be noted, the tendency to alternation of chain units in a copolymer leads to a more noticeable deviation from linearity at the same values of $\Delta T \neq 0$.

The development of a method for synthesis of copolymers with regularly alternating chain units [109] has provided a means for experimental verification and practical application of formula (76). This allowed determining in an independent manner the glass-transition temperature $T_{\alpha\beta}$ for numerous alternating copolymers [110].

The experimental T_g's for a large number of binary copolymers available in the literature are summarized in Fig. 2.17 (borrowed from the work [111]). They are compared to theoretical curves calculated by formula (77) using the kinetic constants for corresponding processes [111]. The values of $T_{\alpha\beta}$ as determined by various authors from independent experimental measurements and supplemented with corresponding kinetic constants are listed in Table 2.1. A comparison of the experimental and theoretical results for T_g (treated without

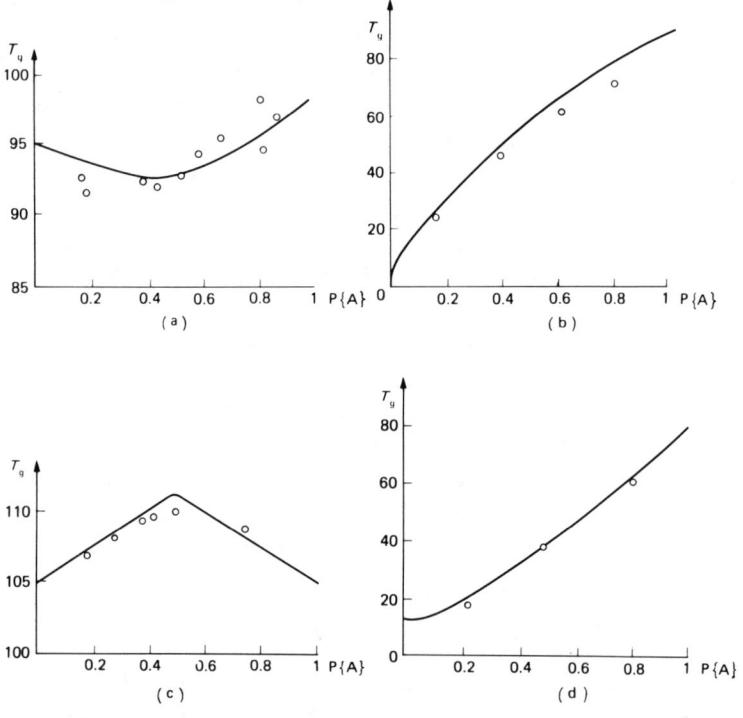

Fig. 2.17 Glass-transition temperature (°C) as a function of average composition for copolymers obtained by bulk copolymerization at initial (denoted ○, x) and high (denoted •) monomer conversions:

(a) methyl methacrylate + styrene [113];
(b) styrene + methyl acrylate [107];
(c) acrylonitrile + styrene [110];
(d) vinyl chloride + methyl acrylate [108];

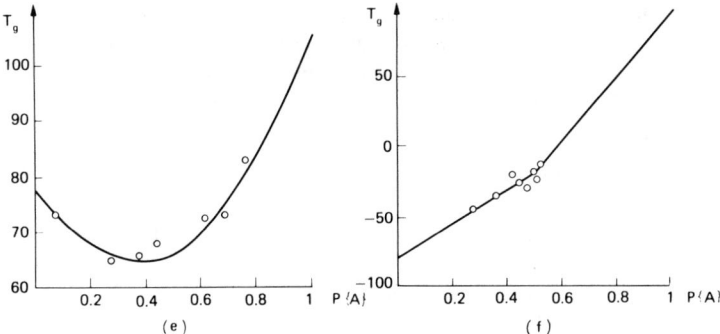

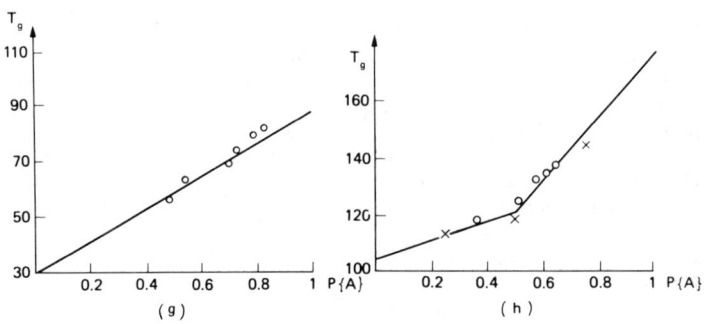

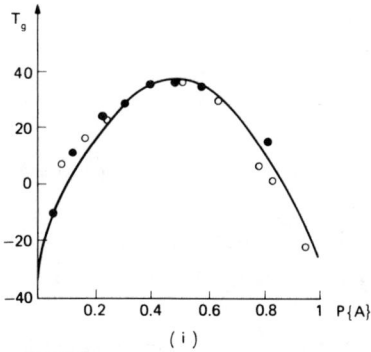

(e) methyl methacrylate + vinyl chloride [110];
(f) acrylonitrile + butadiene [114];
(g) acrylonitrile + vinyl acetate [115];
(h) α-methylstyrene + acrylonitrile [110, 108];
(i) ethyl acetate + vinylidene chloride [116].
Theoretical curves are those computed in [111] by formula (77) within the framework of ultimate model [130]

Table 2.1 Reactivity Ratios of Monomer Pairs (at 60 ± 10°C) and Glass-Transition Temperatures (°C) of Corresponding Amorphous Homopolymers and Alternating Copolymers

M_1	M_2	r_{12}	r_{21}	T_{11}	T_{12}	T_{22}	Reference
Styrene	Methyl methacrylate	0.52	0.46	95.1	91	98.7	[112, 113, 108]
Styrene	Methyl methacrylate	0.75	0.18	91	64.5	6	[112, 107, 108]
Styrene	Acrylonitrile	0.40	0.04	100	111.5	105	[112, 110]
Vinyl chloride	Methyl acrylate	0.07	5.6	80	40	6	[108]
Vinyl chloride	Methyl methacrylate	0.044	11.2	77	49.6	105	[110]
Acrylonitrile	Butadiene	0.04	0.40	100	−25	−80	[112, 114]
Acrylonitrile	Vinyl acetate	5.3	0.04	87	71.5	30	[112, 108, 115]
Acrylonitrile	α-Methyl-styrene	0.088	0.17	105	122	177	[110]
Ethyl acrylate	Vinylidene chloride	0.95	1.11	−27	98	−22	[116]
Butyl acrylate	Methacrylic acid	0.5	1.43				[117]
Methyl acrylate	Methyl methacrylate	1.5	0.3	6	93	103	[113, 107]
Styrene	Methacrylonitrile	0.30	0.16				[112]
Methyl methacrylate	Methacrylonitrile	0.67	0.65				[112]
Styrene	α-Methyl-styrene	2.3	0.38				[112]
Styrene	Vinylidene chloride	1.85	0.085				[112]
Acrylonitrile	Vinylidene chloride	0.91	0.37				[112]

recoursing to fitting parameters) shows that not only a qualitative (viewed as an applicability of the above rule to all systems), but in most cases also quantitative agreements are observed. Likewise, this holds for nonergodic copolymers exhibiting a marked heterogeneity in a number of chain-unit diads of different type (Fig. 2.17i).

Relations of type (76) reflecting the known additivity principle for contributions from various structural units of a macromolecule [104] are also useful in the sense that they permit predicting the properties of a multicomponent copolymer starting from the data on congeneric homo- and bipolymers. For example, a relation similar to (76) has been proposed in Ref. [118] for calculating the contraction factor. Disselhoff has calculated, by making use of the

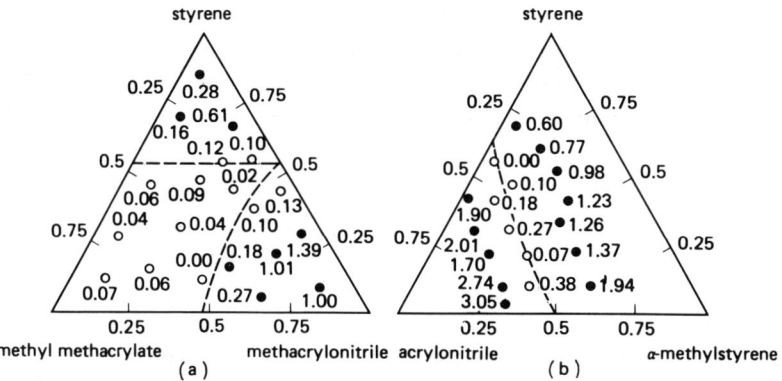

Fig. 2.18 Compositional distribution variance [122] and copolymer transparence [121] as functions of the initial monomer mixture composition for systems: (a) methyl methacrylate + methacrylonitrile + styrene (given are values of $\lambda_{22} \cdot 10^2$); (b) acrylonitrile + α-methylstyrene + styrene ($\lambda_{11} \cdot 10^2$). Open circles refer to transparent and those filled, to nontransparent polymer. The variances have been calculated using formulas (88), (91), (94) at $p = 1$

parameters derived from a dilatometric study of homo- and bipolymerization kinetics [119], the contraction factor also for terpolymerization of the respective monomers, in good agreement with the experiment [120].

The effect of compositional heterogeneity on the copolymer properties may happen to be quite significant, in certain cases even prevalent over that due to the average composition of copolymer. Slocombe [121] has considered experimentally some 40 three-component copolymer systems obtained by bulk copolymerization at limiting conversions, and analyzed the terpolymer transparency as a function of the initial composition of the monomer mixture (apparently, identical with the average composition of the copolymer). Two such systems are represented in the Gibbs phase diagram in Fig. 2.18. The authors [111, 122] provide a plausible explanation to the established empirical relationships [121] assuming a one-to-one correspondence between the copolymer transparency and the variance of its composition distribution in monomer with a polarity different from that of the other two monomers in the triad. In the two systems shown in Fig. 2.18, such species are, respectively, methacrylonitrile and acrylonitrile. As is seen in Fig. 2.18, for either system there exist definite critical values of the variance, $\lambda_{22}^{\text{crit}} = 0.15 \times 10^{-2}$ (a) and $\lambda_{11}^{\text{crit}} = 0.5 \times 10^{-2}$ (b); if these values are exceeded, the copolymer loses its transparency. To be noted, the critical value is independent of the average copolymer composition. The experiment [121] and its theoretical interpretation will be dealt with in greater detail in Sec. 2.3.3.3.

The major feature underlying the effect of compositional heterogeneity of a copolymer on many of its physical and physico-chemical properties is the poor thermodynamic compatibility of the copolymer fractions of different chemical composition, which results in a micro- or a macroscale phase separation in dissolved or in molten state [123, 146]. Such a thermodynamic behaviour in the course of both copolymer synthesis and processing—produces a detrimental effect on operating characteristics of copolymers and their composites.

For a quantitative analysis of the demixing phenomenon we use the Flory-Huggins representation for the volume density of free mixing energy of an arbitrary multicomponent statistical copolymer in molten state (expressed in units of $R_g T$) [124]:

$$\Delta G = \iint \Phi_N f^{(N)}(\vec{\xi}) N^{-1} \ln [\Phi_N f^{(N)}(\vec{\xi})] dN d\vec{\xi} + \varepsilon_{\text{mix}}$$

$$\varepsilon_{\text{mix}} = -\frac{1}{2} \sum_{\alpha,\beta=1}^{m} \chi_{\alpha\beta} \lambda_{\alpha\beta}, \quad \lambda_{\alpha\beta} = \overline{\xi_\alpha \xi_\beta} - \bar{\xi}_\alpha \bar{\xi}_\beta$$

$$\bar{\xi}_\alpha = \int \Phi_N \bar{\xi}_\alpha^{(N)} dN, \quad \overline{\xi_\alpha \xi_\beta} = \int \Phi_N (\overline{\xi_\alpha \xi_\beta})^{(N)} dN \quad (78)$$

$$\bar{\xi}_\alpha^{(N)} = \int \xi_\alpha f^{(N)}(\vec{\xi}) d\vec{\xi} = \bar{\xi}_\alpha, \quad (\overline{\xi_\alpha \xi_\beta})^{(N)} = \int \xi_\alpha \xi_\beta f^{(N)}(\vec{\xi}) d\vec{\xi}$$

where the volume of separate macromolecules N is expressed in units of the least volume v_α of chain units A_α. The augend in (78) reflects the contribution due to combinatorial mixing entropy, whereas ε_{mix}, that due to the energy of $A_\alpha - A_\beta$ interaction; $\chi_{\alpha\beta} = \chi_{\beta\alpha}$ are the Flory-Huggins exchange interaction parameters, and $\chi_{\alpha\alpha} = 0$ at any α. For a binary component, the only parameter remained is $\chi_{12} = \chi$, and $\varepsilon_{\text{mix}} = \chi\lambda$.

From Eq. (78), expressions for the chemical potentials (in units of $R_g T$) of macromolecular fractions with fixed N and $\vec{\xi}$ are easily derived:

$$\mu(N, \vec{\xi}) = \ln [\Phi_N f^{(N)}(\vec{\xi})] + 1 - N/\bar{N} - \frac{1}{2} N \sum_{\alpha,\beta=1}^{m} \chi_{\alpha\beta} \times (\xi_\alpha - \bar{\xi}_\alpha)(\xi_\beta - \bar{\xi}_\beta)$$

$$(\bar{N})^{-1} = \int \Phi_N N^{-1} dN \quad (79)$$

Expressions (78) and (79) are instrumental in studying the thermodynamic properties of an arbitrary m-component copolymer in melt as an object of continuous thermodynamics [125, 126]. Thus, the equality of thermodynamic potentials $\mu^*(N, \vec{\xi}) = \mu(N, \vec{\xi})$ of polymer components in co-existing phases (one of these is designated by an asterisk) lead to a relation

$$\Phi_N^* f^{*(N)}(\vec{\xi}) = \Phi_N f^{(N)}(\vec{\xi}) \exp\left[N\left(P + \sum_{\alpha=1}^{m} Q_\alpha \xi_\alpha\right)\right]$$

$$P = \frac{1}{2} \sum_{\alpha,\beta=1}^{m} \chi_{\alpha\beta}(\bar{\xi}_\alpha^* \bar{\xi}_\beta^* - \bar{\xi}_\alpha \bar{\xi}_\beta)$$

$$+ \frac{1}{N^*} - \frac{1}{N}, \quad Q_\alpha = \sum_{\beta=1}^{m} \chi_{\alpha\beta}(\bar{\xi}_\beta - \bar{\xi}_\beta^*) \quad (80)$$

An essential point is that the quantities P and Q_α in Eq. (80) are independent of the statistical distribution moments $f^{*(N)}(\vec{\xi})$ and $f^{(N)}(\vec{\xi})$ of order higher than one. For this reason, closed equations for calculating the conditions for limit equilibrium demixing (the binodal surface) of a homogeneous system as described by function $\Phi_N f^{(N)}(\vec{\xi})$ can be obtained by substituting expression (80) into the equalities below:

$$\int\int \Phi_N^* f^{*(N)}(\vec{\xi}) dN d\vec{\xi} = 1, \quad \int \Phi_N^* N^{-1} dN = \frac{1}{N^*} \quad (81)$$

$$\int\int \xi_\alpha \Phi_N^* f^{*(N)}(\vec{\xi}) dN d\vec{\xi} = \bar{\xi}_\alpha^*, \quad \alpha = 1, \ldots, m$$

Expressions (81) permit calculating the averages for composition, $\bar{\xi}_\alpha^*$, and macromolecule volume, $\overline{N^*}$, of the new nascent phase.

Using the Gibbs stability criteria [127], the spinodal surface equation is derived from Eq. (78):

$$\sum_{\gamma=1}^{m} (\chi_{\gamma m} - \chi_{\gamma\beta}) \int N\Phi_N \lambda_{\alpha\gamma}^{(N)} dN = \delta_{\alpha\beta}, \quad \alpha, \beta = 1, \ldots, m-1 \quad (82)$$

which, in conjunction with the equation

$$\sum_{\gamma_1,\gamma_2=1}^{m} (\chi_{\gamma_1 m} - \chi_{\gamma_1\alpha})(\chi_{\gamma_2 m} - \chi_{\gamma_2\beta})$$

$$\times \int N^2 \Phi_N \lambda_{\gamma_1\gamma_2\delta}^{(N)} = 0, \quad \alpha, \beta, \delta = 1, \ldots, m-1 \quad (83)$$

allows one to define the conditions for a critical phase. In the case of $m = 2$, formulas (82) and (83) reduce, respectively, to

$$2\chi \int N\Phi_N \lambda^{(N)} dN = 1,$$

$$\int N^2 \Phi_N \int (\xi - \bar{\xi})^3 f^{(N)}(\xi) d\xi dN = 0 \qquad (84)$$

of which the former is the known expression [128] for the spinodal surface of a binary copolymer melt.

It is to be emphasized that condition (82) for spinodal phase separation is dependent only on the second central moments $\lambda_{\alpha\beta}^{(N)}$ of composition distribution $f^{(N)}(\vec{\xi})$ (this moment characterizing the distribution width) and is independent of the copolymer composition average $\vec{\xi}$. By virtue of Eq. (83), any symmetric distribution $f^{(N)}(\vec{\xi})$ satisfying the spinodal equation (82) is capable of forming a critical phase.

Substituting the composition distributions for ergodic and nonergodic copolymers into Eqs. (80)-(84) reveals a basic distinction in thermodynamic behaviour of these copolymer species. Thus, in the former case, irrespective of the copolymer molecular-weight distribution, the conditions for equilibrium demixing (80) and for spinodal demixing (82), (84) in the molten state are the same (in the sense that a metastable state region does not exist) and take the form

$$\sum_{\gamma=1}^{m} (\chi_{\gamma m} - \chi_{\gamma \beta}) D_{\alpha\gamma} = \delta_{\alpha\beta} \; (m \geqslant 2); \; 2\chi D = 1 \; (m = 2) \qquad (85)$$

Moreover, because of the symmetry of distribution $f^{(N)}(\vec{\xi})$ in an ergodic copolymer, Eq. (83) is immediately fulfilled, and the phase separation always occurs at a critical point, that is,

$$f^{*(N)}(\vec{\xi}) = f^{(N)}(\vec{\xi}), \quad \Phi_N^* = \Phi_N$$

In examining to what extent the type of composition distribution can affect the thermodynamic properties of a nonergodic copolymer, one must be aware that the function $f^{(N)}(\vec{\xi})$ and its statistical moments are independent of N, that is, in formula (80) one may set $f^{(N)}(\vec{\xi}) = f(\vec{\xi})$, and Eqs. (82) and (83) reduce, respectively, to

$$\overline{N_W} \sum_{\gamma=1}^{m} (\chi_{\gamma m} - \chi_{\gamma \beta}) \lambda_{\alpha\gamma} = \delta_{\alpha\beta}, \quad \sum_{\gamma_1, \gamma_2 = 1}^{m} (\chi_{\gamma_1 m} - \chi_{\gamma_1 \alpha})(\chi_{\gamma_2 m} - \chi_{\gamma_2 \beta}) \lambda_{\gamma_1 \gamma_2 \delta} = 0$$

$$\overline{N_W} = \int N\Phi_N dN \qquad (86)$$

where the corresponding central moments (59) are calculated by formulas (74). The function $f(\vec{\xi})$ being asymmetric, the equilibrium phase separation in a nonergodic copolymer is not coincident with the spinodal phase separation, and both states are characterized by very small parameters $\chi_{\alpha\beta} \sim (\overline{N_W})^{-1}$, which is typical of polymer mixtures [123, 147-149].

The phase state of a concentrated copolymer solution can also be examined within the framework of the Flory-Huggins theory. However, the large number of intervening exchange interaction parameters (for example, these are six for a binary copolymer in its proper monomers) renders a quantitative analysis rather cumbersome. The qualitative results obtained for a polymer melt remain essentially similar to those for a concentrated solution of this polymer; however, on dilution, the effect of compositional heterogeneity on the copolymer properties becomes less pronounced [124].

2.3.3 Free-Radical Copolymerization

The modelling of a free-radical copolymerization involving two or more monomers includes a solution of two basic sets of problems. On the one hand, this is a description of the rate of the copolymerization process, its thermal regimes and the MWD of reaction products, and on the other hand, this is a quantitative analysis of composition, structure, and compositional heterogeneity of copolymers under various synthetic conditions. Such a two-fold approach to the copolymerization modelling is not accidental. The point is that the composition and the chain structure of free-radical copolymerization products are most commonly determined by the initial monomer concentration, the extent of monomer conversion, and the relative chain-growth rate constants (reactivity ratios), which are little affected by temperature. At the same time, in solving problems of the former type, of primary importance are diffusion factors exerting a marked influence both on the kinetics of pair recombination of polymer radicals and on the temperature dependence of rate constants of various elementary reaction stages, especially of the initiation stage.

In principle, approaches to the calculation of reaction rates and mass-molecular characteristics of homopolymerization products, including a quantitative description of gel effect (see Sec. 2.2) are also transferable onto the solution of an analogous problem for copolymerization. The only specific condition is that one must take into account the dependence of chain termination rate constants $k_t(i, j)$ for interacting macroradicals of size i and j on the macroradical chemical composition (but not on the type of terminal chain units, as is not infrequently assumed [129] failing to take into consideration the diffusion-controlled rate of these reactions). This dependence is associated with the effect of the chemical composition on the chain stiffness and can be a determining factor, for instance, for the effective termination reaction radius at the initial stage of a process. However, the existence of such a dependence makes little difference in the qualitative picture of the results and conclusions summarized in Sec. 2.2. For example, expression (10) remains valid in calculating the polymerization rate in the presence of gel effect, providing that the chain-growth reaction rate constant k_p has been appropriately redefined. Therefore, in what follows we are concerned only with the latter type of problems specific of copolymers, staying within the framework that has been outlined in this Section.

A major advance in the theory of free-radical copolymerization has been accomplished using the ultimate chain model based on the Flory principle for the corresponding reactions [130]. In the copolymerization of M_α monomers of m different types ($m \geq 2$), the chain-growth reaction within the framework of this model can be written in the form

$$A_\alpha^* + M_\beta \xrightarrow{k_p(\alpha, \beta)} A_\beta^*, \quad \alpha, \beta = 1, \ldots, m \qquad (87)$$

where A_α^* is a macroradical whose terminal chain unit has been supplied by an added monomer M_α. We define the relative monomer activity (reactivity ratio) as the ratio $r_{\alpha\beta} = k_p(\alpha, \alpha)/k_p(\alpha, \beta)$; we also introduce a vector $\vec{x}$ with the components

$$x_\alpha = M_\alpha/M, \quad M = \sum_{\alpha=1}^{m} M_\alpha$$

which are equal to the relative monomer fractions of definite type in the reaction mixture. This vector takes definite values at each time t characterized by the degree of conversion $p = 1 - M/M_0$, where M_0 is the total initial monomer concentration. The instantaneous statistical characteristics that describe the ergodic classes (fractions) of a copolymer within the framework of kinetic model (87) with reference to the principle of quasistationary concentrations are dependent on $\vec{x}$ and $r_{\alpha\beta}$. For example, the instantaneous composition average is given by the expression [130]:

$$\bar{\xi}_\alpha = P\{A_\alpha\} = \frac{\Delta_\alpha(1-h_\alpha)}{\sum_{\beta=1}^{m} \Delta_\beta(1-h_\beta)}, \quad h_\alpha = 1 - x_\alpha^{-1} \sum_{\beta=1}^{m} r_{\alpha\beta}^{-1} x_\beta \qquad (88)$$

where Δ_α refers to the minor of the α-th diagonal element h_α in the matrix

$$H = \begin{pmatrix} h_1 & r_{12}^{-1} & \cdots & r_{1m}^{-1} \\ r_{21}^{-1} & h_2 & \cdots & r_{2m}^{-1} \\ \cdots & \cdots & \cdots & \cdots \\ r_{m1}^{-1} & r_{m2}^{-1} & \cdots & h_m \end{pmatrix}$$

Standing in need of calculating instantaneous statistical characteristics of a copolymerization product, it is expedient to make use of the Markovian theory of uniform chains of first order [98]. This theory allows expressing the stationary probability for any chain sequence through the so-called transitional probabilities

$$\nu_{\alpha\beta} = \frac{P\{A_\alpha A_\beta\}}{P\{A_\alpha\}}, \quad \alpha, \beta = 1, \ldots, m \tag{89}$$

For example, $P\{A_\alpha A_\beta A_\gamma A_\delta\} = P\{A_\alpha\} \nu_{\alpha\beta} \nu_{\beta\gamma} \nu_{\gamma\delta}$; to be noted, $P\{A_\alpha\}$ can also be expressed through transitional probabilities [98]. For the ultimate model (87), one has

$$\nu_{\alpha\beta} = r_{\alpha\beta}^{-1} x_\beta \Big/ \sum_{\gamma=1}^{m} r_{\alpha\gamma}^{-1} x_\gamma, \quad \alpha, \beta = 1, \ldots, m \tag{90}$$

At low conversions, when a change in the initial composition $\vec{x}^{(0)}$ of monomer mixture can be safely neglected, an ergodic copolymer is formed, compositionally and structurally characterized by formulas (88)-(90) at $\vec{x} = \vec{x}^{(0)}$. However, as the reaction proceeds with increasing p, the monomer composition in the reaction mixture is seen to vary owing to different relative monomer activities $r_{\alpha\beta}$ by the law [130]

$$(1 - p) \frac{dx_\alpha}{dp} = x_\alpha - \bar{\xi}_\alpha(\vec{x}), \quad x_\alpha(0) = x_\alpha^{(0)}, \quad \alpha = 1, \ldots, m \tag{91}$$

The integration of equations (91) with allowance for Eq. (88) permits determining not only the instantaneous, but also the average copolymer composition at different conversions by the formula

$$\langle \xi_\alpha \rangle = x_\alpha + (x_\alpha^{(0)} - x_\alpha)/p \tag{92}$$

The reaction yields a nonergodic copolymer made up of a continuous set of ergodic classes. The polymer chains that have grown by the time $t' < t$ account for a weight fraction

$$\beta(t') = -\frac{dM(t')}{dt'} \bigg/ (M_0 - M(t))$$

of the whole of the copolymer formed at instant t. By substituting this expression into Eq. (72) and integrating with respect to the variable $dp' = dM(t')/M_0$, we obtain, for the "conversion" (cumulative) component of the copolymer weight distribution (in α-th component), an expression similar to Eq. (73):

$$f_\alpha(p, \xi_\alpha) = \frac{1}{p} \sum_{p^*} \left| \frac{dp}{d\bar{\xi}_\alpha} \right|_{\bar{\xi}_\alpha = \xi_\alpha}, \quad \bar{\xi}_\alpha(p^*) = \xi_\alpha \tag{93}$$

To be noted, the components ξ_α at $m \geq 3$, as distinct from binary copolymerization, can vary nonmonotonically with conversion; therefore, more than one point p^* may occur in the trajectory $\bar{\xi}_\alpha(p)$.

The $f_\alpha(p, \xi_\alpha)$ distribution variance and the probabilities of any chain sequence U for the nonergodic copolymer of interest takes, in accordance with Eqs. (72) and (74), the form:

$$\lambda_{\alpha\alpha} = \frac{1}{p} \int_0^p \bar{\xi}_\alpha^2 \, dp' - (\langle\xi_\alpha\rangle)^2, \quad P\{U\} = \frac{1}{p} \int_0^p P\{p', U\} \, dp' \tag{94}$$

In a similar manner, the expressions for other statistical moments of the multidimensional copolymer composition distribution can be derived.

2.3.3.1 Analysis of Compositional Heterogeneity of Multicomponent High-Conversion Polymerizates

The above equations (91)-(94) are amenable to numerical integration at any $r_{\alpha\beta}$ and $\vec{x}^{(0)}$, taking into account Eq. (88), which enables calculating the statistical and compositional characteristics of a nonergodic copolymer at any degree of conversion. It is to be recalled that the relative activities $r_{\alpha\beta}$ that we have used in our calculations are summarized in Table 2.1. High degrees of conversion ($p \approx 1$) are of prime importance for technology. An essential point is to perform a thorough analysis of the relationship between the compositional heterogeneity of the polymerizate produced at high conversion and the relative activity and the initial composition of monomer mixture. This can be achieved [122, 111, 99] without making resort to a numerical integration of dynamical equations (91). To gain the end, a qualitative examination of the solution of this system will suffice enough. Especially helpful is such an analysis for multicomponent copolymerization ($m \geq 3$), where the search for optimum process conditions is subject to a large uncertainty in the choice of the initial monomer composition $\vec{x}^{(0)}$.

It is necessary, for one thing, to explore the mode in which the monomer mixture composition $\vec{x}(p)$ varies during copolymerization. Each particular composition is characterized by the coordinates of a point in the $(m - 1)$-dimensional phase space representing a domain for feasible values of the vector $\vec{x}$. In binary copolymerization, the phase space is a straight-line segment of unit length. The distances x_1 and x_2 of the segment measured from any of its interior point to its respective ends satisfy the condition $x_1 + x_2 = 1$ and give thus a monomer mixture composition. In ternary polymerization, the phase space is a Gibbs triangle. In order to determine the composition corresponding to an interior point of the Gibbs triangle, three straight lines are drawn through this point parallel to the triangle sides. The portions of triangle sides at the intercept with the straight lines determine the values of x_1, x_2, x_3 for a given composition. Each of the triangle vertices corresponds to homopolymerization of one of the monomers, whereas the triangle side opposite the vertex corresponds to copolymerization of the two other monomers. The phase space of a four-component system is the interior of a tetrahedron. The tetrahedral faces, edges, and vertices correspond to feasible three-, two-, and one-component systems that can be made up of the initial tetrad of monomers.

The compositional evolution of a monomer mixture $\vec{x}$ during copolymerization may be visualized as the movement along a certain trajectory, characterized by the parameter p, within the phase space. An important role is assigned here to the azeotrope compositions $\vec{x}^*$ [131, 132] which are stationary points of a dynamical system of ordinary nonlinear differential first-order equations (91), that is, singular points of the system's phase space:

$$x_\alpha^* = \bar{\xi}_\alpha(\vec{x}^*), \quad \alpha = 1, \ldots, m \tag{95}$$

where the function $\bar{\xi}_\alpha(\vec{x})$ is defined by formula (88). In azeotropic copolymerization, when $\vec{x}^{(0)} = \vec{x}^*$, the compositions of both monomer mixture and copolymer are invariant and coincident at any p. If all the components of $\vec{x}^*$ are positive ($x_\alpha^* > 0$), the azeotrope in question is internal; if at least one of them is zero, $x_\alpha^* = 0$, the azeotrope is a boundary one. In particular, there always exist m boundary azeotropes as all the components of vector $\vec{x}^*$ are equal to zero except one which is unity. Such azeotropes that are involved in a homopolymerization process are called angular azeotropes.

The mathematical problem aimed at calculating the azeotrope compositions $\vec{x}^*$ of a given multicomponent system consists in determining the roots of a system of algebraic equations (95) subject to a constraint $0 \leqslant x_\alpha^* \leqslant 1$ at any α. The problem of boundary azeotrope determination is equivalent to finding internal azeotropes for a system with a smaller number of components, and for this reason hardly requires a special consideration. The internal azeotropes satisfy the following system of linear equations:

$$Q^{(m)} x_\alpha^* = Q_\alpha^{(m)} \sum_{\beta=1}^{m} x_\beta^* r_{\alpha\beta}^{-1}, \quad \alpha = 1, \ldots, m \qquad (96)$$

where $Q^{(m)}$ is the determinant of a matrix H with diagonal unit elements $h_\alpha = 1$, and $Q_\alpha^{(m)}$ is the determinant of a similar matrix in which, along with the diagonal elements, all the elements of the α-th row are set equal to unity. The solution of system (96) takes the form

$$x_\alpha^* = \Delta_\alpha^{(m)} (\omega_\alpha^{(m)})^{-1} \Big/ \sum_{\beta=1}^{m} \Delta_\beta^{(m)} (\omega_\beta^{(m)})^{-1}, \quad \omega_\alpha^{(m)} = Q_\alpha^{(m)} / Q^{(m)} \qquad (97)$$

where $\Delta_\alpha^{(m)}$ are the minors of the diagonal elements of matrix H with $h_\beta = 1 - (\omega_\beta^{(m)})^{-1}$, $(\beta = 1, \ldots, m)$. It follows from Eqs. (96) that, within the ultimate model (87), no system (except those with two or more monomers of the same relative activities) can possess more than one internal azeotrope; the necessary and sufficient condition for its existence is a positive sign of all the quantities $\omega_\alpha^{(m)}$.

Qualitatively, the behaviour of the trajectories (solutions) of the system of equations (95) is defined by the type of its singular points, that is, its azeotropes. The reason for this is that the trajectories behave in a most complicated manner precisely in the vicinity of these points, each type of singular point exhibiting its specific "phase picture" (Fig. 2.19). The trajectories that divide the vicinity of a singular point into four sectors are called the separatrices. With rare exceptions (which will be dealt with somewhat later), all the trajectories originate from unstable singular points and terminate in stable ones. This indicates that during copolymerization, the monomer mixture composition (accordingly, the instantaneous average composition of the copolymer) tends to one of the stable azeotropes (in a total number s) and it is of importance to know to which one. The fact is that, as follows from formula (93), the functions $f_\alpha(\xi_\alpha) = f_\alpha(1, \xi_\alpha)$ at complete conversion and at $\xi_\alpha \to x_\alpha^*$ tend either to infinity, or to zero. The former case, when the azeotrope is "singular" in its α-th component, corresponds to a copolymer markedly heterogeneous in this component. It is typified by a broad $f_\alpha(\xi_\alpha)$-distribution exhibiting at least two maxima at the point $\xi_\alpha = \bar{\xi}_\alpha(0)$ and $\xi_\alpha = \bar{\xi}_\alpha(1) = x_\alpha^*$ and, consequently,

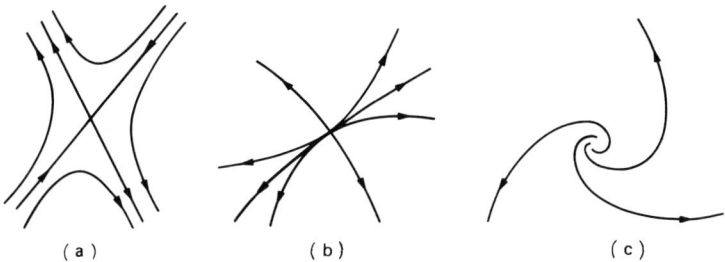

Fig. 2.19 Types of singular points in the phase plane for terpolymerization processes: (a) saddle point; (b) unstable node; (c) unstable focus. The arrowheads indicate the direction of compositional change with conversion. Stable singular points are arrived at from the respective unstable ones through reversal of direction in all the trajectories

a large variance $\lambda_{\alpha\alpha}$ (see curves *1* in Figs. 2.20 and 2.21). The only exceptions here are processes with the initial monomer composition very close to that of a singular point $\vec{x}^*$. If so, both maxima are either almost coincident (when the trajectory terminates in this singular point), or the maximum corresponding to $\xi_\alpha = x_\alpha^*$ is too narrow and does not contribute to the variance $\lambda_{\alpha\alpha}$ (the situation where the trajectory goes to another azeotrope). If $\bar{\xi}_\alpha(p)$ varies monotonically, then, with $f_\alpha(\xi_\alpha)$ tending to zero at $\xi_\alpha \to x_\alpha^*$ (the azeotrope is "regular" in its α-th component), the end product is always relatively homogeneous in this component, since in this particular case the composition distribution $f_\alpha(\xi_\alpha)$ is unimodal, with one maximum corresponding to fractions with $\xi_\alpha = \bar{\xi}_\alpha(0)$ (see curve *3* in Fig. 2.21). If $\bar{\xi}_\alpha(p)$ varies nonmonotonically with increasing conversion (which may occur in multicomponent systems with $m \geqslant 3$), then the function $f_\alpha(\xi_\alpha)$ can exhibit additional maxima corresponding to the points $\xi_\alpha = \bar{\xi}_\alpha(p^*)$ (at these points, $d\bar{\xi}_\alpha/dp|_{p=p^*} = 0$), with the resultant increase in heterogeneity of the product (curves *2* in Figs. 2.20 and 2.21).

We can specify the type of the singular point (97), its stability, and the "singularity" of an azeotrope by performing the conventional analysis of a system of linear differential equations with constant coefficients for small deviations $y_\alpha = x_\alpha - x_\alpha^*$. This system is obtained by linearizing equations (91):

$$(1-p)\frac{dy_\alpha}{dp} = y_\alpha - \sum_{\beta=1}^{m} \frac{\partial \bar{\xi}_\alpha^*}{\partial x_\beta} y_\beta, \quad 1 \leqslant \alpha \leqslant m-1, \quad \sum_{\alpha=1}^{m} y_\alpha = 0 \qquad (98)$$

where $\dfrac{\partial \bar{\xi}_\alpha^*}{\partial x_\beta}$ denotes the derivative $\dfrac{\partial \bar{\xi}_\alpha}{\partial x_\beta}$ at the azeotrope point $\vec{x}^*$. It is necessary

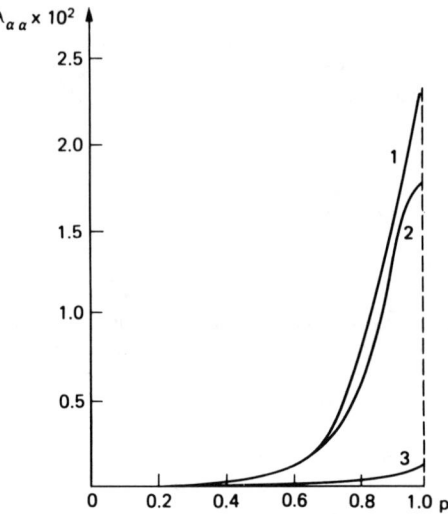

Fig. 2.20 Theoretical curves for variances λ_{11}, (1); λ_{22}, (2), and λ_{33}, (3) as functions of p in a system styrene + methyl acrylate + methyl methacrylate at $x_1^{(0)} = 0.3$, $x_2^{(0)} = 0.6$, $x_3^{(0)} = 0.1$ [111]

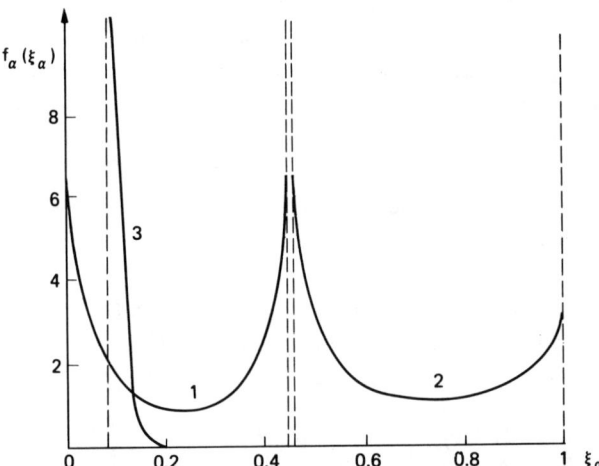

Fig. 2.21 Compositional distribution functions as calculated by (93) for styrene + methyl acrylate + methyl methacrylate terpolymerizates at $p = 1$ ($x_1^{(0)} = 0.3$; $x_2^{(0)} = 0.6$; $x_3^{(0)} = 0.1$) [111]

to calculate the roots of the characteristic polynomial of a matrix $A^{(m)}$,

$$\text{Det } |A^{(m)} - \mu E| = \mu^{m-1} + \sum_{\alpha=1}^{m-1} b_\alpha \mu^{m-\alpha-1} \qquad (99)$$

with elements

$$A_{\alpha\beta}^{(m)} = \delta_{\alpha\beta} - \frac{\partial \bar{\xi}_\alpha^*}{\partial x_\beta} + \frac{\partial \bar{\xi}_\alpha^*}{\partial x_m}, \qquad \alpha, \beta = 1, \ldots, m-1$$

(E is a unit matrix). For example, for an internal azeotrope

$$A_{\alpha\beta}^{(m)} = \delta_{\alpha\beta} + \Lambda_{\alpha\beta}^{(m)} + \sum_{\gamma=1}^{m} \sum_{\delta \neq \gamma}^{m} \frac{\Delta_{\gamma\delta}^{(m)} (x_\alpha^* - \delta_{\alpha\delta})}{\lambda_\gamma^{(m)} \omega_\delta^{(m)}} (\Lambda_{\gamma\beta}^{(m)} - \delta_{\gamma m}) \qquad (100)$$

$$\Lambda_{\alpha\beta}^{(m)} = \delta_{\alpha\beta} + \omega_\alpha^{(m)} (r_{\alpha m}^{-1} - r_{\alpha\beta}^{-1}), \qquad \alpha, \beta = 1, \ldots, m-1$$

where $\Delta_{\gamma\delta}^{(m)}$ is the determinant of $(m-2)$th order of the matrix obtained from the H matrix with $h_\alpha = 1 - (\omega_\alpha^{(m)})^{-1}$ by deleting two rows and two columns numbered γ and δ. (At $m = 2$, one must set $\Delta_{\gamma\delta}^{(2)} = 1$ in formula (100)).

The real parts Re μ of all the roots of polynomial (99) being negative, the azeotrope in question is stable. If they lie within the interval -1 to 0, the azeotrope is "regular" in all its components; in the case of Re $\mu < -1$, the azeotrope is "singular". If several coexistent roots exist, some of these possessing the real part greater, and the others, less than -1, the azeotrope is "singular" in some components, and "regular", in the others. To be noted, the Raus-Hurwitz method [133] allows, without resorting to a calculation of the polynomial roots of (99), specifying simple relationships between the coefficients b_α; these relationships, when fulfilled, provide for the stability of a singular point. For example, in terpolymerization, the stability criterion of an azeotrope is the positiveness of both coefficients b_1 and b_2, whereas a quaternary system requires the additional condition $b_3 < b_1 b_2$.

We now present some results concerned with the studies of the boundary azeotrope stability. Suppose, a boundary azeotrope has its first components of the vector $\bar{x}^*$, in number $m_1 < m$, definitely positive, the other vector components being equal to zero. Then a system of the first m_1 monomers possesses an internal azeotrope. If this azeotrope is unstable, the boundary azeotrope under consideration is also unstable. The conditions that provide for a stability of the boundary azeotrope are: (i) stability of the m_1-component internal azeotrope, and (ii) compliance, at each $\alpha > m_1$, to the inequality

$$\varrho_\alpha^{(m_1)} = \sum_{\beta=1}^{m_1} r_{\beta\alpha}^{-1} \omega_\beta^{(m_1)} - 1 > 0 \tag{101}$$

where the quantities $\omega_\beta^{(m_1)}$ are determined through the internal azeotrope components of the m_1-component system. If α is such that the inequality (101) becomes reverse in sign, the boundary azeotrope in question is necessarily unstable. The procedure of checking the angular azeotropes for stability is quite simple. Suppose, for example, that x_β^* is equal to 1, whereas all the x_α^* are equal to zero ($\alpha \neq \beta$). The angular azeotrope is stable, if all the reactivity ratios are less than unity, $r_{\beta\alpha} < 1$, or unstable, if there is at least one α such that $r_{\beta\alpha} > 1$.

It should be emphasized that the trajectories that originate from and terminate in simple singular points, as shown in Fig. 2.19, do not determine in any means all possible types of solutions to system (91) in the phase trajectory even for a three-component system. For example, one may envisage the existence of limit cycles which are closed trajectories inside the Gibbs triangle and which either attract (stable limit cycles), or repulse (unstable limit cycles) all the trajectories that pass in their close vicinity [134]. In the presence of a stable limit cycle, the copolymer composition distribution may become multimodal, since in such a case the instantaneous copolymer composition and the monomer composition oscillate with rising conversion. A still more complicated dynamic behaviour may be observed in systems with the number of components $m > 3$, where the occurrence of chaotic trajectories of the type of strange attractors [135] may be presumed.

Thus, to be able to analyse the influence of the initial monomer composition $\vec{x}^{(0)}$ on the compositional heterogeneity of the products obtained at limiting conversions in a given multicomponent system, one must specify all the numbers s for its stable boundary and internal azeotropes and draw a conclusion as to the occurrence (or nonoccurrence) of a limit cycle. The value of s determines the number of "domains of influence" into which the phase space of the system is partitioned. If the initial composition $\vec{x}^{(0)}$ is found in a definite domain of influence, then, in the course of copolymerization, the monomer composition $\vec{x}$ remains confined to this domain and, at complete conversion, together with the instantaneous copolymer composition, falls into a stable azeotrope, the only one in this domain. If this azeotrope is singular in some component, then the copolymer exhibits heterogeneity in this particular component. In such a contingency, in order to obtain a copolymerizate compositionally more uniform, one must stop the process at lower conversions. The boundaries of domains of influence themselves consist of "specific" trajectories (separatrices) and can be determined by solving numerically the system of equations (91) (see Fig. 2.18).

2.3.3.2 Universal Classification of Terpolymerization Processes

The application of the outlined concepts to an analysis of terpolymerization processes has been reported in the works [111, 122, 136]. To recall, the phase space in this case is an equilateral triangle whose vertices correspond to angular azeotropes, whereas nonangular boundary azeotropes can lie only in the triangle sides (one azeotrope per triangle side). The compositional evolutions for monomer mixture and terpolymer are demonstrated by trajectories traced within the triangle and arrowed to point the direction of a process.

We now analyse the azeotrope point types. To start with, consider the triangle vertices as exemplified by the upper vertex corresponding to an azeotropic point ($x_1^* = x_2^* = 0$, $x_3^* = 1$). This point may be either a node, or a saddle (Fig. 2.19), and its adjacent triangle sides are separatrices. If the reactivity ratios r_{31} and r_{32} are both greater than unity, then the resultant singular point is an unstable node; if one of the two is greater and the other smaller than unity, then the singular point is a saddle; if the relative activities are both smaller than unity, the singular point is a stable node. We now turn our attention to the lower, subtending triangle side which corresponds to a binary copolymerization in the absence of a third component ($x_3 = 0$). A condition for the existence of a boundary azeotrope corresponding to the subtending side of a composition triangle is imposed by two inequalities $\omega_1^{(2)} > 0$ and $\omega_2^{(2)} > 0$, where $\omega_1^{(2)}$ and $\omega_2^{(2)}$ for binary copolymerization are determined as

$$\omega_1^{(2)} = \frac{r_{12}(r_{21} - 1)}{r_{12}r_{21} - 1}, \quad \omega_2^{(2)} = \frac{r_{21}(r_{12} - 1)}{r_{12}r_{21} - 1} \tag{102}$$

To be noted, this constraint is equivalent to the known conditions for the existence of a binary azeotrope [130]: $r_{12} < 1$, $r_{21} < 1$, or $r_{12} > 1$, $r_{21} > 1$. The former instance corresponds to an unstable, and the latter, to a stable azeotrope. Since in free-radical copolymerization the reactivity ratios, as a rule, never become simultaneously greater than unity, the latter case is not common in practice, and the singular point may be either an unstable node (at $\varrho_3^{(3)} = \omega_1^{(2)}r_{13}^{-1} + \omega_2^{(2)}r_{23}^{-1} - 1 < 0$), or a saddle (at $\varrho_3^{(3)} > 0$). This singular point is always an origin for two separatrices, each lying in a triangle side. In the case of a saddle point, the only incoming trajectory is a separatrix that lies inside the triangle.

A similar analysis can be performed for the internal azeotropic point of a three-component system, and this point can be a saddle point, an unstable node, or a focus. However, an alternative way is to use the azeotropy rule [136] based on the Poincaré index [137], which imposes definite constraints on the number and the type of singular points of a system. By making use of this rule and of the results for boundary azeotropes, a classification has been proposed [136] for potential ternary systems lacking limit cycles analogous to the classification for three-component open-vaporization processes [138]. In copolymerization, two specific features are to be mentioned: (i) the internal azeotrope can also be a focus (which never occurs in open-vaporization processes); (ii) all boundary nonangular azeotropes are necessarily unstable. The number of azeotrope points of all types being known and the trajectory behaviour defined, the above factors ultimately allow specifying 17 various patterns for phase pictures (phase diagrams); these are shown in Fig. 2.22. Any real three-component system with no limit cycles is described, in terms of a properly chosen numbering for monomers, by one of the diagrams in Fig. 2.22 (in diagrams 1, 3, 4, and 17, no distinction is made between internal focuses and nodes). Interestingly, it is only the types of boundary azeotropes that determine unambiguously the appropriate diagram of a system. Exceptions to the rules are two pairs only: 1, 17 and 15, 16. In particular, given the boundary azeotropes only, one can predict the occurrence of an internal azeotrope in the system. Diagram 1 is unique in that it possesses a stable internal azeotrope. Except this case, all stable azeotropes can be located only at the triangle vertices; therefore, $s \leqslant 3$. The lines that demarcate the domains of influence at $s > 1$ are special trajectories originating from unstable nodes and terminating in saddle points. Therefore, for the practical determination of these boundary lines, it suffices to locate only one point lying in a special (or boundary) trajectory in the vicinity of the trajectory origin and to choose it as an initial constraint for numerical integration of system (91). The triangle sides are in fact represented by boundary trajectories. Diagram 17 also needs a brief comment, since it has none of stable azeotropes ($s = 0$), which never occurs in open vaporization processes. The trajectories in such systems are infinite; in binary copolymerization, similar systems are impossible.

The type of diagram in Fig. 2.22 for each particular system is defined by a matrix whose elements are six reactivity ratios $r_{\alpha\beta}$ each assigned a sign deter-

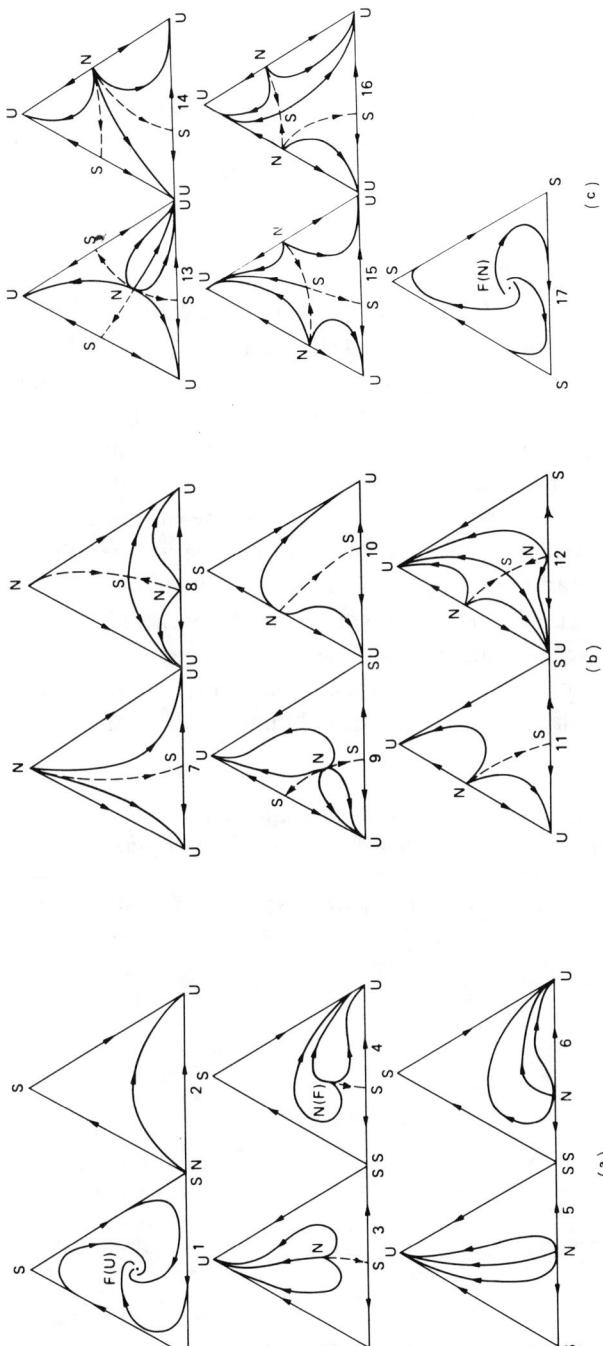

Fig. 2.22 Possible types of phase pictures in terpolymerization: (a) $s = 1$; (b) $s = 2$; (c) $s = 3$ and 0 (diagram 17). F denotes focus; S, saddle point; N, unstable node; U, stable node; dashed lines denote boundary trajectories

mined by one of their three combinations $\varrho_\alpha^{(3)}$ as shown below.

$$\begin{pmatrix} 1 & r_{12} & r_{13} \\ r_{21} & 1 & r_{23} \\ r_{31} & r_{32} & 1 \end{pmatrix} \begin{array}{l} \varrho_1^{(3)} = [r_{32}(r_{23} - 1)r_{31}^{-1} + r_{23}(r_{32} - 1)r_{21}^{-1}](r_{23}r_{32} - 1)^{-1} - 1 \\ \varrho_2^{(3)} = [r_{13}(r_{31} - 1)r_{12}^{-1} + r_{31}(r_{13} - 1)r_{32}^{-1}](r_{13}r_{31} - 1)^{-1} - 1 \\ \varrho_3^{(3)} = [r_{12}(r_{21} - 1)r_{13}^{-1} + r_{21}(r_{12} - 1)r_{23}^{-1}](r_{12}r_{21} - 1)^{-1} - 1 \end{array}$$

(103)

First, we must determine the number n of parameters $r_{\alpha\beta} > 1$, which usually is seldom greater than three. If $n = 3$, and each row and each column of matrix (103) have only one element $r_{\alpha\beta} > 1$, we have a diagram of type 1 or 17; if the matrix has a row or a column accommodating two elements $r_{\alpha\beta} > 1$, we have a type 2 diagram. Suppose, at $n = 2$ both quantities $r_{\alpha\beta} > 1$ are positioned either in a column, or in a row, and a column with the number α. Then, at $\varrho_\alpha^{(3)} > 0$, the diagram types 3, 7, 4, respectively, correspond to such systems and at $\varrho_\alpha^{(3)} < 0$, the diagram types 5, 8, 6. In the case of $n = 1$, only one parameter $r_{\alpha\beta}$ is greater than unity, and the diagram type is determined by the number of positive parameters $\varrho_\alpha^{(3)}$ ($\alpha = 1, 2, 3$). This number being equal to 3, 2, or 1, the diagrams for the system are, respectively, 13, 14, or 15 (alternatively, 16). To have the diagrams for the systems of interest brought in correspondence to those shown in Fig. 2.22, one must number the monomers in accordance with Table 2.2 in whose columns are listed the parametric signs determining the diagram type.

Table 2.2 Parametric Signs Characterizing the Relationship Between the Reactivity Ratios of a System and Its Phase Diagram Type

Diagram	$r_{12} - 1$	$r_{13} - 1$	$r_{21} - 1$	$r_{23} - 1$	$r_{31} - 1$	$r_{32} - 1$	$\varrho_3^{(3)}$	$\varrho_2^{(3)}$	$\varrho_1^{(3)}$
1, 17	−	+	+	−	−	+			
2	+	+	−	−	−	+			
3	−	+	−	+	−	−	+		
4	−	+	−	−	−	+	+		
5	−	+	−	+	−	−	−		
6	−	+	−	−	−	+	−		
7	−	−	−	−	+	+	+		
8	−	−	−	−	+	+	−		
9	−	−	−	+	−	−	+	+	
10	−	−	−	+	−	−	−	+	
11	−	−	−	+	−	−	+	−	
12	−	−	−	+	−	−	−	−	
13	−	−	−	−	−	−	+	+	+
14	−	−	−	−	−	−	+	+	−
15, 16	−	−	−	−	−	−	+	−	−

2 Modelling of Polymerization Processes

Until now, we have been concerned with systems that lacked limit cycles. However, as has been shown in Ref. [134], in principle, the existence of limit cycles may be envisaged in three-component systems with an internal azeotrope and the reactivity ratios corresponding to the first row of Table 4.2. The fact is that in such systems, the triangle boundaries can be either repulsive, like in system 1, the value of q being

$$q = (1 - r_{12}^{-1})(1 - r_{23}^{-1})(1 - r_{31}^{-1}) + (1 - r_{13}^{-1})(1 - r_{32}^{-1})(1 - r_{21}^{-1}) > 0$$

or attractive, like in system 17 with $q < 0$. Given $q > 0$ and the internal azeotrope being stable,

$$\left(2 - \sum_{\alpha=1}^{3} \partial \xi_\alpha^* / \partial x_\alpha = A_{11}^{(3)} + A_{22}^{(3)} < 0\right)$$

we have to deal with a system 1; if $A_{11}^{(3)} + A_{22}^{(3)} > 0$, the internal azeotrope is unstable, and the system must possess at least one stable limit cycle. By contrast, at $q < 0$ and the internal azeotrope being unstable, we have system 17, and for the stable internal azeotrope, a system with unstable limit cycle.

We now present the results for possible azeotropic types ("regular", partially or completely "singular") in three-component systems [122, 136]. Once again, we consider an angular azeotrope as exemplified by the triangle vertex $(x_1^* = x_2^* = 0, x_3^* = 1)$. This azeotrope is "regular" if both parameters r_{31} and r_{32} are greater than 1/2. "Singular" azeotropes (being such completely or partially in the first monomer) correspond to the case of both parameters, or only one parameter of the two (for example, r_{31}) being less than 1/2. To be noted, no azeotrope partially singular in two components is feasible in a three-component system. The above analysis is essential not only in exploring a stable node which determines the degree of heterogeneity of a copolymer produced at high conversion in the respective domain of influence, but also in examining a saddle point, this being an unstable singular point in which only a boundary trajectory can terminate at $p \to 1$. This is due to the fact that in the case of $\vec{x}^{(0)}$ localized in a sufficiently close vicinity of such a trajectory, the system is retained in the saddle-point neighbourhood practically till the reaction has run to completion, and then is allowed to go to a stable azeotrope. For this reason, the compositional heterogeneity of a copolymer produced in such a trajectory is determined by the "regularity" of this particular saddle point, rather than of the stable azeotrope. Since for the saddle point $(x_1^* = x_2^* = 0, x_3^* = 1)$, one of the reactivity ratios, r_{31} or r_{32}, is necessarily greater than unity, this point can be either "regular" (the latter reactivity ratio

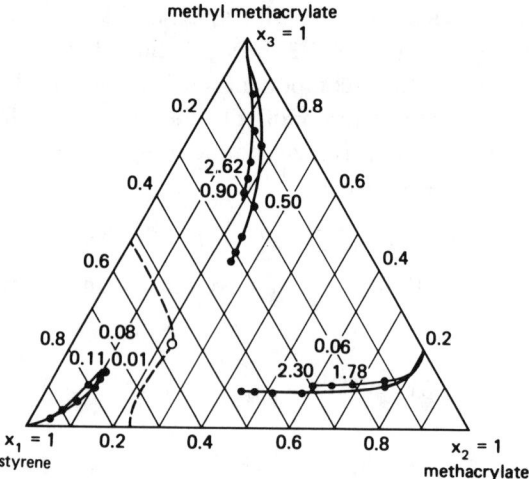

Fig. 2.23 Theoretical curves for terpolymerization of styrene, methyl methacrylate and methyl acrylate as computed by formulas (88), (91), (94) [111]. Variations for monomer mixture composition vs. conversion are denoted by light lines, and those for instantaneous polymer composition, by heavy lines. The trajectory segments between two neighbouring points correspond to a 20% conversion change. Given are the variances $\lambda_{\alpha\alpha} \cdot 10^2$ at $p = 1$ in the order of monomer numbering. The internal azeotrope is denoted by open circle and boundary trajectories, by dashed lines

being greater than 1/2) or, otherwise, partially "singular" in the former or the latter component. The saddle point frequently lies in one of the triangle sides (diagrams 3, 4, 7, 9-11, and 13-16). For example, the saddle point located in the bottom triangle side ($x_3 = 0$) is always "regular" in the second and the third monomer; in the first monomer, it is regular at $\varrho_3^{(3)} < 1$, and "singular" at $\varrho_3^{(3)} > 1$. A similar analysis can also be performed for the saddle point which is an internal azeotrope (diagrams 8, 12, 15, and 16).

Detailed theoretical studies of free-radical terpolymerization of styrene, methacrylate, and methyl methacrylate have been carried out in the works [99, 111] (Fig. 2.23). The calculations have shown that the system in question is described by diagram 9 with an unstable internal node ($x_1^* = 0.57$; $x_2^* = 0.21$; $x_3^* = 0.22$). Among the boundary azeotropes, "regular" are both nonangular saddle points ($x_1^* = 0.77$; $x_2^* = 0.23$, $x_3^* = 0$) and ($x_1^* = 0.53$; $x_2^* = 0$; $x_3^* = 0.47$), as well as the angular stable node ($x_1^* = 1$; $x_2^* = x_3^* = 0$). As to

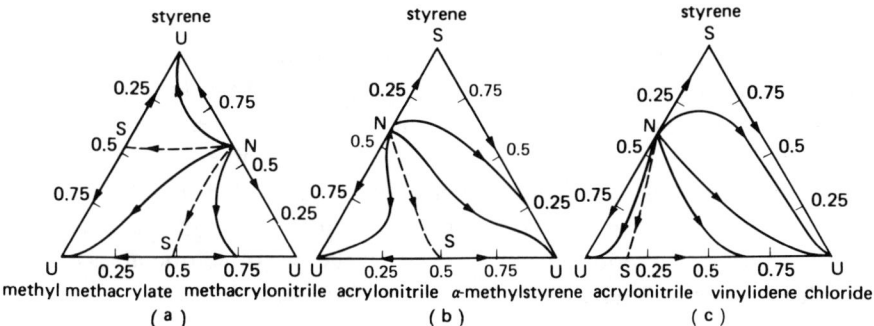

Fig. 2.24 Composition trajectories $\vec{x}$ for the monomer mixture in systems:
(a) methyl methacrylate + methacrylonitrile + styrene; (b) acrylonitrile + α-methylstyrene + styrene;
(c) acrylonitrile + vinylidene chloride + styrene. Boundary trajectories are denoted by dashed lines. Curves have been computed by integrating numerically equations (88), (91) [122]

the remaining angular points, of these the stable node ($x_1^* = x_2^* = 0$; $x_3^* = 1$) is completely "singular", whereas the saddle point ($x_1^* = 0$; $x_2^* = 1$; $x_3^* = 0$) is "singular" in styrene only. In the system in question, there are two domains of influence ($s = 2$) delimited by a boundary line that passes through all nonangular azeotropes. Its shape, constructed by numerical integration, is shown in Fig. 2.23. As evidenced by the above qualitative analysis, a copolymer homogeneous in all its components is expected to form in the given system at the initial monomer-mixture compositions that are confined within the left-hand domain of influence and also within a region adjacent to its boundary. Otherwise stated, in order to provide for obtaining a compositionally homogeneous product at high conversion, the styrene content in the initial monomer mixture must be greater than 50%. The calculated results in Fig. 2.23 lend support to this prediction. To be noted, according to the calculations in Figs. 2.20 and 2.21, the initial monomer mixture composition lies close to the singular trajectory running through the bottom triangle side. For this reason, the compositional heterogeneity of a copolymer is determined by the saddle point ($x_1^* = 0$, $x_2^* = 1$, $x_3^* = 0$) which is partially "singular" in styrene. The substantial heterogeneity in methyl acrylate (the azeotrope being regular in this component) is explained by a nonmonotonic behaviour of the component $\bar{\xi}_2(p)$ in the polymerization process (Fig. 2.23); this is manifested by the occurrence of an additional maximum exhibited by the function $f_2(\xi_2)$ (Fig. 2.21).

Other examples of computed trajectories for ternary systems: (a) methyl methacrylate + methacrylonitrile + styrene, (b) acrylonitrile + α-methylstyrene + styrene, and (c) acrylonitrile + vinylidene chloride + styrene are shown in Fig. 2.24. The first system (a) belongs to type 14, the second (b) and the third (c), to type 10. The variances λ_{22} (in methacrylonitrile) for system (a) and λ_{11} (in acrylonitrile) for system (b) at $p = 1$ are shown in Fig. 2.18. In system (a), all the three angular azeotropes are stable nodes ($s = 3$); of these, two (top and right-hand) are singular, and one (left-hand) is regular. For this reason, the variance λ_{22} in the domain of influence of the latter azeotrope is markedly inferior in value to those in the two other domains. A more detailed analysis of systems (b) and (c) is given below.

2.3.3.3 Experimental Verification of the Theory and Its Use in Predicting the Copolymer Properties

As has already been noted in Sec. 2.3.2, Slocombe [121] studied experimentally a large number (about 40) of three-component systems. He analyzed the transparency of high-conversion copolymers as a function of initial monomer mixture composition. We have referred to these experiments in our discussion of Fig. 2.18. Most of the Slocombe systems possess two nonangular boundary azeotropes and can, in reference to our findings, be described by diagrams 9-12 ($s = 2$) in Fig. 2.22. Based on his experimental evidence, Slocombe divided the systems of interest into two classes: "regular" and "irregular". To the former, he assigned the systems in which the terpolymerization products, at the initial monomer mixture composition lying in a close vicinity to the line joining the two boundary azeotropes (the so-called azeotropic line [121]), were transparent; any other nontransparent system was assigned to the "irregular" class. For the "regular" systems, Slocombe established two interesting experimental rules. The first rule was that, in many such systems, a relatively small deviation of the initial monomer mixture composition from the azeotropic line neighbourhood led to a loss in product transparency. The second rule was that the copolymers with significantly different, but lying close to the azeotropic line, initial compositions $\vec{x}^{(0)}$ were thermodynamically compatible, whereas two monomer mixtures, little differing in composition $\vec{x}^{(0)}$ but lying on the opposite sides of the azeotropic line gave, as a rule, incompatible copolymers. An example of a "regular" system for which both rules hold is the triad acrylonitrile + α-methylstyrene + styrene (diagram 10) shown in Figs. 2.18b and 2.24b. An "irregular" system is exemplified by the triad acrylonitrile + vinylidene chloride + styrene; it can also be described by diagram 10 and is shown in Fig. 2.24c.

The authors of the paper [122] have succeeded, on the basis of literature evidence, to determine the reactivity ratios for eleven systems reported in Ref. [121], of which nine were "regular" systems (styrene + acrylonitrile + α-methylstyrene, styrene + acrylonitrile + methacrylonitrile, styrene + acrylonitrile + vinylmethylketone, styrene + acrylonitrile + methyl methacrylate, styrene + acrylonitrile + n-butyl methacrylate, styrene + α-methylstyrene + methyl methacrylate, acrylonitrile + α-methylstyrene + methacrylonitrile, acrylonitrile + α-methylstyrene + methyl methacrylate, and styrene + α-methylstyrene + methacrylonitrile) and two systems were "irregular": acrylonitrile + vinylidene chloride + styrene and styrene + acrylonitrile + methyl acrylate. The numerical calculations [122] have shown that the boundary trajectories for all these systems lie comparatively close to the azeotropic lines. In all the "regular", according to the Scolombe classifications, systems, the boundary trajectory terminates in a regular boundary azeotrope, and in the two "irregular" systems (including the one shown in Fig. 2.24c), the trajectory terminates in a partially singular saddle point. Consequently, the copolymerization products of a monomer mixture of composition close to that of azeotropic line are more homogeneous in the former case ("regular" systems) than in the latter case ("irregular" systems), which provides a plausible explanation of their different transparencies as observed by Slocombe.

In the light of the expounded concepts, one may attempt to interpret the empirical Slocombe rules for "regular" systems [121]. For example, in the system acrylonitrile + α-methylstyrene + styrene (Figs. 2.18b and 2.24b), the boundary saddle point ($x_1^* = 0.49$m, $x_2^* = 0.51$, $x_3^* = 0$) in which a boundary trajectory terminates is a "regular" point, whereas both stable angular nodes are "singular" in all the components. Therefore, the copolymer produced from a monomer mixture with the composition distant from the azeotropic line (coincident practically with the boundary which demarcates the domains of influence, see Fig. 2.18b) must exhibit a noticeable compositional heterogeneity. In explaining the second empirical rule, it should be pointed out that, if two initial monomer mixture compositions lie on the opposite sides of the boundary line, the trajectories that pass through these composition points tend to different stable "singular" azeotropes. It is clear that the compositional heterogeneity of the mixture of these two copolymers is very high.

The authors of papers [111, 112] have attempted, by making use of the experimental data from Ref. [121], to extend the qualitative features to a quantitative correlation between the degree of compositional heterogeneity and the copolymer transparency (Fig. 2.18). For the two systems in Fig. 2.18, the empirically determined critical variances λ^{crit} of composition distribution in methacrylonitrile and acrylonitrile (the most polar and, for this reason, regarded as the major property-determining species in the respective triads) are independent of the terpolymer average composition. In principle, this fact enables predicting the properties of a copolymer at its synthetic stage. Thus, one may expect that, for the initial monomer mixture compositions with calculated variance $\lambda_{\alpha\alpha}$ (α is the determining monomer) at complete conversion smaller than the critical value, no phase separation of the reaction mass will take place in the course of copolymerization. If a transparent copolymer is to be prepared at another, "unfavourable" monomer mixture composition (Figs. 2.25 and 2.26), the copolymerization should be stopped as a conversion p_{crit} is reached at which the variance $\lambda_{\alpha\alpha}$ has become critical in value: $\lambda_{\alpha\alpha}(p_{crit}) = \lambda^{crit}$.

In a recent work [139], the transparency of bulk bi- and terpolymerizates of styrene with acrylate and methacrylate monomers as a function of the com-

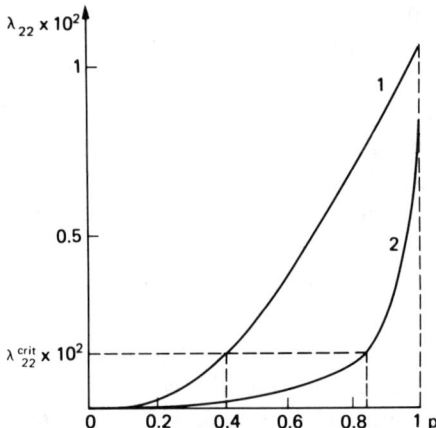

Fig. 2.25 Theoretical curves for the variance λ_{22} as a function of conversion in terpolymerization of methyl methacrylate, methacrylonitrile and styrene at: (1), $x_1^{(0)} = 0.11$, $x_2^{(0)} = 0.78$, $x_3^{(0)} = 0.11$; (2) $x_1^{(0)} = 0.09$, $x_2^{(0)} = 0.27$, $x_3^{(0)} = 0.64$. Shown by dashed line is the critical value $\lambda_{22}^{crit} = 0.15 \cdot 10^{-2}$ at which the copolymer loses its transparence [111]

position distribution variance (in styrene) has been analysed. The critical variances for two binary systems (styrene + methyl acrylate) and (styrene + heptyl acrylate) have been found to be close in value:

$$\lambda^{crit} = (0.55 \pm 0.1)\,10^{-2} \tag{104}$$

Almost the same value has been shown to correspond to the clouding of the reaction mixture of two terpolymers (styrene + methyl acrylate + heptyl acrylate) and (styrene + methyl acrylate + methyl methacrylate) derived from initial monomer mixtures of different compositions. On the basis of this evidence, the authors of Ref. [139] have put forward a supposition that styrene is the determining monomer for compositional heterogeneity in its copolymers with acryl and methacryl monomers. The value of λ^{crit} (104) may be thus used as a universal characteristic for such systems in choosing conditions for preparation of transparent copolymers.

We now make use of the thermodynamic results for the nonergodic multicomponent copolymer from Sec. 2.3.2 as the basis for a theoretical analysis

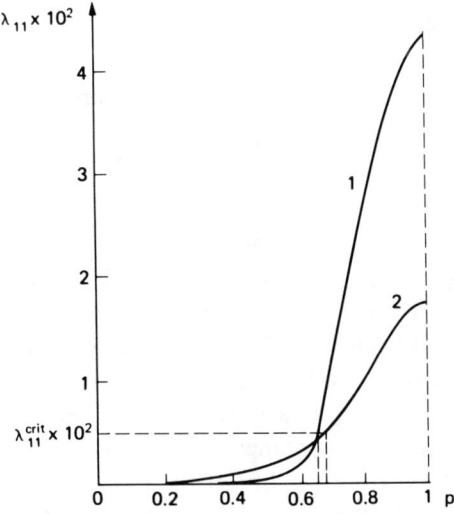

Fig. 2.26 Theoretical curves for the variance λ_{11} as a function of conversion in terpolymerization of acrylonitrile, α-methylstyrene and styrene at: (1) $x_1^{(0)} = 0.7$, $x_2^{(0)} = 0.15$, $x_3^{(0)} = 0.15$; (2) $x_1^{(0)} = 0.2$, $x_2^{(0)} = 0.5$, $x_3^{(0)} = 0.3$. Shown by dashed line is the critical value $\lambda_{11}^{crit} = 0.15 \cdot 10^{-2}$ at which the copolymer loses its transparence

of the empyrical conclusions that have been outlined above. Since the phase transformations conducive to a loss of transparency took place at very high conversions [121, 139], the residual monomers as a solvent are expected to produce little effect on the thermodynamic properties of the system, and the use of the formulas obtained for polymer melts appears to be justified. Taking into account the high viscosity of reaction medium at such conversions, it may be supposed that the clouding is observed only as the system has reached a state close to the spinodal surface (86), that is, when the conditions for stability of metastable states are no longer fulfilled. Consequently, the condition for clouding of a copolymer can be defined in terms of the second central moments $\lambda_{\alpha\beta}$ (59) of the copolymer composition distribution.

Let us consider in some detail relations (86) as applied to a terpolymer ($m = 3$). If equalities (60) are taken into account, only three of the six second-order central moments are independent. The variances $\lambda_{\gamma\gamma}$ ($\gamma = 1, 2, 3$) appear to be a good choice for the three central moments. For the remaining $\lambda_{\alpha\beta}$ ($\alpha \neq \beta$), we have

$$\lambda_{\alpha\beta} = \frac{1}{2}(\lambda_{\gamma\gamma} - \lambda_{\alpha\alpha} - \lambda_{\beta\beta}), \quad (\gamma \neq \alpha, \gamma \neq \beta) \tag{105}$$

With reference to (105), only three Eqs. (86) at $m = 3$ are independent. By solving them for $\lambda_{\gamma\gamma}$, we obtain from (86):

$$\lambda_{\gamma\gamma}^{sp} = \frac{2\chi_{\alpha\beta}}{\Delta \overline{N_W}}, \quad \Delta = (\chi_{12} + \chi_{13} + \chi_{23})^2 - 2(\chi_{12}^2 + \chi_{13}^2 + \chi_{23}^2) \tag{106}$$

To be noted, the subscripts α, β, and γ in equalities (105) and (106) may be taken in any combination of the triad numbers 1, 2, and 3.

Thus, only if each of the variances $\lambda_{\gamma\gamma}$ ($\gamma = 1, 2, 3$) takes a value smaller than the respective value by (106), no spinodal demixing occurs, and a transparent copolymer is thus formed. If no significant stereospecific interactions are operative between any pair of monomers α and β such that $\chi_{\alpha\beta} > 0$, a simple necessary condition for compatibility of the resultant terpolymer follows from (106), viz., $\Delta > 0$, which is equivalent to a simultaneous fulfillment of the three inequalities below:

$$\chi_{12} < \chi_{13} + \chi_{23}, \quad \chi_{13} < \chi_{12} + \chi_{23}, \quad \chi_{23} < \chi_{12} + \chi_{13} \tag{107}$$

Let us now turn attention to the ternary systems in Refs. [121, 139] which have been studied experimentally and shown to exhibit a quantitative relationship between the composition distribution variance in one of the monomer ("determining" monomer) and the copolymer transparency. The monomer of interest is, as a rule, distinct for its polarity and, consequently, the Hilderbrand solubility parameter [123] from the two others, these, in turn, being close to each other in this parameter. Accordingly, two of the three parameters $\chi_{\alpha\beta}$ are close to each other, the third being significantly smaller. For example, in the systems shown in Fig. 2.18, the monomer pairs (methyl methacrylate + styrene) and (styrene + α-methylstyrene) exhibit relatively small Flory-Huggins exchange parameters [140, 141]. By contrast, the interaction of either of the two pairs, respectively, with methacrylonitrile or acrylonitrile is characterized by a significantly larger value of χ [123]. It can easily be shown, using relation (106), that for such systems, the spinodal demixing correlates only with the determining monomer variance $\lambda_{\alpha\alpha}$. Indeed, let us set $\chi_{23} \ll \chi_{12} = \chi_{13}$ in (106). It follows then that

$$\lambda_{11}^{sp} = \frac{1}{2\chi_{12}\overline{N_W}} \ll \lambda_{22}^{sp} = \lambda_{33}^{sp} = \frac{2}{\chi_{23}\overline{N_W}} \qquad (108)$$

The relation (108) shows that the phase separation sets in as the variance in the first monomer has reached a value of λ_{11}^{sp} implying that this monomer is the determining one.

The quantity $\overline{N_W}$ in the right-hand sides of formulas (106) and (108) is, generally speaking, dependent on the initial composition of monomer mixture and, consequently, on the copolymer average composition. A definitive quantitative analysis in this particular case is rather difficult to perform, since the said relationship is affected not only by the chain-growth reaction rate, but also by chain termination rate. Nonetheless, as has been reported in Refs. [111, 122, 139] for a number of systems, the critical values λ^{crit} of composition distribution variance in a determining monomer, independent of the copolymer average composition, are suggestive of minor importance of such a dependence for the phase transformation analysis. It should be noted, however, that the universality of λ^{crit} for a number of binary and ternary copolymers of styrene with acrylate and methacrylate comonomers as alleged in Ref. [139] appears to be rather accidental, considering that the difference in molecular mass of such a copolymer can be quite large.

The theory as it has been expounded in Sec. 2.3.3 is based on the ultimate model of free-radical polymerization. For certain monomers, whose molecules contain bulky or strongly polar substituents, one must take into account also the activity of the next-to-terminal chain unit of a free radical (the so-called penultimate model [130, 150]). However, allowance for this effect does not entail a basic modification of the theory, and the needed improvements can be made within the framework of the earlier concepts [134].

In examining a concrete process, the eventual interference of other factors—multicenter copolymerization, formation of a donor-acceptor complex between monomer molecules, selective sorption, and others—must be taken into consideration. Of special importance are macrokinetic effects arising from the heterogeneity and nonisothermicity of a process. These factors can lead to a nonergodicity of the copolymer and, consequently, affect its compositional heterogeneity. In this respect, the most important problem relevant to process modelling is calculation of the density $\beta(t)$ for distribution of macromolecules over the ergodicity classes, t, considering that the density in such cases is dependent on thermodynamic, diffusive, and hydrodynamic parameters.

2.4 Conclusion

The above examples exhaust by no means the possibilities of mathematical modelling methods in exploring the free-radical polymerization. The macrokinetic approach (Sec. 2.2) and the method for statistical description of molecular structure of copolymers, including the analysis of relationship between molecular structure and thermodynamic properties (Sec. 2.3), provide a good demonstration of the concepts outlined above.

A progress in the development of modelling as applied to polymerization and to other methods for synthesis and chemical modification of polymers is largely dependent on the extent in which the current concepts of polymer physics can gain acceptance in the chemical engineering [20-25]. A strong argument for potential utility of these concepts is that they provide the possibility to construct nonformal and realistic kinetics or macrokinetic models for various chemical processes. In addition, the scaling problems in chemical engineering, primarily concerned with the choice of optimal conditions for effective operation of a chemical engineering process designed for fabrication of certified commercial products can successfully be resolved only using reliable quantitative data that relate the physical properties of polymers to their molecular characteristics.

References

1. Z. Tadmor and C. G. Gogos, *Principles of Polymer Processing*, Wiley, New York, 1979.
2. J. A. Beisenberger and D. H. Sebastian, *Principles of Polymerization Engineering*, Wiley, New York, 1983.

3. K.-H. Reichert and H. Geisler, *Polymer Reaction Engineering*, Hanser-Verlag, 1983.
4. W. H. Ray, in: *ACS Symposium Series No. 226, p. 101*, Amer. Chem. Soc., Washington, D.-C., 1983.
5. W. H. Ray, *Ber. Bunsenges. Phys. Chem.*, **90**: 947 (1986).
6. A. M. North, *Kinetics of Free Radical Polymerization*, Pergamon Press, 1966.
7. R. Jasinghani and W. H. Ray, *Chem. Eng. Sci.*, **32**: 811 (1977).
8. A. D. Schmidt, A. B. Clinch, and W. H. Ray, *Chem. Eng. Sci.*, **39**: 419 (1984).
9. W. Y. Chiu, G. M. Carrat, and D. S. Soong, *Macromolecules*, **16**: 348 (1983).
10. C. J. Kim and A. E. Hamielec, *Polymer*, **25**: 845 (1984).
11. L. Blavier and J. Villermaux, *Chem. Eng. Sci.*, **39**: 87 (1984).
12. G. M. Carrat, C. R. Shervin, and D. S. Soong, *Polym. Eng. and Sci.*, **24**: 442 (1984).
13. P. E. Billagou and D. S. Soong, *Chem. Eng. Sci.*, **40**: 75 (1985).
14. T. J. Tulig and M. Tirrel, *Macromolecules*, **14**: 1501 (1081).
15. P. J. Flory, *Principles of Polymer Chemistry*, Cornell University, Ithaca, 1953.
16. S. M. Benson and A. M. North, *J. Amer. Chem. Soc.*, **81**: 1339 (1959).
17. K. Ito, *J. Polymer Sci.*, A1, **7**: 2995 (1969).
18. K. Ito, *J. Polym. Sci.*, A1, **8**: 1823 (1970).
19. O. Cardenas and K. F. O'Driscoll, *J. Polym. Sci., Polym. Chem. Ed.*, **14**: 883 (1976).
20. M. Doi and S. F. Edwards, *J. Chem. Soc. Faraday Trans.* II, **74**: 1789 (1978).
21. P.-G. De Gennes, *Scaling Concepts in Polymer Physics*, Cornell University Press, Ithaca, London, 1979.
22. W. W. Gressley, *Adv. Polym. Sci.*, **47**: 67 (1982).
23. B. Nyström and J. Roots, *Progr. Polym. Sci.*, **8**: 333 (1982).
24. M. Tirrel, *Rubber Chemistry and Technology*, **57**: 523 (1984).
25. M. Doi and S. F. Edwards, *The Theory of Polymer Dynamics*, Oxford, Clarendon Press, 1986.
26. K. Ito, *Polym. J.*, **12**: 499 (1980).
27. H. M. J. Boots, *J. Polym. Sci. Polym. Phys. Ed.*, **20**: 1695 (1982).
28. S. K. Soh and D. C. Sundberg, *J. Polym. Sci. Polym. Chem. Ed.*, **20**: 1299, 1315, 1331, 1345 (1982).
29. E. B. Brun, V. A. Kaminsky, and G. P. Gladyshev, *Dokl. Akad. Nauk SSSR*, **278**: 134 (1984).
30. V. A. Kaminsky, E. B. Brun, and V. A. Ivanov. *Dokl. Akad. Nauk SSSR*, **282**: 923 (1985).
31. P.-G. De Gennes, *J. Chem. Phys.*, **76**: 3316, 3322 (1982).
32. R. G. W. Norrish and R. R. Smith, *Nature*, **150**: 336 (1942).
33. G. Schulz and G. Harbort, *Macromol. Chem.*, **1**: 106 (1947).
34. K. Horie, I. Mita, and H. Kambe, *Polym. J.*, **4**: 341 (1973).
35. T. Yasukawa, T. Takahashi, and K. Murakami, *Macromol. Chem.*, **174**: 235 (1973).
36. K. Ito, *J. Polym. Sci. Polym. Chem. Ed.*, **12**: 1991 (1974).
37. H. Kh. Mahabadi, K. F. O'Driscoll, *J. Polym. Sci. Polym. Chem. Ed.*, **15**: 283 (1977).
38. O. F. Olaj and G. Zifferer, *Macromol. Chem., Rapid. Commun.*, **3**: 549 (1982).
39. H. Kh. Mahabadi, *Macromolecules*, **18**: 1319 (1985).
40. S. W. Benson and A. M. North, *J. Amer. Chem. Soc.*, **84**: 935 (1962).
41. H. Fujita and A. Kishimoto, *J. Chem. Phys.*, **34**: 393 (1961).
42. J. S. Vrentas, and J. L. Duda, *J. Polym. Sci. Polym. Phys. Ed.*, **15**: 403 (1977).
43. G. Odian, *Principles of Polymerization*, McGraw-Hill Book Company, New York, 1970.
44. E. B. Brun, V. A. Ivanov, and V. A. Kaminsky, *Dokl. Akad. Nauk SSSR*, **291**: 618 (1986).
45. A. L. Efimov. A. I. D'yachkov, S. I. Kuchanov, V. P. Zubov, and V. A. Kabanov, *Vysokomolek. Soed.*, **B24**: 83 (1982).
46. A. I. D'yachkov, A. L. Efimov, L. I. Efimov, T. A. Bugrova, V. P. Zubov, A. F. Samarin, V. M. Artemichev, and V. A. Kabanov, *Vysokomolek. Soed.*, **A25**: 2176 (1983).

47. K. Ito, *Polymer. J.*, **16**: 761 (1984).
48. E. V. Guzeeva, A. L. Efimov et al., *Vysokomolek. Soed.*, **B28**: 587 (1986).
49. Polymer Handbook (Eds. J. Brabdrup and E. M. Immergut), 2nd ed., Interscience, New York, 1975.
50. G. T. Russell, D. H. Napper, and R. G. Gilbert, Macromolecules, **21**: 2133 (1988).
51. G. T. Russell, D. H. Napper, and R. G. Gilbert. Macromolecules, **21**: 2144 (1988).
52. B. A. Korolev, M. B. Lachinov, N. N. Avdeev, V. E. Dreval', A. E. Chalykh, and V. P. Zubov, *Vysokomolek. Soed.*, A **30**: 60 (1988).
53. P. Hayden and H. Melvill, *J. Polym. Sci.*, **43**: 201 (1960).
54. Dispersion Polymerization in Organic Media (Ed. K. E. J. Barret), John Wiley and Sons. London, 1975.
55. W. I. Bengough and R. G. W. Norrish, *Nature*, **163**: 325 (1949).
56. H. S. Mickley, A. S. Michaels, and A. L. Moore, *J. Polym. Sci.*, **60**: 121 (1962).
57. G. Talamini, *J. Polym. Sci.*, A2, **4**: 535 (1966).
58. A. Grosato-Arnaldi, P. Gasparini, and G. Talamini, *Makromolek. Chem.*, **117**: 140 (1968).
59. O. F. Olaj, J. W. Breitenbach, H. Reif, and K. J. Parth, *Angewandte Chem.*, **83**: 370 (1971).
60. J. Ugelstad, H. Flogstad, T. Herzberg, and E. Sund, *Makromolek. Chem.*, **164**:171 (1973).
61. S. I. Kuchanov and D. N. Bort, *Vysokomolek Soed.*, A **15**: 2393 (1973).
62. A. K. Suresh and M. Chanda, *Eur. Polym. J.*, **18**: 607 (1982).
63. W. D. Harkins, *J. Chem. Phys.*, **13**: 381 (1945).
64. W. D. Harkins, *J. Chem. Phys.*, **14**: 47 (1946).
65. W. D. Harkins, *J. Amer. Chem. Soc.*, **69**: 1428 (1947).
66. A. I. Yurzhenko and M. S. Kolechkova, *Dokl. Akad. Nauk SSSR*, **47**: 354 (1945).
67. W. V. Smith and R. H. Ewart, *J. Chem. Phys.*, **16**: 592 (1948).
68. J. L. Gardon, *J. Polym. Sci.*, A1, **6**: 623, 665, 687, 2853, 2859 (1968).
69. J. Ugelstad and F. K. Hansen, *Rubber Chem. Technology*, **49**: 536 (1976).
70. *Emulsion Polymerization* (Ed. I. Piirma), New York, Academic Press, 1982.
71. K. W. Min and W. H. Ray, *J. Appl. Polym. Sci.*, **22**: 89 (1978).
72. I. A. Gritskova, L. I. Sedakova, D. S. Muradyan, B. M. Sinekaev, A. V. Pavlov, and A. H. Pravednikov, *Dokl. Akad. Nauk SSSR*, **238**: 607 (1978).
73. *Micellization, Solubilization, and Microemulsions* (Ed. K. L. Mittal), Plenum Press, New York, 1977.
74. G. A. Simakova, V. A. Kaminsky, I. A. Gritskova, and A. N. Pravednikov, *Dokl. Akad. Nauk SSSR*, **276**: 151 (1984).
75. M. Morton, S. Kaizerman, and M. W. Altier, *J. Colloid. Sci.*, **9**: 300 (1954).
76. V. A. Kaminsky, G. I. Litvinenko, and M. G. Slin'ko, *Dokl. Akad. Nauk SSSR*, **296**: 1393 (1988).
77. N. Friis, Hamielec A. E., *J. Polym. Sci. Polym. Chem. Ed.*, **12**: 251 (1974).
78. M. Khaddazh, G. I. Litvinenko, I. A. Gritskova, V. A. Kaminsky, and A. N. Pravednikov, *Vysokomolek. Soed.*, B **25**: 139 (1983).
79. G. I. Litvinenko, M. Khaddazh, V. A. Kaminsky, A. V. Pavlov, A. G. Davtyan, I. A. Gritskova, and A. N. Pravednikov, *Vysokomolek. Soed.*, B **26**: 683 (1984).
80. Y. Talmon and S. Prager, *J. Chem. Phys.*, **69**: 2984 (1978).
81. Y. Talmon and S. Prager, *J. Chem. Phys.*, **76**: 1535 (1982).
82. P.-G. de Gennes and C. Taupin, *J. Phys. Chem.*, **86**: 2294 (1982).
83. C. Taupin, M. Dvolaitzky, and R. Ober, *Nuovo Cimento*, **30**: 62 (1984).
84. M. J. Grimson and F. Honary, *Phys. Lett.*, A **102**: 241 (1984).
85. D. J. Mitchel and B. W. Ninham, *J. Chem. Soc. Faraday Trans.* II, **77**: 601 (1081).
86. B. Widom, *J. Chem. Phys.*, **81**: 1030 (1984).
87. B. N. Okunev and V. A. Kaminsky, *Kolloidn. Zh.*, **50**: 703 (1988).
88. J. A. Barker, *J. Chem. Phys.*, **20**: 1526 (1952).

89. *Recent Advances in Liquid-Liquid Extraction* (Ed. C. Hansen), Pergamon Press, Oxford (1971).
90. T. S. Sorensen, M. Hennehberg, in: *Dynamics and Instability of Fluid Interfaces* (Ed. T. S. Sorensen), p. 276.
91. G. Meyergoft, R. Soc, and M. Konloumbris, *Polym. Preprint.*, **26**: 203 (1985).
92. D. A. Frank-Kamenetsky, *Diffusion and Heat Transfer in Chemical Kinetics*, Nauka, Moscow, 1987 (in Russian).
93. E. I. Maksimov, *Dokl. Akad. Nauk SSSR*, **191**: 1091 (1970).
94. V. A. Ivanov, E. B. Brun, V. A. Kaminsky, and A. B. Rabinovich, *Khim. Prom.*, No. 8: 468 (1988).
95. E. B. Brun and S. I. Kuchanov, *Theor. Found. of Chem. Engin.*, **14**, No. 2: 148 (1980).
96. E. B. Brun, V. A. Filimonov, and S. I. Kuchanov, *Khim. Prom.*, No. 11: 666 (1981).
97. M. Gordon and W. B. Temple, *The Graph-Like State of Matter and Polymer Science. Chemical Application of Graph Theory*, Academic Press, New York, 1976.
98. *Markov Chains and Monte Carlo Calculations in Polymer Science* (Ed. G. G. Lowry), Marcel Dekker Inc., New York, 1970.
99. S. I. Kuchanov, *Metody kineticheskikh raschetov v khimii polimerov* (Methods for Kinetic Calculations in Polymer Chemistry), Khimiya, Moscow, 1978 (in Russian).
100. E. B. Brun and S. I. Kuchanov, *Dokl. Akad. Nauk SSSR*, **257**: No. 4: 907 (1981).
101. B. D. Coleman and T. G. Fox, *J. Polymer Sci.*, A **1**: 3138 (1963).
102. *Spravochnik po teorii veroyatnosti i matematicheskoi statistike* (Handbook on Probability Theory and Mathematical Statistics) (Ed. V. S. Korolyuk), Naukova Dumka, Kiev, 1978 (in Russian).
103. C. Tosi and G. Catinella, *Makromol. Chem.*, **137**: 211 (1970).
104. D. W. Van Crevelen, *Properties of Polymers. Correlations with Chemical Structure*, Elsever Publ. Company, Amsterdam, 1972.
105. A. A. Dontsov, G. I. Lozovik, and S. P. Novitskaya, *Khlorirovannye polymery* (Chlorinated Polymers), Khimiya, Moscow, 1979 (in Russian).
106. I. Uematsu and K. Honda, *Rep. Progr. Polymer Japan*, **8**: 111 (1965).
107. J. M. Barton, *J. Polymer Sci.*, No. 30: 573 (1970).
108. M. Hirooka and T. Kato, *J. Polymer Sci.*, B **12**: 31 (1974).
109. J. Furukawa, *J. Polymer Sci.* C **51**: 105 (1975).
110. H. W. Jonston, *J. Macromol. Sci., Rev. Macromol. Chem.*, C **14**, No. 2: 215 (1976).
111. E. B. Brun, Cand. Diss. (Phys. Chem.), State Research Institute of Chlorine, Moscow, 1981.
112. H. Mark, E. H. Immergut, L. J. Young, K. J. Beynon, *Reactivity Ratios*, in: *Copolymerization* (Ed. G. E. Ham), Interscience, New York, 1966.
113. R. B. Beevers, *Trans. Faraday Soc.*, **58**: 1465 (1962).
114. J. Furukawa and A. Nishioka, *J. Polymer Sci.*, B **9**: 199 (1971).
115. W. H. Howard, *J. Appl. Polym. Sci.*, **5**, No. 15: 303 (1961).
116. J. Comyn and R. A. Fernandes, *Eur. Polymer J.*, **11**: 149 (1975).
117. N. N. Slavnitskaya, Yu. D. Semchikov, and S. A. Ryabov, *Vysokomolek. Soed.*, B **21**, No. 1: 23 (1979).
118. P. Wittmer, *Angew. Makromol. Chem.*, B **39**: 35 (1974).
119. D. Braun and G. Disselhoff, *Polymer*, **18**, 963 (1977).
120. G. Disselhoff, *Polymer*, **19**, No. 1: 111 (1978).
121. R. J. Slocombe, *J. Polymer Sci.*, **26**, No. 112: 9 (1957).
122. E. B. Brun and S. I. Kuchanov, *Vysokomolek. Soed.*, A **19**, No. 3: 488 (1977).
123. *Polymer Blends* (Eds D. R. Paul and S. Newman), Vol. 1, Academic Press, New York, 1978.
124. E. B. Brun and A. O. Malakhov, *Dokl. Akad. Nauk SSSR*, **313**: 357 (1990).
125. M. T. Rätzsch and H. Kehlen, *Prog. Polymer Sci.*, **14**, No. 1: 1 (1989).
126. M. T. Rätzsch, D. Browarzik, and H. Kehlen, *J. Macromol. Sci.-Chem.*, A **26**, No. 6: 903 (1989).

127. A. Münster, *Chemische Thermodynamik*, Akademie-Verlag, Berlin, 1969.
128. R. Koningsveld and L. Kleintjens, *J. Polymer Sci.*, Part C. Polymer Symposium, No. 61: 221 (1977).
129. *Copolymerization* (Ed. G. E. Ham), Interscience, New York, 1952.
130. T. Alfrey, J. Bohrer, and H. Mark, *Copolymerization, High Polymer Series*, Vol. 8, Interscience, New York, 1952,
131. L. Rios and J. Guillot, *J. Macromol. Sci.*, A **12**, No. 8: 1151 (1978).
132. H. H. Robertson, *J. Inst. Maths. Applics.* **23**, No. 4: 405 (1979).
133. P. Lancaster, *Theory of Matrices*, Academic Press, New York, London, 1969.
134. S. I. Kuchanov, V. A. Efremov, and M. G. Slin'ko, *Vysokomolek. Soed.*, A **28**, No. 5: 964 (1986).
135. J. Guckenheimer, *A Strange, Strange Attractor*, in: Marsden J. and M. McGracken, *The Hopf Bifurcation*. Springer Verlag, New York, 1976.
136. S. I. Kuchanov and E. B. Brun, *Zh. Prikl. Khimii*, **50**, No. 5: 1106 (1977).
137. F. G. Tricomi, *Differential Equations*, Blackie and Son, Ltd., New York, 1961.
138. V. T. Zharov and L. A. Serafimov, *Fiziko-khimicheskie osnovy distillyatsii i rektifikatsii* (Physico-Chemical Fundamentals of Distillation and Rectification), Khimiya, Leningrad, 1975 (in Russian).
139. S. I. Kuchanov, Z. V. Orlova, A. F. Kosheleva, and Yu. P. Gorelov, *Vysokomolek. Soed.*, A, **31**, No. 3: 474 (1989).
140. D. J. Dunn and S. Krause, *J. Polym. Sci., Polymer Lett. Ed.*, **12**: 591 (1974).
141. T. P. Russel, R. P. Hjelm, and P. A. Seeger, *Macromolecules*, **23**: 890 (1990).
142. B. Nyström and J. Roots, Macromolecules, **24**: 184 (1991).
143. V. A. Kaminsky and E. B. Brun, *Vysokomolek. Soed.*, **A32**: 2167 (1990).
144. H.-Kh. Mahabadi, *Macromolecules*, **24**: 606 (1991).
145. V. A. Ivanov, V. A. Kaminsky, and E. B. Brun, *Vysokomolek. Soed.*, **A33**: 1442 (1991).
146. T. Shiomi and K. Imai, *Macromol. Chem., Macromol. Symp.*, **38**: 233 (1990).
147. D. Broseta and G. H. Fredrickson, *J. Chem. Phys.*, **93**: 2927 (1990).
148. C. J. T. Landry, H. Yang, and J. S. Machell, *Polymer*, **32**: 44 (1991).
149. T. Shiomi and K. Imai, *Polymer*, **32**: 73 (1991).
150. T. Fukuda, Y.-D. Ma, K. Kubo, and H. Inagaki, *Macromolecules*, **24**: 370 (1991).

3 Modelling and Macrokinetics of Electrochemical Processes

A. B. Goldberg and L. I. Kheifets

State Research Institute of Chlorine,
Ugrezhskaya 2, Moscow 109088, USSR

3.1 Introduction

The electrochemical processes are receiving of late an ever-increasing attention both owing to their involvement in a straight-forward high-effective conversion of relatively cheap electric energy to chemical energy and to the ability to rationally store and use energy supplied to storage and fuel cells. The application of novel materials and constructions, the development of fundamental principles of electrochemistry provide good grounds for extending the field of practical employment of electrochemical processes and assist in the solution of energetic and ecological problems.

In listing the advantages of electrochemical reactors, one must mention the high efficiency in using materials and energy, high selectivity of processes, purity of end products, potentially low toxicity, and waste-free production. However, the relatively low conductance inherent in conductors of the 2nd kind, which the electrolytes are, entails low yields, both in time and space, of desired products, which necessitates either the use of a large number of reaction units (cells), or of expanded electrode surfaces. For this reason, the problem of developing an all-inclusive and scientifically-based approach to the design and optimization of electrochemical reactors has advanced to the foreground of urgent issues.

Currently, the wide spread of computers has made such an approach practicable. As early as the 1960s, the Hooker Company (USA) outlined a research program aimed at the design of an efficient chlorine diaphragm-type cell [1]. Algorithm and program complexes were developed enabling, on the basis of theoretical and experimental studies of the processes occurring in an electrolytic cell, determination of technical and economical characteristics of a reactor [2, 3], contemporaneously, metal-oxide anodes and ion-exchange membranes made their first appearance in technology. Following the same trend, pioneering studies on modelling a chlorine membrane-type electrolytic cell were started [4]. In the 1970s, the term "chemical engineering approach" gained full recognition as applied to electrochemical reactors [5, 6], implying thereby a universal strategy in the modelling of reactors of any type. The leading chemical companies spared no efforts to make the computer an effective tool in designing efficient commercial electrolytic cell, using the methods of macrokinetic analysis in conjunction with empirical correlations obtained on laboratory- and pilot-scale plants.

At present, both the engineering practice and technology, including electrochemical engineering, have suffered drastic changes vis-à-vis the empirical methods of the turn of the 20th century, when the first steps were attempted at commercial utilization of the chemical effects by electric current in experimental plants. The modern electrochemical technology is, in point of fact, a theory of electrochemical reactors embracing a number of scientific and engineering disciplines such as electrochemistry, physical chemistry, hydrodynamics, mass transport and heat transfer, chemical kinetics, mathematical modelling and process optimization. The electrochemical engineers have elaborated specific experimental methods applicable in various fields of science [7, 8].

Since the systematic consideration of all the effects due to electric and magnetic fields is quite an arduous task, the analysis of a process was mainly focused on separate characteristic phenomena which in many instances bore a remote resemblance of the reality. Hybrid models aimed at practical applications require a comprehensive study of physico-chemical properties of a system in question, a correct statement of the problem, and a large amount of computation. For these reasons, despite the fact that the principles underlying the operation of electrochemical systems are well known and can be applied to mathematical modelling, their straightforward use for the design and optimization of engineering processes becomes occasionally impossible because of a poorly understood theory of electrochemical reactors.

Complex macrokinetic problems have been dealt with in the studies of corrosion, in particular, of anodic dissolution under the conditions of salt passivation when alterations in the geometrical structure of a reactive system should be taken into account [9]. In electrochemical machining, intense electrode processes are involved which require, for their deeper understanding, a revision of the conventional approach to the formulation of transport equations in order to take into account finite mass flow rates at the electrode surface [10]. In recent years, a noticeable progress has been achieved in the simulation of chlorine cells, water electrolysis electrodes, and storage cells of novel type, in the application of porous fluidized-bed and suspended-particle electrodes in hydroelectrometallurgy, electrosynthesis of organic compounds, and nostationary electrolysis [11, 12, 188, 189].

3 Modelling of Electrochemical Processes

Fundamental problems, mathematical computations, and engineering applications in various fields of electrochemistry have been dealt with in the now classical monograph [13] and later works [14, 15]. The present study has been intended to give a systematization of the fundamentals of macrokinetics and modelling of electrochemical processes, currently regarded as an independent field of applied scientific research [16]. An important concern has also been to consider the relationship between charge transfer and hydrodynamic and diffusion effects typical of electrolytic systems only. Along with the classical problems of mass transport and charge transfer in electrolyte solutions, problems related to stability of electrochemical reactors, occurrence of dissipation structures, percolation phenomena, and nonequilibrium thermodynamics of membranes have been considered. Within the framework of a chemical engineering approach, a mathematical model for the process of the production of chlorine and alkali in a solid cathode cell has been developed.

3.2 Mass Transport and Charge Transfer in Ionic Solutions

3.2.1 Phenomenological Transport Equations

The mass transport in a multicomponent liquid mixture obeys linear phenomenological relations between the flux components and their driving forces; the relevant issues have been treated in the thermodynamics of irreversible processes. The classification of transport phenomena in electrochemical systems, the derivation of computation equations, as well as the analysis of limit cases, are convenient to make using the Stefan-Maxwell form for linear relations expressing a balance of driving and resisting forces in the relative motion of transport system components [13]. In an N-component system, one may consider concentrations of components c_i (mol/cm³), macroscopic velocities $\vec{v}_i$ (cm/s), chemical potentials $\bar{\mu}_i$ (J/mol), and interaction coefficients K^0_{ij}. For an isothermal case in the absence of barodiffusion one may write

$$c_i \nabla \bar{\mu}_i = \sum_{j=1}^{N} K^0_{ij}(\vec{v}_j - \vec{v}_i), \quad i = 1, \ldots, N \tag{1}$$

Here, the component flow rates are $\vec{N}_i = c_i \vec{v}_i$. The coefficients K^0_{ij} are dependent on the solution composition and are expressed through the mutual diffusion coefficients as $K^0_{ij} = \dfrac{RT c_i c_j}{c_T D_{ij}}$, where $c_T = \sum_{i=1}^{N} c_i$. From Onsager reciprocity, $D_{ij} = D_{ji}$; therefore, the total number of independent transport characteristics is $1/2N(N - 1)$. On a number of occasions, the validity of equation (1) was a topic for discussion in the literature [193]. Eq. (1), known to be linearly dependent, must be supplemented with hydrodynamic equations and with equations for conservation of mass, which ultimately permits formulating the boundary problem for unknown concentrations c_i, taking into account the specific properties of a given system.

A distinguishing feature of electrochemical processes is the charge transfer involving ionic components; for this reason, the partial potentials $\bar{\mu}_i$ are dependent not only on the solution composition, but also on the local electric state. This necessitates the introduction of an additional variable, the electric potential φ, and an additional equation, Poisson's equation. These conditions result in a significant expansion of the class of problems related to the electrochemical engineering as compared to the issues with which the theory of conventional chemical reactors is concerned. The necessity of deriving volt-ampere characteristics brings the problem of electric potential calculation to the same level of priority as those of calculating component concentrations and component flow rates, with allowance made for migration effects in the electric field—a situation which has no analogue in nonelectrolytic systems. The parameter characterizing the size of a space charge region in an electrolyte is the Debye length

$$\lambda_D = \left[\varepsilon_0 \varepsilon RT / (F^2 \sum_{i=1}^{N} z_i^2 c_{i0}) \right]^{1/2}$$

where $\varepsilon_0 = 8.85416 \times 10^{-14}$ F/cm; ε is the relative permittivity; $R = 8.314$ J/mol·K; T is the absolute temperature; F is the Faraday constant, 96487 C/equiv; z_i is the charge number (an integer); and c_{i0} (mol/cm^3) is the concentration of a component in solution bulk. By way of example, $\lambda_D \approx 10^{-4}$ cm for pure water. If L is the characteristic size of a given system, the electroneutrality disturbance in solution bulk, $\sum_{i=1}^{N} z_i c_{i0}/z_j c_{j0}$ is of the order of $(\lambda_D/L)^2$ [17, 18]. Therefore, outside the double electric layer region at phase interface, it may be assumed that, in place of Poisson's equation, the relation

$$\sum_{i=1}^{N} z_i c_i = 0 \qquad (2)$$

holds, which is a supplementary equation for determining the potential φ in an electrochemical system. The electroneutrality is also observed in a large variety of processes occurring in ionized gases, semiconductors, and solid-state ionic conductors. The effect the charged region at the "electrode-solution" or "membrane-electrode" interface (however small its characteristic value λ_D may appear to be) exerts on the mass transport, and the dynamics of concentrational changes within the double electric layer were extensively studied theoretically [19]. Since the condition (2) in all rigour applies to the simplest case of a binary electrolyte solution, in principle one cannot rule out the existence of charged regions in more complex systems. For example, given a system of ions of three types of which the ions of one type remain immobile in space (the fixed charge of an ion-exchange membrane), the electroneutrality disturbance under definite boundary conditions can be as high as $(10^2\text{-}10^3)\cdot(\lambda_D/L)^2$ [20]. The spatial position, height and shape of the charge peak depend on the current density [194]. However, in concentrated solutions, the small value of λ_D entails a smallness of the above effect.

In the thermodynamics of a variable-composition solution, the problem of defining a quantitative measure for the electric state has been discussed in Ref. [13]. From a formal point of view, any function of the thermodynamically measurable parameters c_i and $\bar{\mu}_i$, that is, $\varphi = \varphi(c_1, \ldots, c_N, \bar{\mu}_1, \ldots, \bar{\mu}_N)$, can serve for such a measure. For example, the function $\varphi = (\bar{\mu}_N - RT \ln c_N)/z_N F$ was named the quasi-electrostatic potential based on the Nth ionic components [13]. This definition is convenient for use in analyzing the relation of the chosen measure to the theory of dilute solutions: as c_i tends to zero, the function φ becomes independent of the particular ion and may reasonably be regarded as an electric potential in homogeneous medium. From a practical standpoint, in order to derive, starting from Eqs. (1) and (2), a closed system of equations with respect to c_i and φ, one must specify the partial potentials $\bar{\mu}_i$ as functions of local-state parameters (c_i and φ). For this reason, it appears expedient to change, in Eq. (1), from $\bar{\mu}_i$ to new variables that would allow a straightforward use of the available experimental data on thermodynamic and transport properties of electrolyte solutions. By fixing the Nth charged component, one can manipulate the quantities $\bar{\mu}_i - \dfrac{z_i}{z_N}\bar{\mu}_N$, which are independent of the electric state and which reflect the overall contribution of oppositely charged ions to the thermodynamic properties of the solution. Introducing the current density $\vec{i} = F \sum_{i=1}^{N} z_i \vec{N}_i$ and using the expression for the velocities from Eq. (1), one obtains [13]:

$$\frac{1}{z_N}\nabla\bar{\mu}_N = -\frac{F}{\varkappa}\vec{i} - \sum_{i=1}^{N}\frac{t_i^0}{z_i}\left(\nabla\bar{\mu}_i - \frac{z_i}{z_N}\nabla\bar{\mu}_N\right) \qquad (3)$$

where $\varkappa$ is the conductivity of the solution; t_i^0 is the transference number relative to the solvent ($\varkappa$ and t_i^0 are uniquely determined via initial parameters D_{ij}). By virtue of the Gibbs-Duhem equation, there exist $(N-2)$ independent driving forces $\nabla\bar{\mu}_i - \frac{z_i}{z_N}\nabla\bar{\mu}_N$. On account of the electroneutrality condition (2) and the thermodynamic state equation for a mixture, they can be expressed through $(N-2)$ independent component concentrations $(N-1)$ independent velocity differences (or diffusion fluxes) can be expressed from Eq. (1) in terms of the driving forces and $\nabla\bar{\mu}_N$ (or $\vec{i}$ by using formula (3)). Conditions $\frac{\partial c_i}{\partial t} + \text{div}\,\vec{N}_i = 0$ ($i = 1, \ldots, N-2$) become then the equations for $(N-2)$ unknown concentrations (to this end, one must resort to the auxiliary use of hydrodynamic equations), providing the current density is known, as is the case with one-dimensional problems. For binary and ternary solutions, the expressions for diffusion fluxes were given in the works [13, 21]. If the current distribution is initially unknown, then the values of $\vec{i}$ must be calculated together with $(N-2)$ concentration fields to satisfy certain specified conditions, for example, a known voltage drop across the electrochemical cell, or a known net current load, or equipotentiality of the current leads.

Thus, one may regard the current density as an adequate characteristic of the electric state of a solution. However, it may be inferred from the above that, in order to calculate the volt-ampere characteristic with an unknown current distribution, it appears convenient to use an alternative measure for an electric state considering that the quantity $\vec{i}$ is not informative as to the ohmic voltage drop in solution. Following Newman [13], the potential φ in solution may be defined as the potential of a hypothetic probe electrode with an equilibrium charge-transfer reaction identical to that operative on one of the cell electrodes. Then the expression for $F\nabla\varphi$ is analogous to Eq. (3), the only distinction being the coefficient t_i^0/z_i replaced by $(s_i/n + t_i^0/z_i)$. Here s_i and n are the coefficients of a reaction represented in the form $\sum_{i=1}^{N} s_i M_i^{z_i} = ne^-$, M_i being the symbolic formula of a component. The above definition allows representing the cell voltage as the sum of the surface overvoltage η_s, concentrational overvoltage η_c, and ohmic voltage drop $\Delta\varphi_{\text{ohm}}$, taking thus into account the kinetic parameters of the electrode reaction and the concentrational changes within the electrode-solution boundary layers.

3.2.2 Determination of Transport Characteristics of Solutions

The title problem may be regarded as an inverse one in relation to the determination of a concentration field and diffusion fluxes, supposing that this is carried out in special experiments, for example, with diaphragm cells [22, 23]. Accordingly, in the data preprocessing determined are the conductivity $\varkappa$, the total of $(N-1)$ transference numbers t_i^0, and the diffusion coefficients for the neutral combinations of ions (which are the proportionality coefficients between diffusion fluxes and concentration gradients). The mutual diffusion coefficients D_{ij} are calculated by using the formulas derived from solving Eqs. (1)-(3) for diffusion fluxes; this procedure in principle presents no problem. For ions of the same sign, $D_{ij} < 0$ is to be expected, which is attributed to the repulsion effect and can be explained by the statistical Onsager theory at infinite dilution. For a concentrated NaCl-KCl-H$_2$O mixture, the condition $D_{\text{Na, K}} < 0$ was checked in Ref. [24] starting from the known transport characteristics. Similar results have been obtained in treating data on ionic isotope diffusion in aqueous solutions. However, for industrial synthetic ion-exchange membranes regarded as polyelectrolytes with spatially fixed charges of ionogen groups, all the D_{ij} coefficients have been found to be positive. This effect may be explained by the resistance to the ion motion exerted by the ordered cluster-channel structure of a polymer membrane matrix. This fact is thus suggestive of the caution one should exercise in treating the quantities K_{ij}^0 in Eq. (1) as partial "friction coefficients" in the presence of long-range electrostatic interactions in solution.

In performing practical calculations, a problem emerges as to the number of equations in system (1) that would suffice for an adequate description of the mass transport as outlined by an adopted modelling strategy. In most of the earlier theoretical works concerned with either the application of Nernst-Planck transport equations, or their generalizations, emphasis was put on the use of strong electrolytes only. Meanwhile, from a macroscopic point of view, the notion "strong electrolyte" is based on an interpretation of experimental data without reference to the actual structure of the solution [25]. On the other hand, in many chemical engineering processes, the working solutions contain compounds prone to association. According to the current concepts [26], the association constants of a number of alkaline metal salts lie within a range of 0.2 to 0.8, which provides the reason to anticipate a significant association at concentrations of the order of 1 mol/dm^3. Therefore, in handling macrokinetic problems, one must specify the independent components of the solution that determine entirely both thermodynamic and transport properties. Commonly, reliable data on the microscopic composition of associated electrolytes and on the diffusion coefficients of solution components are either scarce, or lacking altogether. Therefore, to circumvent this limitation, a simplified solution to the problem was proposed in Ref. [26] as exemplified by the cathodic deposition from a CdI_2 salt solution, in which all the associates of $(CdI_n)^{(n-2)-}$-type were assigned the same mobility; with this condition, the conventional equation of convective diffusion can be applied to the salt as a whole. It will be clear from what follows that the condition for equal ion mobilities is not in fact of essential importance. In considering the diffusion of the label $^{35}SO^{2-}$ in a mixture of chloride solutions [27], the propensity of this anion to pairing with metal ions was taken into account through introducing chemical reaction terms into the transport equation in the dilute solution approximation (Nernst-Planck equations). Assuming the reaction to be at equilibrium, the resultant equation for the label takes the form of an electrodiffusion equation, but with a modified diffusion coefficient. Stockmayer [28] was the first to consider, in a general approach, the diffusion of a weak electrolyte $A_{\nu_A}B_{\nu_B}$ capable of forming a variety of multiply charged complexes $A_{p_i}B_{q_i}$ (p_i, q_i being nonnegative integers) existing in equilibrium with the cation A and the anion B in the absence of electric current. By expressing the diffusion fluxes for all the components through the respective driving forces $\nabla\bar{\mu}_k$, with allowance made for the electroneutrality and the equality of chemical potentials at equilibrium, Stockmayer obtained a law for diffusion of the electrolyte as a whole in the conventional form $N = -\Omega\nabla\mu$. Here Ω is the thermodynamic transport coefficient dependent on the component transport coefficients, and $\mu = \nu_A\bar{\mu}_A + \nu_B\bar{\mu}_B$ is the chemical potential of the salt.

This formal result can be extended to any multicomponent solution keeping in mind, however, that in determining the partial chemical potential of an equilibrium system or in applying the consistent formalism of nonequilibrium thermodynamics [29], no information on the intrinsic structure of a solution is used, given the complete set of state variables. For an N-component system subject to local equilibrium, the number of independent concentrations, by virtue of the Gibbs-Duhem condition, is $(N - 1)$, and N equations (1) for the component fluxes remain valid independent of the occurrence of an equilibrium reaction in the system. It is expedient to take the concentrations of "elementary" ions, that is, of real entities extant in solution at infinite dilution,

3 Modelling of Electrochemical Processes

for the state variables. The formulas for converting the Onsager coefficients of all the extant components to the coefficients of initial reactants in the associated mixture have been reported in Ref. [30]. The works [31, 32] suggest an alternative set for state variables. To say but the least, in the absence of electric current one may introduce fluxes for $(N - 2)$ neutral salts as independent kinetic units, for which the transport equations formally do not differ from those for the ions.

The only constraint here is that associated with the reaction rate. For example, for most sulphuric acid salts reported in Ref. [26], the rate constant values are suggestive of relaxation times of the order of 10^{-8}-10^{-4} s. They define a lower limit for the characteristic times of macrokinetic phenomena which provide for the establishment of a local equilibrium. A more detailed description of ionic pair formation is possible following the known scheme after Eigen for sequential desolvation

$$M^{m+} + L^{n-} \underset{k_{21}}{\overset{k_{12}}{\rightleftarrows}} M^{m+}(S_2)L^{n-} \underset{k_{32}}{\overset{k_{23}}{\rightleftarrows}} M^{m+}(S)L^{n-} \underset{k_{43}}{\overset{k_{34}}{\rightleftarrows}} ML^{(m-n)+} \quad (4)$$

(with M, L symbolizing the ions, and S, the solvent). By linearizing the Nernst-Planck diffusion equations for the five components of reaction (4) and Poisson's equation in the vicinity of equilibrium state and by applying subsequently a normal mode analysis, the authors of the paper [33] have obtained an expression for the diffusion coefficient of the electrolyte as a whole. For a symmetric electrolyte ($m = n$), the diffusion coefficient has been shown to be a linear combination of the diffusion coefficients of the four complexes involved in equilibrium (4); it is therefore dependent on the equilibrium constants. For an asymmetric electrolyte ($m \neq n$), an additional term is introduced to take into account the transference number of charged components. These results might have also been obtained via summarizing directly the ion fluxes, by analogy with an approach used in Refs. [28, 30]. Another conclusion is that for the asymmetric electrolyte, the kinetic relaxation cannot be treated separately from the electrostatic, since the ionic pairs are involved in the build-up of a charge balance in solution. According to the current concepts [34-38], thorough experiments for determining the transport characteristics, including those defined in different reference systems of diffusion fluxes, in conjunction with measurements of diffusion coefficients for isotope labels might have provided a quantitative information about the equilibrium structure of a solution, including the solvation numbers. Taking into account the finite reaction rates leads to a dependence of the transport equations on the dimensionless parameters $\varepsilon_{ij} = D_{ij}c/k_{ij}L^2$ characterizing the kinetic-to-diffusional times ratio. Fine effects due to a boundary layer spaced apart to a distance about $\sqrt{D_{ij}c/k_{ij}}$ from the electrode surface have been analyzed in [39, 40] as exemplified by a binary electrolyte at $\varepsilon_{ij} \to 0$. As the current i tends to a limiting diffusion current $i_{\lim}$, the local equilibrium within the boundary layer breaks down, if $(i_{\lim} - i)/i_{\lim} \leqslant \sqrt{\varepsilon_{ij}}$.

In treating electrolyte mixtures for which experimental data are lacking, semi-empirical rules for correlation with the ion mobilities at infinite dilution, or with the properties of binary components are applied [195]. For ternary electrolytes with a common ion, a complete calculation of transport characteristics can be performed using the so-called LN-approximation after Miller [21, 41] (in the original papers, the equivalent concentration was denoted by N_i); the utility of this approach has been substantiated by practical applications [15]. Let us consider an aqueous NaCl + NaOH solution, for the sake of definiteness. The linear relationships for diffusion fluxes (relative to solvent) can be written in a form equivalent to Eq. (1) as

$$\vec{N}_i^0 = -\sum_{j=1}^{3} l_{ij} \nabla \bar{\mu}_j, \quad l_{ij} = l_{ji}, \quad i = 1, 2, 3 \tag{5}$$

For binary solutions (NaCl or NaOH) we write, in an analogous manner,

$$\vec{N}_+^0 = -l_{++}^* \nabla \bar{\mu}_+ - l_{+-}^* \nabla \bar{\mu}_-$$
$$\vec{N}_-^0 = -l_{-+}^* \nabla \bar{\mu}_+ - l_{--}^* \nabla \bar{\mu}_-, \quad l_{+-}^* = l_{-+}^* \tag{6}$$

The concentration of the binary solution, for example, that of NaCl, being c_1, the quantities $A_{++}^* = l_{++}^*/c_1$, $A_{+-}^* = l_{+-}^*/c_1$, $A_{--}^* = l_{--}^*/c_1$ can be expressed, in the known manner, through the equivalent conductance λ(ohm^{-1} equiv^{-1} cm^2), the diffusion coefficient (with respect to average volume velocity) D^v (cm^2/s), and the transference numbers t_+^0, t_-^0 [21]:

$$A_{++}^* = \frac{(t_+^0)^2 \lambda}{F^2} + \frac{1}{2} \frac{D^v}{RT \left(1 + m \dfrac{d \ln \gamma_\pm}{dm}\right)} \tag{7}$$

$$A_{+-}^* = -\frac{t_+^0 t_-^0 \lambda}{F^2} + \frac{1}{2} \frac{D^v}{RT \left(1 + m \dfrac{d \ln \gamma_\pm}{dm}\right)} \tag{8}$$

$$A_{--}^* = \frac{(t_-^0)^2 \lambda}{F^2} + \frac{1}{2} \frac{D^v}{RT \left(1 + m \dfrac{d \ln \gamma_\pm}{dm}\right)} \tag{9}$$

Here m is the molality (mol/1000 g H_2O), and $\gamma_\pm$ is the average molal ionic activity coefficient. The analogous quantities for NaOH solution we denote by B^*_{++}, B^*_{+-}, and B^*_{--}.

Let a mixed solution has partial concentrations c_1 mol/cm^3 NaCl and c_2 mol/cm^3 NaOH, the total concentration being $c = c_1 + c_2$. After Miller

$$l_{Cl-Cl} = c_1 A^*_{--}(c) \tag{10}$$

$$l_{OH-OH} = c_2 B^*_{--}(c) \tag{11}$$

$$l_{Na-Na} = c_1 A^*_{++}(c) + c_2 B^*_{++}(c) \tag{12}$$

$$l_{Cl-Na} = c_1 A^*_{+-}(c) \tag{13}$$

$$l_{OH-Na} = c_2 B^*_{+-}(c) \tag{14}$$

$$l_{Cl-OH} = -\frac{c_1 c_2}{c^2} \sqrt{l_{Cl-Na} l_{OH-Na}} \tag{15}$$

If, in formulas (10) and (11), the quantities A^*_{--} and B^*_{--} are calculated not per total concentration c, but rather per c_1 and c_2, respectively, the above reactions are referred to as treated within the so-called LNI-approximation. In the work [41], the possibility for correlating the properties of ternary and binary solutions was discussed from the standpoint of the same activity of the solvent, by analogy with the Zdanowski rule for activity coefficients.

3.2.3 Potential-Theory Problems

If the current density in an electrochemical system is small in comparison to the limiting mass-transport current ($i \ll i_{lim}$), variations in concentration can be ignored. The above definitions of potential φ accurate to a constant factor are identical to that of common electrostatic potential. Equation (3) becomes transformed to Ohm's law at constant conductance, with φ obeying Laplace's equation. The relevant range of problems is commonly referred to as the potential-theory problems. Boundary conditions at the electrode surface are formulated on the basis of the electrode kinetics equation $i_n = f(\eta_s)$ and the conditions $i_n = -\varkappa \frac{\partial \varphi}{\partial n}$, $\eta_s = V(x) - \varphi(x)$, where x denotes a coordinate parallel to the electrode surface, and $V(x)$ is the working electrode potential. As a rule, the overall current load is known, and the parameter to be determined is the total voltage applied to the terminals of the electrolytic cell. In insulating parts of the electrode surface, the value of $\partial \varphi / \partial n$ is assumed to be zero. In practice, the anodic and cathodic reactions are different, and for this reason the ohmic losses are considered separately in the anode and cathode compartments of the reactor, with additional measurements or estimations of voltage drop made in intermediate sections of the cell. In the overall cell voltage balance, one must take into account the reversible decomposition voltage corresponding to the free energy of global chemical transformations. Corrections due to concentrational overvoltage are frequently small. Analytical solutions are, as a rule, possible for a limited number of cases under simplifying boundary conditions. The interested reader is referred to the overviews of mathematical methods of potential theory and numerical algorithms in the monographs cited above and also in the paper [42]. A point to be noted is that the potential theory is conceptually analogous to the theory of heat transfer in continuous media, which allows using the already known theoretical results.

The current density distribution function characterizes the effective use of the electrode surface, the current efficiency of end product formation, and the resistance of materials; in addition, it enables one to prognosticate the thermal regime of the cell. The nonuniform distribution of current density is determined by the Wagner number $Wa = \dfrac{d\eta_s}{di} \varkappa/L$ which is the ratio of the effective resistance of an electrode reaction to the ohmic resistance of electrolyte. At $Wa \to 0$ the *primary current distribution* is referred to. In this case, given the overall current, the cell voltage is minimal, and the nonuniformity of current distribution across the electrolyte is markedly pronounced, especially at the electrode surface regions of large curvature. In precision electromachining, the operation at lowered Wa numbers makes it possible to reach a better workmanship of the article profiled to the shape of the cathode tool. From a mathematical standpoint, this case is the simplest one, since it provides the possibility of specifying the equipotential surface of the electrolytic solution in solving a formulated problem.

In the inverse extreme case of $Wa \to \infty$, the total of electric resistance is confined within the interphase boundary, relative to which the electrolyte solution bulk may be regarded as equipotential implying nearly a uniform current distribution throughout the electrode surface. Such conditions are highly welcome in electroplating to provide for a uniform coating thickness. In Ref. [17], a problem for electrochemical metal corrosion has been solved for $Wa \to \infty$ for the case of a heterogeneous small-size inclusion in the electrodeposition surface; an analogous problem has been considered in the case of protective coating. At a finite Wagner number, one must take into account the simultaneous effect due to polarization and ohmic resistance. In such a case, the *secondary current distribution* is referred to. As compared with the primary current distribution, the secondary distribution is more uniform exemplifying the so-called dissipative capability of the electrolyte.

The secondary current distribution is often manifested as a terminal effect in solving electrodeposition problems. Even at $Wa \ll 1$, the distance l from the electrode edge is important, alongside the major size of the reactor L, in current distribution; near the electrode-solution interface, the ohmic resistance $l/\varkappa$ is small, which results in a levelling-off of the interfacial current density. The respective problem has been solved in Ref. [43] to a boundary layer approximation for a disk electrode implanted in an insulating surface. The ratio of the current density at the disk edge, i_{edge}, to the average current density, i_{avg}, is of the order of magnitude of $Wa^{-1/2}$ for a linear electrode kinetics, that is, in this particular case, $i_{\text{edge}} \sim i_{\text{avg}}$. For the Tafel kinetics, the Wagner number is, by definition, $Wa \sim 1/i_{\text{avg}}$, and, as it turns out, $i_{\text{edge}}/i_{\text{avg}} \sim Wa^{-1}$ implying $i_{\text{edge}} \sim i_{\text{avg}}^2$. Thus, the nonlinear kinetics leads, with increasing current load, to a current distribution towards the electrode boundary. This conclusion is general enough. For example, the current increase in a porous electrode operating in an ohmic activation regime produces a displacement of the reaction zone toward the front side of the electrode, whereas at low polarization (linear kinetics), the current load varies proportionally across the layer thickness.

A wide range of problems emerges in developing the methods for prognostication of the galvanic deposition distribution over surfaces with different microgeometry [44]. The two-dimensional problem for Laplace's equation in a semi-space delimited by the surface of a regularly profiled electrode under uniform current supply, distant from the electrode surface, has been studied in Ref. [45] by conformal mapping method. For a complex-geometry reactor, a numerical solution is quite often the only possible; a number of relevant examples with appropriate algorithms have been discussed in recent monographs. Of the recent work, papers [46, 196] are concerned with the practical utility of a computer-aided solution for current distribution in the neighbourhood of a perforated electrode of finite thickness and for resistance in a system analogous to the anode compartment of a chlorine ion-exchange membrane cell. The variational principle of dissipation energy dealt with in Ref. [17] may prove to be helpful in approximate calculations.

In the case where an analytical solution is difficult to perform and numerical methods require much preparatory work or a special study to guarantee the desired accuracy, it appears expedient to develop a mathematical model permitting accounting for major characteristics of the elements of a construction, their arrangement and electrochemical properties. Exactly from this standpoint, a problem concerned with the secondary current distribution and the potential between two plane electrodes has first been considered in [47]. As has been ascertained, a condition for perpendicularity of field lines to the electrode surface is the short interelectrode distance as compared with the electrode size, which is the typical configuration of high-efficiency reactors. The numerical solution of Laplace's equation for a three-electrode system with a wedge-shaped anode and two cathodes adjacent to it on both sides has shown that the current distribution becomes uniform at a distance from the wedge nose about twofold of the interelectrode distance [48]. The *à priori* design of a simple geometry for the field lines provides a means to circumvent the solution of Laplace's equation, given the voltage balance along each field line.

The study of conductance in disperse systems represents an important problem from the standpoint of current distribution modelling of processes with the concomitant formation of a new phase [49, 50] and has been a major concern for the researchers since the pioneering attempts by Maxwell. A generalization of experimental data on gas-evolving electrode processes provides evidence for the existence of a layer of gas bubbles that are retained on the electrode surface during their growth by surface tension forces. The gas layer resistance can be superior to the bulk dispersion resistance, a circumstance that should be taken into account in choosing an electro- and hydrodynamic model for gas-evolving electrodes. Studies on current density microdistribution within the surface layer between the electrode and nonconducting inclusions have been reported [51-55]. Along with rigorous calculations, quasi-one-dimensional models with a varied cross section of solution bulk normal to the electrode surface have been discussed. It should be recalled that the gas-bubble contact angle and the electrode surface coverage are functions of electrode potential.

3.2.4 Convective-Diffusion Problems

The electrolysis regime at currents close to the limiting mass-transport current defines another class of experimental observations. The prediction of limiting currents is important for estimating the maximum efficiency of an electrochemical reactor; in addition, it provides a source of valuable information on the kinetics of electrode processes and transport characteristics in electroanalytical chemistry. Given $i \sim i_{\text{lim}}$, the reactant concentration c at the electrode surface tends to zero; consequently, the concentrational and surface overvoltages grow infinitely by a law $\sim \ln c$. The latter fact is due to the exponential exchange current i_0 vs. concentration relationship [13]. For this reason, the ohmic potential drop in solution is negligeably small and can be excluded from the overall voltage balance providing that no high computation accuracy is required and that the size L of the cell is sufficiently small. In the presence of a supporting electrolyte whose ions are in excess to the reactants but are not involved in the electrode process, the ion migration in electric field may be ignored. Indeed, the migration rate of a reactant relative to the overall flow rate is equal to the transference number of the reactant. Obviously, this conclusion holds not only for the limiting current regime. The presence of a supporting electrolyte allows regarding the whole system quasi-uniform in composition. Then, in equation (1) written for the reactant, we obtain a simple expression for the flow rate by decomposing the potential $\bar{\mu}_i$ into concentrational and electrostatic components and neglecting both the concentrational dependences for the activity coefficient and migration. Finally, it leads to the classical problems of convective diffusion, if we accept that $\vec{v}_1 = \vec{v}_2 = \ldots = \vec{v}_N$ for all the components, except the reactant. The latter condition holds in many cases, since the electric field in the diffusion layer produces no mutual separation of the ionic components of the supporting electrolyte and the solvent (providing that the latter is not involved in the reaction). In addition, for moderately concentrated solutions, one may presume that the only essential interaction is reactant-solvent interaction. In most cases, the velocity of convective motion $\vec{v}$ can be calculated, without specifying its particular sense, from the solution of a hydrodynamic problem. The concentration of a binary electrolyte also satisfies, to a certain degree of accuracy, the convective diffusion equation.

At present, there is a wealth of evidence on the mass-transport coefficient correlations under a variety of geometrical and hydrodynamic conditions. In the monographs [14, 15], the available data have been tabulated in a form enabling one to correlate the Sherwood number $Sh_i = iL/z_i F D_i (c_0 - c_s)$ to the Reynolds number $Re = vL/\nu$ and the Schmidt number $Sc_i = \nu/D_i$ for forced convection (the subscript 0 refers to the volume, and the s, to the surface; the other notations are conventional). For free convection, the Re is replaced by the Grashof number $Gr = gL^3(\varrho_0 - \varrho_s)/\varrho_0 \nu^2$, where ϱ is the density and g, the gravitational acceleration. In addition, the correlations must take into account characteristic geometrical relations. Some of the correlations have been substantiated by ample theoretical and experimental evidence (free convection about vertical planes, diffusion towards a rotating disk). Other correlations have been obtained by regression analysis of the experimental data; such are the correlations for "wire-cloth" electrodes or expanded metals [56, 57]. A point to be noted is that, in electrochemical engineering, the convective diffusion problems are commonly solved under the strict conditions of $c_s = 0$ or $i = i_{\lim}$. Otherwise, at currents smaller than the $i_{\lim}$, the value of c_s is greater over zero and is largely unknown, whereas the surface boundary condition is dependent on an additional variable, the overvoltage η_s. In some detail, this issue will be discussed below.

The mass transport by convective-diffusion mechanism for a gas-evolving electrode has been a subject of interest for the researchers during the two last decades. For example in performing electrodeposition under conditions close to those of limiting current, a rise in the cathode potential to a sufficiently high absolute value initiates hydrogen evolution, which can lead to an increase in the mass-transport coefficient at least by an order of magnitude. However, such an intensified regime is accompanied by two objectionable factors—a diminished current efficiency for the deposited metal and additional ohmic losses. As has been shown, mass transport is enhanced by microconvective agitation of the liquid phase due to the growth and detachment of gas bubbles in a close neighbourhood of the electrode surface. Since the structure of the gas layer is subject to many difficult-to-control conditions, a theoretical approach to the problem is rather cumbersome. Nevertheless, numerous studies (see, for example, the references in the review paper [58]) have allowed describing, at a semi-quantitative level, the dynamics of birth, growth, and detachment of gas bubbles and specifying the effects of technological factors in the simplest systems. This, in turn, has laid the grounds for developing mass-transport models [15, 59]. It has been noted that, owing to the chaotic character of liquid pulsations produced by gas-bubble motion, one may use to advantage an analogy with the heat and momentum transport [59, 60] known in the turbulent motion theory. Despite numerous attempts, none of the proposed models can be accepted for a universal one; however, it must be recognized that currently a popular approach seems to be that suggested in Refs. [15, 61].

Consider an electrode surface divided into small circles of radius L_B, in one-to-one correspondence to the number of the surface-attached gas bubbles, each of these represented by a model hemisphere. Let $R(t)$ be the time-dependent radius of a gas bubble. As is known, $R(t) \sim \sqrt{t}$, and the flow velocity about the free electrode surface within a ring $R(t) \leqslant r \leqslant R_{12}$ ($R_{12} < L_B$), associated with the gas-bubble growth, can be estimated by the formula $v_x = \frac{dR}{dt}\frac{R^2}{r^2}$ (here the parameter R_{12} defines a region within which the microconvective effects are operative). We denote $x = r - R(t)$. Applying the correlation for the local mass-transport coefficient $K_{1x,t}$ to a planar surface, we can write $Sh_{x,t} = kRe_{x,t}^{0.5}Sc^l$, where $Sh_{x,t} = K_{1x,t}x/D$, $Re_{x,t} = v_x x/\nu$ (k and l are constants). The authors of the paper [61], starting from an analogy with boiling-liquid mass transport, have introduced the concept of collective action arising from microconvective and macroconvective effects. The latter are due to the flow of gas-liquid mixture along the electrode in a regime of forced or free convection; the relevant issues have been well studied within the framework of convective diffusion theory. It has been assumed in this model that the macroconvection is operative in a region of $R_{12} \leqslant r \leqslant L_B$, with the mass-transport coefficient K_{2x} being a constant. Denoting the gas-bubble growth time by t_B, the flux of reactant by N_c, and the concentration difference between the electrode surface and the bulk solution by Δc, we obtain according to the model

$$\frac{N_c}{\Delta c} = \frac{1}{t_B}\int_0^{t_B}\int_0^{R_{12}} 2\pi r K_{1x,t}\,dr\,dt + \pi(L_B^2 - R_{12}^2)K_{2x} \tag{16}$$

The overall mass-transport coefficient is thus $K_{12} = N_c/(\Delta c \pi L_B^2)$. In performing the calculations, it should be kept in mind that $K_{1x,t} = K_{2x}$ at $r = R_{12}$. Given $R_{12} = L_B$, then, by definition, $K_{12} = K_1$, the latter quantity being the mass-transport coefficient for neat microconvection. Be R_{12} equal to R_m (mean gas-bubble radius), then, by definition, K_{12} is equal to K_2, the mass-transport coefficient for neat macroconvection. Following this line of reasoning, a correlation between K_{12}, K_1 and K_2 and the gas-bubble surface coverage θ has been obtained in Ref. [61]. The numerical results have lended support to the empirical correlation $K_{12} = (K_1 + K_2)^{1/2}$. The correlation for K_2 (eventually, with a correction introduced for gas-bubble surface "asperity") may be regarded as well-studied. Calculations of K_1 using this model, as well as other models, have led to its correlation with the gas evolution rate per unit surface $\dot{V}_g$ in the form of $K_1 \sim (\dot{V}_g)^m$ ($0.18 \leqslant m \leqslant 0.69$) [15].

An analysis of the literature data on the dynamics of hydrogen and oxygen gas-bubble growth in the electrolysis of aqueous H_2SO_4 solutions as well as on the dissolved gas oversaturation near the electrode surface, with reference to the known mass-transport coefficients, has been performed in Ref. [62].

As has been ascertained, at the current density within a range of $3\text{-}10^4$ A/m^2, some 0.2-0.6 fraction of the total of electrolyzed gas becomes desorbed directly into the interior of growing gas bubbles, whereas the remaining gas portion passes into the solution bulk to create oversaturation. This novel finding may necessitate a revision of the quantitative approach to the mass-transport coefficient and to the entire boundary-layer dynamics. On the basis of simple balance equations, a 2-to-3-fold gas oversaturation in the electrolytic volume of a flow-through cell [63]—a point not to be overlooked in the eventual construction of a mathematical model—has been estimated.

3.2.5 Electric-Field Effects in Mass Transport

As the supporting electrolyte concentration is made to decrease, the electric field starts to exert influence on the reactant fluxes. The emergent therewith problems are often considered under the common title "the effect of migration on limiting currents" [13]. In contrast to the potential-theory problems and convective-diffusion problems, they bear no analogy to any phenomena outside the electrolytic domain. Apart from the binary solution, the electric-field problem has been considered within the formalism of a theory of infinitely dilute solutions, with the potentials represented in the form $\bar{\mu}_i = RT \ln c_i + Fz_i\varphi + \text{const}$. In many instances, such an approach allows one to arrive at simple analytical solutions or to develop effective numerical methods. For a demonstrative purpose, let us consider a three-ion electrolyte, c_i being the concentration of an active ion, c_1 the concentration of an extraneous ion of the same sign and charge, $c_2 = c_i + c_1$, and $z_2 = \pm 1$, $z_i = z_1 = \mp 1$ [14]. Suppose, the ion migration takes place in an immobile Nernst layer of constant thickness δ. The concentration at the electrode surface is $c_i = 0$; the concentrations c_{10} and c_{20} in solution bulk are assumed to be known. The diffusion current is defined as $|i_D| = FD_i \left|\frac{\partial c_i}{\partial n}\right|_s$. For the migration-diffusion limiting current $i_{\lim}$, one has

$$\left|\frac{i_{\lim}}{i_D}\right| = \frac{2}{1 + \sqrt{c_{10}/c_{20}}} \qquad (17)$$

In the presence of a convective flux component, the ratio (17) can reach a very high value. The enhancing effect due to migration ($|i_{\lim}| > |i_D|$) is typical of the cathodic reduction of cations and of the anodic oxidation of anions.

If the electric field polarity is opposite to the direction of reactant migration (cathodic reduction of anions, anodic oxidation of cations), then $|i_{\lim}| < |i_D|$, although the effect is less pronounced [13, 14]. For example, studies of multicomponent system have been reported in Refs. [64, 65].

The effect of ion migration on mass transport is manifested also when the component of interest is present in relatively small amounts, whereas the other ions are involved in the charge transfer, as this takes place in diaphragm and membrane processes. Suppose, there is a cation-exchange membrane (m), in contact with an aqueous binary electrolyte solution. Henceforth, water is denoted by the subscript "0", and mobile ions within the membrane, by "+" or "−". The concentrations are referred to unit membrane volume. For the mass transport to be described by Eq. (1), it must be set $\vec{v}_m = 0$ (for more detail, see Sec. 3.7). We introduce the relative friction coefficients by the formulas

$$f_{+0} = \frac{K_{+0}}{K_{+0} + K_{+m} + K_{+-}}, \quad f_{0+} = \frac{K_{0+}}{K_{0+} + K_{0m} + K_{0-}} \quad (18)$$

$$f_{-0} = \frac{K_{-0}}{K_{-+} + K_{-m} + K_{-0}}, \quad f_{0-} = \frac{K_{0-}}{K_{0+} + K_{0m} + K_{0-}} \quad (19)$$

$$f_{+-} = \frac{K_{+-}}{K_{+0} + K_{+m} + K_{+-}}, \quad f_{-+} = \frac{K_{-+}}{K_{-+} + K_{-m} + K_{-0}} \quad (20)$$

We wish to examine the diffusion of an isotope-labelled cation in the membrane at concentration $c_* \ll c_+, c_0$. Suppose, $c_- \ll c_+, c_0$, which provides for a high selectivity, and an electrostatic potential can be introduced. Disregarding the interaction between the "excess" components and the "small-concentration" components, we can easily solve the equations for the velocities $\vec{v}_+, \vec{v}_-, \vec{v}_0$, and $\vec{v}_*$. The obvious relations $\nabla \mu_+ = z_+ F \nabla \varphi$, $\nabla \mu_* = z_+ F \nabla \varphi + \frac{RT}{c_*} \nabla c_*$, $\nabla \mu_- = z_- F \nabla \varphi + \frac{RT}{c_-} \nabla c_-$, $\nabla \mu_0 = 0$ should also be taken into account (assuming the thermodynamic conditions in the membrane-partitioned solutions to be the same). In addition, we also assume that the kinetic properties of the labelled ions are identical with those of normal ions [66], that is, $K_{*0} = c_*/c_+ K_{+0}$, $K_{*m} = c_*/c_+ K_{+m}$. We introduce $K_{++} = c_+/c_* K_{*+}$. Solving the equations yields expressions for mole fluxes across the membrane:

$$\vec{N}_* = -D_* \nabla c_* - \frac{D_* F}{RT} z_+ c_* \left(1 + \frac{\lambda_+ + f_{+0} f_{0+}}{1 - f_{+0} f_{0+}}\right) \nabla \varphi \qquad (21)$$

$$\vec{N}_- = -D_- \nabla c_- - \frac{D_- F}{RT} z_- c_- \left(1 + \frac{z_+ c_+}{z_- c_-} \frac{f_{+-} + f_{+0} f_{0-}}{1 - f_{+0} f_{0+}}\right) \nabla \varphi \qquad (22)$$

In expressions (18)-(20), the respective small coefficients for the f_{ij} must be dropped. The diffusion coefficients are thus given by expressions

$$D_* = \frac{RTc_+}{K_{+0} + K_{++} + K_{+m}}, \quad D_- = \frac{RTc_-}{K_{-0} + K_{-+} + K_{-m}} \qquad (23)$$

The parameter $\lambda_+ = K_{++}/(K_{+0} + K_{+m})$ provides a relative quantitative measure for the interaction of ions of the same sign in solution. For the opposite case of a counterion tracer and a low-concentration coion in a high-selective anion-exchange membrane, we must set $K_{*0} = \frac{c_*}{c_-} K_{-0}$, $K_{*m} = \frac{c_*}{c_-} K_{-m}$, $K_{--} \equiv \frac{c_-}{c_*} K_{*-}$, $\lambda_- = K_{--}/(K_{-0} + K_{-m})$ and the expressions (21)-(23) equally hold for the signs "+" and "−" replaced by their opposites.

Since the coefficients f_{ij} can be obtained from membrane permeability measurements under various conditions, the experiments with radioactive tracers enable determining the quantity K_{++}. Comparing the K_{++} to its analogue in free solution provides an indirect information on the degree of order for the membrane structure. On the other hand, the K_{++} can be dismissed from consideration through combining Eqs. (21) and (23). Therefore, the results by two independent experiments on diffusion and radioactive tracer migration provide a quantitative characterization of the interaction coefficients K_{+0}, K_{+m}, and K_{0m}. The parenthetic expression in Eqs. (21) and (22) are indicative of an inequality of the migratory U_M and diffusive U_D mobilities, a fact that has been observed on a number of occasions in ion-exchange materials [67, 68]. Based on the above evidence, we have estimated the ratio $U_M/U_D = 1.5$-2.5 for most commonly employed membranes in experimental studies on diffusion of labelled counterions; to be definite, we have used the f_{ij} for a Nafion-type membrane in NaCl solution [69] and thus obtained $0 \leqslant \lambda_+ \leqslant 0.9$. Recalling that the quantity $\lambda_+^0 = K_{++}^0/(K_{+0}^0 + K_{+-}^0)$ deter-

3 Modelling of Electrochemical Processes

mined on the basis of self-diffusion data for free solution has a negative sign, the above estimate lends an indirect support to the conclusion on a cluster-channel structure of ion-exchange membranes.

3.2.6 Concentrational Overvoltage

At currents below the limiting value but, nonetheless, attended by significant changes in concentration, no general approach to the statement of problem of their solution has been developed, even to a dilute-solution approximation. If the problem of current density distribution is coupled to the voltage balance calculation, the concentrational component included, the so-called *tertiary* current distribution is occasionally referred to [14]. We now consider a situation with excess supporting electrolyte presuming the diffusion boundary layer thickness to be small as compared to the characteristic size L of a cell with a well-stirred bulk solution. The concentration part of the overvoltage based on the probe electrode potential [13] is

$$\eta_c = \frac{RT}{nF} \sum_{i=1}^{N} s_i \ln \frac{c_{i0}}{c_{is}} \tag{24}$$

Let the ohmic potential drop be neglected; then the net potential $\eta = \eta_c + \eta_s$ is known and is the same at any point of the equipotential electrode surface. Consequently, equations $i_n = f(\eta_s)$ and (24) provide a one-to-one correspondence between i_n and c_{is}. The stoichiometric proportionality $\frac{\partial c_i}{\partial n} = \frac{s_i i_n}{nFD_i}$ at the electrode surface is assumed to be known. As shown in the monograph [13], the $\partial c_i / \partial n$ is related to a surface integral containing $\partial c_{is}/\partial x$; this follows from considering the convective diffusion equation within the boundary layer. The above relations lead to an integral equation with respect to c_{is}, which is reducible to a numerical solution. If the ohmic potential drop $\Delta\varphi_{ohm}$ cannot be ignored, then, by choosing an appropriate distribution for $i_n(x)$, the $\Delta\varphi_{ohm}$ is found by solving Laplace's equation; its solution permits defining the term $\eta - \Delta\varphi_{ohm} = \eta_c + \eta_s$ and repeating once again the above procedure for forming the integral equation. Further, an iterative process can be suggested as designed for convective diffusion around an axisymmetric body. On considering the tertiary current distribution throughout regular (for example, corrugated) surfaces [14], the following specific features are to be noted. If the ratio of the diffusion layer thickness to the size of a surface irregularity is large, then the concentration polarization contributes to an increased nonuniformity of current distribution as compared to secondary current distribution. Otherwise, a tendency to current distribution uniformity is observed.

3.2.7 Interdependence of Various Factors in a Cell Model

The development of any adequate model requires forming balance equations for conservation of fundamental properties of matter. We specify, in a somewhat arbitrary manner, three major classes of phenomena to be considered: hydrodynamic (conservation of momentum and/or energy), diffusive (conservation of mass of components in their relative motion), and electric (conservation of charge). In essence, these phenomena can be studied on a single mechano-statistical basis, but the proposed classification appears to be convenient for characterizing electrochemical reactor models. Nonelectrolytic processes are modelled by the two former classes. The momentum transfer and the diffusion being treated separately, the formulation of boundary conditions for the equations of interest does not present a major limitation. However, for a deeper insight into the processes occurring in an electrochemical cell, one needs to resort to a more sophisticated hybrid model which cannot be a mere additive construct of models for different phenomena [197].

Interrelation of Diffusive and Hydrodynamic Phenomena

There are known systems in which an intense mass exchange causes a substantial change in the hydrodynamic conditions. One may cite, for example, the condensation of vapour mixture on a cooled wall, the vaporization of liquids, crystallization or dissolution of salts, and precision electrochemical machining. Such systems are intractable by the conventional mathematical methods based on a separate solution of equations of hydrodynamics and convective mass transport [10]. A major impediment to a physically consistent model here is the mechanical momentum which can be transferred from one phase to the particles of another phase involved in the mass exchange. Since the transferable momentum is proportional to the flux of particles, the velocity field at the interface becomes thus dependent on the concentration field. To formulate boundary conditions, it suffices to write the law of mass conservation for each component within a reference system fixed to the phase interface surface.

During the anodic dissolution of a metal at a given current density, the true cation concentration within the anode boundary turns out to be smaller than that calculated in neglect of the effects arising from the interactive transport of both mass and momentum. At the current density rising to infinity, a "saturation" effect for the boundary anode concentration has been observed [10]. Regimes for electrodeposition onto the surface of a rotating disk electrode and for a number of other electrochemical processes have also been reported. A well-known example of the effect produced by a diffusion flux on the flow velocity is free convection initiated by electrode reactions as, for example, those occurring in metal-electrorefining baths. The free-convection problems, with account taken for the nonisothermicity of a reactor, have been considered in Ref. [15]. In magnetoelectrolysis (see Sec. 3.3), the Lorentz force acting upon the liquid in the presence of an external magnetic field and affecting the flow velocity around the electrode is proportional to the limiting diffusion current which, in turn, is determined by hydrodynamic conditions.

Relation Between Diffusive and Electric Phenomena

The electric field within an electrolyte is not merely an external factor that exerts an independent action upon the motion of ions; the field itself is determined by concentrational variations. A well-known fact is the so-called diffusion potential which persists in a variable-composition solution after the initially applied electric current has been switched off. In the foregoing, difficulties have been discussed arising in formulating and solving the mass-transport problems in concentrated solutions. Even if the electromigration of a component in solution bulk may be neglected, a dependence of the boundary conditions on the electrode potential is retained. For example, electrodiffusion phenomena at the electrode surface shielded by a gas-bubble layer with a definite contact angle and hexagonal packing [70] can be studied. The diffusion of electrolytically evolved gas molecules towards the gas-bubble surface is determined by the intensity of gas molecule production on the electrode surface, that is, by the current density distribution. The latter factor is dependent on the electrode surface overvoltage which is directly related to the concentration of molecules on the electrode surface. It follows, therefore, that the buildup of both the electric field and the current density is inalienable from the concentration field formation. A numerical solution has been given for the evolution of hydrogen from a 30% KOH solution at the current density of 0.3 A/cm^2 under typical conditions of chlorine-alkali electrolysis, which enabled estimation of the effect of electrode surface gas shielding on the reaction overvoltage.

Relation Between Electric and Hydrodynamic Phenomena

The relation between electric and hydrodynamic phenomena is manifested in the processes leading to the formation of a new phase and to a change in local electric resistance. For example, in gas-evolving circulating cells with vertically positioned electrodes, the local gas content is a determining factor for both the current density distribution with height and the circulation rate. In turn, the gas content is dependent on the current distribution (that is, the gas-source density) and the balance of convective fluxes. It follows, therefore, that the calculation of current distribution in such a cell is dependent on the hydrodynamic conditions, and vice versa.

Depending on the designated purpose of an electrochemical cell, conditions for external and internal circulation of the electrolyte are to be distinguished. The former circulation is typical of the systems equipped with an external circuit and designed for a complete separation of the evolved gas from the liquid; a design version of such systems is a cell where the upstream and downstream electrolyte flows are directed to compartments partitioned mechanically from each other. The hydraulic calculation of circulation circuits requires a knowledge of local resistance coefficients. If these are not available, the equations for overall balance of potential and dissipation energies, based on the characteristic velocity gradients, may come profitable for the purpose [71, 72]. The internal circulation occurs spontaneously within the cell, and the emergent flow is, as a rule, a two-dimensional vortex difficult to visualize or control. In a vertical two-electrode cell, typical is the superimposition of external and internal circulations. The flow rate of the latter determines the length of the input region, that is, the initial stretch of the cell height where the gas evolving at one of the electrodes is distributed nonuniformly across the channel section. Occasionally, the boundary region between one-phase and two-phase flows may appear distinctly visible, with a sudden upstream funnelling of the gas-bubble track at a certain height; starting from this point, the flow in the narrow interelectrode gap may be regarded as a quasi-one-dimensional one. Such a change in the flow regime is explained by the interference of a downstream flow in the expanded region, that is, by internal circulation. In Ref. [73], a simple hydraulic model for the input region of a cell with a single gas-evolving electrode has been considered, taking into account the balance of both, the gas and the liquid across the channel cross section. This made it possible to estimate a critical gas flow rate which, once reached, makes the gas stream at a certain height expand through the entire cross section.

If the two-dimensional character of an electrolyte flow cannot be ignored, an exact solution of the Navier-Stokes equations is required. However, as distinct from the classical problems of natural convection with the electroactive component obeying the diffusion equation, currently no generalizing equations for gas fraction distribution have been formulated. One can *à priori* form a function for gas content distribution and then calculate the convective velocity [74] estimating the internal circulation contribution to the mass transport towards the electrode. A sophisticated model of turbulent flow and current distribution in the bulk of an electrolytic gas-evolving anode bath for hydro-electrometallurgy deposition of metals has been proposed [75]; however, this model requires an experimental verification.

To briefly summarize, the interrelation of diffusive-electric and electric-hydrodynamic phenomena constitutes a distinctive feature of the theory of electrochemical reactors.

3.3 Magneto- and Electrohydrodynamic Approximations in the Theory of Transport Phenomena in Electrochemical Systems

The thermodynamic theory of multicomponent transport in electromagnetic field [29] is not at present sufficiently elaborate for its practical applications. For this reason, in estimating the effect of magnetic forces, simplified models are most commonly used in practice; however, in each particular case, the chosen model must be examined for the range of its applicability. In constructing a model, the Lorentz force $\vec{F}_L$ acting upon the unit volume of charge density ϱ_e in an electromagnetic field of strengths $\vec{E}$ and $\vec{H}$ must be introduced into the hydrodynamic Navier-Stokes equations (from now and on, only nonpolarized media are taken into consideration):

$$\vec{F}_L = \varrho_e \vec{E} + \frac{1}{c} \vec{i} \times \vec{H} \tag{25}$$

Ohm's law is written in the form

$$\vec{i} = \varrho_e \vec{v} + \varkappa \left(\vec{E} + \frac{\vec{v}}{c} \times \vec{H} \right) \tag{26}$$

with allowance made for the convection current $\varrho_e \vec{v}$ and the conduction current (the addend). Occasionally, the Hall current arising from a diffusive motion of charged particles relative to the mass-average velocity $\vec{v}$ in magnetic field is included in Eq. (26). The Hall effect has been well studied in conductors of the first kind. It develops as a transverse electric field in a current-carrying conductor placed in a magnetic field; a measure of the Hall effect, the so-called Hall coefficient, is the strength of transverse electric field divided by the product of the current density and the magnetic induction. In an electrolyte solution, the Hall effect is proportional to the difference of the products of the squared ion mobility and the ion concentration. Since the ion mobility is by a few orders of magnitude smaller than the electron mobility in metals, and the concentrations of oppositely charged ions are comparable, the experimental observation of the Hall effect in solution is an arduous task. An exception to the rule is the possibility to record the Hall potential and the change in surface conductance by applying crossed electric and magnetic fields, respectively, in a longitudinal and transversal direction to the charged interface, owing to a predominance of ions of the same charge in the double electric layer [76]. It has been hypothesized that such an approach allows studying in greater detail the composition and the fine structure of the surface layer of an electrolyte. A macroscopic theory has recently been developed in Ref. [77] based on the inclusion of the Lorentz force into the Nernst-Planck equation and on the assumption of nonoccurrence of ion flows in the direction of the induced Hall field. The degree of electrolyte dissociation, responsible for the number of charged species, has been shown to be of prime importance in the process.

The field strengths $\vec{E}$ and $\vec{H}$ are related to each other through the known Maxwell equations. We use one of these for our estimations:

$$\operatorname{rot} \vec{H} = \frac{4\pi}{c} \vec{i} + \frac{1}{c} \frac{\partial \vec{E}}{\partial t} \tag{27}$$

The dimensionless parameter $\varkappa L/c$ being $\gg 1$, we have to deal with a magnetohydrodynamic (MHD) approximation [78] (L is the characteristic scaling factor of the system). This is typical of electrochemical reactors with the conductance of working solutions (in the CGSE system) of the order of $\varkappa \sim 10^{11}\text{-}10^{12}$ s^{-1} (at $L \geqslant 1$ cm). In such an approximation, the Coulomb force in Eq. (25), the convection current in Eq. (26), and the displacement current in Eq. (27) may be ignored. We assume that the imposition of an external magnetic field produces no effect on the volume of the electrolyte under study (henceforth, the external magnetic field is meant to be the one generated by

3 Modelling of Electrochemical Processes

a source of current extraneous to the given system). The Lorentz force is commonly neglected as compared to other forces, for example, gravitational. Let us define conditions under which the current induced by Lorentz force is not taken into consideration in Eq. (26). On the basis of Eqs. (27) and (26), the order of magnitude for $\overline{H}$ is estimated to be $H \sim \frac{4\pi}{c} Li \sim \frac{4\pi}{c} L \varkappa E$. Then the ratio $\left|\frac{\vec{v}}{c} \times \vec{H}\right| / |\vec{E}| \sim \frac{4\pi}{c^2} \varkappa L v$ is the magnetic Reynolds number Re_m. As is easily seen, in the typical case Re_m is $\ll 1$ owing to the small value of v/c. This may serve as an argument for the conventionally accepted form of Ohm's law $\vec{i} = \varkappa \vec{E}$ for electrolytic processes.

For heavy-duty commercial electrolytic cells operated at large current loads, one must make allowance for external magnetic fields that are induced in the electric buses, mainly owing to the action of an additional force (25) on the liquid, rather than to the general law (26). With an external field of strength H_0 imposed on a solution, the field strength, induced by the current in the solution bulk, becomes changed by a value of ΔH. As can easily be shown, $\Delta H/H_0 \ll 1$ at $Re_m \ll 1$. Therefore, the magnetic forces that can be neglected in the absence of an external field, are taken into account in a system exposed to the external field by just substituting a given quantity H_0 into Eqs. (25) and (26). The problem is further simplified with the supporting electrolyte taken in excess, when the conduction current may be neglected and, in place of Eq. (26), one may write $\vec{i} = -Fz_1 D_1 \nabla c_1$ for the concentration c_1 of an electroactive component within the diffusion layer. Then, owing to the applied external field, H_0, the hydrodynamic flow around the electrode surface and the diffusion of components become interdependent through the intervention of the Lorentz force in the system. Complex effects are observed arising from the interaction of MHD-convection and natural convection in the gravitational field. According to whether the magnetic force makes the convection increase or, conversely, decrease, the limiting diffusion current either increases, or decreases [79]. With the electrode or the ion-exchange membrane in question positioned horizontally, the MHD-convection as effected in the direction parallel to the electrode (membrane) surface can stabilize the flow and thus prevent the emergence of dissipative vortices.

The effects due to the electrodynamic forces operative in liquid-cathode cells used for electrowinning of metals from the melt are well known [80]. The magnetic field, on its interaction with the current of the melt, elicits a circulation within the melt bulk and distorts the melt surface to a curvature, which finally results in a reduced metal yield and an irregular wear of the anode. In electrolytic cells for production of chlorine and alkali by the mercury method, the magnetic forces play a decisive role in the formation of mercury electrode surface. Currently, metal anodes are gaining an ever-increasing application in the electrochemical engineering. These anodes are effectively and economically exploited at large current densities, which necessitates the use of interelectrode gaps as small as possible. This incurs the risk of destructive short-circuit, unless appropriate precautionary measures are taken against the interference of magnetic forces. Another objectionable feature associated with the magnetic forces is the deposition of the so-called "amalgam oil", which is a colloid suspension exhibiting a weak ferromagnetism, at the bottom of an electrolytic cell [81]. The deposited suspension impedes the normal conditions of amalgam flow and leads to frequent off-schedule maintenance cleanings of the cell. These facts should be taken into account in designing the bus system of the electric supply circuit of mercury cells.

In recent years, searches are being carried out for a purposeful action by external magnetic forces on the intensity of electrochemical processes; in other words, one may anticipate the possibility of a close control over the electrolysis regime via combining appropriately the strengths of electric and magnetic fields in the cell. This new area in electrochemical engineering has been named the magnetoelectrolysis [82]. Magnetoelectrolytic methods, applied in practice, have proved to be effective in increasing the mass-transport rate, improving the quality of electrolytic deposits, and inhibiting corrosion. MHD-effects in an electrolyte on its passage through a narrow electrode gap [83], also in electrochemical polishing of orifices [84] and in melt electrolysis using a laboratory cell with horizontally positioned electrodes [85] have been studied. The electric current passing through an electrolytic system can be used for generating a magnetic field of required strength, whereas the additional energy expenditure is believed to arise only due to ohmic losses in the magnetizing coils. Analytical solutions for a number of MHD problems have been reported in the works [82, 86, 87, 198].

In the inverse extreme case of $\varkappa L/c \ll 1$, the equations of continuous-media electrodynamics admit solutions for $H \ll E$; otherwise stated, the effects due to magnetic forces in expressions (25) and (26) may be neglected, whereas the field strength $\vec{E}$ obeys the conventional equations of electrostatics [78]. This is the so-called electrohydrodynamic (EHD) approximation. Similar conditions can be realized in the bulk of poorly conducting liquids (for example, organic) in which the space charges and ponderomotive forces arise from the passage of current and also from the medium inhomogeneity in conductivity [88]. Since the concentration of mobile charge carriers is very low in such media, one must apply a voltage of several thousand volts to any noticeable current or hydrodynamic flow, which is difficult for practical implementation in conventional electrochemical systems. On the other hand, the EHD conditions can be satisfied by going down to the L scale as low as 1 mm or less for common electrolytic solutions. This class of conditions embraces processes occurring in porous electrodes and membranes. It stands to reason that, if within the pore space the charge density $\varrho_e = 0$, we have to deal with the earlier considered problem of electrodiffusion transport under electroneutrality conditions (a similar transport problem in the MHD approximation is associated with a small effect due to magnetic field). The response of the charge region of a double electric layer at phase interface to an external electric field fits into the EHD scheme considering the small scale of the Debye length as a characteristic measure for the interaction region. The theory of electrokinetic phenomena has found a wide application in technology and scientific research [17, 89]. It may be inferred, therefore, that the EHD approximation in electrochemical systems is used as wide as the MHD approximation.

As an important application, one may refer to the electric logging in geoexploration, in particular, oil prospecting. The main idea of this method is a study of the relaxational decay of current strength in a disperse layer containing hydrocarbon droplets and electrolyte solution after the direct-current electric field to which the system was exposed is switched off. In analysing the shape of the measured current decay curve one can obtain information on the composition and physico-chemical properties of the oil pool explored. An approach to the development of an adequate theory has been outlined, for example, in the works [90, 91], where there have been considered models for relaxation effects in a thin cylindric capillary filled with an electrolyte solution and containing a single nonconducting droplet extended through a certain length within the capillary. An essential point is that, in formulating the boundary conditions for equations describing phase transport by the action of an electric field, we must include the components of Maxwellian tension tensor

$$T_{ij} = \frac{1}{4\pi} \left[\frac{1}{2} (E^2 + H^2)\delta_{ij} - E_i E_j - H_i H_j \right]$$

into the momentum balance at the charged interface (in this particular case, $\vec{H} = 0$). The neglect of this condition leads to a 50% error in the calculated stationary velocity of interface displacement under uniform field [90].

Undoubtedly, in electrochemical systems, phenomena may occur whose characteristics are intermediate with respect to those of the two extreme approximations considered.

3.4 Dynamics and Stability of Electrochemical Reactors

The use of pulsating voltage calls into play a larger number of control parameters as compared to that under stationary voltage conditions. A variety of parameters are amenable to control, for example, peak voltage, pulse duration, pulse periodicity, reverse voltage, and so forth. Nonstationary methods are also used in electroorganic synthesis, in electrochemical precision machining, in electrodeposition [5, 11, 92]. The introduction of a temporal function in modelling electrochemical reactors is essential for a proper control of transient on-off regimes, for the design of batch-flow cells and must constitute a basis for developing the software support to computer-aided production control. An essential point is the stability analysis of stationary states of the cell operation as a whole, or at separate stages. By way of example, in a thermokinetic instability [93], because of the electrode surface warm-up, the current density and the mass transfer rate tend to increase, and a stationary state can be provided for only by good conditions for heat removal.

An approach to modelling nonstationary regimes from a standpoint of the theory of dynamic systems has been discussed in the literature [6, 14], and a mathematical method has been developed for the signal-response representation using transition functions. These are easy to determine for the simplest linear models of a reactor. If no information on the processes operative in a reactor is available, the transition functions can be devised through studying the experimental response of the system in question to stepwise or sinusoidal perturbative signals of the input variables.

Typical nonstationary problems of convective-diffusion type with regard for the specificities of electrochemical processes were first considered in the 1960s by T. Z. Fahidy [14]. This author has proposed a classification of liquid flow characteristics and boundary conditions for equations and finally came, in a close analogy with the chemical reactor theory, to two extreme models: continuous stirred tank electrochemical reactor (CSTER) and plug flow electrochemical reactor (PFER). In stability analysis, the electric current load, I, is the control parameter mediated in the source (efflux) of matter and Joule heat release within the reactor space.

We now consider a model for the CSTER flow reactor with two state variables: concentration c and temperature T. For a stationary state, they are designated by a superscript asterisk *, that is c^* and T^* at the current load I^*; the respective input variables are denoted c_0^* and T_0^*. Both the input variables and the current are, in the general case, functions of time ($c_0 = c_0(t)$, $T_0 = T_0(t)$, $I = I(t)$) and may be regarded as external control factors. The system of equations for mass and heat balance is represented in a dimensionless form according to Refs. [94-96] as

$$\begin{cases} \dfrac{dx_1}{d\tau} = -\psi x_1 + R(x_1, x_2) - S + \xi_1 \\ \dfrac{dx_2}{d\tau} = -x_2 + Q + \xi_2 \end{cases} \quad (28)$$

where $x_1 = \dfrac{T - T^*}{T_0^*}$, $x_2 = \dfrac{c - c^*}{c_0^*}$, $\tau = t/t_r$ (t_r is the average residence time of a particle in the reactor). The quantities $\xi_1 = \dfrac{T_0 - T_0^*}{T_0^*}$ and $\xi_2 = \dfrac{c_0 - c_0^*}{c_0^*}$ are, respectively, the dimensionless input temperature and concentration. The quantities ψ, R, S, and Q are dependent on the geometrical and physicochemical properties of a system; in addition, they are functions of I^* and I. The particular expressions have been specified in the cited papers. The function $R(x_1, x_2)$ is an expression for Joule heat, and it includes the electrolyte solution conductance dependent on both c and T. If the cell voltage V, rather than the cell current load I, has been chosen for a control factor, one must use the volt-ampere characteristic equation $I = I(V)$ for current-to-voltage conversion.

Three classes of perturbations may be envisaged:

(a) perturbations of internal variables x_1, x_2, regarded as random variations of the initial conditions;

(b) exponentially decaying perturbations of external variables $|\xi_i| < A_i e^{-\lambda_i \tau}$, $A_i > 0$, $\lambda_i > 0$, $\tau \geq 0$, $i = 1, 2$;

(c) sustained (for example, sinusoidal) perturbations of current or voltage.

The Lyapunov function method has proved to be particularly effective in studying the stability of a stationary CSTER state towards perturbations of any type. For example, the cell for copper electrowinning from sulphuric acid solutions is stable towards perturbations of types (a) and (b). For perturbations of type (c), a stability in the sense of Malkin theory is reached: small external perturbations are attended by small variations in parameters c and T. Moreover, if the response amplitude decreases with increasing current or voltage oscillation frequency [96], and the system in question functions as a low-pass filter.

The PFER model is described by the equations below [97, 98]:

$$\begin{cases} \dfrac{\partial c}{\partial t} = -v\dfrac{\partial c}{\partial z} - K_1 I \\ \dfrac{\partial T}{\partial t} = -v\dfrac{\partial T}{\partial z} - K_2(T - T_a) + K_3(c_1 T)I^2 - K_4 I \end{cases} \quad (29)$$

where v is the flow velocity, and z is the longitudinal coordinate. The second, third, and fourth terms of the temperature equation characterize, respectively, the release of heat into the surroundings at temperature T_a, the Joule heat, and the electrochemical reaction heat. In [97], small deviations of the variables c and T from the stationary solution of system (29) (obtainable in the form of a wave propagating at the velocity v) have been analyzed for stability. Conditions for a local stability of the reactor have been defined. In the paper [98], a transition-function method (temporal Laplace transformation) has been used to study the frequency response to external factors (input concentration $c(t, 0)$, input temperature $T(t, 0)$, and current $I(z, t)$). For simplicity, the linearization of the heat-balance equation has been performed at small current amplitudes, assuming K_3 to be constant. Analytical results have been obtained for the output reactor variables $y_1(t, L) = c(t, L) - c^*(L)$, $y_2(t, L) = T(t, L) - T^*(L)$, with the constants $c(t, 0) = c^*(0)$, $T(t, 0) = T^*(0)$ corresponding to a stationary regime $c^*(z)$, $T^*(z)$, $I^*(z)$ and imposed sinusoidal current perturbations $x(t, z) = I(t, z) - I^*(z) = x_0 \sin \omega t$. The reactor residence time is $t_r = L/v$. By denoting $\beta = \omega t_r$, the concentration oscillation amplitude $y_1(t, L)$ is seen proportional to the factor $\psi_c(\beta) = (\sqrt{1-\beta})/\beta$. An infinite set of resonance frequencies $\beta_1, \beta_2, \ldots$, is thus defined corresponding to local maxima of the function $\psi_c(\beta)$, which is typical of a system with distributed parameters and a perturbation imposed along the spatial coordinate. An experimental study of the frequency response can easily be implemented because of its simplicity. A comparison with the theory can provide the basis for validation of the PFER model in each particular case. In the works [99, 100], attempts have been made to apply some elements of stochastic process dynamics. However, realistic models must, to a certain extent, take into account the dynamic dispersion of a component in the reactor, based, for example, on experimental measurements of the residence time distribution of a labelled species.

The dynamics and stability of definite classes of electrochemical processes have also been studied within the framework of a sequential macrokinetical analysis. Numerical methods have been developed for solving a system of electrodiffusion equations within a nonstationary boundary layer taking into account the concentrational dependence of transport properties, nonisothermicity, and specific features of reactions at the electrode surface and in the solution bulk [101, 102]. The effect of statistical characteristics of a turbulent flow on the current pulsation of an electrochemical reaction has been studied and interpreted in terms of the cell impedance [103].

The stability of any continuous system is determined by a hydrodynamic Glensdorf-Prigogine criterion. Let S be the entropy per unit mass in a nonequilibrium system, and v, the convective velocity of the centre of mass. For a locally stable state, the condition $\delta^2 S < 0$ holds. The Lyapunov function is $Z = S - T^{-1}v^2/2$, and the inequality $\partial(\delta^2 Z)/\partial t \geq 0$ corresponds to a decay of local fluctuations of the variables and, consequently, of the process stability. In the electromagnetic field, a generalized criterion applies, $\partial[\delta^2(\varrho\varsigma)]/\partial t \geq 0$, where $\delta^2(\varrho\varsigma) = \delta^2(\varrho z) - \frac{1}{2\pi} T^{-1}[(\delta E)^2 + (\delta H)^2]$ and ϱ is the density of the medium [104]. An explicit equation for $\delta^2(\varrho\varsigma)$ contains the product of the variations of thermodynamic forces and fluxes, including $\delta(\vec{i}T^{-1})\delta\vec{E}$. The direction of the imposed electric field opposite to that of the electrolysis current is a destabilizing factor, since the variation $\delta^2(\varrho\varsigma)$ may in time acquire a negative sign. Consequently, from general considerations one may anticipate a loss of stability in electrolyte solutions of varied concentration, if the diffusion potential counteracts the ohmic component. Such an effect for potentiostatic conditions has been predicted in Ref. [105] from an analysis of the diffusion problem in the immobile Nernst layer. Under stationary conditions,

$$d\varphi/dx = -RTi/(F^2 D^* c^*)$$

where $c^* = \Sigma z_i^2 c_i$, $(D^*)^{-1} = (1/n)\Sigma \nu_i z_i D_i^{-1}$, $n = \Sigma z_i \nu_i$, ν_i being the stoichiometric numbers of an electrode reaction ($\nu_i > 0$ for reaction products, $\nu_i < 0$ for reactants), and D_i are the diffusion coefficients. The quantity D^* may be termed the stoichiometric diffusivity. As has been shown in [105], for the Tafel electrode kinetics of type $i \sim \exp(-\alpha F\varphi/RT)$, the condition $D^* < 0$ is a necessary criterion for the multiplicity of stationary states. This multiplicity is typical of anodic reactions in which negatively charged low-mobility organic ions are generated. Further, attempts at generalization have been made using the linear analysis methods for the case of an arbitrary kinetic function including concentration and degree of adsorption, also taking into account the resistance of electric circuit and the properties of double electric layer [106, 107]. This has allowed, in a number of cases, classification of the kinetic relations potentially conducive to instability. For example, in adsorption-free systems, the instability is manifested in the region of drooping volt-ampere characteristics. This behaviour is borne out experimentally by anionic electroreduction or by electrooxidation of cations. If the convective velocity normal to the diffusion layer surface is too large to be ignored, additional factors arise that can lead to a loss of stability [108]. In a binary electrolyte, if the sufficiently intense convection becomes coincident in direction with the flow of discharging ions, the stationary state may fail to arise.

The results that have been reviewed above provide evidence for the interplay of a large number of factors determining the dynamics of electrochemical reactors, and much work needs to be done to get a deeper insight into the stability problems of electrolyte systems.

3.5 Origination of Dissipative Structures

Under the notion "dissipative structures" a variety of events are united, including both self-sustained oscillatory regimes (temporal structures) and regularly recurrent spatial formations (spatial structures) that emerge under stationary boundary conditions. According to the Glensdorf-Prigogine theory, the dissipative structures may be regarded as arising from a loss of stability by highly nonequilibrium stationary states in which the relation between flux characteristics and thermodynamic forces becomes nonlinear. One will have easily perceived a close connection with the theory of stability (Sec. 3.4); however, the problem of modelling dissipative structures extends far beyond the scope of this theory, and for electrochemical systems it remains a practically unexplored area.

The phenomena of self-organization in electrochemistry have become apparently an object of extensive studies since the 1950s, when Theorell discovered and described in detail "a membrane oscillator" [109]. A thin membrane of porous slag glass or porcelain was used to partition a vessel into two compartments containing aqueous NaCl solutions of different concentration. Electric current was applied to the membrane to create a drop in potential and in pressure across the membrane; the resultant effect was recorded as a difference between the liquid levels in the anodic and cathodic compartments. If the strength of applied current was low, within a short period of time, an electroosmotic pressure difference could be recorded between the two compartments. As the current strength was allowed to exceed a threshold value, periodic oscillations of the liquid levels, of the potential difference, and of the membrane electric resistance were observed to occur. Theorell explained these effects as arising from a nonlinear current load vs. potential difference relationship and calculated the membrane resistance as a function of the electrolyte flow rate on the basis of a convective-diffusion equation in the membrane pores. However, the equation for temporal resistance variation was not rigorously substantiated and was written rather by analogy with the equation for oscillation in electron-tube circuits. In Ref. [110], an attempt was undertaken to specify the oscillatory effect in more strict terms via studying the stability of hydrodynamic equations. Subsequently, oscillations in membrane systems were observed on a number of occasions. For example, there was reported an oscillation of the electrolyte flow in an oxygen-oxygen macropore-membrane cell filled with a 50% KOH solution [111]. The mechanism of resistance oscillations observed in a synthetic dioleylphosphate-based cation-exchange membrane in a chloride solution [112] was explained by competing sorption of K^+, Na^+, and Li^+ ions and low-mobility doped Ca^{2+} ions acting to block the hydrophilic ion-exchange groups. Also, an example of the redox system is known capable of generating current oscillations in a dropping mercury electrode at a sufficiently high circuit voltage [113], which has been explained by oscillatory tangential movements of the mercury drop surface.

The potential oscillations at a constant current were often observed for anodic dissolution in the active-passive transition region for a variety of metal-electrolyte systems. However, there remains much to be understood in regard to elementary electrochemical acts and passivation conditions, and for these reasons, none of the proposed mathematical models can be accepted as quite satisfactory. It appears to be universally recognized that any model must take into account the dynamics of the proper electrochemical stage for formation of a passive film, also the varied coverage of metal surface and the nonstationary ion transport affecting the surface pH and, consequently, the passivation potential. In the general case, it is necessary to consider the eventual formation of porous salt deposits between the oxide film and the electrolyte solution, whose thickness and structure are likewise variable in time. By applying the linear theory of stability to an analysis of typical models, the authors of Ref. [114] have identified a bifurcation point within the parametric space β, θ_s (where β is the Temkin interaction parameter for the adsorbed metal ions as oxide film constituents, and θ_s is the equilibrium surface coverage). A region of multiple (in number three) stationary states has been identified. For a model with transport limitations in solution, the conditions for formation of a stable limit cycle (the Hopf bifurcations) within the θ, φ-space have been determined, which provides for periodic oscillations of both quantities (here θ is the surface coverage, and φ is the potential). In particular, dissolution of iron in sulphuric acid has been considered.

As has recently been noted [115], many oscillatory systems—those known in heterogeneous catalysis and those reported for electrochemical reactions—are by nature chaotic rather than periodic. It was not clear, however, should the chaos be considered determinate, or rather originating from random noises. In the work [115], the experimental results for dissolution of iron in 1 M H_2SO_4 on a rotating disk were treated using a temporal series method. Under potentiostatic conditions, chaotic current oscillations were observed at a period τ of about 1/129 s. Denoting the time-dependent current strength by $I(t)$, we can consider a trajectory for the system in d-dimensional vector space $I(t)$, $I(t + \tau)$, ..., $I(t + (d - 1)\tau)$. We define a correlation integral

$$C(r) = \lim_{N \to \infty} \frac{1}{N^2} \sum_{\substack{i \neq j}}^{N} \theta[r - |\vec{I}_i - \vec{I}_j|] \tag{30}$$

where θ is the Heaviside function and $|\vec{I}_i - \vec{I}_j|$ is the distance between two points in a trajectory within the d-dimensional space. Performing summation over all the t_i's and t_j's such that $|\vec{I}(t_i) - \vec{I}(t_j)| < r$, we write expression (30) in the form $C(r) = Ar^\nu$. Here A is a constant, and ν is termed the correlational dimension and is the lower bound of the fractal trajectory dimension. Calculations have been carried out for $4 \leqslant d \leqslant 10$ at characteristic electrode potentials. The maximum of correlational dimension has been shown to be $\nu = 6$; it follows therefore that the dissolution must be described by at least seven ordinary differential equations of the 1st order or, possibly, by partial-derivative equations.

The available data on spatial organization of electrolyte systems are rather scarce. Results for liquid metal electrodes studied in the electrodeposition of mixed melts of alkali metal chlorides have been reported in Ref. [116]. The recorded abnormally high current densities under potentiostatic conditions have been ascribed to the interface instability conducive to the formation of dissipative structures which may be visualized as circulatory electrolyte cells of size about 0.5 mm migrating from the centre to the periphery of the electrode. With the negative potential rising, the circulatory cells grow in number and, at a maximum of current density, an ordered structure builds up in the form of a cellular convection. The mass-transport coefficient is observed to increase by more than an order of magnitude. In mercury polarography, maxima of the first kind are recorded as peaks on the ascending portion of a volt-ampere curve. As is commonly believed, they are caused by liquid movements stimulated by the surface tension difference across the surface of the mercury drop, which, in turn, is due to a nonuniform potential distribution in solution. In other words, the effect was ultimately explained as originating in a geometrical nonuniformity of the system. A theoretical analysis [117] has provided the reason to believe that the occurrence of maxima is associated with the hydrodynamic and electrochemical instability inherent in two-phase systems with a variable surface tension. Thus, the Marangoni effect, modified by the interfacial charge transfer, can be observed here, irrespective of the practical implementation of the experiment.

It has been shown recently [118] that in a system where at least one of the contact phases is an electrolyte solution and the interface surface accommodates adsorbed thereon ionic charges, the system stability may break down because of the interaction of viscous and electric forces. One may refer, by way of example, to the contact surface between an aqueous phase containing organic acids or complex ions and an organic phase, which is common in liquid extraction processes [119]. Let us consider a planar interface $y = 0$ between two immiscible solutions exposed to electric field E; the interface surface has specifically adsorbed charges of surface density $ez\Gamma$ (ez is the charge on a single adion, and Γ is the surface charge density. Let M be the molecular mass of adsorbed particles; η_1 and η_2, the dynamic phase viscosities; σ, the surface tension; v_s, the surface displacement velocity; and v_x, the projection of the liquid velocity vector onto a directional axis x (parallel to the surface). The balance of forces for an elemental area of the surface, taking into account tangential motions only, takes the form:

$$\frac{\partial \sigma}{\partial x} + ez\Gamma E_x = \Gamma M \frac{\partial v_s}{\partial t} + \eta_1 \frac{\partial v_x^{(1)}}{\partial y} + \eta_2 \frac{\partial v_x^{(2)}}{\partial y}, \quad \text{at } y = 0 \tag{31}$$

As follows from this equation, the action of a longitudinal electric field on the adsorbed charges must necessarily elicit convective flows both within the interface itself and the layers adjacent to it in the two phases. In the work [118], the simplest model system "hard metal wall/two-dimensional planar horizontal electrolyte layer/infinite bulk of the second liquid phase" has been studied. Along with the well-known Marangoni instability arising from the surface tension vs. concentration dependence, in this particular case, the destabilizing action is associated with an electric factor proportional to the current density.

Dissipative structures can also form at the interface "solid electrode—electrolyte", for example, as arising from the compositional inhomogeneity of a composite layer in electrodeposition [120], or at the ion-exchange membrane interface [121]. With reference to the concept of negative conductance under strongly nonequilibrium conditions (see above Sec. 3.4), the structurization can also occur in the bulk of the solution. Attempts to take into account the effect of Coulomb forces on the development of instability have been reported in Refs. [122, 123]. Since the space charge in solution is rather small, its effect can show up only if the difference in ion mobility is very large (by two orders of magnitude according to Ref. [122]). Uncommon and little yet understood effects have been observed in electrochemical cells with horizontal or inclined electrodes, where the Bénard instability is complicated by specific electrode reactions and experimental conditions. Under potentiostatic conditions for dissolution-deposition in a copper sulphate solution [124], the current density was significantly higher with the cathode located below, whereas the observation of current hysteresis provides evidence for the formation of steady-state convective cells. The linear analysis of stability of stationary states in Refs. [125, 126] has allowed explaining the diminution of the critical Rayleigh number in comparison to the classical Bénard problem (in which the boundary concentrations are assumed to be constant); also, the dependence of this number on the electric circuit parameters has been explained. In [126, 199], an attempt has been made to treat, in general terms, the stability of an "electrode-electrolyte" system at any number of reactions and under the conditions of specific adsorption, double-layer effects, and electrodiffusion kinetics.

In the domain of magnetoelectrolysis, an experimental study of the influence of magnetic field on the current density of redox reactions at a vertical electrode under natural convection was carried out [127]. In magnetic fields of strength up to 8000 G, the effects due to gravitational forces could be ignored. The plotted graphs showing the current density vs. field strength relationship exhibited markedly pronounced sinusoidal fluctuations and multiple values of current density; however, no attempt was made to interpret the observed effects.

One may therefore draw a conclusion that the electrochemical process exhibits a greater tendency, in comparison to others, to forming dissipative structures owing to the intervention of an additional factor of Coulomb instability. In practice, either the removal, or intentional sustenance of these effects is determined by the purpose of a given electrochemical engineering process.

3.6 Percolation Models as Applied to Electrochemical Systems

There are two reasons suggestive of the potential utility of percolation models in the theory of electrochemical processes. First, the percolation concepts constitute the basis for quantitative "structure-property" correlations as long as the properties of the medium are determined by the geometric order of its structural units. This allows a purposeful search for the optimal structure (porous membrane, composite electrode coating, conducting polymer films, and so forth). Second, they constitute the basis for a theoretical substantiation and define the limits to the applicability of a macroscopic description of any system, if the relations between its constituents admit of being treated in terms of a percolation theory. Viewed from these two standpoints only, percolation models can claim a wider application as compared to the range of a conventional problem on distribution of elements with mutually exclusive properties in an infinite lattice.

The said may be illustrated by the following examples. Ruthenium-titanium oxide electrode coatings deposited as solid-solution films $Ru_xTi_{1-x}O_2$ are obtained by thermal decomposition of ruthenium and titanium chlorides. The catalytic activity of electrode reactions and the coating conduction are attributed to RuO_2. A comparison of the infrared absorption spectra, exchange reaction current, electroconductance, light absorbance and X-ray data have allowed drawing a conclusion on the formation, at $x \geqslant 0.25$, of an infinite cluster RuO_2, that is, a bound system of conducting elements [128]. In modelling the nonuniform inner surface of a porous electrode, each element constitutive of a cubic lattice was assigned, with a definite probability, to one of the four states: metal, electrolyte, surface (semiconductor), and air [129]. Imparting an electric conductance to each element and, additionally, a capacitance to the surface element, one obtains thus a random electric circuit whose impedance is tractable by calculation. In Ref. [130], the problem of an optimal filling of the electrode layer with small platinum particles imbedded in a conducting polymer film has been considered with a view of reaching a maximum of effectively operating specific surface per particle. A necessary condition for normal operation of such a system is the electric contact of agglomerated particles with the current-conducting substrate as well as the occurrence of current-conducting regions in the film accessible to free electrolyte solution. The layer structure, that is, the mode for particle arrangement was modelled on three-dimensional lattices of different type. The optimum specific surface has been determined at volume fraction of particles about 0.20-0.25, which corresponds to the percolation threshold. The conductance of synthetic ion-exchange membranes as a function of moisture content U is described by a power law providing that U is greater than a critical value U_c; otherwise, at $U \leqslant U_c$, the conductance goes to zero [131]. These findings, based on the concept of a cluster-channel structure of hydrated functional groups, allowed us to conclude that the conductance exhibits a percolation character. In [132], asymptotic formulas for calculating the conductance in a close vicinity to the percolation threshold have been derived, and the scaling relations in the fractal region for scales smaller than the correlation length l_{corr} have been analyzed. In particular, for l_{corr} greater than the membrane thickness L, the membrane resistance varies with L by a law $R_{ohm} \sim L^{1+\alpha}$, where $\alpha > 0$. This implies the nonapplicability of a quasi-homogeneous macroscopic description at $U \to U_c$. In film electrodes, the electron conduction is effected via jump mechanism between localized electron states. Along with the redox process at the film/electrolyte interface, the electron-transfer stage also contributes to the reaction kinetics. A problem has also been considered for the overall conduction of a random set of localized states, where each pair of states was assigned a definite conductance vs. distance relationship [133]. The electron transfer can be effected only via the trajectories in which the conductance of each constituent element is greater than a threshold value.

The list of the available evidence on percolation phenomena can be extended still further, but we feel that the above examples suffice to reveal the characteristic features of problems of this class. These are: (a) spatial confinement of a studied region and the associated therewith asymmetry of the boundary conditions (for example, film electrodes); (b) multiplicity of states for the elements of a model lattice (porous electrodes); and (c) nonlinearity of transport properties of the constituent elements (electrochemical kinetics). These limitations exclude practically the application of analytical methods. For this reason, the most widespread and effective approach to percolation phenomena in electrochemistry is the modelling by Monte-Carlo method.

3.7 Nonequilibrium Membrane Thermodynamics

In this section, our major goal has been to show that, in a thermodynamic approach to transport phenomena, there is no fundamental difference in the description of a free electrolyte solution and a membrane system. However, certain aspects of the internal structure of membranes pose a number of problems that require further substantiation. Abstracting from physicochemical properties, we understand, in what follows, by the membrane any partition which separates two solutions from each other and which may happen to be selectively permeable to certain components of these solutions. Many of the issues that are discussed below apply with equal right to processes not necessarily electrochemical ones.

3.7.1 Discontinuous Systems

Consider a membrane separating two N-component solutions and permeable to the solution components; then, under certain specified conditions, it is possible to relate the component fluxes across the membrane to the chemical potential difference of these solution components. Such conditions are the quasistationarity of transmembrane processes and small deviations from the state of equilibrium. Then, neglecting inertial forces, one can construct a linear nonequilibrium thermodynamics for a discontinuous system [29] within which the membrane zone is modelled by a set of phenomenological Onsager coefficients. For an electrochemical system including membrane-separated salt solutions, Kedem and Katchalsky [134] have derived phenomenological equations relating experimentally measurable quantities. Differences in concentration, pressure, and electric potential or variables equivalent to them have been shown to be the driving forces operative in solutions [200].

The state of a transition zone (membrane) being close to the state of local equilibrium is a fundamental requirement of linear thermodynamics. By way of illustration, we now consider the simplest event of diffusion of one compound in an immobile medium between two regions of concentration c_1 and c_2 in a layer of thickness L. If the chemical potential is $\mu = \mu^\theta + RT \ln c$, then, according to the thermodynamics of discontinuous media, the flux of matter is $N = l(\mu_2 - \mu_1)/L = lRT(c_2 - c_1)/\bar{c}L$, with $\bar{c} = (c_2 - c_1)/\ln(c_2/c_1)$. The thermodynamic coefficient l is thus defined in terms of the average concentration. Let us consider variations of the boundary values of c_1 and c_2 such that the average $c_0 = \dfrac{1}{2}(c_1 + c_2)$ remains invariant. At small deviations from the average, that is, $c_2 - c_1 \to 0$, we arrive at Fick's law $N = lRT(c_2 - c_1)/c_0 L$. Since it holds for any deviation from equilibrium, the latter formula is an exact one, and the true diffusion coefficient is $D_0 = lRT/c_0$. The diffusion coefficient as defined in terms of discontinuum thermodynamics is $\overline{D} = lRT/\overline{c}$; consequently, in all cases, $\overline{D} > D_0$. The error associated therewith is not large at $c_1/c_0 > 0.5$. For example, the value of c_1/c_0 being 0.1 and 0.05, the ratio $\overline{D}/D_0$ is, respectively, 1.64 and 1.93.

The membrane model can also be conceived as a sequence of energy barriers [135], in which the probability for a particle jumping over the barrier is determined by the transition-state theory. The intervention of external forces is taken into account via introduction of appropriate potential differences into the energy of activation of transition state. Thermodynamic description of diffusion in a membrane regarded as a continuous medium is practically applicable, if $\Delta\varphi < 1.1RT$, $\Delta\varphi$ being the potential difference between two successive sites for localization of a particle within the membrane. For a local equilibrium to remain valid on the scale of a single jump, this condition is not restrictive.

The problem is much less clear with the mutual diffusion of several compounds. A wealth of experimental evidence on nonlinear effects in membrane systems sets definite limits to the practical applicability of the classical thermodynamics of discontinuous systems. For example, in the monograph [136] on mass transport in porous media, a description of the mutual diffusion of neon and argon across a graphite membrane is given. It has turned out that the Onsager reciprocal relations, considering this system as a discontinuous one (neglect of the finite filter thickness), remain valid only within a very narrow region near the equilibrium state. Recently, the classical thermodynamic analysis of de Groot and Mazur [29] has been extended in [137] via introduction of corrections for center-of-mass acceleration, kinetic energy of diffusion, and electric polarization in the membrane region. This has removed the constraints imposed by quasiequilibrium conditions. Assuming that a mixture inside the membrane pore moves under continuous-medium conditions, the dissipative function (that is, the production of entropy in the system) for isothermal conditions may be written in the form

$$T\sigma = -\sum_k N_k [(\Delta\overline{\mu}_k)_p - \alpha_3 m_k^2 N_k^2/2c_k]$$
$$- \sum_k V_k N_k [\Delta p - \alpha_1 \varrho v^2/2 - \alpha_2(\Delta\varphi)^2] \quad (32)$$

where Δp is the pressure difference, ϱ is the density, $\Delta\varphi$ is the electric potential difference, m_k is the molecular mass, V_k is the specific volume, and v is the volume flow velocity. Parameters α_1, α_2, and α_3 are characteristics of the porous membrane structure. By constructing phenomenological equations for the transmembrane fluxes N_k via linear combinations of the "forces" (enclosed in square brackets in Eq. (32)), we finally obtain nonlinear relations between the fluxes and thermodynamic potentials $\Delta\mu_k$, Δp, and $\Delta\varphi$. However, the extent of applicability of the linear combinations remains unclear. In a region distant from the state of equilibrium, more constructive appears to be the macrokinetic analysis of concentrational changes within the membrane based on the differential "force vs. flux" correlations in any physically small volume.

3.7.2 "Dusty Gas" Model

The title model was initially developed on the basis of a kinetic theory for the gas-mixture flux in a porous medium. This medium can be visualized as an equipotent gas component composed of heavy immobile particles, uniformly distributed within the space. Conversion to macroscopic parameters is performed through solving the Boltzmann equation by Enskog method, which ultimately leads to the known Stefan-Maxwell equations, equivalent to the Onsager formalism in nonequilibrium thermodynamics. Simultaneously, diffusion coefficients are calculated as functions of molecular collision parameters. The conclusions derived from kinetic theory cannot be applied in a straightforward manner to liquid membrane systems; however, two possible routes can be suggested. The first route consists in a phenomenological generalization of the results ensuing from the gas-kinetic equations. The second route postulates the *à priori* applicability of nonequilibrium thermodynamics formalism to a combined system "solution-membrane". The "dusty gas" model has been substantiated and worked out to completion mainly owing to the works by Mason and coauthors [136, 138].

Without going into particulars, the major concepts of the model may be outlined in the following manner. We consider the membrane as an equipotent $(N + 1)$th component of the mixture. Then, under isotropic conditions, the $(N + 1)$ Stefan-Maxwell equations (1) hold. If the system is not at mechanical equilibrium [29] and, in addition, is nonisothermal, then forces $\nabla\mu_i$ are replaced by forces

$$\vec{d}_i = (\nabla\mu)_T - \frac{m_i}{\varrho}\left(\nabla p - \sum_{j=1}^{N+1} c_j\vec{X}_j\right)$$

where $\vec{X}_i$ is an external potential force acting upon the ith component (this force determines also the mechanical part of $\bar{\mu}_i$), ϱ is the total density, and m_i is the molecular mass. The velocity difference $\vec{v}_i - \vec{v}_j$ must be supplemented by a thermodiffusion term proportional to $\nabla \ln T$. However, the introduction of the temperature gradient does not entail any fundamental change, and the system, as previously, is assumed to be isothermal. The coefficients K_{ij}^0 are replaced by K_{ij} to emphasize their distinction from those for free solution. Thus, formally a supplementary set of interaction parameters is introduced, $K_{i,N+1}$ ($i = 1, \ldots, N$). For an immobile membrane, $\vec{v}_{N+1} = 0$. Further, we may assume that an inflexible membrane, on its interaction with the incoming components of solution, behaves, in a sense, analogously to the rigid walls of a vessel within which the system in question is confined; for this reason, the chemical potential $\mu_{N+1} = 0$. Expressing the force $c_{N+1}\vec{X}_{N+1}$ from the condition for mechanical equilibrium, $\nabla p = \sum_{j=1}^{N+1} c_j \vec{X}_j$, the $(N + 1)$th Stefan-Maxwell equation may be written in the form

$$\sum_{j=1}^{N} K_{N+1,j}\vec{v}_j = \sum_{j=1}^{N} c_j\vec{X}_j - \nabla p \tag{33}$$

Equation (33) is a generalization of Graham's law for diffusion of a gas mixture through a porous medium. In the presence of a pressure gradient and external forces, Eq. (33) may also be referred to as a generalization of Darcy's law.

For microporous materials, the solution phase is, within the framework of this theory, inseparable from the membrane phase, and the $(N + 1)$-component mixture must be regarded as homogeneous and the coefficients K_{ij} be treated as purely empirical ones. In coarse-grained materials, the size of pores is substantially larger than the molecule diameter. Therefore, a problem arises to express K_{ij} through K_{ij}^0 and structural properties of the medium so as to be able to make use of the known transport characteristics of free solution. To this end, a transformation has been introduced in Ref. [136], which in our notations takes the form

$$K_{i,N+1} = K'_{i,N+1}\left(1 + k\sum_{\gamma=1}^{N} K'_{\gamma,N+1}\right)^{-1}, \quad K_{ij} = K'_{ij} + kK_{i,N+1}K'_{j,N+1} \tag{34}$$

for $i, j = 1, \ldots, N$. The substitution of (34) into Stefan-Maxwell equations for the permeative components yields, with reference to Eq. (33):

$$c_i \vec{d}_i = \sum_{j=1}^{N} K'_{ij}(\vec{v}_j - \vec{v}_i) - K'_{i,N+1}\left[\vec{v}_i - k\left(\sum_{j=1}^{N} c_j \vec{X}_j - \nabla p\right)\right] \quad (35)$$

If all the velocities are the same and the driving forces $\vec{d}_i$ are absent, one refers to a "viscous flow" moving at a velocity of $\vec{v}_i = k\left(\sum_{j=1}^{N} c_j \vec{X}_j - \nabla p\right)$. For a capillary-pore model, the permeability k may be represented, in accordance with Poiseuille's law, as $k = f_\varepsilon \langle r^2 \rangle / 8\eta$. Here f_ε is the porosity-tortuosity factor [136] which is a measure for attenuation of mass transport by a porous structure (membrane) and is a function of porosity; $\langle r^2 \rangle$ is the statistical mean square of the pore radius; and η is the viscosity of the solution contained in the pores. For coarse-grained materials, the concentrations c_i in Eq. (35) are referred to unit pore volume, and the K'_{ij} can be expressed through the coefficients K^0_{ij} that have been introduced in Sec. 3.2.1, namely, $K'_{ij} = K^0_{ij}/f_\varepsilon$, at $i, j = 1, \ldots, N$.

A detailed study of the gas diffusion mechanism in capillaries has shown that precisely the coefficients K'_{ij}, rather than K_{ij}, provide a measure for the interaction of components, whereas $K'_{i,N+1}$ characterizes the wall resistance, that is, the Knudsen slip. The occurrence of this resistance is an essential feature that distinguishes the membrane transport kinetics as expressed by Eq. (35) from that by Eq. (1) in free solutions. The relative contributions due to the viscous flow and the Knudsen slip are estimated by quantities k and $\left(\sum_{j=1}^{N} K'_{j,N+1}\right)^{-1}$ according to Eqs. (33)-(34). In liquids, the slip is insignificant, and $k \gg \left(\sum_{j=1}^{N} K'_{j,N+1}\right)^{-1}$, whereas it follows from (34) that $K_{i,N+1} \approx K'_{i,N+1} / \left(k \sum_{j=1}^{N} K'_{j,N+1}\right)$. Assuming $K'_{i,N+1} \sim c_i$, we estimate $K_{i,N+1} \sim x_i/k$, where x_i is the mole fraction, and $K_{ij} \sim x_i x_j$ within a constant factor. Estimates of such a kind simplify the expression for transport equations. In dilute solutions, we can neglect the interaction between solvent and other components, that is, we can omit the diffusion terms in Eq. (35) for solvent. For electrolyte solutions, the inequality $K'_{i,N+1} \ll K'_{ij}$ ($i \neq$ solvent) is expected to hold for mutual diffusion within the pores under continuous medium conditions. Then the last summand in Eq. (35) for $i \neq$ solvent is eliminated, and the only distinction of this expression from the system of equations (1) consists in K^0_{ij} being replaced by K'_{ij}, recalling that $K'_{ij} = K^0_{ij}/f_\varepsilon$. This implicit exclusion of the viscous flow from the analysis of electrochemical processes in porous media is a common expedient, although its strict justification has as yet been given only for a hydrodynamic model of the flow of suspended spherical particles in a capillary [139] at infinite dilution limit.

3.7.3 System Analysis Under Strongly Nonequilibrium Conditions

The fundamental hypothesis of local thermodynamic equilibrium constituting the basis of nonequilibrium thermodynamics [29] appears to be justified by a good comparison between theoretical predictions and experimental evidence for the fluxes in a nonequilibrium system. Invoking the simplest examples in Sec. 3.7.1, one can conclude that, as the membrane as a whole is made to shift from the state of equilibrium, nonlinear "flux" vs. "force" relationships arise. In a continuous "dusty gas" model, like in free solution, the elementary physical volume of the medium is commonly believed to persist in a state of local equilibrium. In coarse-grain materials, the elementary volume is in size comparable to the pore diameter and, presumably, cannot be smaller than a certain limiting value. Therefore, if the intensive properties are observed to vary markedly on the elementary volume scale to which the equations of thermostatics can no longer be applied, we have good reasons to suspect a breakdown of linearity. In homogeneous media, such a behaviour is an exception rather than a rule, since the thermal relaxation time for them has the same order of magnitude as the mean free-path time of molecules, which is beyond comparison smaller than the macroscopic characteristic time. Relaxation of an element constitutive of a coarse-grain medium exhibits a macroscopic nature (as described by the laws of Newtonian viscous friction, Fourier heat conduction, and Fick diffusion) and proceeds at an incomparably slower rate. Therefore, a macroscopic description of strongly nonequilibrium conditions in heterogeneous media can include, in principle at least, generalization of local equilibrium hypothesis and nonlinear dependencies.

Our theory [140] that we wish to briefly outline below has emerged from an attempt to give a rational explanation of the experimental results concerned with the determination of electrolyte diffusion coefficients in electrochemical matrix cells [141]. A membrane, soaked with a concentrated KOH solution, of pore size ranging from several tenths of a micron to several microns and of thickness about 1 mm was tightly sandwiched between two layers of catalyst on which the reaction $H_2 + 2OH^- = 2H_2O + 2e^-$ was allowed to run. The stationarity of the process was maintained owing to a specially designed circulation circuit for gaseous hydrogen. Under the conditions of limiting mass-transfer current, the measured "electrochemical" diffusion coefficients $D_\pm^{el}$ have been found to be several times smaller than the coefficients $D_\pm$ determined at the same concentration, but in the absence of current, by porous diaphragm method. These findings failed to be explained within the framework of classical nonequilibrium thermodynamics where, by definition, the transport coefficients are postulated to be independent of the velocity of a process. However, evoking once again the simplest example in Sec. 3.7.1, one will easily see that, under conditions strongly shifted from a state of equilibrium, the linear thermodynamics predicts an augmented diffusion flux. Consequently, in solving the inverse problem, the experimental values of l (or D_0) as measured thermodynamically are found to be smaller than their true values, which necessitates applying corrections to bring them to an agreement.

We have extended the above concepts to the case of an $(N + 1)$-component isotropic mixture within the so-called formalism of extended irreversible thermodynamics [142]. It can be shown, in neglect of the viscous tension and the heat flux due to heat conduction, that the entropy per unit mass S under strongly nonequilibrium conditions is dependent, by definition, on the internal energy per unit mass U, density ϱ, mass fraction ϱ_i/ϱ, and diffusive mass fluxes $\vec{j}_i = \varrho_i(\vec{v}_i - \vec{v})$ relative to the center-of-mass velocity $\vec{v} = \sum_{k=1}^{N+1} \varrho_k \vec{v}_k/\varrho$, with $\varrho = \sum_{k=1}^{N+1} \varrho_k$, $i = 1, \ldots, N$. The extended Gibbs equation takes the form

$$\varrho dU = \varrho T^* dS - \varrho p^* d\frac{1}{\varrho} + \varrho \sum_{i=1}^{N} M_i^* d\frac{\varrho_i}{\varrho} + \sum_{i=1}^{N} \vec{B}_i^* d\vec{j}_i \tag{36}$$

For complete thermodynamic description of the system we must define equations of state, that is, specify the partial derivatives for S as functions of the same intensive variables. In the asymptotic limit of small fluxes, this can be done on the basis of the tensor function representation theorems in an isotropic medium keeping in mind that the properties of a material are independent of the frame of reference chosen. In addition, the specified conditions must correspond to thermostatic requirements as $\vec{j}_i = 0$. Accurate to the second order in flux parameters, this yields

$$T^* = T + O(2) \tag{37}$$
$$P^* = P + O(2) \tag{38}$$

$$M_i^* = \mu_i - \mu_N + O(2) \tag{39}$$

$$\vec{B}_i^* = \sum_{k=1}^{N} \beta_{ik}\vec{j}_k + O(2) \tag{40}$$

where the matrix β_{ik} is symmetric. The entropy S may be represented as the sum of an equilibrium entropy S_{eq} dependent on U, ϱ, ϱ_i/ϱ, and an additional entropy S'. It can readily be shown, by introducing a quadratic-form matrix $M_S = \|\beta_{ik}\|$ and an N-dimensional vector $\vec{f} = (\vec{j}, \ldots, \vec{j}_N)$, that

$$2\varrho TS' = -\vec{f}(M_S\vec{f}) \tag{41}$$

For locally stable states, the additional entropy is $S' < 0$, and the matrix M_S is positively definite. The expression for entropy production, σ, is derived by equations for conservation of mass, momentum, and energy. The "heat flux", originating in diffusion [29] and making part of the energy equation, as well as the entropy flux are defined (assuming the absence of viscous tension) up to the third order terms by the known expressions of linear thermodynamics. Finally, one obtains $\sigma = \sigma_0 + \sigma'$, where σ_0 is given by the classical expression (under isothermal conditions)

$$\sigma_0 = -\frac{1}{T}\sum_{i=1}^{N}\vec{j}_i\nabla(\bar{\mu}_i - \bar{\mu}_N) \tag{42}$$

The additional summand σ' is

$$\sigma' = -\frac{1}{T}\sum_{i=1}^{N}\vec{j}_i\frac{d\vec{B}_i}{dt} \tag{43}$$

By definition, the symbol d/dt refers to a substantial derivative, and

$$\frac{d\vec{B}_i}{dt} = \sum_{k=1}^{N}\beta_{ik}\frac{d\vec{j}_k}{dt} \tag{44}$$

Linear phenomenological equations for diffusion fluxes with a symmetric coefficient matrix can be postulated on the basis of Eqs. (42)-(43), by analogy with the procedure suggested in Ref. [137] for a discontinuous system. Further, it can be shown that the only distinction between the Stefan-Maxwell equation and its generalization is that the forces $\vec{d}_i$ in the former case are to be replaced by forces $\vec{g}_i = \vec{d}_i + m_i\vec{\Delta}_i$ with $\sum_{i=1}^{N+1}\vec{c}_i\vec{g}_i = 0$. The summands $\vec{\Delta}_i$ are defined in the following manner:

$$\vec{\Delta}_i = \frac{d\vec{B}_i}{dt} - \frac{1}{\varrho} \sum_{k=1}^{N} \varrho_k \frac{d\vec{B}_k}{dt}, \quad i = 1, \ldots, N \tag{45}$$

$$\vec{\Delta}_{N+1} = -\frac{1}{\varrho} \sum_{k=1}^{N} \varrho_k \frac{d\vec{B}_k}{dt} \tag{46}$$

The next stage is a generalization of the "dusty gas" model [140]. Along with the assumptions made in Sec. 3.7.2, we take the density of the $(N+1)$th component to be uniform: $\varrho_{N+1} = $ const. A strongly nonequilibrium state can be characterized not only by the vector $\vec{f} = (\vec{j}_1, \ldots, \vec{j}_N)$, but also by any vector related to the former through a nondegenerate linear transformation. We define now a set of variables $\vec{f}_0 = (\vec{N}'_1, \ldots, \vec{N}'_N)$ and a set $\vec{f}_* = (\vec{j}^*_1, \ldots, \vec{j}^*_{N-1}, \vec{v}^*)$, where $\vec{v}^* = \left(\sum_{k=1}^{N} \varrho_k \vec{v}_k\right) \bigg/ \varrho_*$ is the baricentric velocity of all the mobile components; $\varrho_* = \sum_{k=1}^{N} \varrho_k$, $\vec{j}^*_i = \varrho_i(\vec{v}_i - \vec{v}^*)$, $i = 1, \ldots, N-1$, is the diffusive mass flux of the ith component in the center-of-mass coordinate system of $\vec{v}^*$; and $\vec{N}'_i (i = 1, \ldots, N)$ are the total mass fluxes within the laboratory coordinate system. The variables are related to each other in a matrix form as

$$\vec{f} = D_* \vec{f}_*, \quad \vec{f} = D_0 \vec{f}_0 \tag{47}$$

The entropy S' likewise can be expressed in a variety of ways, for example, in accordance with (41):

$$-2\varrho TS' = \vec{f}(M_S \vec{f}) = \vec{f}_*(M_S^* \vec{f}_*) = \vec{f}_0(M_S^0 \vec{f}_0) \tag{48}$$

It is evident that

$$M_S^* = D_*^T M_S D_*, \quad M_S^0 = D_0^T M_S D_0 \tag{49}$$

(the superscript T denotes transposition). The transformation matrices D_* and D_0 are dependent on the densities ϱ_i. We define now a vector of forces $\vec{\Delta} = (\vec{\Delta}_1, \ldots, \vec{\Delta}_N)$. With reference to equations (44), (45), we obtain

$$\vec{\Delta} = D_0^T M_S \frac{d\vec{f}}{dt} \tag{50}$$

Going over to the vector $\vec{f}_0$ by Eq. (47) and taking into account Eq. (49), we perform differentiation of the matrix D_0 subject to $\varrho_N = $ const to obtain, after appropriate transformation,

$$\vec{\Delta} = -M_S^0 \left[\begin{pmatrix} d\varrho_1/dt \\ \cdots \\ d\varrho_N/dt \end{pmatrix} \vec{v} + \frac{d\vec{f_0}}{dt} \right] \tag{51}$$

In the case of a one-dimensional stationary flow in the direction x, the total fluxes N_i' must be coordinate-independent and, as follows from (51),

$$\vec{\Delta} = -M_S^0 \begin{pmatrix} d\varrho_1/dx \\ \cdots \\ d\varrho_N/dx \end{pmatrix} v^2 \tag{52}$$

In considering coarse-grained filtering media, that is, those in which the solution contained in the pore voids forms a bulk phase, we must emphasize their specific properties resembling the transport properties of free solutions. The relation of the resistance coefficients K_{ij} to the coefficients K_{ij}^0 for free solutions has been discussed within the framework of the classical theory in Sec. 3.7.2. The said formulation requires, in addition, an assumption about the entropy S'. Our assumption is that the entropy S' for coarse-grained media, when expressed through $\vec{f}_*$, is independent of $\vec{v}_*$, that is, it is a quadratic form only with respect to $\vec{j}_1^*, \ldots, \vec{j}_{N-1}^*$, as this should be expected for a free solution. Mathematically, this is expressed by the respective matrix elements equated to zero:

$$(M_S^*)_{iN} = (M_S^*)_{Ni} = 0; \quad i = 1, \ldots, N \tag{53}$$

By virtue of the conversion formulas (49), one can obtain constraints for the matrix elements as expressed in terms of other variables. The quadratic form S' becomes, therefore, degenerate.

At this stage, the above assumption should be regarded as an *ad hoc* hypothesis, validated, nonetheless, by the "microscopic" argumentation. Indeed, if the pore size is much larger than the molecular diameter and the free-path length, then any averaged property is in principle amenable to determination through solving the local (classical) equations for a transport within the pores of physical elementary volume. Since any theory is constructed in a small-flux approximation, accordingly, the local baricentric velocity for the solution contained in the pores, v_{loc}, is also small. Accordingly, the Peclet number as estimated by pore size must not be large: $v_{\text{loc}} \cdot r/D_{ij} < 1$. If so, the solution of macroscopic problem becomes independent of v_{loc}. Since $\vec{v}_*$ is an integral of $\vec{v}_{\text{loc}}$ over an elementary volume, we arrive at the earlier formulated hypothesis. The condition for smallness of the Peclet number is typical of most electrochemical processes.

Now, going back to the interpretation of the measured diffusion coefficients, we set $N = 3$ for a binary solution. Corrections to be applied to thermodynamic forces are given by expression (52).

In considering the Stefan-Maxwell equations, we use a condition for mechanical equilibrium $\nabla p = \sum_{k=1}^{N+1} c_k \vec{X}_k$ in neglect of viscous flow effect (Section 3.7.2). Suppose, an electrolyte molecule dissociates into ν_1 cations and ν_2 anions of respective charges z_1 and z_2, so that $\nu_1 z_1 + \nu_2 z_2 = 0$. The ion concentrations are $c_1 = \nu_1 c$ and $c_2 = \nu_2 c$. The chemical potential per g-mol electrolyte is $\mu = \nu_1 \mu_1 + \nu_2 \mu_2$, and the current density is $\vec{i} = F(z_1 \vec{N}_1 + z_2 \vec{N}_2)$. Simple transformations lead to the expressions for fluxes

$$\vec{N}_1 = -\frac{\nu_1}{\nu} \frac{Dc}{RT} \frac{c_T}{c_0} (\nabla \mu + \nu_1 m_1 \vec{\Delta}_1 + \nu_2 m_2 \vec{\Delta}_2) + \frac{\vec{i} t_1^0}{F z_1} + c_1 \frac{\vec{N}_0}{c_0} \quad (54)$$

$$\vec{N}_2 = -\frac{\nu_2}{\nu} \frac{Dc}{RT} \frac{c_T}{c_0} (\nabla \mu + \nu_1 m_1 \vec{\Delta}_1 + \nu_2 m_2 \vec{\Delta}_2) + \frac{\vec{i} t_2^0}{F z_2} + c_2 \frac{\vec{N}_0}{c_0} \quad (55)$$

Here $\nu = \nu_1 + \nu_2$; $D = \dfrac{(z_1 - z_2) D_{10} D_{20}}{z_1 D_{10} - z_2 D_{20}}$ is the electrolyte diffusion coefficient; $t_1^0 = \dfrac{z_1 D_{10}}{z_1 D_{10} - z_2 D_{20}}$ and $t_2^0 = 1 - t_1^0$. Eqs. (54) and (55) are general expressions for diffusion processes in a binary electrolyte. Introducing the concentrational driving force ∇c, we write in a conventional manner $\dfrac{c c_T}{\nu R T c_0} \nabla \mu = \beta(c) \nabla c$, where $\beta(c) = \dfrac{c_T}{c_0} \left(1 + \dfrac{d \ln \gamma_\pm}{d \ln m}\right) \left(1 - \dfrac{d \ln c_0}{d \ln c}\right)$, m is the molality, and $\gamma_\pm$ is the average ionic activity coefficient. We thus arrive at an equality similar to that used in [141] in measurement results handling:

$$\frac{\vec{N}_1}{\nu_1} t_1^0 + \frac{\vec{N}_2}{\nu_2} t_2^0 = \frac{\vec{N}_0}{c_0} c - D_\pm \left[\nabla c + \frac{c c_T}{\nu R T c_0 \beta(c)} (\nu_1 m_1 \vec{\Delta}_1 + \nu_2 m_2 \vec{\Delta}_2)\right] \quad (56)$$

where $D_\pm = D \beta(c)$. The use of conditions (53) leads to a system of algebraic equations expressing the matrix elements M_S^* equal to zero:

$$\begin{cases} \beta_{11}^\circ \varrho_1 + \beta_{12}^\circ \varrho_2 + \beta_{10}^\circ \varrho_0 = 0 \\ \beta_{21}^\circ \varrho_1 + \beta_{22}^\circ \varrho_2 + \beta_{20}^\circ \varrho_0 = 0 \\ \beta_{01}^\circ \varrho_1 + \beta_{02}^\circ \varrho_2 + \beta_{00}^\circ \varrho_0 = 0 \end{cases} \qquad (57)$$

For electrolyte solutions, it can be accepted, to a satisfactory accuracy, $\dfrac{dc_0}{dx} \approx -\dfrac{dc}{dx}$. By making use of this condition and expressing β_{10}°, β_{20}° in Eq. (57) with a view to (52) for one-dimensional stationary flows, we obtain

$$\nu_1 m_1 \Delta_1 + \nu_2 m_2 \Delta_2 = -v^2 \left(1 + \frac{c}{c_0}\right) [(m_1\nu_1)^2 \beta_{11}^\circ + 2(m_1\nu_1)(m_2\nu_2)\beta_{12}^\circ$$

$$+ (m_2\nu_2)^2 \beta_{22}^\circ] \frac{dc}{dx} \qquad (58)$$

A comparison of Eqs. (56) and (58) shows that the transport coefficient $D_\pm^{el}$ to be measured is determined, with reference to the driving force $\dfrac{dc}{dx}$, as

$$D_\pm^{el} = D_\pm \left\{ 1 - v^2 \frac{cc_T}{\nu RTc_0 \beta(c)} \left(1 + \frac{c}{c_0}\right) \left[(m_1\nu_1)^2 \beta_{11}^\circ \right.\right.$$

$$\left.\left. + 2(m_1\nu_1)(m_2\nu_2)\beta_{12}^\circ + (m_2\nu_2)^2 \beta_{22}^\circ \right] \right\} \qquad (59)$$

The matrix M_S° being positive determinate, it follows that $\beta_{11}^\circ > 0$, $\beta_{11}^\circ \beta_{22}^\circ - (\beta_{12}^\circ)^2 > 0$. Considering the evident inequalities $(m_1\nu_1)^2 \beta_{11}^\circ + (m_2\nu_2)^2 \beta_{22}^\circ \geqslant 2(m_1\nu_1)(m_2\nu_2)\sqrt{\beta_{11}^\circ \beta_{22}^\circ} > 2(m_1\nu_1)(m_2\nu_2)|\beta_{12}^\circ|$, we come to a conclusion that the nonequilibrium correction to Eq. (59) is negative in sign, that is, $D_\pm^{el} < D_\pm$ (the positiveness of $\beta(c)$ for a KOH solution has been shown by concrete calculations in the work [141]).

Thus, an analysis of strongly nonequilibrium conditions in electrochemical systems has shown that the measurable coefficient $D_\pm^{el}$ becomes identical to the electrolyte diffusion coefficient $D_\pm$ only if no center-of-mass motion occurs in the system. The porous-diaphragm method with no current applied provides, at a small concentration gradient, for conditions under which the velocity v is actually zero. In the limiting current regime, the finite value of v is provided for by specially designed conditions, namely, the circulation of gaseous hydrogen. Starting from the value of limiting current, the overall mass flow in the cell, ϱv, can be determined. Formula (59) is also suggestive of a coincidence of the measurable coefficients in the limit of $c \to 0$, in agreement with the experiment. It should be noted that the above analysis is essentially qualitative by character, since a number of new parameters, introduced in nonequilibrium thermodynamics, remain as yet indefinite. Further improvements are possible both via a more thorough analysis of experimental conditions in systems with intense mass-exchange and via clarification of the physical sense of suggested hypotheses. Regrettably, apart from the references cited above, the authors are unaware of similar attempts to measure transport characteristics of solutions with large compositional gradients in porous media. It is felt, however, that the suggested approach throws light on certain specific problems of membrane transport.

3.8 Macrokinetics of Chlorine Solid-Cathode Cells

This and subsequent Sections are concerned with demonstrating the potential utility of mathematical modelling as applied to the problems of intensified production of chlorine and alkali in solid-cathode cells (by diaphragm and membrane methods), one of the most large-scale and energy consumptive branches of electrochemical engineering. Two major indices—current efficiency and voltage drop—are necessarily present in the feasibility study of any designed electrochemical construction. They are also taken into account in the stage-to-stage development of a mathematical model including [11]:

systematization of physico-chemical data;
analysis of separate steps of electrolysis (macrokinetic study of processes occurring in the electrolyte and diaphragm; study of electrode processes; determination of material and heat balances, also of voltage balance and current density distribution; study of hydrodynamic factors);
development of algorithms for computing mass and heat flow rates, voltage and current balance;
design of computer programs;
input information specification (specification of design objectives and characterization of construction elements);
computer-aided computations and data processing.

3.8.1 Thermodynamic and Transport Properties of NaCl-NaOH-H$_2$O

This Section is an introduction to a theory of secondary chlorine-alkali mixing processes by diaphragm electrolysis method. Since the filtering diaphragm separating the catholyte and analyte solutions has a small thickness, it develops high compositional gradients. Consequently, apart from convective and migration components, one must take into account diffusion fluxes, which leads to a problem of determining the concentrational dependence for all the coefficients of Eq. (1). For ternary chloride-alkali solutions with a common ion, the fluxes, expressed through concentrational driving forces, take the form [21]:

$$\vec{N}_1 = -t_1^o \frac{\vec{i}}{F} + c_1 \vec{v}_0 - D_{11}^o \nabla c_1 - D_{12}^o \nabla c_2$$

$$\vec{N}_2 = -t_2^o \frac{\vec{i}}{F} + c_2 \vec{v}_0 - D_{21}^o \nabla c_1 - D_{22}^o \nabla c_2 \quad (60)$$

$$\vec{N}_3 = t_3^o \frac{\vec{i}}{F} + c_3 \vec{v}_0 - (D_{11}^o + D_{21}^o)\nabla c_1 - (D_{12}^o + D_{22}^o)\nabla c_2$$

(v_0 is the solvent velocity; for other notations, see Sec. 3.2.2).

Unfortunately, the available data on the properties of solutions at elevated temperatures at which the electrolysis is commonly carried out are rather scarce, and for this reason a full description of the system of interest will be given only for 25°C. As shown below (see Sec. 3.8.3), this does not impose a major constraint from the standpoint of modelling the reactor as a whole. Calculations, performed by using the exact equations at 25°C, allow estimating the adequacy of the approximative models whose parameters are also amenable to assessment at higher temperatures.

If there are the activity coefficients available, we can pass from thermodynamic driving forces to concentrational ones. In accordance with the Pitzer model [143-145], the following expressions for activity coefficients of chloride (γ_1) and alkali (γ_2), and for the practical osmotic coefficient φ can be written in terms of the total molality $m = m_1 + m_2$ and chloride mole fraction $x = m_1/m$:

$$\varphi = 1 - 0.392 \frac{\sqrt{m}}{1 + 1.2\sqrt{m}} + m[xB_1^\varphi + (1-x)B_2^\varphi + x(1-x)\theta]$$

$$+ m^2[xC_1^\varphi + (1-x)C_2^\varphi + x(1-x)\psi] \quad (61)$$

$$\ln \gamma_1 = -0.392 \left[\frac{\sqrt{m}}{1 + 1.2\sqrt{m}} + \frac{2}{1.2} \ln(1 + 1.2\sqrt{m}) \right]$$

$$+ m[B_1^\gamma + (1-x)(B_2^\varphi - B_1^\varphi + \theta)]$$
$$+ m^2[1.5C_1^\varphi + (1-x)(C_2^\varphi - C_1^\varphi + 0.5\psi) + 0.5x(1-x)\psi] \quad (62)$$

$$\ln \gamma_2 = -0.392 \left[\frac{\sqrt{m}}{1 + 1.2\sqrt{m}} + \frac{2}{1.2} \ln(1 + 1.2\sqrt{m}) \right]$$
$$+ m[B_2^{\gamma} + x(B_1^{\varphi} - B_2^{\varphi} + \theta)]$$
$$+ m^2[1.5 C_2^{\varphi} + x(C_1^{\varphi} - C_2^{\varphi} + 0.5\psi) + 0.5x(1-x)\psi] \tag{63}$$

where

$$B_1^{\varphi} = \beta_1^{(0)} + \beta_1^{(1)} \exp(-2\sqrt{m}) \tag{64}$$
$$B_2^{\varphi} = \beta_2^{(0)} + \beta_2^{(1)} \exp(-2\sqrt{m}) \tag{65}$$
$$B_1^{\gamma} = 2\beta_1^{(0)} + \frac{1}{2} \frac{\beta_1^{(1)}}{m} \{1 - (1 - 2\sqrt{m} - 2m) \exp(-2\sqrt{m}\} \tag{66}$$
$$B_2^{\gamma} = 2\beta_2^{(0)} + \frac{1}{2} \frac{\beta_2^{(1)}}{m} \{1 - (1 - 2\sqrt{m} - 2m) \exp(-2\sqrt{m}\} \tag{67}$$

The coefficients β and C characterize the properties of binary solutions. According to the data in [144], for NaCl, $\beta_1^{(0)} = 0.0765$, $\beta_2^{(1)} = 0.2664$, $C_1^{\varphi} = 0.00127$, and for NaOH, $\beta_2^{(0)} = 0.0864$, $\beta_2^{(1)} = 0.253$, $C_2^{\varphi} = 0.0044$. The coefficients θ and ψ characterize interactions in the ternary system and must be regarded as fitting parameters. In Ref. [145], the values $\theta = -0.050$ and $\psi = -0.006$ were given as estimated from the reported results on e.m.f. measurements for a circuit containing hydrogen and silver-chloride electrodes [146]. However, since those measurements had been made on solutions with $m_2 = 0.01$ and $m_1 < 0.3$, we were of opinion that supplementary data on solutions of higher concentrations were needed. To this end, we used potentiometric data borrowed from [147, 148] which were obtained by techniques combining hydrogen, amalgam, and silver-chloride electrodes in chloride-alkaline solutions of different composition. The original results were represented as isomolal series with variable x: γ_1 and γ_2 at $m = 0.5$ and 1.0 [147], $\gamma_2' = \gamma_2 \sqrt{a_w}$ at $m = 3$ and 5 [148] (a_w is the water activity). These results have been supplemented by reference handbook data on NaCl solubility in aqueous NaOH solution; the NaCl activity coefficient along the saturation line at saturation concentration m_1^* can be calculated by the formula

$$\gamma_1 = \exp\left(\frac{\mu_{\text{NaCl}}^{\ominus}(s) - \mu_{\text{NaCl}}^{\ominus}(aq)}{2RT} \right) \bigg/ \sqrt{m_1^*(m_1^* + m_2)} \tag{68}$$

At 25°C, the standard potentials are: for the solid salt, $\mu_{\text{NaCl}}^{\ominus}(s) = -384.04$ kJ/mol; and for the solute salt, $\mu_{\text{NaCl}}^{\ominus}(aq) = -393.04$ kJ/mol. Parameters θ and ψ have been determined by the least-squares method, comparing the experimental γ_1, γ_2, γ_2' and those calculated by Pitzer equations. The parameters obtained are: $\theta = -0.0546$ and $\psi = -0.0066$. The advantage of this refinement for activity calculations is demonstrated by Table 3.1. Given in this Table are the relative maximum errors of calculated activity coefficients estimated in comparison to the corresponding experimental series.

Table 3.1 Relative Errors (%) for Calculated Activity Coefficients of Chloride-Alkali Solutions at 25°C

Source	$\theta = -0.500$, $\psi = -0.0060$			$\theta = -0.0546$, $\psi = -0.0066$		
	max $|\delta\gamma_1|$	max $|\delta\gamma_2|$	max $|\delta\gamma_2'|$	max $|\delta\gamma_1|$	max $|\delta\gamma_2|$	max $|\delta\gamma_2'|$
[147] $m = 0.5$	1.8	1.1	—	2.1	1.1	—
[147] $m = 1.0$	2.5	1.9	—	2.9	2.4	—
[148] $m = 3$	—	—	4.1	—	—	3.8
[148] $m = 5$	—	—	6.7	—	—	5.8
Saturation line ($m_2 \leqslant 11.5$)	14.2 ($m_2 \leqslant 11.5$)	—	—	8.3 ($m_2 \leqslant 11.5$)	—	—

The Sources of Experimental Data and an Approximation to Transport Properties of Binary Solutions. Given the values of λ, t°, and D^V as a function of concentration (see Sec. 3.2.2) and the multiplier $(1 + d \ln \gamma_\pm / d \ln m)$ calculated by formulas (61)-(67) at $x = 0$ and $x = 1$, one can proceed to specifying the characteristics A^*_{++}, A^*_{+-}, A^*_{--} and B^*_{++}, B^*_{+-}, B^*_{--} for binary chloride and alkali solutions, respectively.

(a) *NaCl solution.* Out of the wealth of available experimental data, only those obtained for concentrations not exceeding 3 mol/dm³ [21] have been used for calculating the Onsager coefficients.

The data on electric conductance are available up to a concentration of 5.4 mol/dm³ [149]; for approximating λ (ohm^{-1} equiv^{-1} cm²), the following formula can be used:

$$\lambda = \begin{cases} 0.09885\ m(1 - 0.1443\ m + 0.008107\ m^2)/c, & \text{at } m > 0.7 \\ 126.3 - 85.01\sqrt{m} + 85.39\ m - 42.41\ m\sqrt{m}, & \text{at } m \leqslant 0.7 \end{cases} \quad (69)$$

The diffusion coefficients for $c \leqslant 5$ mol/dm^3 have been given in Refs. [25, 150]; approximation to the quantity $\hat{D} = 10^5 D^V$ (cm^2/s) takes the form

$$\hat{D} = \begin{cases} 1.464 + 0.005816\ c + 0.01151\ c^2 - 0.001378\ c^3, & \text{at } c > 0.3 \\ 1.61 - 0.7785\sqrt{c} + 1.61\ c - 1.177\ c\sqrt{c}, & \text{at } c \leqslant 0.3\ (\text{mol/dm}^3) \end{cases} \quad (70)$$

The transference number t_+^o has been borrowed from Ref. [151]; its approximation as a function of m and temperature, t °C, has been given in [3]:

$$\begin{aligned} t_+^o &= 0.3754377 + 4.864677 \times 10^{-4}\ (t\ °C) \\ &+ [1.455249 \times 10^{-4}\ (t\ °C) - 0.03151785]m \\ &+ [0.011552 - 3.588446 \times 10^{-5}\ (t\ °C)]m^2 \\ &+ [3.044053 \times 10^{-6}\ (t\ °C) - 1.297191 \times 10^{-3}]m^3 \end{aligned} \quad (71)$$

for a concentration region $c \leqslant 5$ mol/dm^3.

(b) *NaOH solution*. The electric conductance up to a concentration of 12 mol/dm^3 [149, 152] is described by the expression

$$\lambda = \begin{cases} \dfrac{1478 - 104.56\ c}{6.572 + c}, & \text{at } c > 0.5\ \text{mol/dm}^3 \\ 245 - 79.59\sqrt{c} + 4.08\ c + 1495\ c\sqrt{c}, & \text{at } c \leqslant 0.5\ \text{mol/dm}^3 \end{cases} \quad (72)$$

The diffusion coefficient data are available only for concentrations $c \leqslant 3.9$ mol/dm^3 [149]; however, for KOH solution, they extend over a larger concentration range, up to 10 mol/dm^3 [153]. Also in [153], a semiempirical relationship was proposed based on the generalized Eyring theory of activated transition state, with the diffusion visualized as a replacement of solvent molecules by solute molecules. Owing to a basically similar behaviour of the alkali solutions studied, it is presumed that, in extending the NaOH properties to a higher-concentration region, the results from [153] may be used:

$$D^V = D^\infty(1 + 0.018\ m)\left(1 + \frac{d\ln \gamma_\pm}{d\ln m}\right)\exp\left(-\frac{0.018\ m}{1 + 0.036\ m}\frac{A}{RT}\right) \quad (73)$$

The diffusion coefficient at infinite dilution is $D^\infty = 2.1 \times 10^{-5}$ cm^2/s. The adjustable parameter A has been determined by the least squares method using the experimental data from [149] and found to be $A = 33.2463$ kJ/mol.

The transference numbers t_-^o up to a concentration of 10 mol/dm^3 within a 18-25°C temperature range were reported in the works [152, 154-157];

however, the accuracy of measurements was not high and, at $c \geq 2$ mol/dm³, could amount to 0.05-0.1. A mean least squares treatment of these data has yielded the relationship

$$t_-^o = 0.8088 + 0.04864\, c - 0.004295\, c^2 \tag{74}$$

Chloride-Alkali Solutions. The transport properties have been calculated according to Miller's LN-approximation (see Sec. 3.2.2) using the formulas for conversion from the Onsager coefficients to the equivalent conductance λ, transference numbers t_1^o, t_2^o, and diffusion coefficients D_{ij}^o borrowed from [21]. The chemical potential derivatives with respect to composition can be calculated using expressions (62)-(67). Finally, in determining the mixed-solution density, the approximation procedure as reported in [3] has been used; its applicability at 25°C has been checked using the known reference data on the density of cell liquors [158] and saturated solutions [152].

The results are convenient to represent in terms of relative quantities defined in the following manner. At infinite dilution, the limiting equivalent ionic conductances are $\lambda_1^\infty = 76.35$, $\lambda_2^\infty = 198.3$, $\lambda_3^\infty = 50.1$ ohm^{-1} equiv^{-1} cm². For the solution as a whole, $\lambda^\infty = \lambda_1^\infty x_1 + \lambda_2^\infty x_2 + \lambda_3^\infty$, where $x_1 = c_1/c$, $x_2 = c_2/c$, $c = c_1 + c_2$. For the coefficients in Eqs. (60), we have

$$t_i^\infty = x_i \lambda_i^\infty / \lambda^\infty, \quad (i = 1, 2) \tag{75}$$

$$D_{11}^\infty = \frac{RT}{F^2} \lambda_1^\infty [(1 + x_1)\lambda_3^\infty + x_2 \lambda_2^\infty]/\lambda^\infty \tag{76}$$

$$D_{12}^\infty = \frac{RT}{F^2} x_1 \lambda_1^\infty (\lambda_3^\infty - \lambda_2^\infty)/\lambda^\infty \tag{77}$$

$$D_{21}^\infty = \frac{RT}{F^2} x_2 \lambda_2^\infty (\lambda_3^\infty - \lambda_1^\infty)/\lambda^\infty \tag{78}$$

$$D_{22}^\infty = \frac{RT}{F^2} \lambda_2^\infty [x_1 \lambda_1^\infty + (1 + x_2)\lambda_3^\infty]/\lambda^\infty \tag{79}$$

One may use, as a possible approximation, simple formulas relating the diffusion coefficients and transference numbers in solutions of finite concentration with the same mole ratios x_1, x_2 to those of infinitely dilute solutions:

$$\overline{D_{ij}^o} = \frac{\lambda}{\lambda^\infty} D_{ij}^\infty, \quad \overline{t_i^o} = t_i^\infty \tag{80}$$

We have calculated, starting from the properties of binary solutions, dimensionless quantities defined as $f^\lambda = \lambda/\lambda^\infty$, $f_1^t = t_1^o/t_1^\infty$, $f_2^t = t_2^o/t_2^\infty$, $f_{ij}^D = D_{ij}^o/\overline{D_{ij}^o}$ ($ij = 1, 2$). The calculations have been carried out for concentrations

Table 3.2 Regression Coefficients for Transport Properties of Chloride-Alkali Solutions

	Property						
	$f^\varkappa$	f_1^t	f_2^t	f_{11}^D	f_{12}^D	f_{21}^D	f_{22}^D
a	−24.31	1.34	20.85	5.726	3.45	29.32	0
b	102.196	−10.6368	556.702	−9.45	−10.5	−9.8415	0
b_1	0	0	0	0	0	0	2.027
b_2	0	0	0	0	0	0	−1.116
a_1	−121.4	−2.924	65	157.7	−65.8	−234.2	96.02
a_2	−120.1	−73.45	52.68	−136.1	309.4	−936.9	122.8
a_{11}	8139	−11570	−14880	−7435	−17170	−338500	−18700
a_{12}	11080	−2461	−15940	−11290	31330	−409400	14070
a_{22}	5732	8840	−4585	−1264	−110500	−28680	14400
a_{111}	0	0	0	0	-5227×10^3	2114×10^4	−106200
a_{112}	0	0	0	0	-2526×10^4	2478×10^4	-1621×10^4
a_{122}	0	0	0	0	7639×10^3	-4101×10^4	-4686×10^3
a_{222}	0	0	0	0	2145×10^4	-2392×10^4	3281×10^3
	Approximation error (%)						
	4	7	4	7	10	—	6

$c \leqslant 7$ mol/dm^3, which in fact covers the concentration range of working electrolyte solutions. At need, as required by the LN-approximation (see formulas (10), (12), and (13)), the concentration range could be extended beyond the NaCl solubility limit to an oversaturation region by extrapolating formally expressions (69)-(71). In the final analysis, it is only a comparison with experimental properties that can lend support to the correctness of this technique. Any of the calculated dimensionless characteristics can be represented by the formula below (with c_1 and c_2 expressed in mol/cm^3):

$$f = 1 + \frac{a\sqrt{c}}{1 + b\sqrt{c}} + b_1\sqrt{c_1} + b_2\sqrt{c_2} + a_1 c_1 + a_2 c_2 + a_{11} c_1^2 \\ + a_{12} c_1 c_2 + a_{22} c_2^2 + a_{111} c_1^3 + a_{112} c_1^2 c_2 + a_{122} c_1 c_2^2 + a_{222} c_2^3 \quad (81)$$

The regression coefficients are listed in Table 3.2. The approximation error is indicated at the bottom for each of the functions, except f_{21}^D. Function f_{21}^D, within a definite argument region, changes the positive sign to the negative on passing from low to high concentrations, which may cause relatively large errors. However, this is not expected to interfere significantly with the mass-transport calculations, since in most cases, the coefficient D_{21}^o is smaller than the other diffusion coefficients, the more so in the region of sign reversal, where $D_{21}^o \to 0$. With the exception of this particular case, the error range remains much the same as shown in Table 3.2 for the other characteristics. It should be pointed out that the largest errors are associated with dilute or binary solutions (that is, as $c_1 \to 0$ or $c_2 \to 0$). In mixed solutions of moderate and high concentrations, the errors are, on the average, smaller by a factor of 1.5-3 than those in Table 3.2.

Experimental data on transport properties of chloride-alkali solutions are quite scarce in the literature. Measurements of the cell liquor conductance were carried out by Angel on four compositions [159]. The approximation errors showed a 5 to 11% overestimation relative to the calculated values. In [160], the results for conductance of solutions obtained by electrolysis of aqueous 290, 300, and 310 g/dm^3 sodium chloride at different degrees of decomposition (neglecting water vaporization losses during electrolysis) have been reported. The discrepancy between these data and the calculated results was not higher than 3.9%. Finally, three measured electroconductances in Ref. [157] have been approximated by a regression equation to an accuracy better than 4.5%. The transference numbers t_2^o (a total of 8), that we calculated to an accuracy better than 6%, have also been reported in [157] for cell liquors. As to other properties, the diffusion coefficients for hydroxyl ions in a supporting NaCl solution were measured by capillary tube method in [157]. A strict interpretation of the results in terms of D_{22}^o coefficients was valid only in the limit of $c_2 \to 0$; therefore, out of the four compositions dealt with in [157] only one, namely, that with $c_1 = 4.34$ mol/dm^3, $c_2 = 0.24$ mol/dm^3, should be taken into consideration. The experimental coefficient $D_{22}^o = 2.58 \times 10^{-5}$ cm^2/s has been fitted by the above regression to an accuracy of 0.6%. Considering that the measurement error was about 7%, this agreement should be recognized as exceptionally good.

3.8.2 Macrokinetics of Chlorine Dissolution in Brine

Interactions of chlorine in salt solutions have been extensively studied, which was necessitated by a wide-spread use of bubble-type cells in commercial electrolysis with no metal electrodeposition. The disperse phase in this technique is composed of chlorine gas bubbles mixed to solvent vapours. The dissolution rate determines also the rates of side reactions, the current efficiency for the formation of alkali and chlorine, and the accumulation of chloro-oxide contaminants in electrolyte solution.

In modelling a chlorine cell, the problem of quantitative expression for anolyte chlorine and hypochlorous acid compositions is solved to a first approximation by using the thermodynamic equilibrium relations between the gaseous and liquid phase components and the equilibrium hydrolysis reaction [11]. However, as evidenced by the data in Ref. [161], in a chlorine-alkali horizontal-diaphragm cell, the best agreement between the calculated and experimental concentrations of active chlorine and sodium chlorate is observed under the conditions of insignificant chlorate formations when the cell performance parameters are close to the nominal. As the alkali concentration c_{alk} is raised to 160-170 g/cm^3, the experimental values tend to become lower than those calculated. Commercial electrolytic cells of the same type always exhibit a certain spread in performance characteristics, which may be manifested by increased chlorate contents in certain baths. Deviations from the nominal characteristics may occur under nonstationary conditions due to a sudden change in current load or in brine feed. The eventual interference of these factors has stimulated a deeper study of certain theoretical aspects of chlorine dissolution in weakly acidic solutions.

The higher the active chlorine concentration in the anodic space of the cell, the shorter is the characteristic time for reaction of sodium chlorate formation. At pH $\geq$ 4.5, this time becomes comparable to the time for passage of chlorine from the gas-bubble space into the bulk of the solution as the process is being carried out in a stirred bubbling cell. With increasing pH, the consumption of active chlorine to produce NaClO$_3$ will not only decrease the growing concentration of active chlorine, but even result in its drop as soon as the maximum point is reached.

Chlorine dissolution is attended by reversible reactions of hydrolysis and dissociation:

$$Cl_2 + H_2O \rightleftarrows HClO + H^+ + Cl^- \tag{82}$$

$$HClO \rightleftarrows H^+ + ClO^- \tag{83}$$

and also by the irreversible decomposition of hypochlorite:

$$2HClO + ClO^- \rightarrow ClO_3^- + 2H^+ + 2Cl^- \tag{84}$$

$$HClO + 2ClO^- \rightarrow ClO_3^- + 2Cl^- + H^+ \tag{85}$$

The rate constants for reactions (84) and (85), K_{dec}, are quite close. The quantitative contribution due to reaction (85) becomes sizeable at pH $\sim$ 6.

We denote by c'_1, c'_2, and c'_3 the respective concentrations of Cl_2, $HClO$, and ClO^- in solution, and by γ, the ratio of liquid volume to interfacial area. The driving force in the process is the difference between the thermodynamically equilibrated Cl_2 concentration, c'_{1s} (corresponding to the partial pressure of chlorine in the gas phase, p_{Cl_2}), and the volume concentration c'_1 at a given time. We introduce the mass transport coefficient k and the rate constants for reactions (82) and (83); thus, the balance equations for solution components are written in the form:

$$\frac{dc'_1}{dt} = \frac{k}{\gamma}(c'_{1s} - c'_1) - (k_1 c'_1 - k_2 c'_2) \tag{86}$$

$$\frac{dc'_2}{dt} = (k_1 c'_1 - k_2 c'_2) - (l_2 c'_2 - l_3 c'_3) - 2K_{dec}(c'_2)^2 c'_3 - K_{dec} c'_2 (c'_3)^2$$

$$\frac{dc'_3}{dt} = (l_2 c'_2 - l_3 c'_3) - K_{dec}(c'_2)^2 c'_3 - 2K_{dec} c'_2 (c'_3)^2$$

Summing up, we obtain an equation for the active chlorine concentration, $c' = c'_1 + c'_2 + c'_3$:

$$\frac{dc'}{dt} = \frac{k}{\gamma}(c'_{1s} - c'_1) - 3K_{dec}[(c'_2)^2 c'_3 + c'_2 (c'_3)^2] \tag{87}$$

Our objective is to determine the mass-transport coefficient on the basis of the known experimental relationships for the stationary concentration of active chlorine [162]. Since the hydrolysis and dissociation reactions are fast, it may be assumed that the concentration proportions of species Cl_2, $HClO$, and ClO^- in the liquid bulk are the same as those under the conditions of a complete gas-liquid equilibrium at given electrolyte pH. Otherwise stated, the proportion $\dfrac{c'_1}{c'_{1s}} = \dfrac{c'_2}{c'_{2s}} = \dfrac{c'_3}{c'_{3s}} = \dfrac{c'}{c'_s} = x$ is a quantitative measure for deviation of the solution composition from the thermodynamically equilibrated concentration c'_{is}. In a dimensionless form, we have the expression

$$\frac{dx}{dt} = \frac{k}{\gamma}\frac{c'_{1s}}{c'_s}(1-x) - 3K_{dec}x^3 \frac{1}{c'_s}[(c'_{2s})^2 c'_{3s} + c'_{2s}(c'_{3s})^2] \tag{88}$$

in place of Eq. (87).

Under stationary conditions, the equation

$$x^3 + A(x - 1) = 0 \tag{89}$$

where $A = kc'_{1s}/[3\gamma K_{dec}\{(c'_{2s})^2 c'_{3s} + c'_{2s}(c'_{3s})^2\}]$ has a single root within an interval $0 \leqslant x \leqslant 1$:

$$x_0 = \left[\frac{1}{3} + \left(\sqrt{\frac{1}{4} + \frac{1}{27}A} + \frac{1}{2}\right)^{2/3}\bigg/A^{1/3} + \left(\sqrt{\frac{1}{4} + \frac{1}{27}A} - \frac{1}{2}\right)^{2/3}\bigg/A^{1/3}\right]^{-1} \quad (90)$$

The volume concentration of active chlorine is defined as $c' = c'_s x_0$. The quantity c'_s increases monotonically with pH, whereas the quantities A and x_0 tend to decrease; consequently, the product of the two latter quantities can be a nonmonotonic function of pH.

Since the specific surface of a gas-liquid contact is not, to our knowledge, accessible by experiment, the quantity we have estimated is in fact a rate constant $K = k/\gamma$. To obtain A, we used the most reliable literature data on equilibrium properties of the system chlorine-solution [163, 164]. The rate constant K_{dec} at 40°C was assumed, in accordance with [165], to be equal to 0.2 (dm^6)/(mol^2 s). At temperatures above 70°C, a formula from [3], was used,

$$K_{dec} = 5.555 \times 10^{-3} \exp(0.6 + 0.056 \, t \, °C) \quad (91)$$

In Fig. 3.1, the calculated curves (shown in solid lines) are compared with the experimental concentrations of active chlorine (circles) for a stirred bub-

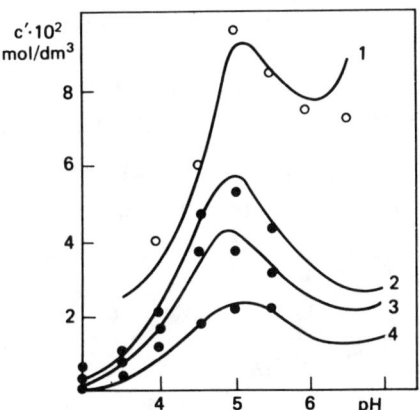

Fig. 3.1 Stationary concentration of active chlorine as a function of solution pH:

1, $K = 1 \cdot 10^{-3} \, c^{-1}$, 40°C, 150 g/dm^3 NaCl;
2, $K = 2 \cdot 10^{-3} \, c^{-1}$, 75°C, 270 g/dm^3 NaCl;
3, $K = 3 \cdot 10^{-3} \, c^{-1}$, 85°C, 270 g/dm^3 NaCl;
4, $K = 3 \cdot 10^{-3} \, c^{-1}$, 95°C, 270 g/dm^3 NaCl

bling cell. The values of K have been calculated by the mean least squares method for a range of $3 \leq \text{pH} \leq 6$. A satisfactory agreement with the experimental measurements at 40°C and 150 d/dm³ NaCl has been arrived at, starting from the equilibrium relationship in [163]. In high-temperature experiments ($t > 70°C$, 270 g/dm³ NaCl), the best agreement has been obtained using the data in [164] within a range of $4 \leq \text{pH} \leq 5.5$. In the case of pH ≤ 3.5, a somewhat better agreement has been achieved with the results reported in [163]. On passing through a maximum of chlorate yield at pH ≈ 6.5, the free chlorine concentration is observed to increase.

With a view to further verification of the model, formula (90) has been used in calculating the side reaction characteristics in a diaphragm-type cell. The best agreement with the commercial electrolyzer control analysis data for active chlorine and sodium chlorate contents in the anolyte has been obtained using the relationships in Ref. [164] and a value of $K = 2.5 \times 10^{-3} \text{s}^{-1}$. A good agreement with the K's calculated for independent experimental conditions lends support to the proposed mass-transport model of the gaseous chlorine dispersed in a NaCl solution. To make the picture complete, we give below correlations for the data tabulated in Ref. [164], that were used here with allowance made for nonequilibrium dissolution effects at temperatures above 70°C ($p_{Cl_2} = 1$ atm, NaCl concentration in mol/dm³):

$$\log c'_{1s} = \frac{1748}{t\,°C + 273.16} - 0.1189[\text{NaCl}] - 7.073 \tag{92}$$

$$\log c'_{2s} = \text{pH} - 0.25 \times 10^{-2} t\,°C - 0.248[\text{NaCl}] - 4.34 \tag{93}$$

$$\log c'_{3s} = 2\text{pH} + 0.45 \times 10^{-2} t\,°C - 0.24[\text{NaCl}] - 11.63 \tag{94}$$

Let us consider a nonstationary dissolution. Neglecting the eventual decomposition of hypochlorite at pH < 4.5, we find by Eq. (88)

$$x = 1 + B \exp\left(-K \frac{c'_{1s}}{c'_s} t\right) \tag{95}$$

where B is a constant dependent on the initial conditions. The characteristic dissolution time is estimated as $t_d \sim K^{-1} \cdot c'_s / c'_{1s}$, and its value is greater over the physical adsorption time $t_d^0 \sim K^{-1}$. Indeed, an increase in t_d with rising pH was observed occasionally in a number of experiments. Assuming the process to occur in the vicinity of a stationary state, we write a solution to the linearized equation (88):

$$x = x_0 + B \exp\left[-K \frac{c'_{1s}}{c'_s}\left(1 + 3\frac{x_0^2}{A}\right) t\right] \tag{96}$$

whence $t_d \sim K^{-1} \dfrac{c'_s}{c'_{1s}} \left(1 + 3\dfrac{x_0^2}{A}\right)^{-1}$. A thermodynamic equilibrium is reached as $A \to \infty$, $x_0 \to 1$, and the solution (96) reduces to Eq. (95). In the case of strongly nonequilibrium conditions, for example, at pH > 5, we may set $A \to 0$, $x_0 \to 0$ and obtain, on the basis of Eq. (89), another estimate $t_d \sim \dfrac{1}{3} K^{-1} \dfrac{c'_s}{c'_{1s}} A^{1/3}$. These expressions show that chlorate formation favours the tendency to diminution of t_d.

Thus, the model that has been proposed in this Section nicely agrees with the available experimental results and provides a more reliable quantitation of the side reactions occurring in chlorine cells.

3.8.3 Current Efficiency Theory for Diaphragm-Type Cell

According to the experimental relationship reported by Angel [159], the alkali current efficiency initially rises with catholytic alkali concentration as high as 120-160 g/dm³ (depending on the cell construction and electrolysis conditions) and then, on passing a maximum, sharply decreases. Simultaneously, the chlorate concentration is increased in the cell liquor. At low alkali concentration, the accumulation of hypochlorite is possible.

Theoretical calculations were attempted on a number of occasions, but they lacked generality and argumentation and could not provide an adequate description of an electrochemical process with parameters varied within a wide range. A point to be noted is that the tradiditonal "current efficiency theory" is concerned with calculations of the current efficiency B_{OH^-} for the formation of alkali at given catholyte and anolyte compositions and, in principle, must be included, as an essential stage of side-reaction calculations, within a generalized algorithm. In some detail, this is shown below.

In Ref. [149], procedures for estimating the alkali losses due to convective transport of chlorine and acid at low alkali concentrations have been systemized. In that particular case, the current efficiency is expressed by the formula

$$B_{OH^-} = \dfrac{c^{cat}_{OH^-}}{c^{cat}_{OH^-} + R^{an}(c^{an}_{Cl_2+HClO} + c^{an}_{H^+})} - E_{H^+} - D - K \qquad (97)$$

Here $c^{cat}_{OH^-}$ is the catholyte alkali concentration, mol/cm³; $c^{an}_{Cl_2+HClO}$ is the concentration of chlorine and hypochloric acid in solution; $c^{an}_{H^+}$ is the concentration of H⁺ ions in the anolyte; R^{an} is the anolyte volume contraction coefficient [149]; E_{H^+} is the current efficiency loss due to hydrogen ion migration; and D and K are, respectively, the current efficiency losses due to the diffusion of chlorine and acid and to the mechanical agitation of the solution. At low alkali concentrations, with the current efficiency determined by chlorine solubility, the quantities D and K may be neglected, with no loss in the predictive value of the above formula. An increase in $c_{OH^-}^{cat}$ leads to a sizeable decrease in B_{OH^-} which should be attributed to the migration of hydroxyl ions into the anodic space. In this case, the loss in D due to the diffusion of chlorine and hypochloric acid increases sharply; considering that the loss estimation is difficult to perform, the utility of relation (97) appears thus to be somewhat uncertain. On the other hand, the alkali transport on the alkaline side of the diaphragm can be analyzed.

Quite a number of theoretical models were suggested treating specific aspects of the hydroxyl ion transport into the anodic space (see the literature reviewed in the papers [11, 166, 167]). A a rule, preference was given to a single driving force involved in transport, for example, concentrational or electromigratory, and attempts were made to relate this force to the anolyte filtration rate across the diaphragm. The relationships thus obtained were in a qualitative accord with the general trend of the current efficiency diminishing with alkali concentration, but failed to provide a satisfactory quantitative correlation within wider concentration ranges. A typical presumption in allowing for the simultaneous involvement of diffusion, electromigration, and countercurrent was the validity of Ohm's law in the diaphragm region with high concentration gradients and different characteristics for current conduction in solution. Such an approach allowed treating the hydroxyl ion transport independent of the transport characteristics for Na^+ and Cl^- ions present in concentrations comparable to that of OH^- ions. However, it can be shown that the potential drop across the diaphragm, $d\varphi_{ohm}/dx$, calculated at the catholyte interface with allowance made only for the ohmic component, is, under typical conditions, about twice as large as the $d\varphi/dx$ calculated by more strict equations including the diffusion potential. Since the electric field strength is the driving force for ion transport, we come to a conclusion on the nonapplicability of Ohm's law to this particular system.

Another drawback inherent in most models was the arbitrary assessment of the alkaline-reaction zone width which in most cases was assumed to be identical with the diaphragm thickness. However, in a real process under the conditions of maximum current efficiency, only part of the diaphgram shows alkaline reaction. At present, the most consistent appears to be the method for concordant solution of a mass-transport problem in two regions within the diaphragm: alkaline region (species Na^+, Cl^-, OH^-; $x < f$) and acidic region (species Cl_2, HClO, H^+; $x > f$), first put forward in [168]. At the interface f, the ion OH^- reacts with the dissolved chlorine and acid:

$$2OH^- + Cl_2 \rightarrow ClO^- + Cl^- + H_2O$$
$$OH^- + HClO \rightarrow ClO^- + H_2O \qquad (98)$$
$$OH^- + H^+ \rightarrow H_2O$$

The equilibrium of these reactions is shifted towards the formation of hypochlorite ion and water, and the kinetic neutralization time, as has been estimated, is not greater than $\approx 10^{-3}$ s, which allows regarding the reactions as instantaneous, with the reactant concentrations equal to zero at $x = f$. The flux of OH^- ions towards the anode is determined by solving a mass-transport problem for the specified regions under the known boundary conditions (anolyte and catholyte compositions) and flux balance:

$$N_{OH^-} + 2N_{Cl_2(soln)} + N_{HClO} + N_{H^+} = 0$$

The anolyte and catholyte compositions with respect to chloride and alkali are determined from the cell material balance for these species. The concentrations for dissolved chlorine and hypochlorous acid as functions of pH are treated by the theory of delayed mass transport of chlorine from the gas bubbles (Sec. 3.8.2). Within the framework of "dusty gas" model (Sec. 3.7.2), the role of a porous diaphragm is merely to reduce the mass transport, which is characterized by a factor of f_ε^{-1}. Indeed, the available experimental data and the experience with industrial electrolyzers provide no evidence for an explicit correlation between the current efficiency and the chemical composition of diaphragm material. In view of a poor knowledge of the diaphragm structure under electrolysis conditions, the factor f_ε is actually regarded as a fitting parameter in the current efficiency theory.

In the alkaline region, Eqs. (60) must be solved with the current density $\vec{i}$ and the solvent filtration velocity $\vec{v}_0$ known for a given part of the diaphragm. In the acidic-reaction zone, the fluxes can be estimated starting from the solution for a one-dimensional diffusion problem (in the x-direction across the diaphragm). Given the diaphragm thickness d, the x-components of both the solvent velocity, v_{0x}, and the current density, i_x, are assumed to be known at the anolyte interface, that is, at $x = d$. Thus, we obtain

$$N_i = \frac{[v_{0x}(d) + (z_i D_i F/RT\varkappa)i_x(d)]c_i^{an}}{\exp\{[v_{0x}(d) + (z_i D_i F/RT\varkappa)i_x(d)][f(y) - d]/D_i\} - 1} \qquad (99)$$

with the subscript i denoting species Cl_2, $HClO$, and H^+. Here $\varkappa$ is the effective conductivity of the solution, and D_i is the effective diffusion coefficient. It has been assumed that, in the general case, the location of the neutralization zone boundary is determined by the longitudinal coordinate y.

The above model allows calculating the back migration of OH^- ions, given the anolyte pH value. Considering, however, that the value of pH is, for its part, dependent on the technology, one is confronted with the necessity of developing a generalized algorithm for calculating the side reactions in the electrolytic cell; conceptually, this algorithm must include the conventional "current efficiency theory" as its constitutive part.

To this end, the overall electricity balance is made up for the cell (the sum of the current efficiencies for the formation of chlorine and side products is taken equal to unity):

$$B_{Cl_2(g)} + B_{O_2} + B_{ClO^-} + B_{ClO_3^-} = 1 \tag{100}$$

Taking into account that the current efficiencies for the formation of alkali and chlorine satisfy the acid balance in the cell, $B_{OH^-} = B_{Cl_2(g)} + b_0$ (b_0 is the input flow of alkali or acid fed in the brine, expressed in fraction of the overall current; with HCl fed in, $b_0 < 0$), we obtain

$$B_{OH^-} - b_0 + B_{O_2} + B_{ClO^-} + B_{ClO_3^-} = 1 \tag{101}$$

As is seen, each of the terms in expression (101) is a unique function of anolyte pH. Indeed, since the pH-dependent concentrations $c_{Cl_2}^{an}$, c_{HClO}^{an}, and $c_{H^+}^{an}$ are the boundary conditions for a problem of diaphragm mass transport, the solution to this problem (that is, the flux N_{OH^-} and, consequently, B_{OH^-}) is also a function of pH. The current efficiency for the formation of oxygen, B_{O_2}, is determined by the properties of electrode material and can be described by an appropriate regression equation in a manner similar to that used in [3]. The current efficiency for the formation of hypochlorite, B_{ClO^-}, roughly corresponds to the convective flux of active chlorine from the anolyte. The electric-field migration correction taken into account [166], a formula for the ClO^- flux in the alkaline diaphragm zone at a given height y may be proposed:

$$N_{ClO^-}(y) = \frac{D_a F}{RT} \frac{i_x(d)}{\varkappa} \left\{ (c_{Cl_2}^{an} + c_{HClO}^{an}) e^{-\frac{v_{ox}(d)[d-f(y)]}{D_a}} \right. \\ \left. + c_{ClO^-}^{an} \right\} - v_{ox}(d)[c_{Cl_2}^{an} + c_{HClO}^{an} + c_{ClO^-}^{an}] \tag{102}$$

Here D_a is the diffusion coefficient for active chlorine in the diaphragm.

Formation of sodium chlorate in an electrolytic cell can proceed both by chemical reactions (84), (85) and by the known electrochemical mechanism [149]:

$$6ClO^- + 3H_2O \rightarrow 2ClO_3^- + 4Cl^- + 6H^+ + 3/2O_2 + 6e^- \quad (103)$$

The kinetics of this chemical process has been studied in some detail; however, the quantitative contribution of reaction (103) to the total of chlorate formed remains a controversial issue [11, 166]. In practice, even at a low concentration of the alkali product, when the anolyte pH is within a range of 2-4, that is, under complete inhibition of reactions (84) and (85), the alkali inevitably contains a certain amount of contaminant $NaClO_3$. With laboratory electrochemical cells, this might have been attributed to the eventual occurrence of stagnation zones close to the anodic diaphragm interface, unless appropriate measures for more vigorous stirring of the solution are taken. Under commercial production conditions, the residual concentration $[NaClO_3]$ within a range of 0.1-0.2 g/dm^3 in the catholyte can hardly be attributed to the above cause in view of a circulation and vigorous stirring of the solution by evolving gas bubbles in the bath. It has been shown in Ref. [161] that the formation of chlorate in any noticeable amount in a close vicinity to the neutral layer in the diaphragm is little probable. In Ref. [169], no difference in the rate of chlorate accumulation in chlorine bubbling experiments run with the current "on" and "off" has been observed, which argues for the nonoccurrence of an electrochemical reaction (103). If so, a reasonable explanation is that design faults and in-process defects of the electrode unit and diaphragm deposition cause a mechanical mixing of the electrolyte solutions, a fact pointed out in the classical monograph [149]. Owing to the same reason, the actual current efficiency, even under optimal conditions, never reaches as high as 98-99% with respect to the chlorine solubility.

However, in recent years, an attempt was made to revise these conclusions in favor of a predominant role of the electrochemical mechanism. The studies of hypochlorite solutions by UV- and Raman spectroscopies have led the authors of the papers [170-172] to a conclusion that the reaction (103) proceeds through an electrodic stage of chlorite formation over the entire pH range, from acidic to weakly alkaline solutions. The concentrational changes for both platinum and catalytic metal-oxide anodic coatings are described by a system of equations:

$$\begin{cases} \dfrac{d[ClO_2^-]}{dt} = k_1[ClO^-] - k_2[ClO_2^-] \\[2ex] \dfrac{d[ClO_3^-]}{dt} = k_2[ClO_3^-] + K_{dec}[HClO]^2[ClO^-] \end{cases} \quad (104)$$

Under stationary conditions, $[ClO_2] = k_1/k_2[ClO^-]$, and the reaction rate is a function of [HClO] and [ClO$^-$] only. We have estimated, by making use of the rate constants reported in [171], that at low pH (about 2.5), the electrochemical process yields, under diaphragm electrolysis conditions, 0.02-0.04 g/dm^3 NaClO$_3$ in the electrolytic alkali. This result agrees fairly well with independent laboratory experiments, but fails to explain the occurrence of higher concentrations under operating conditions. Simultaneously, according to Eqs. (104) at pH $\approx$ 4.5, the electrochemical reaction is predominant [171] (regrettably, no comparison was made in [171] with experimental data). In view of the conflicting experimental evidence, we deemed it justified to ignore the electrochemical process in our model. This assumption turned out quite acceptable from a practical standpoint in defining, by numerical experimentation, the operating conditions capable of initiating noticeable side reactions. To reach a complete agreement between the experimental and computed concentrations of NaClO$_3$, the "electrochemical" corrections should be applied following the procedure that has been outlined above.

Thus, in neglect of reaction (85), the current efficiency for the formation of chlorate can be calculated by the formula

$$B_{ClO_3^-} = 6FV_a K_{dec}(c_{HClO}^{an})^3 c_{ClO^-}^{an}/Si^{avg}$$

where V_a is the volume taken by the solution in anodic compartment, S is the electrode surface area, and i^{avg} is the size-average current density in the cell.

We now wish to outline a general procedure enabling one to take into account electrode construction, for the filtering rate and current density in different parts of the diaphragm. The diaphragm area is divided into N elementary areas, not necessarily equal among themselves; it is assumed that in each elementary area, the average current density i_j^{avg} and the average filtering rate v_{0j}^{avg} remain invariant. In allowing for the microinhomogeneities arising from the network structure of a mesh cathode [11], the elementary diameters must be much larger than the mesh spacing. We denote the elementary fractional areas by α_j^s, so that $\sum_{j=1}^{N} \alpha_j^s = 1$. In solving the problem of mass transport across the diaphragm for each elementary area, we find the local fluxes N_{OH^-j} and N_{ClO^-j}, and then the overall losses $N_{OH^-}^{avg} = \sum_{j=1}^{N} \alpha_j^s N_{OH^-j}$, $N_{ClO^-}^{avg} = \sum_{j=1}^{N} \alpha_j^s N_{ClO^-j}$. Subsequently, the current efficiencies are calculated, $B_{OH^-} = 1 - FN_{OH^-}^{avg}/i^{avg}$, $B_{ClO^-} = 2FN_{ClO^-}^{avg}/i^{avg}$. These are substituted, jointly with $B_{ClO_3^-}$ and B_{O_2}, into (101), which becomes thus the equation that must be solved for the anolyte pH. The equation has a unique solution within a range of $-1 < \text{pH} < 5.5$. The computational procedure has been implemented in appropriate algorithms and computer programs.

Now we should come back to the problem of transport coefficients in chloride-alkali solutions. For high temperatures, the only possibility at present is to formulate reasonably acceptable assumptions of type (80). In fact, we have ignored the D_{ij}^o coefficients and preferred to start from the ionic diffusion coefficients at infinite dilution, $D_i^\infty = RT\lambda_i^\infty/F^2$ ($i = 1, 2, 3$), to replace these subsequently by the quantities D_i^o. The latter must satisfy the following conditions: (i) they must obey, to a certain accuracy, the rule (80) for diffusion coefficients; (ii) they must agree, when manipulated by the formula for transference number conversion $t_n^o = z_n^2 D_n^o c_n / \sum z_i^2 D_i^o c_i$, with the known experimental data. As is easily seen, these requirements are in fact equivalent to the conditions for application of equations for infinitely dilute solutions, however, with the coefficients corrected for the real conductance and the transference numbers. Here the option of choosing the coefficients is somewhat larger than that for the coefficients $\overline{D_{ij}^o}$ in Sec. 3.8.1, since the inequality $t_n^o \neq t_n^\infty$ is admitted. For the typical electrolysis conditions, the ratio λ/λ^∞ is varied from 0.35 to 0.40. The transference numbers, estimated by formula (71) to a certain accuracy, are known for the chloride solution of anodic space. The temperature coefficient for the hydroxyl transference number in cell liquor has been estimated in [157] to be about -0.1%/deg. Taking into account this evidence, we have estimated at 90°C the coefficients $D_1^o = 2.34 \times 10^{-5}$, $D_2^o = 5.15 \times 10^{-5}$, and $D_3^o = 1.67 \times 10^{-5}$ cm^2/s. It is noteworthy that authors of the paper [173] have obtained, by treating the experimental data on diffusion diaphragm permeability, an effective coefficient $D_2 = 1.4 \times 10^{-5}$ cm^2/s at 90°C, which gives, with reference to $f_\varepsilon^{-1} = 3.42$-$3.62$ in the same paper, $D_2^o = (4.79$-$5.07)10^{-5}$ cm^2/s. The latter value agrees, within a few percent, with our previous estimate of this quantity.

The only fitting parameter in the "current efficiency theory", the membrane porosity-tortuosity factor $f_\varepsilon = 0.40$, has been determined by treating the available literature results on side reactions in a horizontal diaphragm-type cell [161]. The probable value of f_ε for technological baths lies within a narrow range of 0.35-0.40. Shown in Fig. 3.2 are the current efficiency and anolyte pH curves calculated using the approximation (80) and the Miller transport rule conducive to formula (81). Through adjusting the value of f_ε one can describe, with an accuracy amply sufficient for the technology conditions, the

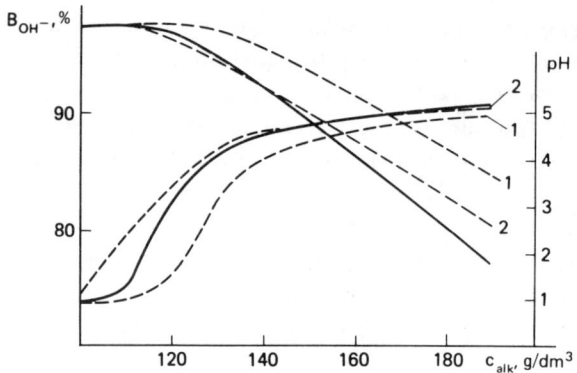

Fig. 3.2 Current efficiency B_{OH^-} and anolyte pH vs. alkali concentration c_{alk} curves for a diaphragm-type cell (d = 2.5 mm) at uniform current density i = 0.14 A/m² (25°C).

Solid lines (calculated using rule (81)): f_ε = 0.4; dashed lines (calculated using rule (80)): 1, f_ε = 0.4; 2, f_ε = 0.6

rate of side reactions by making use of the approximate estimation (80). This conclusion remains valid also as the process parameters are varied. Simultaneously, it should be inferred that the value of f_ε obtained by a fitting procedure within the above interval of 0.35-0.40 for technological baths would have been with certainty higher, had the experimental results been treated by the improved rule (81); however, a more detailed study of this problem is beyond the scope of the present work.

3.8.4 Electrolytic Cell as Electro- and Hydrodynamic System

Design faults and the lack of required precision in fabrication or assembly of an electrolytic unit or its separate parts may be the cause of a significant electrode height-scale nonuniformity in filtering rate and current density distribution throughout the electrode surface. An incomplete filling of the cathodic space leads to a reduced penetration and, consequently, to a higher cell-liquor strength in the underfilled areas as compared to the average value of the diaphragm-type cell as a whole. The cell construction, commonly exploited in commercial production, provides for about 85% of catholyte filling capacity. The nonuniform filtration across the diaphragm may also arise owing to a difference between the catholyte and anolyte densities.

The large linear size of an electrode unit and the relatively small electrode gap are a major reason of the fact that even a slight deviation from the vertical position of the electrodes or their displacement relative to the symmetry axis lead to an electrode misalignment with the gap widened at the one end, and narrowed at the other. The evolving gas makes the electrolyte resistance vary with electrode height. The electrode nonequipotentiality also plays a role. These factors are mainly responsible for a macroscopic nonuniformity of current density distribution. In calculating the voltage regime at the electrolytic cell terminals, one must also take into account the manner in which the current leads are connected to the electrolysis circuit.

Along with the cathode height H, the cathode mesh spacing $h \ll H$ is also a characteristic length scale for an electrolytic cell. The cathode-mesh geometry is, by itself, a cause of nonuniformities in filtering rate and current density, respectively, on small scales d_F and d_C (the so-called microscopic nonuniformities) within the diaphragm (membrane) adjacent to the cathode-mesh elements.

The overall filtration flow in a diaphragm-type cell is determined both by the overall current load and by the concentration of alkali product. In the areas of augmented (in comparison to the average) current density or reduced convection, the effect due to the migration component of the hydroxyl ion flux is enhanced, since the countercurrent is no longer capable of offsetting this component. Presumably, this must result in a reduced current efficiency in those areas. For ion-exchange membranes, the filtration regime is determined by other factors; nonetheless, the above statements remain in force. Apparently, the thinner is the diaphragm (membrane), the stronger are the microscopic nonuniformity effects. If $d_F > d$ and $d_C > d$, the electrolytic reaction proceeds under completely nonuniform conditions.

3.8.4.1 General Method for Calculating the Volt-Ampere Characteristic and Current Density Distribution Along the Cell Height

Owing to the basically similar construction of diaphragm-type and membrane-type electrolytic cells, the voltage balance and current distribution for them can be considered from the same standpoint. If, as most commonly is the case, the membrane is tightly held against the anodic plane surface, the voltage drop across the gas-filled catholyte is to be taken into account, as distinct from the chloride solution in a diaphragm-type cell. To be definite, we consider the diaphragm-type cell assuming that both the concept and the mathematics are equally applicable to the membrane type.

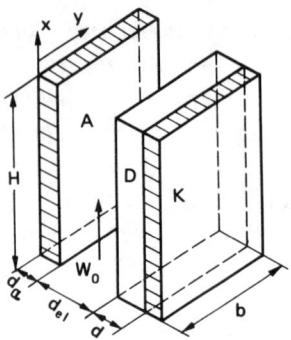

Fig. 3.3 Geometry of a diaphragm-type cell with properly aligned anode.

A, anode; D, diaphragm; K, cathode; w_0, anolyte circulation velocity vector at cell inlet; d_a, effective anode thickness; d_{el}, electrolyte layer thickness; d, diaphragm thickness; H, cell height; b, cell width

Major contributions to the overall electrolytic cell voltage E are due to the sum of chlorine and hydrogen generation potentials, $E_{el\text{-}ch}$ (electrochemical component), to the electrolyte voltage drop, E_{el}, to the diaphragm voltage drop, E_d, and to the anode voltage drop, E_a. The voltage drop in other components of the electrode unit (cathode, cell bottom, buses, etc.) ΔE may be of an order of 0.2 V; however, we will not go into details of the current distribution circuit. The sum

$$U = E_{el\text{-}ch} + E_{el} + E_d + E_a \tag{105}$$

is dependent on the cell geometry and the electrolyte circulation conditions and can be determined by calculation. Further, we obtain $E = U + \Delta E$, where ΔE is assumed to be a constant of known value.

The reference coordinate system and the main dimensions of a symmetric vertical electrode unit are shown in Fig. 3.3. The parameters characterizing a displaced position of the anode are shown in Fig. 3.4. An analysis of the available experimental data on electrode reactions has provided evidence for the existence of a stationary gas-enriched layer contingent to the electrode surface, not participating in the gas-liquid balance but contributing in some measure to the electric resistance. We denote the thickness of this layer by d_{gas}. Contingent to the stationary layer is a mobile electrolyte layer whose thickness varies with electrode height and is determined, at small deflections θ from

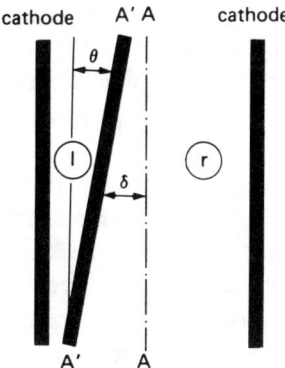

Fig. 3.4 Cathode-anode geometry. AA, correct alignment of anode symmetry axis; A'A', misaligned position; δ denotes the positive direction of anode displacement; θ is the deflection angle

the vertical (Fig. 3.4), by the formula

$$a(x) = d_{el} - d_{gas} \mp \delta \pm \left(x - \frac{H}{2}\right)\theta$$

Here the upper sign "+" (or "−") refers to the region on the left-hand side (l), and the lower sign, to the right-hand side (r) of the anode. Let us take $a(x) > 0$. Denoting $\nu_1 = (d_{el} - d_{gas} - \delta)/(d_{el} - d_{gas})$, $\nu_2 = H\theta/(d_{el} - d_{gas})$, we obtain $1 - |1 - \nu_1| > 1/2\nu_2$.

We assume, for the sake of simplicity, both the cathode and the anode leads to be equipotential. Indeed, the voltage loss in either is known to be about 0.02 V, which is an order of magnitude smaller in comparison to any summand in formula (105). With a lead connected to the anode top or to the anode bottom, the respective top end-face and the bottom end-face are assumed to be equipotential; with the lead connected to the anode side, equipotential is the anode side-face. We assume also that the gas is uniformly distributed across the channel section, measuring $a(x)$ in width. For the top (bottom) lead connection, the field lines within the anode are assumed to be directed strictly parallel to the vertical x-axis, and for the side lead connection, parallel to the y-axis.

In deriving the equations for current density distribution along an arbitrary field line, the voltage balance (105) is to be written using the components expressed by Ohm's law. Also, the elementary balance for anodic and electrolytic currents must be taken into account. The size-average current density in the cell, i^{avg}, is assumed to be known.

In constructing a model, it appears expedient to use the potentials linearized with respect to deviation of the current density from the average in place of those expressed by the conventional Tafel relations. Then one has $E_{el-ch} = m + ni(x, y)$ for each field line. We understand by $i(x, y)$ a local current density in the electrolyte bulk. In modelling a perforated (hole) electrode, the $i(x, y)$ refers to a current density averaged over an elementary area, in size much larger over the perforation spacing. Since the true geometric surface is different from the apparent one, the absolute term m in the above relationship is dependent on the perforation characteristics. In accordance with Sec. 3.8.4.4, the true current densities at the parallel and facing each other anode and cathode surfaces are estimated as $i_a = \frac{1}{2} i(x, y)(2 - \alpha^a)/(1 - \alpha^a)$, $i_c = \frac{1}{2} i(x, y)(2 - \alpha^c) /(1 - \alpha^c)$, where α^a and α^c are the free cross-section areas (in fractional units) for the respective electrodes. The dependences for m and n (under diaphragm electrolysis conditions) as functions of i^{avg}, α^a, α^c and electrolysis temperature for dimensionally stable anodes (DSA) can be derived using the data in [174]. Also, the experimentally measured resistivities for neat NaCl solution (ϱ_{el}, ohm·cm), titanium and graphite (ϱ_a) should be taken into account. We now introduce a coefficient $\lambda_{el} = \lambda_{el}(\varepsilon)$ to specify the electrolyte resistance increment arising from the occurrence of gas (ε is the gas content) and analogous coefficients for an electrolyte-soaked diaphragm, λ_d, for the stationary gas layer, λ_{gas}, and for a perforated anode, λ_a.

With the electrode unit exhibiting an asymmetry (Fig. 3.4), two equations for current density distribution corresponding to the voltage balances in the left-hand (l) and right-hand (r) regions of the anodic space are obtained. The area-average current densities i_l^{avg} and i_r^{avg} in these regions are different; however, the overall voltage drop along each field line remains the same. In addition, one must also take into account specific features due to the anodic box ($d'_{el} > 0$, independent current balance for either region), solid anodic sheet, or anodic plate ($d'_{el} = 0$, coupled current balance between the two regions); here d'_{el} denotes a half of the electrolyte layer thickness in the anodic box.

We set $\xi_l = [n_l + \varrho_{el}(d_{el} - d_{gas} + \lambda_d d + \lambda_{gas} d_{gas})]i_l(x, y)/(U - m_l)$, $\xi_r = [n_r + \varrho_{el}(d_{el} - d_{gas} + \lambda_d d + \lambda_{gas} d_{gas})]i_r(x, y)/(U - m_r)$, which are in fact dimensionless current densities, with n_l and m_l being functions of i_l^{avg}, and n_r and m_r, functions of i_r^{avg}. Then, for the side anode lead in the left-hand region, Eq. (105) takes the dimensionless form

$$\left\{1 + \omega_l\left[\lambda_{el,l}\nu_1 - 1 + \lambda_{el,l}\nu_2\left(X - \frac{1}{2}\right)\right]\right\}\xi_l = 1 + \alpha_{bl}\Phi_S \quad (106)$$

$$\Phi_S = \int_0^y d\eta \int_0^\eta \tau(X,\xi)d\xi - Y\int_0^1 \tau(X,\xi)d\xi$$

Here $X = x/H$, $Y = y/b$, $\omega_l = \varrho_{el}(d_{el} - d_{gas})[n_l + \varrho_{el}(d_{el} - d_{gas} + \lambda_d d + \lambda_{gas}d_{gas})]^{-1}$, $\alpha_{bl} = \lambda_a\varrho_a b^2/\{d_a[n_l + \varrho_{el}(d_{el} - d_{gas} + \lambda_d d + \lambda_{gas}d_{gas})]\}$,

$$\tau = \xi_l + \left(\frac{U - m_r}{U - m_l}\right)\frac{n_l + \varrho_{el}(d_{el} - d_{gas} + \lambda_d d + \lambda_{gas}d_{gas})}{n_r + \varrho_{el}(d_{el} - d_{gas} + \lambda_d d + \lambda_{gas}d_{gas})}\xi_r$$

for the coupled current balance. In Eq. (106), the effect of gas content on current distribution is manifested by a functional λ_{el} vs. ε dependence. The coefficient α_{bl} is a characteristic of electrode resistance. Accordingly, an equation for the right-hand region is obtained through formally replacing the subscripts l or r by r or l in Eq. (106). In addition, the ν_1 must be replaced by $2 - \nu_1$, and the ν_2, by $-\nu_2$.

Quantities ξ_r and ξ_l having been defined by the above equations as functions of the coordinates x, y, one can proceed to the determination of quantities $V_l = 1/\int_0^1\int_0^1 \xi_l\,dx\,dy$, and $V_r = 1/\int_0^1\int_0^1 \xi_r\,dx\,dy$; as is easily seen

$$V_l = (U - m_l)/\{i_l^{avg}[n_l + \varrho_{el}(d_{el} - d_{gas} + \lambda_d d + \lambda_{gas}d_{gas})]\},$$
$$V_r = (U - m_r)/\{i_r^{avg}[n_r + \varrho_{el}(d_{el} - d_{gas} + \lambda_d d + \lambda_{gas}d_{gas})]\}$$

Further, according to the overall current balance, $i_l^{avg} + i_r^{avg} = 2i^{avg}$. The three latter relations define implicitly the unknowns U, i_l^{avg}, and i_r^{avg}. In the simpler case of independent current balance, one must set $\tau = \xi_l$ for the left-hand, and $\tau = \xi_r$ for the right-hand region: if so, the equations for ξ in different regions are independent. In the particular case of a symmetric anode ($\nu_1 = 1$, $\nu_2 = 0$), evident are the equalities $i_l^{avg} = i_r^{avg} = i^{avg}$, $\xi_l = \xi_r = \xi$; if so, then either $\tau = 2\xi$, or $\tau = \xi$ (for independent balance). In this case, $U_{min} = m + [n + \varrho_{el}(d_{el} - d_{gas} + \lambda_d d + \lambda_{gas}d_{gas})]i^{avg}$ is a theoretical minimum voltage, providing that the anode is equipotential, and the anolyte is degassed. Then $U = V(U_{min} - m) + m$, where the coefficient V characterizes the net effect due to electrolyte gas content and anodic resistance.

The balance equations for the top and bottom anode leads are written in a similar manner, the only distinction being the vanish of y-coordinate dependence. The coefficient α_{bl} must be replaced by $\alpha_{Hl} = \lambda_a \varrho_a H^2 / \{d_a[n_l + \varrho_{el}(d_{el} - d_{gas} + \lambda_d d + \lambda_{gas} d_{gas})]\}$. In place of Φ_S, one must write for the bottom anode lead

$$\Phi_B = \int_0^X d\eta \int_0^\eta \tau(\xi)\,d\xi - X \int_0^1 \tau(\xi)\,d\xi$$

and for the top anode lead,

$$\Phi_T = \int_0^X d\eta \int_0^\eta \tau(\xi)\,d\xi - \int_0^1 dX \int_0^X \tau(\xi)\,d\xi$$

In what follows, we chiefly consider a case of the side anode lead, since the analysis of other anode-lead configurations presents no problem.

Following the idea adopted in [175], we use a step-by-step approximation in our calculations. To this end, we initially set $\xi_l = V_l^{-1}$, $\xi_r = V_r^{-1}$ in the left-hand side of Eq. (106) (enclosed within the square brackets) and in the expression for τ. The physical meaning of this operation is that, in a zero approximation, the current density remains uniform within the confines of either region and is defined by the unknowns i_l^{avg} and i_r^{avg}. The $\xi_l(X, Y)$ is expressed by Eq. (106) as

$$\xi_l(X, Y) = \frac{1 - \bar{\alpha}_{bl} V_l^{-1} Y + \frac{1}{2} \bar{\alpha}_{bl} V_l^{-1} Y^2}{1 + \omega_l \left[\lambda_{ell}^\circ \nu_1 - 1 + \lambda_{ell}^\circ \nu_2 \left(X - \frac{1}{2}\right)\right]} \qquad (107)$$

Recalling the conditions $\int_0^1 \int_0^1 \xi_l\,dX\,dY = V_l^{-1}$ and $\int_0^1 \int_0^1 \xi_r\,dX\,dY = V_r^{-1}$, the V_l and V_r are defined by the formula

$$V_l = \left(\int_0^1 \frac{dX}{1 + \omega_l \left[\lambda_{ell}^\circ \nu_1 - 1 + \lambda_{ell}^\circ \nu_2 \left(X - \frac{1}{2}\right)\right]} \right)^{-1} + \frac{\bar{\alpha}_{bl}}{3} \qquad (108)$$

Note that λ_{ell}° is a function of the height X and corresponds to the electrode-gap gas content $\varepsilon_l^\circ = \varepsilon_l^\circ(X)$ as obtained for a uniform current distribution of density i_l^{avg} or $\xi_l^\circ \equiv V_l^{-1}$. In formulas (107), (108), we have $\bar{\alpha}_{bl} = \alpha_{bl}(1 + i_r^{avg}/i_l^{avg})$ for the coupled current balance, and $\bar{\alpha}_{bl} = \alpha_{bl}$ for the independent balance. Analogous formulas for ξ_r and V_r are derived from (107) and (108) through replacing l by r, r by l, ν_1 by $2 - \nu_1$, and ν_2 by $-\nu_2$.

3.8.4.2 Gas Content in Circulation Electrolyte Flow

To make further calculations practicable, we must relate the gas content to the current density distribution with height. Suppose, for one thing, that the gas is uniformly distributed across the ascending flow cross section. The perturbative effect on the ascending flow buildup at the cell bottom feed inlet is regarded as negligible. With a perforated electrode, the width of ascending flow is the sum of the distances d_{el} (diaphragm-electrode gap) and d'_{el} (electrolyte layer half-thickness of the anodic box). For solid electrode sheets, it is assumed that $d'_{el} = 0$.

The gas evolution rate is proportional to the local current density, and the ideal-gas laws are assumed to hold. The liquid volumetric flow rate is assumed to be constant with height by virtue of a large difference between the liquid and gas densities. The balance equation for the liquid and gas volumetric flow rates along the height of the cell includes the phase flow rates averaged over the channel cross section. The electrolyte velocity at the cell inlet (at $x = 0$) is assumed to be known and equal to w_0; the gas content in the general case is different from zero: $\varepsilon = \varepsilon_0$ (at $x = 0$). Models for the relative motion of phases in the gravitational field have been discussed in [59]. Most commonly, the bubble ascent velocity w_b relative to the liquid velocity is ignored, although in principle a case of $w_0 \sim w_b$ should not be excluded. For the chlorine bubbles in the anolyte of a diaphragm-type cell, the velocity w_b is accepted to be 4-5 cm/s [176]. For catholyte circulation in the cathodic compartment of a membrane-type cell, the relative velocity is to be taken a value of the same order of magnitude owing to the coalescence of hydrogen bubbles [177]. Henceforth, it is accepted that $w_b = $ const.

For compact anodes, the gas and liquid balance equations in the regions (l) and (r) are independent (Fig. 3.4). Written in a dimensionless form:

$$\frac{d}{dX}\left\{\left[\nu_1 + \nu_2\left(X - \frac{1}{2}\right)\right](W_l(X) + \sigma)\varepsilon_l(X)\right\} = \frac{V_l}{M_l}\int_0^1 \xi_l(X, Y)dY \qquad (109)$$

$$\frac{d}{dX}\left\{\left[\nu_1 + \nu_2\left(X - \frac{1}{2}\right)\right]W_l(X)[1 - \varepsilon_l(X)]\right\} = 0$$

The initial conditions are $W_l = 1$ (at $X = 0$), $\varepsilon_l = \varepsilon_0$ (at $X = 0$). Here $W_l = w_l/w_0$, $\sigma = w_b/w_0$, $M_l = F(p - p_w)(d_{el} + d'_{el} - d_{gas})w_0/(4.9346RTHi_l^{avg})$;

w_l is the liquid-phase ascent velocity, variable along the height X, cm/s; p is the average pressure within the cell, atm; p_w is the water vapour pressure over solution, atm; $F = 96487$ C/g-equiv; $R = 8.314$ J/(deg·mol); and T is the absolute temperature. It is assumed that the current efficiency for the gaseous chlorine formation is but slightly smaller than 100%. Having set $\xi_l = V_l^{-1}$ in the adopted step-by-step procedure, we arrive at the solution

$$\varepsilon_l^o = 2\left[\left(\nu_1 - \frac{1}{2}\nu_2\right)(1+\sigma)\varepsilon_0 + XM_l^{-1}\right]\left\{\left[\left(\nu_1 - \frac{1}{2}\nu_2\right)[1+\sigma(1+\varepsilon_0)]\right.\right.$$

$$\left. + (\sigma\nu_2 + M_l^{-1})X\right] + \sqrt{\left[\left(\nu_1 - \frac{1}{2}\nu_2\right)[1+\sigma(1+\varepsilon_0)] + (\sigma\nu_2 + M_l^{-1})X\right]^2}$$

$$\left.- 4\left[\left(\nu_1 - \frac{1}{2}\nu_2\right)(1+\sigma)\varepsilon_0 + XM_l^{-1}\right]\sigma\left[\nu_1 + \nu_2\left(X - \frac{1}{2}\right)\right]\right\}^{-1}, \qquad (110)$$

$$W_l^o = (1-\varepsilon_0)\left(\nu_1 - \frac{1}{2}\nu_2\right)\left\{\left[\nu_1 + \nu_2\left(X - \frac{1}{2}\right)\right](1-\varepsilon_l^o)\right\}^{-1}$$

The value of V_l is obtained by integrating by the use of formula (108). For the right-hand region (r), the respective quantities are obtained, as previously, through replacing l by r, r by l, ν_1 by $2 - \nu_1$, and ν_2 by $-\nu_2$. For solid electrode sheets, the distance d'_{el} in the expression for M_l must be set to zero. A solution of the problem for perforated electrodes is obtained by formulas (110) written for a symmetric electrode unit, that is $\nu_1 = 1$, $\nu_2 = 0$, with $d'_{el} > 0$.

For convenience of performing the voltage calculation, the Bruggeman relationship for the electrolyte resistance coefficient $\lambda_{el} = (1+\nu_\varepsilon)^{3/2}$ [49], where $\nu_\varepsilon = \varepsilon/(1-\varepsilon)$, is replaced by a function $1 + A_1\nu_\varepsilon$, linear within the interval $0 \leqslant \varepsilon \leqslant 0.5$. Having set $\int_0^1 [1 + A_1\nu_\varepsilon - (1+\nu_\varepsilon)^{3/2}]d\nu_\varepsilon = 0$ we find $A_1 \approx 1.7$. Thus, the final result is $\lambda_{el}(\varepsilon) = 1 + 1.7\varepsilon/(1-\varepsilon)$. Analytical expressions for voltage can be derived from relation (110) for a symmetric electrode unit (or a perfect-mixing perforated anode), on condition that $\varepsilon_0 = 0$ in the limit case of $\sigma \to 0$, that is, for the gas transported in the liquid flow. Then one has $\varepsilon^o = X/(X+M)$. For the side anode-lead configuration, a very simple formula is thus obtained:

$$V = \frac{\Gamma}{\ln(1+\Gamma)} + \frac{\overline{\alpha_b}}{3}$$

(where $\Gamma = 1.7\omega/M$), which enables one to calculate the voltage in volts, as has been shown in the foregoing Section. Turning to the limit case of $\Gamma \to 0$

(low gas content), one has $V \approx 1 + \dfrac{\overline{\alpha_b}}{3} + \dfrac{\Gamma}{2}$; for the top and bottom anode leads, $V \approx 1 + \dfrac{\overline{\alpha_H}}{3} + \dfrac{\Gamma}{2}$. It can easily be shown that the same asymptotic expressions stem out of the general integral equation (106) at $\sigma = 0$, $\nu_1 = 1$, $\nu_2 = 0$, $\alpha_b \to 0$ (or $\alpha_H \to 0$), $\Gamma \to 0$, even if a method other than the step-by-step approximation has been applied.

We now consider the hydrodynamic conditions for a perforated electrode with no mixing of the solutions contained in the spaces as defined by d_{el} and d'_{el} distances. Let α be the free cross-sectional area (in fractions of a unity); we denote the gas contents and the liquid phase velocities for the d_{el}- and d'_{el}-spaces, as defined above, by ε, w and ε', w', respectively. In the limit case of low free cross-sectional area ($\alpha \to 0$), the mass transport between the two regions is inhibited or absent altogether. At $\alpha \to 1$, the electrode is free of any hydraulic resistance. Then, at sufficiently high flow turbulence, it may be accepted that the gas is uniformly distributed across the ascending flow cross section. This is exactly the case that has been considered above. The simplest approach to the finite mixing velocity is to introduce a coefficient D to characterize the mass transport between the two regions. We retain, for simplicity, that in either region, the gas is uniformly distributed throughout the flow cross section. The volumetric gas (liquid) flux for exchange transport from one region into the other is equal to $D(\varepsilon - \varepsilon')$ (per second per cm^2).

Thus, the free cross section exerts influence not only on the effective electrode surface (with the resultant change in the overvoltage of a chemical reaction), but also on the gas content ε producing a change in the electrolyte voltage drop. In addition, the electrode front side (the one facing the counterelectrode) and the electrode back side are not equivalent with respect to the gas evolution rate, the fact accounted for in the model via introducing the coefficient β as a measure for the back-side current. According to theoretical estimates, $\beta = 0.5\,\alpha$ (see Sec. 3.8.4.4).

With the lateral exchange transport taken into account, one can easily obtain mass balance equations throughout the height of a symmetric electrode configuration. These are, in a dimensionless form:

$$\frac{d}{dX}[(W+\sigma)\varepsilon] = \frac{1-\beta}{\mu M} V \int_0^1 \xi\, dY - \frac{1}{\mu\,\mathrm{Pe}}(\varepsilon - \varepsilon')$$

$$\frac{d}{dX}[W(1-\varepsilon)] = \frac{1}{\mu\,\mathrm{Pe}}(\varepsilon - \varepsilon')$$

$$\frac{d}{dX}[(W' + \sigma)\varepsilon'] = \frac{\beta}{(1 - \mu)M} V \int_0^1 \xi \, dY + \frac{1}{(1 - \mu)\,\text{Pe}}(\varepsilon - \varepsilon')$$

$$\frac{d}{dX}[W'(1 - \varepsilon')] = -\frac{1}{(1 - \mu)\,\text{Pe}}(\varepsilon - \varepsilon')$$
(111)

Here $W = w/w_0$, $W' = w'/w_0$, $\mu = \dfrac{d_{el} - d_{gas}}{d_{el} - d_{gas} + d'_{el}}$, $\text{Pe} = (d_{el} - d_{gas} + d'_{el})w_0/(DH)$ is an analogue of the Peclet number for diffusion processes. The initial conditions are $\varepsilon = \varepsilon' = \varepsilon_0$ (at $X = 0$) and $W = W' = 1$ (at $X = 0$).

The limit $\text{Pe} \to 0$ defines an ideal mixing. The product $(d_{el} + d'_{el})D$ acts for a turbulent diffusion coefficient. Under turbulent mixing conditions, assuming $D \sim w_0$, we obtain $\text{Pe} \sim 10^{-2}$, a value that determines the regime of intense lateral mass exchange. The effect due to a finite mixing velocity becomes manifest at $\text{Pe} \geqslant 1$. By contrast, at $\text{Pe} \to \infty$, the system (111) decomposes into independent pairs of equations for the d_{el}- and d'_{el}-spaces with no mass exchange between them. Assuming the mass-exchange coefficient D to be proportional to α, one must, in modelling the extent of interference due to free cross section, choose correlated values of Pe and α, such that $\text{Pe}\,\alpha = \text{const}$. Further progress in simulation and optimization of the gas-liquid flow structure in electrolytic cells may be accomplished on the basis of liquid flow dynamics, with reference to realistic resistance coefficients for electrolyte flows both in the longitudinal and transverse directions.

3.8.4.3 Similarity Criteria for Electric Fields

The Tafel relation for the electrochemical voltage component taking the form $E_{el\text{-}ch} = A + B \log i(x, y)$, the coefficients of the linearized form $E_{el\text{-}ch} = m + ni(x, y)$ are expressed by the relations

$$m_{l,r} = A + B \log i_{l,r}^{avg} - 0.4343B; \quad n_{l,r} = 0.4343B/i_{l,r}^{avg} \quad (112)$$

We now set $\alpha_b = \lambda_a \varrho_a b^2 / \{d_a[n + \varrho_{el}(d_{el} - d_{gas} + \lambda_d d + \lambda_{gas} d_{gas})]\}$,

$$M = F(p - p_w)(d_{el} - d_{gas} + d'_{el})w_0/(4.9346 RTHi^{avg}),$$

$$\omega = \varrho_{el}(d_{el} - d_{gas})/[n + \varrho_{el}(d_{el} - d_{gas} + \lambda_d d + \lambda_{gas} d_{gas})]$$

where n is determined by the size-average current i^{avg}. By expressing the $\omega_{l,r}$ and $\alpha_{l,r}$ in Eq. (106) through ω and α and referring to (112), (106), (109), and

(111), the quantities $V_{l,r}$ are shown to be dependent on the dimensionless parameters α_b, ω, M, σ, ε_0, Pe, μ, β, $\dfrac{\lambda_d d + \lambda_{gas} d_{gas}}{d_{el} - d_{gas}}$, ν_1, ν_2, i_l^{avg}/i^{avg}, i_r^{avg}/i^{avg}.
In turn, one can form, through applying the above method for voltage calculation and using formula (112), a system of equations for the ratios i_l^{avg}/i^{avg} and i_r^{avg}/i^{avg}:

$$V_l \frac{i_l^{avg}}{i^{avg}} - V_r \frac{i_r^{avg}}{i^{avg}} = \Lambda \left[0.4343(V_r - V_l) + \log\left(\frac{i_r^{avg}}{i^{avg}}\right) \right]$$

$$i_l^{avg}/i^{avg} + i_r^{avg}/i^{avg} = 2$$

where $\Lambda = B/[\varrho_{el}(d_{el} - d_{gas} + \lambda_d d + \lambda_{gas} d_{gas})]i^{avg}$. Consequently, the ratios i_l^{avg}/i^{avg}, i_r^{avg}/i^{avg} are dependent both on the aforementioned parameters and on Λ.

Thus, the dimensionless current distributions ξ_l and ξ_r in the anolyte, the dimensionless voltages V_l, V_r, and the dimensionless ratios i_l^{avg}/i^{avg}, i_r^{avg}/i^{avg} are, in the general case, determined by twelve similarity criteria α_b, ω, M, σ, ε_0, Pe, μ, β, $\dfrac{\lambda_d d + \lambda_{gas} d_{gas}}{d_{el} - d_{gas}}$, ν_1, ν_2, and Λ.

For the voltage U (in volts), formula $U = A - 0.4343 B + B \log i^{avg} + B \log (i_{l,r}^{avg}/i^{avg}) + V_{l,r} B[0.4343 + (i_{l,r}^{avg}/i^{avg})/\Lambda]$ can easily be derived, that is, we have a supplementary relation for U as a function of i^{avg} and electrochemical parameters A and B. The cell voltage, in using electrodes with specified electrochemical properties and configuration (defined by the constants A and B) and with a fixed average current density, is thus entirely determined by the aforementioned twelve parameters.

As can easily be verified, the V and ξ for a symmetric electrode unit are both dependent only on eight parameters α_b, ω, M, σ, ε_0, Pe, μ, and β, whereas the voltage U, on additional four parameters Λ, A, B, and i^{avg}. If the lateral mixing is ideal (or if $d'_{el} = 0$), in the above relations the parameters Pe, μ, and β are eliminated. The parameters remaining for the symmetric electrode unit are α_b, ω, M, σ, and ε_0, or an equivalent set of parameters α_b, Γ, M, σ, and ε_0, where $\Gamma = 1.7\omega/M$. At $\sigma = 0$, the dependence of voltage and current on ω and M is effected only through Γ, that is, only three parameters remain: α_b, Γ, and ε_0. The quantity ε_0 can be taken equal to zero with a good reason. In an infinite-circulation system ($w_0 \to \infty$, $\Gamma \to 0$), of essential importance is only the electrode resistance as expressed by α_b. For the top (bottom) anode lead, the said above remains valid with the α_b replaced by α_H.

3.8.4.4 Ion Transport as Effected on Perforated Electrodes

Perforated electrodes (also known as hole electrodes, or mesh electrodes) exhibit a number of specific features as regards the current density distribution and filtering rate, the major reason being that their respective nonuniformity scales d_C and d_F (the so-called microscopic nonuniformities) are comparable to the perforation spacing.

For an adequate mathematical description of electrochemical cells with microscopic nonuniformities, any proposed electrode model must meet two requirements: (i) it must reflect the essential features of the perforation pattern; (ii) it must be relatively simple to be accessible to an analytical examination.

Let us consider a model made up of an infinite vertical array of equidistant infinitely thin conducting elements of equal length exhibiting the same potential (Fig. 3.5). For simplicity, we confine ourselves to a planar two-dimensional problem. Suppose that the model electrode is imbedded in a continuous homogeneous isotropic medium. In moving towards an infinitely distant electrode of opposite polarity, the velocity $\vec{v}_0$ and the current density $\vec{i}$ are uniform and equal, respectively, to v_0^{avg} and i^{avg}. In moving in the opposite direction, the current tends to zero, $\vec{i} \to 0$, and the filtration flow remains uniform. By virtue of symmetry of the problem, the band of width h_2 delimited by the symmetry axes AB and CF is a completely representative one for description

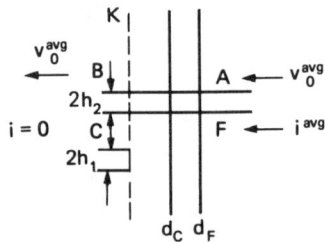

Fig. 3.5 Two-dimensional model of an electrode (cathode K). Horizontal arrows indicate the direction for filtration flow and electric current; d_F and d_C denote the microscopic nonuniformity scale for filtration velocity and current density in the diaphragm; mesh spacing is $h = 2h_2$; free cross-sectional area is $\alpha = 1 - h_1/h_2$.

3 Modelling of Electrochemical Processes

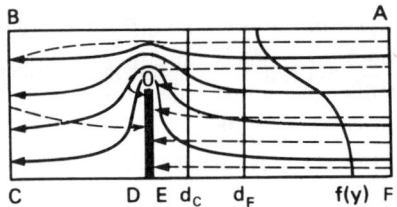

Fig. 3.6 Schematic picture of filtration velocity and current density field lines near a cathode element DOE.

The line $f(y)$ denotes a boundary separating the regions of alkaline and acidic reaction. Solid lines denote filtration field lines, and dashed lines denote current field lines

of the velocity and current fields. The perforation spacing h and the fraction of free cross-sectional area α are expressed through geometric dimensions h_1 and h_2:

$$h = 2h_2, \quad \alpha = 1 - h_1/h_2$$

We consider a potential attached flow with the imposed condition for flow slip at the electrode surface. Indeed, if the electrode process is accompanied by gas evolution, the contact between the solution and the hard electrode surface is weakened due to the buildup of a gas cushion, which leads to a reduction of the viscous drag. Furthermore, to obtain an approximate estimate of the value of d_C, we assume Ohm's law to hold, in neglect of the effects due to the overvoltage vs. current density relationship. Then the solution surface contiguous to the electrode surface has the same electric potential φ_0 throughout. Given below is a solution to the problem within the band *ABCF*. A scheme for the filtration and current fields is given in Fig. 3.6. We place the origin of coordinates at the point $D \equiv E$, with the x-axis pointed along the line *EF*.

Filtering Rate Distribution. In view of the assumptions made, we must find a solution to the boundary problem for velocity potential φ within the band *ABCDOEF* (Fig. 3.6):

$$\Delta \varphi = 0, \quad \left.\frac{\partial \varphi}{\partial n}\right|_{ABCDOEF} = 0; \quad \frac{\partial \varphi}{\partial x} \to v_0^{\text{avg}}, \quad \frac{\partial \varphi}{\partial y} \to 0 \quad (\text{at } x \to \pm \infty)$$

with $\vec{v}_0 = -\nabla \varphi$. We make use of the conformal mapping method in [178] introducing, to this effect, a complex variable $z = x + iy$ and a complex potential $\Phi = \varphi + i\psi$. An essential point is that, by virtue of symmetry, the boundaries AB and $CDOEF$ are the stream lines. The complex velocity is expressed as $\hat{v}_0 = v_{0x} + iv_{0y} = -\dfrac{d\Phi}{dz}$. Actually, the potential in question has already been considered in [178] in the form of

$$\Phi = \frac{h_2 v_0^{\text{avg}}}{\pi} \ln \frac{1 + \chi \cos\left(\dfrac{\pi}{2}\dfrac{h_1}{h_2}\right)}{1 - \chi \cos\left(\dfrac{\pi}{2}\dfrac{h_1}{h_2}\right)} \qquad (113)$$

where $\chi = \sqrt{\tanh^2\left(\dfrac{\pi}{2}\dfrac{z}{h_2}\right) + \tan^2\left(\dfrac{\pi}{2}\dfrac{h_1}{h_2}\right)}$. To estimate the dimension of d_F, let us consider limit cases:

(a) $h_1/h_2 \to 0$. Then, within the terms of the order $\left(\dfrac{h_1}{h_2}\right)^2$ we have

$$\Phi \approx v_0^{\text{avg}}\left[z + \frac{\pi}{4}\frac{h_1^2}{h_2}\coth\left(\frac{\pi}{2}\frac{z}{h_2}\right)\right]$$

$$\frac{d\Phi}{dz} \approx v_0^{\text{avg}}\left[1 - \frac{\pi^2}{8}\left(\frac{h_1}{h_2}\right)^2 \frac{1}{\sinh^2\left(\dfrac{\pi}{2}\dfrac{z}{h_2}\right)}\right]$$

These expansions are valid at $|z| \geqslant h_1$. The second summands in the square brackets characterize the perturbation of flow uniformity by the electrode element DOE and, as can easily be shown, become apparent at $|z| \sim h_1$. In this limit case, it should be assumed that $d_F = h_1$.

(b) $h_1/h_2 \to 1$. Considering the major terms only, it can be shown that

$$\Phi \approx v_0^{\text{avg}}(\operatorname{sgn} x)\left[z + \frac{2}{\pi} h_2 \ln\left(1 + e^{-\pi \frac{z}{h_2}}\right)\right],$$

$$\frac{d\Phi}{dz} \approx v_0^{\text{avg}}(\operatorname{sgn} x)\left[1 - \frac{2e^{-\pi \frac{z}{h_2}}}{1 + e^{-\pi \frac{z}{h_2}}}\right]$$

The expansions hold at $|z - ih| \geqslant h_2 - h_1$. We make estimation of $\dfrac{d\Phi}{dz}$ at the field lines $z = x$ and $z = h_2 i + x$. If $x = h_2/\pi$, then $\left.\dfrac{d\Phi}{dz}\right|_x \approx 0.46 v_0^{\text{avg}}$, $\left.\dfrac{d\Phi}{dz}\right|_{h_2 i + x} \approx$

3 Modelling of Electrochemical Processes

2.2 v_0^{avg}. At large values of x, the flow perturbations are apparent to a lesser extent. Therefore, in this limit case, we can accept $d_F = h_2/\pi$.

Current Density Distribution. Let us consider a problem with mixed boundary conditions for the electric potential φ:

$$\Delta\varphi = 0, \; \frac{\partial\varphi}{\partial n}\bigg|_{AB,CD,EF} = 0, \; \varphi = 0 \text{ at } x \to -\infty$$

$$\varphi|_{DOE} = \varphi_0, \; \frac{\partial\varphi}{\partial x} \to i^{avg}, \; \frac{\partial\varphi}{\partial y} \to 0 \text{ at } x \to +\infty$$

The condition at $x \to -\infty$ determines an electric potential reference level. At the first outset, the value of φ_0 is unknown, and it is to be determined by the problem solution; however, as follows from physical considerations, $\varphi_0 < 0$. Under the condition of the problem $\vec{i} = -\nabla\varphi$. The complex current density is expressed, commonly, through a complex potential $\Phi = \varphi + i\psi$: $\hat{i} = i_x + ii_y = -\overline{\dfrac{d\Phi}{dz}}$. A solution to the problem has been found in [179] by conformal mapping method and shown to be

$$\Phi = \varphi_0 + \frac{h_2 i^{avg}}{\pi} \text{arc cosh} \frac{\sin\left(\frac{\pi}{2}\frac{h_1}{h_2}\right) - \left[\cot\left(\frac{\pi}{2}\frac{h_1}{h_2}\right)\right]W_1}{\left[\cos\left(\frac{\pi}{2}\frac{h_1}{h_2}\right)\right]\sqrt{W_1 - 1}} \quad (114)$$

where

$$\cosh\frac{\pi\varphi_0}{h_2 i^{avg}} = \frac{1}{2}\left[\sin\left(\frac{\pi}{2}\frac{h_1}{h_2}\right) + \frac{1}{\sin\left(\frac{\pi}{2}\frac{h_1}{h_2}\right)}\right], \; \varphi_0 < 0$$

$$W_1 = \sqrt{\tanh^2\left(\frac{\pi}{2}\frac{z}{h_2}\right) + \tan^2\left(\frac{\pi}{2}\frac{h_1}{h_2}\right)}$$

Let us consider limiting cases:

(a) $h_1/h_2 \to 0$. Then $\varphi_0 \approx \dfrac{h_2 i^{avg}}{\pi} \ln\left(\dfrac{\pi}{2}\dfrac{h_1}{h_2}\right)$. In addition,

$$\Phi \approx i^{avg}\left[z + \frac{h_2}{\pi}\ln\left(1 - e^{-\pi\frac{z}{h_2}}\right)\right]$$

$$\frac{d\Phi}{dz} \approx i^{avg}\left[1 + \frac{e^{-\pi\frac{z}{h_2}}}{1 - e^{-\pi\frac{z}{h_2}}}\right]$$

The expansions are valid at $|z| \gg h_1$. Let us estimate the value of $\dfrac{d\Phi}{dz}$ at field lines $z = x$ and $z = h_2 i + x$, $x > 0$. If $x = \dfrac{h_2}{\pi}$, then $\left.\dfrac{d\Phi}{dz}\right|_x \approx 1.6 i^{\text{avg}}$, $\left.\dfrac{d\Phi}{dz}\right|_{h_2 i + x} \approx 0.73 i^{\text{avg}}$. At larger values of x, the flow perturbations become vanishingly small. Consequently, one may take $d_C = \dfrac{h_2}{\pi}$, by analogy with the case (b) of hydrodynamic problem.

(b) $h_1/h_2 \to 1$. If so, $\varphi_0 \approx 0$. At $x < 0$, we have $\Phi \approx i h_2 i^{\text{avg}}$; at $x > 0$, we have, within the terms of the order $\left(1 - \dfrac{h_1}{h_2}\right)^2$,

$$\Phi \approx i^{\text{avg}}\left[z + \frac{\pi}{8} h_2 \left(1 - \frac{h_1}{h_2}\right)^2 \left(\tanh\left(\frac{\pi}{2}\frac{z}{h_2}\right) - 1\right)\right],$$

$$\frac{d\Phi}{dz} \approx i^{\text{avg}}\left[1 + \frac{\pi^2}{16}\left(1 - \frac{h_1}{h_2}\right)^2 \frac{1}{\cosh^2\left(\dfrac{\pi}{2}\dfrac{z}{h_2}\right)}\right]$$

The above expansions are valid at $|z - ih_2| \gg h_2 - h_1$. The latter condition specifies also the uniformity region of electric field; therefore, it may be assumed that $d_C = h_2 - h_1$.

For intermediate values of α, a parabolic interpolation can be suggested, in accordance with the above limiting cases:

$$d_F = \frac{h}{2}\left[\left(\frac{1}{\pi} - 1\right)\alpha^2 + \left(1 - \frac{2}{\pi}\right)\alpha + \frac{1}{\pi}\right] \tag{115}$$

$$d_C = \frac{h}{2}\alpha\left[\left(\frac{1}{\pi} - 1\right)\alpha + 1\right]$$

Let us consider conclusions sequent to the general solution (114). First, $\Phi(ih_1) = \varphi_0 + \dfrac{h_2 i^{\text{avg}}}{2}\left(1 + \dfrac{h_1}{h_2}\right)i$, therefore, $\psi(ih_1) = \dfrac{h_2}{2} i^{\text{avg}}\left(1 + \dfrac{h_1}{h_2}\right)$ determines the current supplied to the electrode element at $x = +0$, $0 \leqslant y \leqslant h_1$. Then the current of strength $h_2 i^{\text{avg}} - \psi(ih_1) = \dfrac{h_2}{2} i^{\text{avg}}\left(1 - \dfrac{h_1}{h_2}\right)$ and of average density $\dfrac{1}{2} i^{\text{avg}}$ at the gap between the electrode elements passes through the free cross-sectional area into the region of $x < 0$. Accordingly, the average current density taken up by the portion OE of electrode surface is equal to $\frac{1}{2} i^{avg} \left(1 + \frac{h_2}{h_1}\right)$.

The nonuniformity in current density distribution must produce a greater voltage drop across the electrolyte layer adjacent to the electrode surface as compared to a uniform distribution. This is tantamount to the increment of layer thickness by an effective value Δl in a uniform electric field. An additional electrolyte potential drop, equal to $\varphi_{add} = i^{avg} \cdot \Delta l$, is dependent on the electrode shape. Apparently, $\varphi_{add} = \lim_{\substack{y=0 \\ x \to \infty}} (\Phi - \varphi_0 - i^{avg} x)$, and it can easily be shown, using (114), that $\Delta l = -\frac{h}{2\pi} \ln \cos\left(\frac{\pi}{2}\alpha\right)$. For typical electrodes exploited in the chlorine production, the value of Δl constitutes several fractions of a millimeter. The additional voltage is dependent on the electric conduction of the medium; it is rather small in diaphragm-type cells. However, this effect should be taken into account in designing the shape of an electrode with an ion-exchange membrane retained against the electrode surface, since the membrane resistance is two orders of magnitude greater over that of neat solution.

The above estimates of the electric characteristics of perforated electrodes are essential for an adequate description of the mass transport towards the electrode, especially a gas-evolving one, as well as for the calculation of ion transport in nonuniform fields. Of practical importance are also analysis and optimization of the voltage balance of the cell, with allowance made for ohmic losses in the electrolyte.

General Mathematical Formulation of Ion-Transport Problem for Diaphragm. As distinct from an idealized electrolytic cell with uniform filtering velocity $\vec{v}_0$ and uniform current density $\vec{i}$, in practice one must take into account the cathode geometry, which in fact signifies that the ion transport in electric and hydrodynamic fields is to be considered on a scale comparable to the cathode perforation spacing.

Let us consider the diaphragm zone $0 < y < h_2$, $0 < x < f(y)$ exhibiting an alkaline reaction (Fig. 3.6).

In Sec. 3.8.3, the transport properties have been expressed through the diffusion coefficients D_1^0, D_2^0, and D_3^0 for individual ions (by analogy with equations in the theory of infinitely dilute solutions). By substituting expressions (60) for the fluxes $\vec{N}_1$ and $\vec{N}_2$ into the law of conservation of matter, div $\vec{N}_1$ = div $\vec{N}_2$ = 0, we obtain two equations of ellyptic type with respect to concentrations c_1 and c_2. In performing the transformations, it should be recalled that div $\vec{i}$ = 0, and div $\vec{v}_0$ = 0.

In expressing the Laplace operator in an explicit form, we obtain equations of the type $\Delta c_1 = f_1(\nabla c_1, \nabla c_2, c_1, c_2, \vec{i}, \vec{v}_0)$ and $\Delta c_2 = f_2(\nabla c_1, \nabla c_2, c_1, c_2, \vec{i}, \vec{v}_0)$ (in an expanded form, these equations have been given in [180]). It appears expedient to convert the original variables to dimensionless variables $\xi = \frac{1}{c^{cat}}(c_1 + c_2)$, $\eta = \frac{1}{c^{cat}}\left(\frac{D_1^0}{D_3^0} c_1 + \frac{D_2^0}{D_3^0} c_2\right)$, where c^{cat} is the Na$^+$ ion concentration in the catholyte ($c^{cat} = c_3$, at $x = 0$). We write the equations:

$$\Delta\xi + \frac{1}{2D_3^0}\left(\frac{D_3^{02}}{D_1^0 D_2^0}\nabla\eta - \frac{D_3^0 D_1^0 + D_3^0 D_2^0 + D_1^0 D_2^0}{D_1^0 D_2^0}\nabla\xi\right)\vec{v}_0 = 0 \quad (116)$$

$$\Delta\eta - \frac{1}{2D_3^0}(\nabla\xi + \nabla\eta)\vec{v}_0 = \frac{D_3^0(D_1^0 - D_2^0)}{\xi(\xi + \eta)}\left\{\frac{\xi + \eta}{2D_1^0 D_2^0 D_3^0}\left[D_2^0\right.\right.$$

$$\times (D_3^0 - D_1^0)\overline{\eta}\nabla\overline{\xi} - D_1^0(D_3^0 - D_2^0)\overline{\xi}\nabla\overline{\eta}\left.\right]\vec{v}_0 + (\overline{\xi}\nabla\overline{\eta} - \overline{\eta}\nabla\overline{\xi})$$

$$\times \left(\nabla\overline{\xi} - \nabla\overline{\eta} + \frac{\vec{i}}{FD_3^0 c^{cat}}\right)\right\}$$

Here

$$\overline{\xi} = \frac{D_2^0 \xi - D_3^0 \eta}{D_3^0(D_2^0 - D_1^0)}, \quad \overline{\eta} = \frac{D_3^0 \eta - D_1^0 \xi}{D_3^0(D_2^0 - D_1^0)}$$

The boundary conditions belong to a mixed type: at $x = 0$, the concentrations c_1 and c_2 are those of their respective components of the catholyte; at $x = f(y)$, one has $c_2 = 0$, whereas c_1 is equal to the concentration of chloride in the anolyte (the variations of NaCl concentration in the acidic zone of the diaphragm can be ignored). At $y = 0$, $y = h_2$, we have, by virtue of symmetry,

$$\frac{\partial c_1}{\partial y} = \frac{\partial c_2}{\partial y} = 0$$

The concentration field having been defined, we can calculate the local alkali losses at a given height y (assuming that the relation $x = f(y)$ is known) by making use of the equality $\vec{N}_2 \vec{n} = -D_2 \vec{n}\nabla c_2$, where $\vec{n}$ is the unit vector normal to the neutralization surface. In solving an analogous problem for the acidic diaphragm zone, the flux $\vec{N}_a = 2\vec{N}_{Cl_2} + \vec{N}_{HClO} + \vec{N}_{H^+}$ is calculated acting to neutralize the alkali at $x = f(y)$:

$$(\vec{N}_2 + \vec{N}_A)\vec{n} = 0 \quad (117)$$

This condition determines, in an implicit manner, the location of the otherwise unknown neutralization surface $x = f(y)$. We have deemed it justified to ignore the general formulation of the mass transport in the acidic zone and to take for simplicity that the flux $\vec{N}_A$ has only a horizontal component which is dependent on y and can be estimated by formula (99).

To obtain a numerical solution of Eqs. (116), one must define the coordinate relationship between quantities $\vec{i}$ and $\vec{v}_0$. This can be done using formulas (113) and (114); however, in view of the model character of the problem in question, certain simplifications may be suggested. Suppose, outside the nonuniformity scales $x \geqslant d_F$ and $x \geqslant d_C$, both the filtration and the current are uniform and are equal, respectively, to v_0^{avg} and i^{avg}, that is: $v_{0x} = v_0^{avg}$ (at $x \geqslant d_F$), $v_{0y} = 0$ (at $x \geqslant d_F$); $i_x = i^{avg}$ (at $x \geqslant d_C$), $i_y = 0$ (at $x \geqslant d_C$). At $x < d_F$ and $x < d_C$, we assume a linear dependence of the horizontal components v_{0x} and i_x on x, recalling (see above) that

$$v_{0x} = 0 \text{ (at } x = 0, y < h_1\text{)}, \quad v_{0x} = \frac{h_2}{h_2 - h_1} v_0^{avg} \text{ (at } x = 0, y > h_1\text{)}$$

$$i_x = \frac{i^{avg}}{2}\left(1 + \frac{h_2}{h_1}\right) \text{ (at } x = 0, y < h_1\text{)}$$

$$i_x = \frac{i^{avg}}{2} \text{ (at } x = 0, y > h_1\text{)}.$$

If so, the components v_{0y} and i_y are uniquely determined by continuity equations div $\vec{v}_0 = 0$, div $\vec{i} = 0$ and by symmetry constraints $v_{0y} = 0$ (at $y = 0$, $y = h_2$), $i_y = 0$ (at $y = 0$, $y = h_2$). The quantities v_0^{avg} and i^{avg} are macroscopic parameters and are determined by the material balance of the cell. In considering the filtration and current macrodistributions, they must be replaced by local quantities v_{0j}^{avg} and i_j^{avg} for the j-th elementary area of diaphragm surface. The d_F and d_C as functions of h and α are determined by Eq. (115).

The system of equations (116) has been solved by iterative finite-difference method: each next iteration has been determined by solving a system of linear equations, whereas the nonlinear right-hand side has been computed on the basis of preceding approximation. This procedure has been applied at each fixed position $f(y)$ of the boundary; preferable was a boundary best satisfying condition (117). In practice, this has been accomplished by minimizing a discrete analogue of the functional

$$\int_{y=0}^{y=h_2} \left(N_{2x} + N_{2y}\frac{n_y}{n_x} + N_{Ax}\right)^2 dy \to \min$$

Other computational details and the check on accuracy have been described elsewhere [180]. The boundary having been defined, the net alkali loss in the cell (per unit of cathode height) can be estimated through integrating the local losses along the line $x = f(y)$, or, as can easily be verified, via

$$N_2^{\text{avg}} = \frac{1}{h_2} \int_0^{h_2} \left(N_{2x} + \frac{n_y}{n_x} N_{2y} \right) dy$$

3.8.5 Features of an Electrochemical Process Using Solid Polymer Electrolyte

At present, various modifications of solid-cathode electrolysis are being extensively studied with a view to eliminate certain drawbacks inherent in this method, such as low alkali concentration, the necessity of regular cleaning or replacing the asbestos diaphragm, sizeable ohmic losses, nonuniformity of current density distribution, and others. To this end, porous-layer electrocatalysts, along with ion-exchange membranes, are used. The former have already gained a wide application in chemical sources of electric energy, in fuel elements, and in water electrolysis. The solid-polymer electrolyte (SPE) technology is currently known as an electrolysis method employing an ion-exchange membrane with a layer of catalyst applied to its surface and serving for an electrode. This method can be coupled with the conventional minimization of the electrode-membrane distance, which is achieved via mechanical apposition of the membrane to the activated electrode surface—the so-called zero gap (ZG) cell technology. In this case, the system membrane—electrode surface is dismountable and provides for a contact of the fresh electrolyte with the membrane. A number of design versions are possible, with due account of the specific processes occurring at the anode (A) and cathode (K), separated by a membrane: ZGA/ZGK, ZGA/SPEK, SPEA/ZGK, SPEA/SPEK. As attested by the majority of sources (for the most part, the patented ones), the version ZGA/SPEK is preferred in chlorine technology; for this reason, in a narrow sense it is referred to as the SPE-technology [181]. Simultaneously, other variants are also being studied. Empirical models for chlorine-alkali membrane-type cells have been described [182] employing the regression analysis of measured current efficiencies, voltage drop and a number of other parameters with a view to determining real physico-chemical conditions at the membrane surface and assessing the effectiveness of various design versions. The authors of the paper [182] have come, on the basis of their findings, to

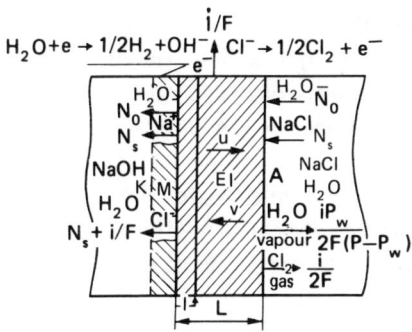

Fig. 3.7 Schematic picture of the anodic section of an electrolytic cell with porous catalytic layer (El) and ion-exchange membrane (M).

A refers to anolyte, and K, to catholyte

a conclusion that the chloride solution in the course of electrolysis becomes significantly depleted, with the alkali concentrated within the active layers. However, the inadequacy of the adopted model has narrowed the predictive value of this conclusion only to a series of performed experiments. We intend, in what follows, to show that a macrokinetic analysis of transport limitations in porous electrode layers can constitute the basis for a rational choice of electrode design and technology [11, 183, 190].

To start with, we consider an anodic process. A schematic diagram of such a process within the SPE framework is shown in Fig. 3.7. On its front side, the electrode layer is limited by a permselective membrane, permeable chiefly to Na^+ ions and, partly, to water molecules. At the inlet, NaCl and water are fed into the cell. Chlorine, produced by a reaction, is transported jointly with water vapours across the electrode body to be further removed via anolyte.

The electrode is regarded as a quasihomogeneous medium composed of three interpenetrant phases: solid, liquid, and gaseous. Since the electrode thickness L is small in comparison to the electrode height, we can confine ourselves to the search for solution of a one-dimensional problem, in which all the variables are functions of the distance X to the front electrode surface. It is easily seen that the modelling of such a system requires a study of four interrelated phenomena attended by chlorine evolution: passage of current across the liquid phase; gas-liquid countercurrent within the electrode body;

NaCl concentration distribution across the electrode thickness; and thermal effect in the electrode-electrolyte system. Below, given are the major macrokinetical equations and analytical results obtained on the model.

Porous Electrode Polarization. Since the metal base of a porous electrode possesses a larger conductance relative to that of the solution, the former is assumed to be equipotential. In the simplest case, assuming the validity of Ohm's law and neglecting the concentrational polarization, the polarization η at any point of the electrode body obeys the equation

$$\delta^2 \frac{d^2\eta}{dX^2} = e^{\alpha\eta} - e^{-\beta\eta} \tag{118}$$

where $\delta = l/L$, $l = \sqrt{\dfrac{RT\varkappa}{FSi_0}}$ (henceforth, the potentials are normalized to RT/F, and the coordinate, to the electrode thickness L). We assume here the conventional exponential dependence for electrode kinetics; S is the inner specific surface of the electrode accessible by electrochemical reaction; i_0 is the exchange current; $\varkappa$ is the conductivity; and l is the characteristic length for electrochemical reaction penetration.

The total current i passes through the solution at $X = 0$; at $X = 1$, no current is observed in solution. Hence, the boundary conditions to Eq. (118) are:

$$\frac{d\eta}{dX} = -A \text{ (at } X = 0\text{)}, \quad \frac{d\eta}{dX} = 0 \text{ (at } X = 1\text{)} \tag{119}$$

where

$$A = \frac{FLi}{RT\varkappa}$$

Changing, in problem (118)-(119), to variables η, $\dfrac{d\eta}{dX}$, we obtain implicit relations for determining $\eta(X)$, $\eta(0)$, $\eta(1)$:

$$\frac{\sqrt{2}}{\delta}(1 - X) = \int_{\eta(1)}^{\eta(X)} \frac{dy}{\sqrt{\dfrac{1}{\alpha}(e^{\alpha y} - e^{\alpha\eta(1)}) + \dfrac{1}{\beta}(e^{-\beta y} - e^{-\beta\eta(1)})}} \tag{120}$$

$$\frac{1}{2}(A\delta)^2 = \frac{1}{\alpha}(e^{\alpha\eta(0)} - e^{\alpha\eta(1)}) + \frac{1}{\beta}(e^{-\beta\eta(0)} - e^{-\beta\eta(1)})$$

To be noted in the case of low polarization, $\eta \ll 1$, the linearization of the exponent in (118) allows obtaining an explicit expression

$$\eta(X) = A \frac{\delta}{\sqrt{\alpha+\beta}} \frac{\cosh\left[\sqrt{\frac{\alpha+\beta}{\delta}}(1-X)\right]}{\sinh\sqrt{\frac{\alpha+\beta}{\delta}}} \qquad (121)$$

Formulas (120) and (121) suggest simple estimates of the front-side and back-side polarizations in certain limit cases: if $\delta \ll 1$, $\delta^2\eta(0) \ll 1$, then

$$\frac{1}{2}(A\delta)^2 + \frac{1}{\alpha} + \frac{1}{\beta} = \frac{1}{\alpha}e^{\alpha\eta(0)} + \frac{1}{\beta}e^{-\beta\eta(0)}, \quad \eta(1) \ll \eta(0) \qquad (122a)$$

if $\delta \gg 1$, $A \ll 1$, then

$$e^{\alpha\eta(0)} - e^{-\beta\eta(0)} = A\delta^2, \quad \eta(1) \sim \eta(0) \qquad (122b)$$

if $\delta \sim 1$, $\eta(0) \ll 1$, then

$$\eta(0) = \frac{A\delta}{\sqrt{\alpha+\beta}}, \quad \eta(1) \sim \eta(0) \qquad (122c)$$

To be noted, formula (122c) is derived from Eq. (122a) in the limiting case of $\eta(0) \to 0$.

Convective Motion in Gaseous and Liquid Phases. Vaporization of water in the electrochemical reaction zone requires a continuous replenishment of anolyte solution, which leads to the occurrence of countercurrents of liquid and gas of respective flow rates v and u (Fig. 3.7) whose intensities are determined by the mass flow in the electrolytic cell. In addition, there occur pressure drops, both in the liquid, Δp_{liq}, and the gas, Δp_{gas}, phase, in the direction of electrode thickness; these drops determine the volume fraction of each of the phases and their connectedness. According to Darcy's law,

$$u = \frac{k_1 K_{gas}}{\mu_{gas} L} \frac{dp_{gas}}{dX}, \quad v = \frac{k_1 K_{liq}}{\mu_{liq} L} \frac{dp_{liq}}{dX} \qquad (123)$$

Here μ_{gas} and μ_{liq} are the gas and liquid viscosities; K_{gas} and K_{liq} are the permeability coefficients; k_1 (and, subsequently, k_2) are dimensional constants. With reference to (123), equations for pressures p_{gas} and p_{liq} are written as:

$$-\frac{k_1 K_{liq}}{\mu_{liq} L} c_0 \frac{dp_{liq}}{dX} - \frac{k_2 K_{gas}}{\mu_{gas} L} \frac{p_w}{RT} \frac{dp_{gas}}{dX} = N_0 \qquad (124a)$$

$$-\frac{d}{dX}\left(\frac{k_1 K_{liq}}{\mu_{liq} L} c_0 \frac{dp_{liq}}{dX} + \frac{k_2 K_{gas}}{\mu_{gas} L} \frac{p_{gas}}{RT} \frac{dp_{gas}}{dX}\right) = \frac{S i_0 L}{2F}(e^{\alpha\eta} - e^{-\beta\eta}) \qquad (124b)$$

$$2\sigma/r_{cr} = P_{gas} - P_{liq} \qquad (124c)$$

The first of equations (124) expresses the constancy of the total water flow distributed in the electrode body between the phases: $N_0 = \text{const}$ (c_0 is the water concentration). The second equation is obtained as the sum of the conservation equation for chlorine and the condition of either the absence of a water source, or water efflux in the electrode $\left(\dfrac{dN_0}{dX} = 0\right)$. The permeability coefficients K_{gas} and K_{liq} as well as the vapour pressure p_w are presumed to be independent of the coordinates. It can be shown that the latter condition holds, if the temperature drop across the electrode, ΔT, is sufficiently small: $\Delta T \ll p_w / \dfrac{dp_w}{dT}$, provided $\Delta T \ll 30\text{-}40°C$. A third equation, which is Laplace's capillary equation, presupposes a complete wetting and allows thus to estimate the critical radii r_{cr} for the pores that constitute an interface between the two phases. The gas-phase diffusion transport can be safely ignored, if, as can be shown, the gas-pore radius is $r_{gas} \geq 10^{-4}$ cm.

At the electrode back side, that is, at $X = 1$, the liquid-pressure continuity is fulfilled implying that the $p_{liq}(1)$ is equal to the known anolyte pressure p. The $p_{gas}(1)$ is defined as a breakdown pore pressure; it is determined only by the porous structure properties. This condition is equivalent to the absence of a resistance to the gas transport from the electrode back side into the solution bulk. Assuming the membrane to be gas-impermeable, we can write boundary conditions to equations (124):

$$p_{liq}(1) = p; \quad p_{gas}(1) = p + \Delta p_{break}; \quad dp_{gas}/dX = 0 \text{ (at } X = 0) \quad (125)$$

The system of equations (124), (125) considered jointly with (118), (119) allows one to express the pressure through polarization in an explicit form. Under typical conditions, the relative changes in pressure,

$$\pi_{gas}(X) = \frac{p_{gas}(X) - p_{gas}(1)}{p_{gas}(1)} \quad \text{and} \quad \pi_{liq}(X) = \frac{p_{liq}(X) - p_{liq}(1)}{p_{liq}(1)}$$

are small, and the problem solution is thus simplified. We obtain, in a dimensionless form:

$$\pi_{gas}(X) = P_1[\eta(1) - \eta(X) + A(1 - X)]$$

$$\pi_{liq}(X) = -P_2\left[\frac{Pe}{Q}(1-x) + \frac{(NQ)^{-1} - P_3}{\gamma_a}(\eta(1) - \eta(X))\right] \quad (126)$$

The notations used are:

$$P_1 = \left(\frac{RT}{F}\right)^2 \frac{\varkappa\mu_{\text{gas}}}{2k_2 K_{\text{gas}} p_{\text{gas}}(1)(p_{\text{gas}}(1) - p_w)}$$

$$P_2 = \frac{D_s \mu_{\text{liq}}}{K_1 K_{\text{liq}} p}, \quad P_3 = k\frac{c^{\text{an}}}{c_0}$$

$$\text{Pe} = \frac{Li}{FND_s c^{\text{an}}}, \quad Q = c_0(Nc^{\text{an}})^{-1}\left[k + \frac{p_w}{2(p_{\text{gas}}(1) - p_w)}\right]^{-1}$$

(c^{an} is the NaCl anolyte concentration). The salt diffusion coefficient D_s and the degree of chloride decomposition N have been introduced here to make easier further estimations of NaCl concentration. By definition, $N = -i/FN_s$, where $N_s < 0$ is the salt flow rate fed into the cell. The coefficient k characterizes the membrane permeability with respect to water, $k = -FN_0/i$; the coefficient γ_a is defined as $\gamma_a = F^2 D_s c^{\text{an}}/(RT\varkappa)$.

Further, by making use of Laplace's equation, we can estimate the variation of the critical pore radius across the electrode thickness. Under typical conditions, one has $|d\,r_{\text{cr}}| \leqslant 10^{-4}$ cm for a gross overestimate, whereas for most gas-evolving electrodes, commonly exploited in electrochemistry, $r_{\text{cr}} \approx 10^{-4}\text{-}10^{-3}$. Therefore, the assumption of a constancy of the liquid-to-gas fraction ratio and, consequently, the transport coefficients in equations seem to be justified: these are dependent only on the structure of a porous body and are in value the same as those characteristic of the phase distribution under gas breakdown conditions [12]. In addition, the small value of change in r_{cr} makes a breakdown of liquid continuity within the electrode body little probable.

Chloride Distribution Within the Electrode Body. For an efficient performance of the system in question, it is essential to provide for a uniform distribution of NaCl concentration across the layer thickness. The electrochemical reaction having been taken into account, an equation for concentration distribution is readily obtained:

$$\frac{d}{dX}\left(-\frac{D_s}{L}\frac{dc}{dX} + vc\right) = -\frac{t_+ RT\varkappa}{F^2 L}\frac{d^2\eta}{dX^2} \tag{127}$$

(t_+ is the cation transference number).

Let the salt concentration in anolyte, c^{an}, and the ion fluxes at $X = 1$ be known:

$$c = c^{\text{an}} \text{ (at } X = 1\text{)}, \quad \left(-\frac{D_s}{L}\frac{dc}{dX} + vc\right) = N_s \text{ (at } X = 1\text{)} \tag{128}$$

By integrating Eq. (127), with allowance made for the salt flux at $X = 1$ and recalling that $d\eta/dX = 0$ (at $X = 1$), a first-order equation is derived. Passing to the dimensionless concentration $\xi = c/c^{an}$ and making use of Darcy's law (123) and formula (126) for the liquid pressure, we obtain an equation

$$\frac{d\xi}{dX} = \text{Pe} - \frac{\text{Pe}}{Q}\xi + \frac{1}{\gamma_a}\left[\left(P_3 - \frac{1}{NQ}\right)\xi + t_+\right]\frac{d\eta}{dX}, \quad \xi = 1 \text{ (at } X = 1) \quad (129)$$

which is more suited for numerical solution. The general solution to problems (127), (128) or, equivalently, to problem (129), takes the form:

$$\xi(X) = \left[1 + \int_1^X \left(\frac{t_+}{\gamma_a}\frac{d\eta}{dX} + \text{Pe}\right)e^{\frac{\pi_{liq}(X)}{P_2}}dX\right]e^{-\frac{\pi_{liq}(X)}{P_2}} \quad (130)$$

Assumed $\xi \sim 1$, the conditions $\left|P_3 - \frac{1}{NQ}\right| \ll t_+$ or $\frac{c^{an}}{2c^0}\frac{p_w}{p_{gas}(1) - p_w} \ll t_+$ are quite probable to hold for moderate temperatures. This condition is tantamount to the neglect of the polarization component in formula (126) for $\pi_{liq}(X)$. It follows in this case from (130) that

$$\xi(X) \approx \xi^0(X) + \frac{t_+}{\gamma_a}[\eta(X) - \eta(1)]e^{\frac{\text{Pe}}{Q}(\overline{X}-X)} \quad (131)$$

where

$$\xi^0(X) = Q + (1-Q)e^{\frac{\text{Pe}}{Q}(1-X)} \quad (132)$$

and $\overline{X}$ is a parameter, $X < \overline{X} < 1$.

The solution to (132) has a simple physical meaning: it determines the salt concentration profile in a hypothetical case of negligibly small electric-field effect on mass transport and is in fact a solution to the ordinary equation for convective salt diffusion with the diffusion coefficient D_s at a given salt feed flux $N_s = i/FN$ and anolyte concentration c^{an} (or a solution to Eq. (129) at $d\eta/dX = 0$). The quantity Pe has a meaning of Peclet number as defined by the characteristic velocity of salt transport, $-N_s/c^{an}$. The quantity Q is the ratio of the salt transport velocity $-N_s/c^{an}$ to the water flow rate fed into the cell, $-\overline{N}_0/c_0$. To be certain of this, it suffices to represent the cell water balance in the form

$$\overline{N}_0 = N_0 + \frac{ip_w}{2F(p_w - p)}$$

with N_0 expressed through k.

The solution to Eq. (131) shows that the electrode polarization always leads to an increased NaCl concentration as compared to that predicted by $\xi^0(X)$-distribution. Since $t_+/\gamma_a \sim 1$, the magnitude of this effect reaches at least $\eta(X) - \eta(1)$. If the ratio Pe/Q is large, the respective correction may be quite significant despite the low polarization. By contrast, at a sufficiently small Pe/Q one may anticipate the formula (132) to be suited for practical calculations of concentration.

Thermal Effects in Electrode-Electrolyte System. The evolution of heat within a given system and its removal through a control surface Σ enclosing the system is determined, under steady-state conditions, by the total enthalpy flux:

$$Q_{total} = -\int_\Sigma \sum_i \bar{h}_i \vec{N}_i d\vec{\Sigma} \qquad (133)$$

The intensities of mole fluxes $\vec{N}_i$—those coming into and those going out of the system—are shown in Fig. 3.7.

Recalling the thermodynamic identity $\bar{h}_i = \bar{\mu}_i - T\dfrac{\partial \mu_i}{\partial T}$ (where $\bar{\mu}_i$ and μ_i denote the electrochemical and chemical potentials, respectively), the definitions of overvoltage and interfacial equilibrium potential drop and using formula (133), one may show, without the loss of generality, that

$$Q_{total} = Q_{vap} + Q'_{el\text{-}ch} + Q^o_{el\text{-}ch} \qquad (134)$$

where $Q_{vap} = -\dfrac{q_{vap} i}{2F}\dfrac{p_w}{p - p_w}$, $Q'_{el\text{-}ch} = \dfrac{RT}{F} i\eta(0)$, $Q^o_{el\text{-}ch} = -\dfrac{RT^2}{F} i \dfrac{\partial E^o}{\partial T}$, and $E^o = \varphi_s^o - \varphi_M$. Here q_{vap} denotes the specific vaporization heat; φ_s^o and φ_M refer, respectively, to the equilibrium solution potential at concentration c^{an} and to the potential of electrode metal base.

The first summand in Eq. (134) is the vaporization heat in the electrode-electrolyte system. The second summand is the irreversible thermal effect due to current passage. The third summand determines reversible heat losses due to the discharge of Cl^- ions in solution at concentration c^{an} to yield gaseous chlorine molecules (the so-called Peltier heat).

Finally, for the Joule heat produced by the electrode metal one has

$$Q_M < \dfrac{Li^2}{\varkappa_M}$$

where $\varkappa_M$ is the effective conductivity of the metal base. However, its contribution is negligibly small and can be safely ignored.

Parametric Response of Anodic Process in a Membrane-Type Cell. To perform calculations for a given process, one needs a knowledge of design and technologic characteristics of the electrolytic cell and specific features of the electrode discharge. Regrettably, the available experimental evidence on salt electrolysis in the cell of interest is quite scarce, and for this reason we turn to better studied characteristics of electrodes used in water electrolysis. In principle, this allows assessing the potential utility of such electrodes in commercial production of chlorine. The supposed characteristics of such electrodes are summarized in Table 3.3.

To provide a quantitative estimate of the polarization effect on the chloride concentration distribution, the system of equations (118), (119), (126), and (129) has been solved numerically. The accepted technologic and design parameters did not differ basically from those of a working electrode; the only exceptions were arbitrarily overestimated values of k (the reason of this will be explained below). The polarization (η) and concentration (ξ) curves as functions of coordinate X for different exchange current densities are shown in Fig. 3.8. The temperature was chosen equal to 90°C.

The calculated results show the polarization effect to play a minor role within a range of Pe/$Q \leqslant 1$. As is explained in greater detail below, a value of $Q \sim 1$ is required to provide for a uniform concentration distribution; otherwise stated, the constraint to be imposed is Pe $\leqslant 1$. Under these conditions, the system remains stable in response to varied electrochemical parameters, and the chloride concentration obeys, to a sufficient accuracy, the dependence for $\xi^\circ(X)$ as described by formula (132) (see Fig. 3.8, curves 4). By contrast, with Pe increasing, the alterations in volume exchange current Si_0 as produced, for example, by a disordered microstructure of the active electrode coating may lead to the undesirable concentration and precipitation in the reaction zone; this, however, exerts little effect on the electric strength $d\eta/dX$ (Fig. 3.8).

At large Peclet numbers, the system becomes unstable also with respect to other process parameters making part of the parametric Q complex. Shown in Fig. 3.9 are the curves $\xi^\circ(0)$ as functions of $\Delta Q = Q - 1$ at different Pe numbers. For convenience, each Pe number has been assigned its particular ΔQ scale: at Pe = 0.1 and 0.4, the abscissa axis is scaled to $1\Delta Q$; at Pe = 1, it is scaled up to $4\Delta Q$; at Pe = 4, to $40\Delta Q$; and at Pe = 10, to $2 \times 10^4 \Delta Q$. As is seen, with growing Pe, the value of $\xi^\circ(0)$ becomes increasingly more sensitive to ΔQ, which imposes strict constraints on the process parameters of the electrolytic cell.

3 Modelling of Electrochemical Processes

Table 3.3 Putative Characteristics of Porous Anode Layer

Notation	Characteristic	Numerical value	Dimension
L	Thickness	$\leqslant 10^{-1}$	cm
ε	Total porosity	0.4–0.7	nil
ε_{liq}	Volume fraction of liquid-filled pores	0.3–0.4	nil
r	Typical pore radius	10^{-6}–10^{-2}	cm
S	Inner specific surface	$\sim 10^4$	cm^{-1}
α	Transfer coefficient for forward chlorine reaction	3 (DSA)	nil
β	Transfer coefficient for reverse chlorine reaction	0.55 (DSA)	nil
i_0	Exchange current for chlorine reaction	$\sim 10^{-2}$ (DSA) $\sim 10^{-3}$ (platinum)	A/cm^2 A/cm^2
Δp_{break}	Laplace gas breakdown pressure	0.1–1	atm
K_{gas}	Gas-phase permeability coefficient	$\geqslant 10^{-10}$	cm^2
K_{liq}	Liquid-phase permeability coefficient	$\sim 10^{-10}$	cm^2
i	Performance current density	0.2–0.3	A/cm^2
c^{an}	Anolyte salt concentration	3.4–5.1	M
t	Electrolysis temperature	>70	°C
N	Degree of salt decomposition	1	nil
k	Transmembrane water permeability factor	3–5	nil
$\eta(0)$	Front-side electrode polarization	$\sim 10^{-2}$	V
l	Characteristic depth of reaction penetration	$\sim 10^{-2}$	cm
Δp_{gas}	Gas-pressure drop	$\leqslant 10^{-2}$	atm
Δp_{liq}	Liquid-pressure drop	$\leqslant 10^{-2}$	atm
$Q^o_{el\text{-}ch}$	Reversible heat absorption for Cl$^-$ ion discharge	$\leqslant 3 \cdot 10^{-2}$	Wt/cm^2
$Q'_{el\text{-}ch}$	Irreversible in-electrode heat release	10^{-3}–10^{-2}	Wt/cm^2
Q_{vap}	Vaporization heat absorption	$>2 \times 10^{-2}$ (at $t > 70$ °C)	Wt/cm^2
Q_M	Heat produced by current in metal base	$\leqslant 10^{-6}$	Wt/cm^2

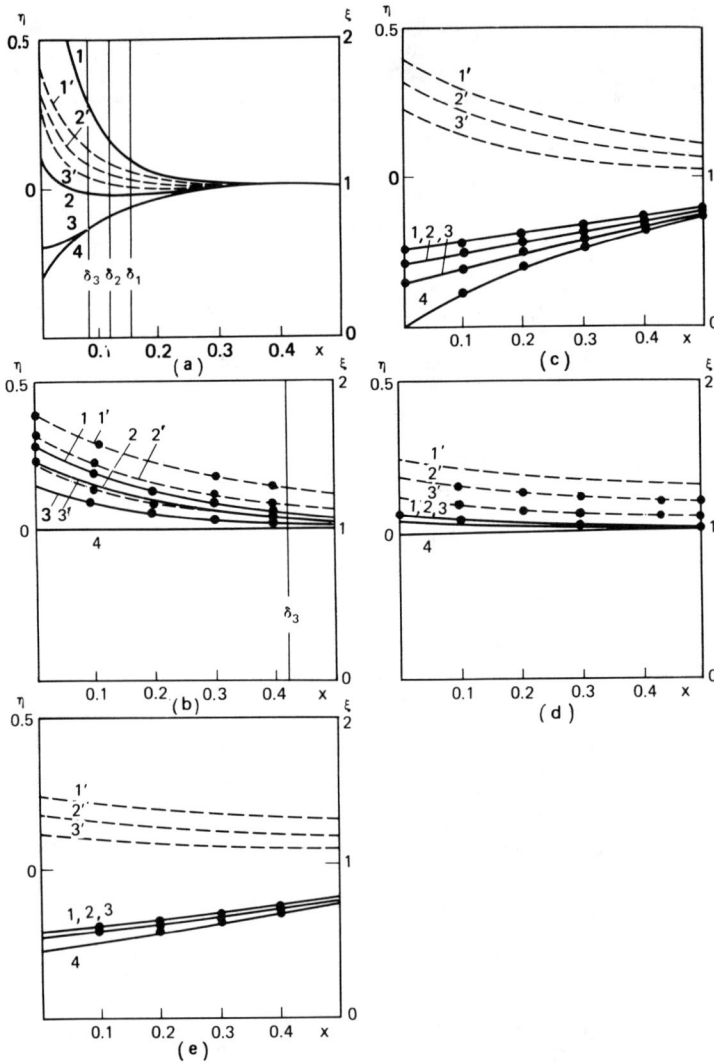

Fig. 3.8 Polarization (η) and concentration (ξ) distribution curves (shown in dashed and solid lines, respectively) for different process parameters and exchange currents:

curves 1, 1', $i_0 = 3 \cdot 10^{-3}$ A/cm^2; curves 2, 2', $i_0 = 5 \cdot 10^{-3}$ A/cm^2; curves 3, 3', $i_0 = 2 \cdot 10^{-2}$ A/cm^2; curve 4 denotes $i_0 = \infty$ ($\xi^0(X)$); δ_1, δ_2, δ_3 are characteristic thicknesses of reaction zones continuous to the electrode front side $X = 0$, for the respective current densities i_0.

	L, cm	k	Pe	Q	i, A/cm^2
a	0.05	10.42	11.8	1.00	0.2
b	0.01	10.42	2.36	1.00	0.2
c	0.01	0.48	2.36	1.15	0.2
d	0.005	10.42	0.59	1.00	0.1
e	0.005	1.31	0.59	6.6	0.1

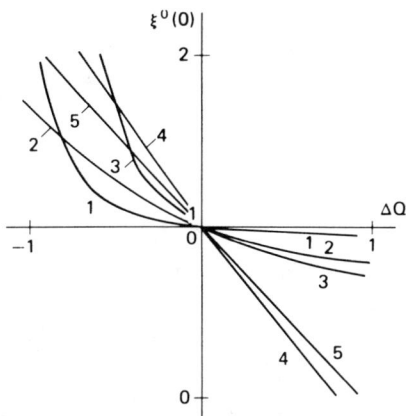

Fig. 3.9 Relationship $\xi^0(0)$ vs. ΔQ for different Peclet numbers in electrolysis using a solid-polymer electrolyte anode (SPEA):

curve 1, Pe = 0.1; curve 2, Pe = 0.4; curve 3, Pe = 1.0; curve 4, Pe = 10 (for scale change, see text)

In commercial electrolyzers, at current densities of 0.2-0.3 A/cm^2, a value of Pe $\sim$ 1 corresponds to the electrode thickness L of the order of 30-70 μm, in agreement with the evidence available in the patent literature.

Denoting $G = k + \dfrac{p_w}{2(p_{gas}(1) - p_w)}$ we can write $Q = \dfrac{c_0}{Nc^{an}G}$; consequently, the conditions $Q \approx 1$, $N = 1$, $c_0/c^{an} \approx 10$ determine a value of $G \approx 10$. However, within a temperature interval of 70-90°C, the p_w is 0.3-0.6 atm, with the value of k being 3-5 for the studied types of membranes. Since the pressure $p_{gas}(1)$ is $>$ 1 atm, we estimate the G to be $\leqslant$ 4.5, that is, $Q \approx 2$, which must produce a sharp depletion of the reaction zone solution and a decrease in current efficiency. Indeed, the available evidence in the patent literature recommends, in the SPE system, either to feed in an acidified brine, or to use an anodic unit specially designed to ensure a direct contact between the membrane and free anolyte solution. Such measures allow one to reduce the current loss due to oxygen formation. If a direct membrane-anolyte contact is impracticable, a membrane with the permeability factor k of the order of 10 is required under conventional process conditions. An alternative version ensuring the value of $G \approx 10$ subject to $k \approx 3 \div 5$ is the implementation of a process at elevated temperature and pressure ($p > 6.5$ atm, $t > 160$°C).

In the general case, to assess the response of a process to varied nonelectrochemical parameters, we expand the quantity Q in terms of parameters c^{an} and G:

$$\Delta Q = -\frac{\Delta c^{an}}{c_0^{an}} - \frac{\Delta G}{G_0} \qquad (135)$$

where $\Delta Q = Q - 1$, $\Delta c^{an} = c^{an} - c_0^{an}$, $\Delta G = G - G_0$, taking into account that $c_0/(Nc_0^{an}G_0) = 1$. Omitting the mathematics, we merely wish to point out that at Pe $\leqslant$ 1 and at $t = 70\text{-}90°C$, the admissible variations Δc^{an}, Δk, Δt (at the value of Q, in accordance with formula (135) close to unity) are large enough, that is, the process is stable to be run. By contrast, at $p > 6.5$ atm and $t > 160°C$, the temperature variation as small as 1°C produces a near 10 % change in NaCl concentration in the reaction zone. For this reason, such conditions should be recognized as impracticable.

Thus, the use of a SPE system as applied to the technology of chlorine membrane electrolysis may be recommended, with regard to the constraints imposed on the thickness of porous anode (30-70 μm) and in conjunction with the use of membranes with water permeability factor k at least twice as large as that of Nafion or MF-4 SK-type membranes. The said above does not extend to electrolysis methods in which the membrane is in partial contact with the fresh anolyte solution (that is, in the case of ZGA).

Estimation of Electrode Layer Acidity. Whereas the variation in NaCl concentration affects the relative rates of chlorine and oxygen evolution, the decrease in pH below a certain level leads to a greater voltage drop, to a reduced current efficiency for the alkali formation, and to a damage of bilayer membranes. The source of acidity is the side reaction $2H_2O \rightarrow O_2 + 4H^+ + 4e^-$. As is easily seen, both the direction of convective flow and the migration of hydrogen ions by the action of electric field favor their accumulation near the membrane surface. To estimate this effect, a transport equation for H^+ ions has been solved. For a condition most appropriate to NaCl concentration, that is, at $Q = 1$, and with allowance made for an estimate similar to that used in deriving equation (131), one may write, in a dimensionless form:

$$\frac{d}{dX}\left[\frac{d\xi_H}{dX} + \left(\frac{D_s}{D_H}\text{Pe} - \frac{d\eta}{dX}\right)\xi_H\right] = -\frac{k_H L^2}{D_H(c_H^{an})^3}\frac{1}{\xi_H^2} \qquad (136)$$

Here ξ_H denotes the ratio of the variable concentration of hydrogen ions to their concentration in the bulk of the anolyte, $\xi_H = c_H/c_H^{an}$; D_H is the diffusion coefficient. The kinetic term in the right-hand side of (136) is that borrowed from [184], at the rate constant $k_H = 2 \cdot 10^{-11}$ (mol/cm^3)$^3 \cdot$s^{-1} for DSA-type anodes. The boundary conditions take the form

$$\left.\frac{d\xi_H}{dX}\right|_{X=0} = N\text{Pe}\,\frac{D_s}{D_H}\,\frac{c^{an}}{c_H^{an}}(1-B_{OH^-}) - \text{Pe}\left(\frac{D_s}{D_H}+N\gamma_a\right)\xi_H\bigg|_{X=0} \qquad (137)$$

and $\xi_H = 1$ at $X = 1$.

Condition (137) expresses the neutralization of H^+ ions by the hydroxyl ions supplied from the cathodic compartment by back migration. The numerical computation has been carried out for the layer thickness of 50 μm and at Pe = 0.5, $i_0 = 10^{-6}$, 10^{-4}, and 10^{-2} A/cm^2, $0.5 \leqslant B_{OH^-} \leqslant 1$. The value of ξ_H (at $X = 0$) is little affected by electrolysis conditions, and in any computation version the value of pH within the layer adjacent to the membrane surface never exceeded 0.6. It is to be inferred, therefore, that the SPEA design scheme, even under the optimum NaCl concentration conditions, does not preclude from a loss in process effectiveness.

Cathodic Process. The macrokinetics analysis within the SPEK system is carried out in a manner very much similar to that applied to the SPEA case. The coordinate $X = 0$ corresponds to the membrane surface, and $X = 1$, to the bulk of catholyte solution. All the notations and equations derived for the anodic process retain their original meaning; however, these characteristics of the solution are related to NaOH. The only distinctive feature is that the total water flow within the layer is not constant but rather liable to variation because of the water decomposition by reaction $H_2O + e^- \to 1/2 H_2 + OH^-$. Accordingly, the quantity Q is defined as

$$Q = c_0(Nc^{cat})^{-1}\left[1+k+\frac{p_w}{2(p_{gas}(1)-p_w)}\right]^{-1}$$

In accordance with the direction of water flow within the membrane, one must take $N_0 > 0$ and, consequently, $k < 0$. The positive flux N_s is determined by the total amount of alkali supplied; therefore, for the quantity N to remain positive, it must be defined for the SPEK as $N = i/FN_s$.

The said above having been taken into account, the form of equations for $\pi_{gas}(X)$ and $\pi_{liq}(X)$ remains unchanged, whereas the equation for dimensionless alkali concentration within the layer, $\xi = c/c^{cat}$ takes the form

$$\frac{d\xi}{dX} = -\text{Pe} - \frac{\text{Pe}}{Q}\xi + \frac{1}{\gamma_K}\left[\left(P_3 - \frac{1}{NQ}\right)\xi - t_+\right]\frac{d\eta}{dX} \qquad (138)$$

where c^{cat} is the alkali concentration in catholyte, and $\xi = 1$ (at $X = 1$). Considering both the definition of Q distinct from that in the anodic problem,

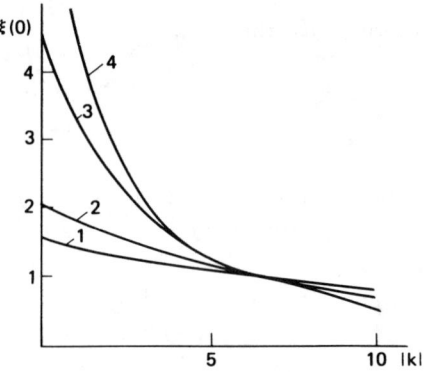

Fig. 3.10 Relative alkali concentration at membrane interface as a function of membrane permeability factor in electrolysis using SPEK at different Peclet numbers:

curve 1, Pe = 0.5; curve 2, Pe = 1.0; curve 3, Pe = 3.0; curve 4, Pe = 5.0

and the smallness of t_+ in alkaline solution, the estimate $\left| P_3 - \dfrac{1}{NQ} \right| \ll t_+$ is no longer actual, and this precludes the derivation of an approximate formula of type (131). Still, the problem solution may be estimated in qualitative terms. As can be easily verified, the parameter Q has a physical meaning of the rate ratio for the efflux of water and alkali from the boundary diffusion layer at $X = 1$, taken with a negative sign. For this reason, at $Q > 0$ (low membrane permeability), the convection prevents the alkali efflux providing thereby for a sharp increase in alkali concentration within the layer. At $Q < 0$ (moderately high permeability), the convection facilitates the alkali efflux, and the concentration effect shows up less notably. This is achieved, in particular, at $-k = 3\text{-}5$. At large values of $-k$, an alkali depletion of the solution is possible. In neglect of polarization, a solution to the problem is obtained, by analogy with (132):

$$\xi°(X) = -Q + (1 + Q)e^{\frac{\text{Pe}}{Q}(1-X)} \tag{139}$$

The negative exponent at $Q < 0$ specifies here a weaker, as compared to the anodic process, dependence on the Pe number, that is, on the process technology. Furthermore, the current efficiency is to a greater extent determined by the permselective properties of an ion-exchange membrane rather than by the catholyte composition. All this predetermines a greater stability of the process vis-à-vis the SPEA.

Since $\left[\left(P_3 - \dfrac{1}{NQ}\right)\xi - t_+\right]\dfrac{d\eta}{dX} > 0$, the polarization effect is manifested by a lower alkali concentration in comparison to $\xi^o(X)$, as distinct from its increase relative to $\xi^o(X)$ in the anodic layer.

By way of example, calculations have been performed for the layers of thickness 30 and 100 μm, and for exchange currents from 10^{-6} to 10^{-3} A/cm², the porous structure characteristics being analogous to those of the anodic layer. The process conditions for the membrane-type cell have been chosen $1/NQ - P_3 \sim 0.16$ and $\gamma_K \sim 0.5$. The membrane permeability effect was accounted for by varying the value of Q in Eq. (138). Within the variation ranges chosen for layer thickness and exchange current, no notable effects were observed. The relevant results are represented in Fig. 3.10.

Briefly summarized, the conclusion of this Section is that the system SPEK, as distinct from the SPEA, poses no basic difficulties for stable process performance from a macroscopic standpoint. For the currently employed types of ion-exchange membranes, conditions can be provided enabling to minimize the concentration effects.

3.9 Diaphragm-Type Chlorine Cells Studied by Computer Experiment Method

In performing computations, one must specify fitting parameters in the chosen model for current density distribution and gas content. Such parameters are the thickness of the gas boundary layer d_{gas} and the resistance coefficients λ_d and λ_{gas}. We have assumed $d_{gas} = 0$ for a diaphragm-type cell in which the electrode gap is not too narrow. However, the respective contribution to the voltage, proportional to $\lambda_{gas}d_{gas}$, is not small, and this is formally accounted for by the parameter λ_d. An agreement with the commercial electrolyzer voltage was obtained at a value of λ_d about 10. The estimate of parameter f_ε within the current efficiency theory has been given above in Section 8.3. More detailed data as well as discussions concerning the choice of an acceptable range of circulation velocity w_0 and the effect due to the exchange coefficient D have been dealt with elsewhere [161, 167, 176]. The agreement between the experimental and computational findings provides evidence for the eventual practical utility of a computer-aided optimization of the chlorine-alkali cell processes [191, 192].

3.9.1 Requirements for Design Parameters

Requirements for Accuracy of Electrode Unit Assembly. Shown in Fig. 3.11 are the calculated curves for current efficiency B_{OH^-} (%) and cell-liquor chlorate concentration (g/dm³) for an asymmetric configuration of the anode unit modelled after a commercial electrolytic cell with bottom anode lead. The anode displacement Δ is given as a fraction of the maximum displacement (with the anode pressed against the diaphragm), in the absence of a skewness. The alkali concentration was varied as 110, 130, and 150 g/dm³ at the beginning, mid-point, and end of the diaphragm's working cycle. In the case of a skewness (that is, with the anode deflecting from the vertical), the c relationships remain similar. The assembly accuracy requirement is such that the relative anode displacement and skewness must not exceed 0.5-0.75 in order not to reduce the cell performance characteristics. The same applies to the anode design with side leads.

Effect of the Cell Height on Performance Efficiency. Let us consider a commercial electrolytic cell designed for a current load of $i^{avg} = 0.143$ A/cm², the height of the electrode working section being $H = 76$ cm. Suppose, the cell height can be varied in such a manner as to allow the overall current load kept at a constant level, that is $Hi^{avg} = $ const. Under these conditions, the voltage decreases with increasing H, and an intensification of side reactions is to be anticipated. By contrast, at small H, the voltage is high, and the side

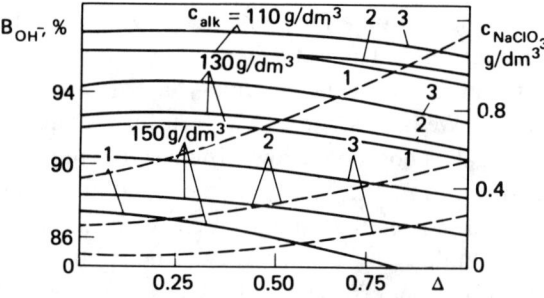

Fig. 3.11 Effect of anode displacement Δ on NaOH current efficiency (solid lines) and catholyte sodium chlorate concentration (dashed lines) at catholyte alkali concentration 130 g/dm³.

Curve 1, temperature 90°C; curve 2, 100°C; curve 3 refers to the regime of additional anolyte saturation with NaCl to concentration level of 300 g/dm³

reactions are essentially suppressed. Thus, a computer experiment was set up with a view to determine optimum parameters of the cell design, with account taken of the overall mass and heat balances in the cell, as well as of the current density distribution and voltage balance. Along with the cell height H, other parameters were allowed to vary: the concentration of alkali produced, brine feed temperature (70-80°C), heat loss factor (to account for the heat lost into the surroundings), circulation velocity w_0 (1-40 cm/s), initial gas content ε_0 (0-0.1), membrane porosity-tortuosity factor f_ε (0.35-0.40), lateral electrolyte flow mixing (see Section 3.8.4.2). Some of the results are summarized in Table 3.4. Since the specific electric energy consumption η_{el} is a performance characteristic, we have plotted $\eta_{el} = 0.67\, E/B_{OH^-}$ as a function of height H (Fig. 3.12); B_{OH^-} is expressed in fractions of unity. At a given alkali concentration, the region between two neighbouring curves corresponds to the calculated voltage and current efficiency intervals as represented in Table 3.4.

Thus, the following conclusions can be drawn: (i) the design of an electrode unit of greater height with the current load kept unchanged is not expedient because of a decreased current efficiency and the formation of chlorate; (ii) the design height smaller than 50-60 cm would result in a sharply augmented energy consumption; (iii) the height $H = 76$ cm seems to be optimal from the standpoint of energy consumption, but it is close to the limit of intense chlorate formation at higher concentration of alkali product. In order to make the technology less susceptible to side reactions, a somewhat smaller design height appears to be preferable. In varying other parameters, for example, current load, additional calculations are needed to reach a more complete optimization of the electrolytic cell design.

Effect of Cathode Mesh Pattern on Current Efficiency; Applicability Range of Thin Modified Diaphragms. Summarized in Table 3.5 are the calculated results for commercial electrolysis technology (alkali concentration 130 g/dm³; current density 0.14 A/cm²; temperature 90°C; diaphragm thickness $d = 2.25$ mm). The greater cathode mesh spacing h results in a marked decline of the performance characteristics as the non-uniformity microscale becomes commensurate with the diaphragm thickness; however, this trend is less pronounced as the free cross-sectional area α is made to increase.

In the electrolysis using thin modified diaphragms, one must control the process conditions that provide for a high current efficiency. One of the similarity criteria for mass-exchange processes within the diaphragm is the ratio $id/(D_2^0 c_{alk})$, where c_{alk} is the catholyte alkali concentration. Indeed, as has been shown by calculations, the current efficiency is kept unchanged with

Table 3.4 Performance Characteristics of a Vertical-Electrode
Diaphragm Cell at Different Lengths of Electrode Working Section

Electrode working section length, cm	Electrolysis temperature, °C	Cell voltage, V	Alkali product concentration, g/dm³									
			115		130		135		140		145	
			B_{OH^-}, %	chlorate, g/dm³	B_{OH^-}, %	chlorate, g/dm³	B_{OH^-}, %	chlorate, g/dm³	B_{OH^-}, %	chlorate, g/dm³	B_{OH^-}, %	chlorate, g/dm³
30	101 ± 1	4.20-4.44	99.2	—	—	—	—	—	—	—	—	—
50	97 ± 1.5	3.62-3.74	99.0-99.1	—	98.2-98.7	—	97.8-98.4	—	97.3-98.0	—	99.1-99.2	—
65	94.5 ± 1.7	3.42-3.51	98.1-98.7	—	96.5-97.4	—	95.8-96.9	—	95.1-96.3	—	96.8-97.6	0-0.03
76	93 ± 2	3.31-3.44	97.1-98.0	—	94.9-96.2	—	94.1-95.5	0-0.05	93.1-94.7	0.01-0.23	94.3-95.6	0.1-0.6
90	92 ± 2	3.22-3.30	95.5-96.8	—	92.7-94.5	0-0.4	91.6-93.6	0.16-0.81	90.5-92.6	0.5-1.47	92.2-93.9	1.0-2.2
120	90 ± 2.8	3.12-3.20	91.8-93.9	0.1-0.7	87.8-90.5	1.4-2.7	86.3-89.2	2.14-3.76	84.8-87.9	3.0-4.66	83.2-86.5	3.9-5.8
150	88 ± 3	3.06-3.13	87.3-90.2	1.3-2.6	82.1-85.8	3.9-6.1	80.2-84.2	4.84-7.06	78.2-82.4	6.39-8.75	76.2-80.7	7.4-10.8

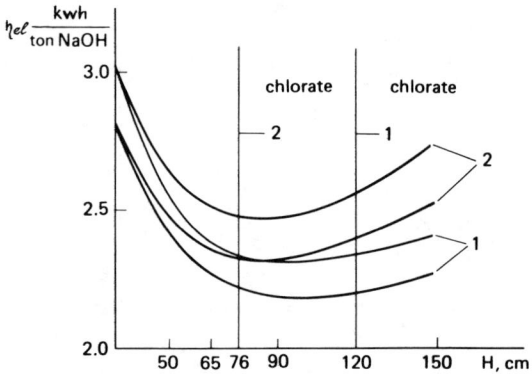

Fig. 3.12 Energy consumption η_{el} as a function of the working length H of a vertical electrode, with overall current load kept constant.

Curve 1, alkali concentration 115 g/dm³; curve 2, 145 g/dm³. In either case, the respective solid line delineates regions with low rates (on the left-hand side) and high rates (on the right-hand side) of side reactions

Table 3.5 Current Efficiency (%, in parentheses) and Catholyte Sodium Chlorate Concentration (g/dm³) as Calculated for Different Cathode Mesh Spacing h (cm) and Free Cross-Sectional Area α

h	α				
	0.05	0.39	0.60	0.90	0.95
0.10	0.15 (93.5)	0.15 (93.4)	0.17 (93.4)	0.19 (93.4)	0.19 (93.4)
0.32	0.24 (93.3)	0.19 (93.3)	0.19 (93.3)	0.22 (93.2)	0.23 (93.1)
0.70	1.35 (90.1)	0.69 (91.7)	0.46 (92.4)	0.38 (92.4)	0.11 (93.5)
1.00	3.65 (85.5)	1.73 (89.3)	1.00 (91.1)	1.13 (90.8)	0.53 (92.3)
1.50	7.74 (76.7)	4.41 (84.3)	2.76 (87.2)	1.93 (89.0)	0.75 (91.8)

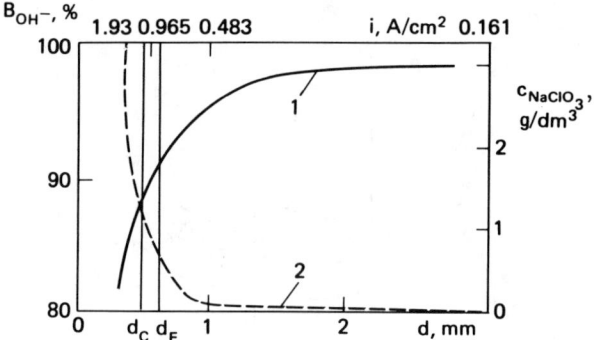

Fig. 3.13 Effect of microscopic current density and filtration velocity nonuniformities on current efficiency (curve 1) and catholyte NaClO$_3$ concentration (curve 2) in thin-diaphragm electrolysis

decreasing d if the current density i is made simultaneously increase in a manner providing for id = const, on condition that both the filtration and the current are uniform. The mesh structure of the cathode entails a dependence of performance characteristics on two other parameters, α and h/d; for the latter, the relevant correlations are shown in Fig. 3.13 (temperature 82.5°C; c_{alk} = 130 g/cm^3; f_ε = 0.40 α = 0.39; h = 0.32) (standard cathode). For simplicity, the change in i (on the scale of electrode height) was ignored, providing thus for $i = i^{avg}$ (size-average current density). The d-to-i ratio was chosen such as to provide for a high ideal-cell current efficiency (close to 98% at id = const).

A significant decline of performance characteristics has been observed at the diaphragm thickness about twice as large as the microscopic nonuniformity scale, which for the standard cathode makes up about 1 mm. Thus, for a diaphragm operating in direct contact with the cathode mesh, there should be specified an acceptable minimum thickness d_{min}, below which the filtration and current nonuniformities both lead to a sharp decline in current efficiency and alkali product quality.

3.9.2 Performance Characteristics as a Function of Current Load

The practical experience in commercial use of chlorine-alkali cells provides evidence that, for accidental reasons independent of the operating personnel, the current load may sharply change. Then to stabilize the cell parameters, the cell control system must, in accordance with the maintenance rules, regulate the anolyte height in correspondence with the altered current load. However, such variations in cell performance should be possibly avoided since they lead to undesirable consequences, in particular, reduced current efficiency and contamination of alkali with sodium chlorate. Regrettably, under workshop conditions in many instances the relevant operational data have not been sufficiently systemized to enable a reliable analysis of emergency situations. Meanwhile, a theoretical prognostication of the eventual extent of disturbance effects would have been a helpful guidance in avoiding the most "hazardous" conditions arising from current load change.

We have calculated the steady-state characteristics of industrial electrolyzers as a function of current density within a range of $0.1 \text{ A/cm}^2 \leqslant i^{\text{avg}} \leqslant 0.25 \text{ A/cm}^2$ and degree of chloride decomposition within a range of $0.25 \leqslant n \leqslant 0.65$.

We have examined the effect of 5, 10, 20 and 50% load drops on the performance characteristics of an electrolytic cell operated at normal loads of 0.2 and 0.25 A/cm². Accordingly, the calculations have been carried out for two series of perturbed current load: (i) 0.19, 0.18, 0.16, 0.10 A/cm² and (ii) 0.2375, 0.225, 0.2, 0.125 A/cm². The temperature of electrolytic process has been calculated as a function of current density and degree of chloride decomposition, given the following initial conditions: brine feed temperature, 70°C; NaCl brine concentration, 300 g/dm³; average pressure, 1 atm; ambient temperature, 23°C; anolyte circulation velocity, 10 cm/s. The calculated curves are represented in Fig. 3.14. The temperature error arising from lateral mixing in the anolyte compartment was not large and did not exceed 0.5-1.0°C. It has been assumed, in examining the side reactions, that the HCl brine concentration is 1 g/dm³; the diaphragm thickness, 2.5 mm; and the membrane porosity-tortuosity factor, $f_\varepsilon = 0.4$. The current efficiency for the alkali formation and the sodium chlorate catholyte concentration as functions of n under different process conditions are shown in Figs. 3.15 and 3.16. In the chlorate concentration vs. degree of decomposition curves, one will have observed a characteristic inflexion with the occasional trend to extend to a plateau. Such a behaviour conforms to the theory of delayed mass transport from the gas bubbles (see Section 3.8.2).

The plotted graphs provide a means to estimate the effect due to current load variation. Suppose, the cell has been run under conditions A_1 close to the optimal ones ($i^{\text{avg}} = 0.2 \text{ A/cm}^2$, $n = 0.5$, $t = 96°C$, $B_{\text{OH}^-} = 97.5\text{-}98.0\%$,

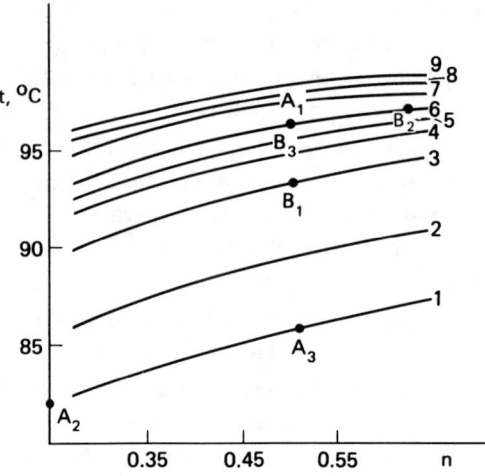

Fig. 3.14 Electrolysis temperature t as a function of the degree of salt decomposition n at current densities

0.1 A/cm² (curve 1); 0.125 A/cm² (2); 0.16 A/cm² (3); 0.18 A/cm² (4), 0.19 A/cm² (5), 0.20 A/cm² (6), 0.225 A/cm² (7), 0.2375 A/cm² (8), 0.25 A/cm² (9). For details, see text

$c_{NaClO_3} = 0$). The 50% load drop leads to conditions A_2 ($i^{avg} = 0.1$ A/cm², $n = 0.25$, $t = 82°C$, $B_{OH^-} = 97.0$-97.5%, $c_{NaClO_3} = 0$). Keeping the current density unchanged and raising the alkali concentration to its specified value, we come to conditions A_3 ($i^{avg} = 0.1$ A/cm², $n = 0.5$, $t = 86°C$, $B_{OH^-} = 93\%$, $c_{NaClO_3} = 0.3$ g/dm³). The $A_1 \rightarrow A_3$ transition results in a 4.5-5.0% current efficiency drop and in accelerated chlorate formation. Stressing the role of temperature, the conditions A_3 may be compared to conditions A_3' ($i^{avg} = 0.1$ A/cm², $n = 0.5$, $t = 96°C$, $B_{OH^-} = 88$-90%, $c_{NaClO_3} = 0.8$-1.17 g/dm³), if the temperature is assumed to be the same in the initial and final states. At a fixed degree of decomposition, the rise in temperature leads to a more concentrated catholyte owing to the carryover of moisture with vapor, which finally results in impaired performance characteristics. On the other hand, at constant alkali concentration, the rise in temperature is known to improve both the current efficiency and product quality. As an example of 25% load increase, one may consider the transition from conditions B_1 ($i^{avg} = 0.16$ A/cm², $n = 0.5$, $t = 92.5°C$, $B_{OH^-} = 96.5$-97.0, $c_{NaClO_3} = 0$) to

3 Modelling of Electrochemical Processes

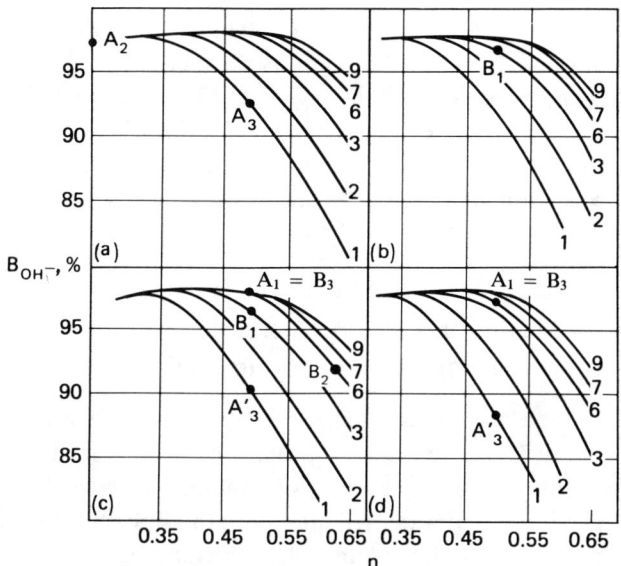

Fig. 3.15 Current efficiency for alkali formation B_{OH^-} (%) as a function of the degree of salt decomposition n and current density at 85 °C (panel a); 90 °C (b); 95 °C (c); 100 °C (d). Current densities are the same as in Fig. 3.14

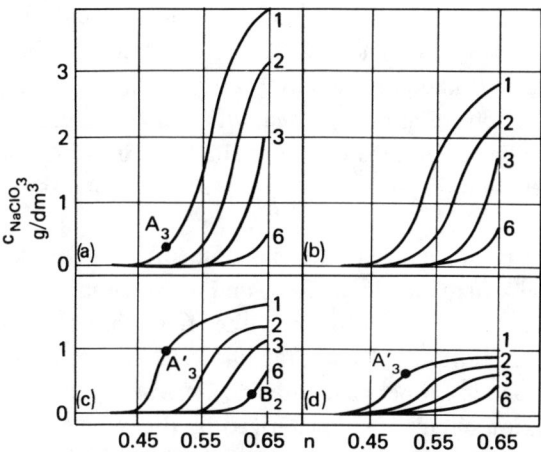

Fig. 3.16 Sodium chlorate concentration c_{NaClO_3} in electrolytic alkali solution as a function of the degree of salt decomposition n, current density, and temperature (for notations, see Figs. 3.14 and 3.15)

conditions B_2 ($i^{avg} = 0.2$ A/cm^2, $n = 0.625$, $t = 97°C$, $B_{OH^-} = 92\%$, $c_{NaClO_3} = 0.35$) and further to conditions $B_3 \equiv A_1$.

A general conclusion is therefore that the most "hazardous" are technology operations leading to an enhanced salt decomposition. The mathematical modelling of an electrolytic cell enables one to predict effects arising from the combined action of essential process conditions.

3.10 Certain Technologic Aspects in Modelling the Performance of Membrane-Type Recycle Cells

The electrolysis technology using ion-exchange membranes provides a possibility of more versatile process control as compared to the diaphragm method. It enables, by way of example, varying the rate of brine fed into the anodic, and of water fed into the cathodic compartment, to control the HCl acidification of electrolytic cells (operated both singly or in series), to alter the process temperature, to provide for cell solution recycling, and a number of other operations.

The maintenance of anolyte pH and sodium chloride anolyte concentration within their required ranges is an essential factor in the process of brine electrolysis: an increase in pH leads to a reduced anodic current efficiency and to a higher content of both sodium chlorate and hypochlorite, whereas the decrease in pH below a definite level (determined by the membrane type) leads to a larger voltage drop and may therefore cause damage to the modified membranes. In this Section we are concerned with a mathematical model for the anodic material balance of cells operated singly or in series, taking into account the side reactions attendant to the dissolution of gaseous chlorine in brine [185].

We consider the performance of a series of membrane-type cells, or a single cell with a parallel scheme for brine feed, and a sequential scheme for current supply (Fig. 3.17). It is assumed that the fresh brine of concentration c^0 mol/cm^3 NaCl is uniformly fed to all the cells at a rate of V^0 cm^3/s per cell. The current load is I A per element. The anolyte recycling as implemented through a circulation circuit is characterized by the recycle ratio k; therefore, the recycle flow rate is kNV^{an}, where NV^{an} is the overall anolyte flow rate (Fig. 3.17).

The difference in current efficiencies observed for the cell elements is primarily due to different service life of the membranes or to inherently differ-

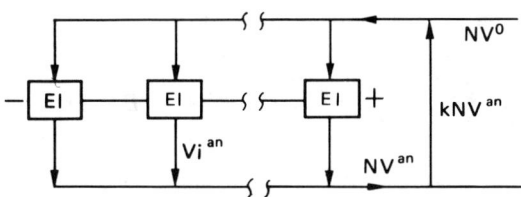

Fig. 3.17 Brine feed diagram for elements (El) of a series of N electrolyzers or N working cells of an electrolyzer

ent membrane performance characteristics. Suppose, there are M groups of cell elements; within each group, the performance characteristics are identical, whereas the groups differ among themselves in the current efficiency for the alkali production, B_{OH^-i}. The relative contribution of the respective elements within a group is denoted by f_i; in other words, this is the probability to identify an element with the given efficiency B_{OH^-i}. Trivially, the f_i is determined by the performance characteristics of a membrane; by definition, $\sum_{j=1}^{M} f_j = 1$.

Estimation of NaCl Concentration in the Anolyte. We denote by V_i^{an} (cm^3/s) the anolyte flow rate at the outlet of an element of the i-th group. The water balance equation for the i-th group element is written as

$$V^0 + k \sum_{j=1}^{M} f_j V_j^{an} = V_i^{an} + \frac{9I}{F\varrho_0} \frac{p_{wi}}{p - p_{wi}} + \frac{18 I B_{OH^-i} k_i}{F\varrho_0} \quad (140)$$

The difference in water content for solutions of different concentration may be ignored; in the concentration range of interest, the density for all the solutions may be taken equal to $\varrho_0 = 0.87$ g H$_2$O/cm^3 solution. The equilibrium vapor pressure p_{wi} can be expressed through temperature and solution composition; the second term in the right-hand side of (140) characterizes the anolyte water loss by vaporization. The third term in (140) determines the transmembrane water transport rate, provided that k_i water molecules comigrate with a single Na$^+$ ion (the Na$^+$ ion flux is IB_{OH^-i}/F mol/s, in neglect of the Cl$^-$ ion transport into the cathodic space). Equation (140) disregards the supply of water as produced by side reactions. By the order of magnitude, this water production does not exceed $I(1 - B_{OH^-i})/F$. Commonly, $k_i = 3\text{-}5$ and, taken $B_{OH^-i} \sim 0.9$, the respective correction would have been a mere 1/30-1/50 of the third term value in Eq. (140). An approximation of typical concentration dependences for k_i has been given in [185].

We represent the Na$^+$ ion balance in the form

$$V^0 c^0 + k \sum_{j=1}^{M} f_j V_j^{an} c_j^{an} = V_i^{an} c_i^{an} + \frac{I}{F} B_{OH^-i} \qquad (141)$$

Through introducing a dimensionless velocity $v_i = V_i^{an}/V^0$ and a dimensionless concentration $\xi_i = c_i^{an}/c^0$, equations (140) and (141) are rewritten as

$$1 + k \sum_{j=1}^{M} f_j v_j = v_i + \varphi_i \qquad (142)$$

$$1 + k \sum_{j=1}^{M} f_j v_j \xi_i = v_i \xi_i + \psi_i \qquad (143)$$

Here $\varphi_i = \dfrac{9I}{F\varrho_0 V^0} \dfrac{p_{wi}}{p - p_{wi}} + \dfrac{18 I B_{OH^-i}}{F\varrho_0 V^0} k_i$, $\psi_i = \dfrac{I B_{OH^-i}}{F V^0 c^0}$. Explicit expressions are thus derived from equations (142), (143)

$$v_i = \dfrac{1 - k \sum_{j=1}^{M} f_j \varphi_j}{1 - k} - \varphi_i, \qquad (144)$$

$$\xi_i = \dfrac{1 - k \sum_{j=1}^{M} f_j \psi_j - (1-k)\psi_i}{1 - k \sum_{j=1}^{M} f_j \varphi_j - (1-k)\varphi_i} \qquad (145)$$

Expressions (144) and (145) are meaningful at $\xi_i > 0$ and $v_i > 0$. Under these conditions, supplemented with $k < 1$, one can arrive at inequalities $\varphi_i < 1$, $\psi_i < 1$, which thus impose constraint on the admissible parametric values. Since the φ_i is a non-linear function of ξ_i, a numerical algorithm has been developed for solving the system of equations (144), (145) ($i = 1, \ldots, M$).

Shown in Fig. 3.18 is the anolyte NaCl concentration vs. brine consumption rate relationship calculated for a series of 40 anolyte cells) equipped with MF-4SK membranes. The following performance characteristics have been assumed: $I = 350$ kA, $c^0 = 300$ g/dm^3, temperature 80°C. The narrow space between the two curves characterizes the spread in values on condition that the current efficiency for the production of alkali varies within the cell series from 85 through 95%, with no electrolyte recycling involved in the process. Assuming that for a steady and economical run of the process the concentration must fall within a range of 200 to 240 g/dm^3, the overall brine

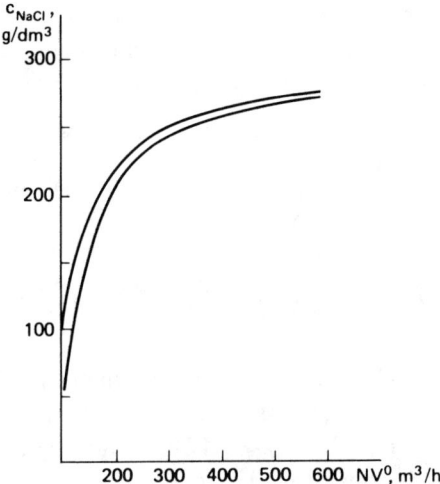

Fig. 3.18 Anolyte chloride concentration as a function of brine consumption for a series of 40 electrolyzers at $85\% \leqslant B_{OH^-} \leqslant 95\%$ (no recycling)

consumption rate must be 200 to 260 m³/h, or 5.0-6.5 m³/h per cell. According to Eqs. (142), (143), the anolyte composition is determined by the I/V^0 ratio. Therefore, in order to comply with the above concentration requirements, this ratio must be maintained within a range of 54 to 70 kA h/m³ per cell. It can be shown that these conclusions remain valid also for Nafion-type membranes.

Anolyte Acidity Estimation. The acid balance for the ith element can be represented in the form

$$\frac{I}{F}(B_{Cl_2 i} - B_{OH^- i}) = V^0 c_H^0 + k \sum_{j=1}^{M} f_j V_j^{an}(c'_{1j} + c'_{2j} + 2c'_{3j})$$

$$- V_i^{an}(c'_{1i} + c'_{2i} + 2c'_{3i}) \tag{146}$$

Here $B_{Cl_2 i}$ is the current efficiency for the formation of chlorine; $c'_{1i}, c'_{2i}, c'_{3i}$ are the concentrations for HCl, HClO, and Cl₂ in the liquid anolyte phase, mol/cm³; c_H^0 is the HCl concentration in brine feed, mol/cm³. Equation (146) is in accord with the conclusions concerning the electrolytic cell material balance as reported in [186]. Since the efficiency $B_{Cl_2 i}$ is to a significant extent dependent on anolyte acidity, its relationship must be treated in terms of electricity balance, taking into account that the sum of the current efficiencies for the formation of chlorine, oxygen, and active chlorine is equal to 1, that is,

$$\frac{I}{F}(1 - B_{Cl_2 i}) = \frac{I}{F}B_{O_2 i} - 2k \sum_{j=1}^{M} f_j V_j^{an}(c'_{2j} + c'_{3j}) + 2V_i^{an}(c'_{2i} + c'_{3i})$$

This, added together with (146), yields

$$\frac{I}{F}(1 - B_{OH^- i}) = \frac{I}{F}B_{O_2 i} + V^0 c_H^0 + k \sum_{j=1}^{M} f_j V_j^{an}(c'_{1j} - c'_{2j}) - V_i^{an}(c'_{1i} - c'_{2i}) \quad (147)$$

In the region of interest (at pH < 4), we neglect both the formation of sodium chlorate and the eventual migration of hypochlorite toward the cathode. Apart from this, it is evident that $[ClO^-]_i \ll [HClO]_i$. Turning to the dimensionless variables, Eq. (147) is transformed into

$$\eta_{1i} - \eta_{2i} = \frac{1 - k \sum_{j=1}^{M} f_j \chi_j - (1 - k)\chi_i}{1 - k \sum_{j=1}^{M} f_j \varphi_j - (1 - k)\varphi_i}, \quad (148)$$

where

$$\eta_{1i} = c'_{1i}/c_H^0, \quad \eta_{2i} = c'_{2i}/c_H^0, \quad \chi_i = \frac{I(1 - B_{OH^- i} - B_{O_2 i})}{FV^0 c_H^0}.$$

In turn, the current efficiency for the formation of oxygen is dependent on the anolyte pH, the electrode type, the cell design, and can be modelled by means of an appropriate regression equation. As attested by calculations, the major conclusions remain in force irrespective of the actual expression for the function $B_{O_2} = f(pH)$. The system of equations (148) has been solved for the unknowns pH_i ($i = 1, \ldots, M$) using numerical methods.

Shown in Fig. 3.19 is the pH vs. B_{OH^-} relationship calculated for an uncoupled electrolytic cell, that is, at $k = 0$. The admissible pH range being $2.5 \leq pH \leq 3.5$, the spread in membrane selectivities (B_{OH^-} values) is expected to be quite narrow, not exceeding 2%. The realistic values of B_{OH^-} for an electrolytic cell may vary from 85 to 95%. In implementing recycle conditions, all the cells of an electrolytic cascade can in principle be brought to the same performance characteristics; however, this is reached only for the limit case of large k-values.

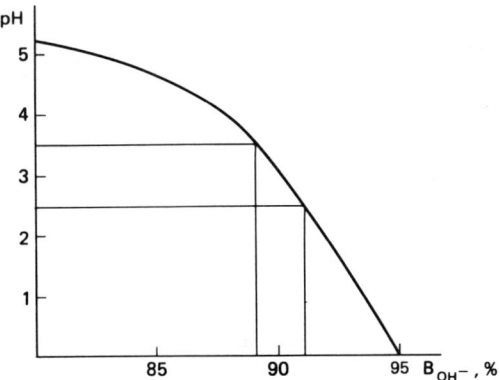

Fig. 3.19 Effect of selective properties of a membrane on anolyte pH (no recycling)

By way of example, model calculations have been performed on a single 56-cell membrane-type electrolyzer with a Gaussian distribution of membrane selectivity over the cells with the average $B^*_{OH^-}$ value of 90%. Standard deviations σ within an interval of 0.3 through 2% have been used. A typical relationship between the cell pH spread interval and the recycle ratio is shown in Fig. 3.20. The membrane nonuniformity is characterized by a spread in current efficiency ΔB over the cells. For example, it is only at $k > 0.95$ that the condition $\Delta pH \sim 1$ starts to hold. For acid-resistant membranes, the admissible interval is $1 < pH < 4$; in such a case, the recycle ratio is subject to less stringent constraints.

The above analysis has provided evidence that, in feeding an acidified brine of constant composition into a cascade of electrolytic cells with different current efficiency, conditions arise conducive to pH values deviating from the required range. Therefore, an adequately designed mathematical model can be helpful in optimizing the entire analytic cycle of brine feed preparation, including the stages of dechlorination, purification, and salt resaturation in the commercial production of chlorine and alkali in membrane cells.

Theoretical Probabilistic Approach to Characterization of Industrial Ion-Exchange Membranes on the Basis of Laboratory Experiment. It is to be inferred from the foregoing that the maintenance of an optimum anolyte composition is the major condition for the process efficiency, which imposes strict constraints on the performance characteristics of membranes intended for use in commercial electrolytic cells. The role of laboratory experiments in reaching

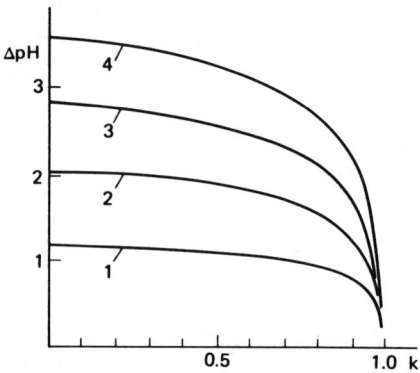

Fig. 3.20 Spread in anolyte pH as a function of recycle ratio for a 56-cell membrane-type electrolyzer.

Curve 1, at current efficiency spread for alkali formation 1.71%; curve 2, 2.84%; curve 3, 4.0%; curve 4, 5.69%

this goal is of prime importance. The membrane properties are studied using laboratory specimens of about 4 cm in diameter. Different results of membrane selectivity (current efficiency) measurements obtained on membranes of the same type provide evidence for a non-uniformity of the membrane properties.

In fact, the problem of estimating the permselective properties of a macroscopic specimen of a membrane on the basis of laboratory measurements performed on a smaller test specimen is a problem of scale-up prognostication [11].

The current efficiency obtained on a specimen of the membrane designed for use in an electrolytic cell is the average of local characteristics of separate parts of the membrane. Therefore, there are good reasons to assume that the normal distribution law holds for the current efficiency B with the average B^* and the variance D_B:

$$f(B) = \frac{1}{\sqrt{2\pi D_B}} \exp\left[-\frac{(B - B^*)^2}{2 D_B}\right] \tag{149}$$

The probable spread in B values is known to be $B^* \pm 3\sqrt{D_B}$.

We denote by N the number of elementary specimens of current efficiency B_i, constitutive of a membrane macrospecimen. First, we assume the B_i quantities to be pair-wise independent, including any two neighboring elements. The current density is assumed to be uniform over the entire membrane area. If so, the average value of a random variable B_i for a set of membranes of the same type is equal to B^*, and the respective variance is $d_B = ND_B$.

To provide for an optimum anolyte pH range throughout all the cells, $2.5 < \mathrm{pH} < 3.5$, the maximally admissible spread in current efficiency for the formation of alkali must not exceed 2%, or $\sqrt{D_B} < 0.33\%$. For a membrane of surface area about $1\,\mathrm{m}^2$, we may take $N = 900$, and, consequently, $\sqrt{d_B} < 10\%$. The quantities $\sqrt{D_B}$ and $\sqrt{d_B}$ are standard deviations; of these, the latter is easily accessible by experiment.

One might have assumed from the above example that the working cell membranes exhibit a sufficient uniformity, since the inequality $\sqrt{d_B} < 10\%$ imposes no strict constraints on the membrane performance characteristics. However, the above considerations are based on a hypothesis, yet to be verified, that the measurement results obtained on the microspecimens are independent.

We consider, in the general case, the local current efficiency $B(\vec{r})$ for the given elementary area to be a random coordinate function with respect to both the height (x-axis) and the width (y-axis) of a cell, characterized by a vector $\vec{r} = \vec{r}(x, y)$. The laboratory measurements have provided evidence for a relationship between the current efficiency and the membrane resistance; therefore, the local current density $i(\vec{r})$ is also a random function which is, however, dependent on $B(\vec{r})$ function. We denote by $d^2\vec{r}$ the elementary area; by i^*, the average-size current density in the cell; by S, the total membrane area in the cell. The value of i^* is assumed to be known and coincident with the area-averaged current density per cell, for example, in a bipolar circuit.

By definition, the current efficiency for a macrospecimen is

$$B = \frac{1}{Si^*} \int B(\vec{r}) i(\vec{r}) d^2\vec{r}$$

The mean statistical value is $\langle B \rangle = \dfrac{1}{Si^*} \int \langle B(\vec{r}) i(\vec{r}) \rangle d^2\vec{r}$ and the variance is

$$D_B = \langle B^2 \rangle - \langle B \rangle^2 = \frac{1}{(Si^*)^2} \iint K(\vec{r}, \vec{r}\,'') d^2\vec{r} d^2\vec{r}\,''.$$ Here $K(\vec{r}, \vec{r}\,'') = \langle B(\vec{r}) i(\vec{r}) \times B(\vec{r}\,'') i(\vec{r}\,'') \rangle - \langle B(\vec{r}) i(\vec{r}) \rangle \langle B(\vec{r}\,'') i(\vec{r}\,'') \rangle$ is the correlation function for the random function $B(\vec{r}) i(\vec{r})$.

The mean statistical current efficiency $\langle B \rangle$ can be estimated from the on-site operational measurements of current efficiency for an industrial electrolytic cell equipped with new membranes. To be noted, it must not necessarily be coincident with the value of B^* from laboratory experiments in view of a nonuniform current distribution. If so, the B^* should be replaced by $\langle B \rangle$ in the formula (149) for normal distribution.

Now we wish to show a way for estimating $\langle B \rangle$ from laboratory measurements. First, we derive an approximate expression for current density distribution. Let us write an overall voltage balance (E_M, the voltage drop across the membrane, is introduced in place of E_d):

$$E = E_{\text{el-ch}} + E_{\text{el}} + E_M + E_{\text{gas}} + \Delta E$$

For simplicity, the voltage drop across the electrodes is ignored. By analogy with the conventions in Section 3.8.4.1, we introduce the following notations: ϱ_{el}, the resistivity for a neat electrolyte; d_{el}, d_M, d_{gas}, the respective thicknesses for electrolyte layer, membrane, and stationary gas layer; λ_{el}, λ_M, λ_{gas}, the resistance increment factors for the respective regions relative to the neat electrolyte layer resistance; m and n, the electrochemical component characteristics. Assuming $E_{\text{el-ch}} = m + ni$, $E_{\text{el}} = \varrho_{\text{el}}\lambda_{\text{el}}d_{\text{el}}i$, $E_M = \varrho_{\text{el}}\lambda_M d_M i$, $E_{\text{gas}} = \varrho_{\text{el}}\lambda_{\text{gas}}d_{\text{gas}}i$ we arrive at an expression

$$i = \frac{E - m - \Delta E}{n + \varrho_{\text{el}}(\lambda_{\text{el}}d_{\text{el}} + \lambda_M d_M + \lambda_{\text{gas}}d_{\text{gas}})} \tag{150}$$

Known the gas content $\varepsilon(\vec{r})$ (measured or calculated) at a given point of the electrolyte bulk, the λ_{el} can be expressed in an explicit form through $\vec{r}$. The product $\lambda_M d_M$ characterizes the membrane resistance and, by assumption, is a dependent random variable: $\lambda_M d_M = \varphi(B_i)$. For this reason, the current $i(\vec{r})$ is also a random variable related to B_i through expression (150).

By performing a series of N measurements for current efficiency B_i on arbitrary membrane microspecimens under laboratory conditions, we can calculate $i(B_i)$ by formula (150) for the given coordinate $\vec{r}$ and thus estimate $\langle B(\vec{r})i(\vec{r}) \rangle \approx \sum_{i=1}^{N} B_i i(B_i)/N$, provided that N is sufficiently large. This having been done, the value of $\langle B \rangle$ is obtained by integrating with respect to the coordinate. Admittedly, such a procedure is rather cumbersome as compared to the direct measurement of $\langle B \rangle$, but it elucidates the next step, namely, the determination of variance D_B.

In accordance with the ergodic property of random functions, the statistical averages of a function can be estimated by a single realization of the function. The current densities as calculated for the coordinate $\vec{r}$ are henceforth denoted by the respective subscript, for example, $i_{\vec{r}}$. The random variable $B_{i_{\vec{r}}}$ is parametrically dependent on $\vec{r}$, since the gas content distribution and, consequently, $\lambda_{\text{el}} = \lambda_{\text{el}}(\vec{r})$ are assumed to be known. For a pair of points $\vec{r}$ and $\vec{r}\,'$,

3 Modelling of Electrochemical Processes

the properties of an isotropic membrane are determined solely by the distance $|\vec{r} - \vec{r}'|$ between the points. Suppose, this distance can accommodate an array of n elementary membrane specimens. The size of an elementary specimen being Δl, one has $|\vec{r} - \vec{r}'| \approx n\Delta l$.

We consider now an array of elementary specimens constitutive of a portion of a macrospecimen of the membrane; these are accommodated within a straight "section" of the macrospecimen and are numbered in the order of succession 1 through N, N being a sufficiently large number. In view of the assumed isotropy of the membrane, the direction of the said "section" may be chosen arbitrarily. For a pair of points $\vec{r}$ and $\vec{r}'$, we define an integer n by the formula $n \approx |\vec{r} - \vec{r}'|/\Delta l$. If so, an estimate

$$\langle B(\vec{r})i(\vec{r})B(\vec{r}')i(\vec{r}')\rangle \approx \frac{1}{N-n} \sum_{i=1}^{N-n} B_i i_{\vec{r}}(B_i) B_{i+n} i_{\vec{r}'}(B_{i+n}) \quad (151)$$

can be applied to the random function $B(\vec{r})i(\vec{r})$. In performing calculations, both the gas content and the current density may be regarded as roughly constant within the microspecimen.

For practical applications, it has been recommended to choose $N \geq 100$ [187]. Consequently, some 100 measurements of current efficiency B_i are to be made on the elements orderly arrayed along a straight line. It is precisely this basic feature that makes the suggested method distinct from those currently employed for measuring the current efficiency without recourse to any correlation between the measured results.

The estimate as obtained by formula (151) is dependent on the variables $\vec{r}$ and $\vec{r}'$; by integrating with respect to these variables, one can determine $\langle B^2 \rangle$ and then D_B. The procedure is simplified providing that (i) the effect of gas content on current density distribution can be ignored; if so, the correlation function $K(\vec{r}, \vec{r}')$ is dependent only on the distance $|\vec{r} - \vec{r}'|$; (ii) variations in $\lambda_M d_M = \varphi(B_i)$ (see formula (150)) are comparatively small, and the effect of membrane non-uniformity on current density distribution can be ignored; if so, the current density is not a random variable.

The suggested method gives the variance D_B somewhat larger than that estimated within the hypothesis of independent laboratory measurements. This imposes more strict constraints on the uniformity of the membrane's performance characteristics as compared to the foregoing numerical solution. Note that a single series of laboratory experiments on microspecimens suffices enough in practice to obtain a full information on the uniformity of membranes of a given type.

3.11 Conclusion

To briefly summarize the experience in mathematical modelling of a chemical reactor of any type, two stages can be singled out in the solution of the problem. The first stage includes a formal division of the reactor, in a broader sense, into its constituent elements or into a variety of physicochemical processes that can be studied independently. Conceptually, this independent approach is based either on the spatial separation of reactor elements, or on the distinctly different spatial and temporal scales of the processes involved, which actually determine the hierarchic structure of a model. In the second stage, the mathematical models of separate processes are synthesized into a unified scheme, with the eventual optimization of a specified objective function of control parameters. The macrokinetic analysis that has been carried out in this paper is thus suggestive of the possibility of a comprehensive study of electrochemical processes in the first modelling stage. Nonetheless, the electrochemical reactor theory stands in need of an in-depth investigation of hybrid models relating the charge transfer to diffusive and hydrodynamic phenomena. Currently, the issues concerned with the stability of electrochemical processes, the emergence of dissipative structures, and the nonlinear effects in membranes are, for the most part, poorly understood. However, one can draw an optimistic conclusion that the electrochemical macrokinetics which has arisen as a merger of several fundamental disciplines offers at present vast opportunities for applied scientific research.

Viewed from a practical standpoint, the suggested approach to reactor modelling allows to be confined to a minimum of fitting parameters. As distinct from adaptive-heuristic models that must be corrected, in the course of process measurements, for a number of reactor performance characteristics, our approach allows an independent prognostication of any characteristics that may be important for the practical improvement of a process. The required number of fitting parameters is determined by the extent of reliability of physico-chemical information available for each step of the process. The lack of such information can be replenished by special experiments run either under laboratory conditions, or on a pilot plant. Then, a computer experiment, based on the developed mathematical model, helps to resolve the scale-up problem in the design of innovatory commercial-sized electrolytic cells. Thus, the union of macrokinetic methods and computer experiment constitutes a solid scientific basis for a large-scale research in the field of applied electrochemistry.

3 Modelling of Electrochemical Processes 283

References

1. R. B. MacMullin, *Electrochem. Technol.*, 1, No. 1: 5-17 (1963).
2. R. B. MacMullin, *J. Electrochem. Soc. Jap. Engl. Ed.*, 38, No. 8: 570-579 (1970).
3. Z. Nagy, *J. Electrochem. Soc.*, 124, No. 1: 91-95 (1977).
4. R. R. Chandran and D. T. Chin, *Electrochim. Acta*, 31, No. 1, 39-50 (1986).
5. T. Z. Fahidy, *Can. J. Chem. Eng.*, 51, No. 5: 521-535 (1973).
6. D. J. Pickett, *Electrochemical Reactor Design*, Elsevier, Amsterdam, 1979, p. 536.
7. J. R. Selman, *AIChE Symp. Ser.*, 77, No. 204: 88-102 (1981).
8. P. Sh. Nigmatulin, B. A. Kader, V. S. Krylov, and L. A. Sokolov, *Uspekhi Khimii*, 44, No. 11, 2008-2034 (1975).
9. Ya. M. Kolotyrkin, Yu. A. Popov, and Yu. V. Alekseev, in: *Itogi nauki i tekhniki. Korroziya i zashchita ot korrozii* (Advance in Science and Technology. Corrosion and Corrosion Protection), VINITI, Moscow, 1982, pp. 88-138 (in Russian).
10. V. S. Krylov, *Uspekhi Khimii*, 49, No. 1: 118-146 (1980).
11. *Intensifikatsiya elektrokhimicheskikh protsessov* (Intensification of Electrochemical Processes) (Ed. A. P. Tomilov), Nauka, Moscow, 1988, p. 214.
12. L. I. Kheifets and A. V. Neumark, *Mnogofaznye protsessy v poristykh sredakh* (Multi-Phase Processes in Porous Media), Khimiya, Moscow, 1982, p. 320 (in Russian).
13. J. S. Newman, *Electrochemical Systems*, Englewood Cliffs, Prentice-Hill, New York, 1973, p. 432.
14. T. Z. Fahidy, *Principles of Electrochemical Reactor Analysis*, Elsevier, Amsterdam, No. 18, 1985.
15. I. Rousar, K. Micka, and A. Kimla, *Electrochemical Engineering*, Academia, Praha, Part 1; Part 2, 1986.
16. L. I. Kheifets and A. B. Goldberg, *Elektrokhimiya*, 25, No. 1: 3-33 (1989).
17. V. G. Levich, *Fiziko-khimicheskaya gidrodinamika* (Physico-Chemical Hydrodynamics), Fizmatgiz, Moscow, 1959 (in Russian).
18. A. A. Chernenko, *Dokl. AN SSSR*, 153, No. 5: 1129-1131 (1963).
19. V. V. Nikonenko, V. I. Zabolotsky, and N. P. Gnusin, *Elektrokhimiya*, 25, No. 3: 301-306 (1989).
20. Yu. Ya. Gurevich, A. V. Noskov, and Yu. I. Kharkats, *Elektrokhimiya*, 25, No. 5: 708-711 (1989).
21. D. G. Miller, *J. Phys. Chem.*, 71, No. 3: 616-632 (1967).
22. R. Francesconi, *Ingeg. Chim. Ital.*, 6, No. 4: 72-84 (1970).
23. G. Kosanovich and H. T. Gulliman, *Can. J. Chem. Eng.*, 49, No. 6: 753-757 (1971).
24. N. G. Pinto and E. E. Graham, *AIChE Journal*, 33, No. 3: 436-443 (1987).
25. R. A. Robinson and R. H. Stokes, *Electrolyte Solutions. The Measurement and Interpretation of Conductance, Chemical Potential and Diffusion in Solutions of Simple Electrolytes*, Butterworths, London, 1970.
26. *Comprehensive Treatise of Electrochemistry* (Eds. E. Yeager, J.O.M. Bockris, B. E. Conway, and S. Sarangapani), Plenum Press, New York, London, v. 5, v. 6, 1983.
27. J.-P. Simonin, J.-F. Gaillard, P. Turq, and E. Soualhia, *J. Phys. Chem.*, 92, No. 6: 1696-1700 (1988).
28. W. H. Stockmayer, *J. Chem. Phys.*, 33, No. 3: 1291-1292 (1960).
29. S. R. de Groot and P. Mazur, *Nonequilibrium Thermodynamics*, North-Holland Publ. Co., Amsterdam, 1962.
30. D. G. Leaist, *Can. J. Chem.*, 66, No. 5: 1129-1134 (1988).
31. D. G. Leaist, *Can. J. Chem.*, 65, No. 10: 2489-2494 (1987).
32. A. Ekman, S. Liukkonen, and K. Kontturi, *Eletrochim. Acta*, 23, No. 3: 243-250 (1978).

33. P. Turq, L. Orcil, J. Chevalet, M. Chemla, and R. Mills, *J. Phys. Chem.*, **87**, No. 20: 4008-4012 (1983).
34. J. Tamas and K. Ujszaszy, *Acta Chim. Acad. Sci. Hung.*, **49**, No. 4, 377-393 (1966).
35. P. C. Carman, *J. Phys. Chem.*, **72**, No. 5: 1713-1721 (1968).
36. A. S. Cukrowski, *J. Nonequilib. Thermodyn.*, **2**, No. 2: 69-84 (1977).
37. M. Halla, P. Turq, and M. Chemla, *J. Chem. Soc. Faraday Trans. I.*, **77**, No. 3: 465-481 (1981).
38. H. Schonert, *Z. Phys. Chem.*, **138**, No. 1: 17-30 (1983).
39. M. A. Vorotyntsev, *Elektrokhimiya*, **24**, No. 9: 1239-1243 (1988).
40. A. V. Sokirko and Yu. I. Kharkats, *Elektrokhimiya*, **25**, No. 3: 331-335 (1989).
41. D. G. Miller, *Faraday Disc. Chem. Soc.*, No. 64: 295-303 (1977).
42. N. P. Gnusin, N. P. Poddubnyĭ, and A. I. Masliĭ, *Osnovy teorii rascheta i modelirovaniya elektricheskikh polei v elektrolitakh* (Fundamentals of Theoretical Calculation and Modelling of Electric Fields in Electrolytes), Nauka, Novosibirsk, 1972 (in Russian).
43. W. H. Smyrl and J. Newman, *J. Electrochem. Soc.*, **136**, No. 1: 132-139 (1989).
44. S. S. Kruglikov, M. M. Yarlykov, and N. N. Starkova, *Elektrokhimiya*, **21**, No. 10: 1372-1376 (1985).
45. A. P. Koryushin, *Elektrokhimiya*, **21**, No. 9: 1180-1184 (1985).
46. K. Aoki, Y. Nishiki, K. Tokuda, H. Matsuda, *J. Appl. Electrochem.*, **17**, No. 3: 552-558 (1987).
47. Ch. W. Tobias and R. Wijsman, *J. Electrochem. Soc.*, **100**, No. 10: 459-467 (1953).
48. R. F. Savinell and G. G. Chase, *J. Appl. Electrochem.*, **18**, No. 4: 499-503 (1988).
49. D. A. G. Bruggeman, *Ann. Phys.*, **24**, No. 7-8: 636-679 (1935).
50. R. E. De La Rue and Ch. W. Tobias, *J. Electrochem. Soc.*, **106**, No. 9: 827-833 (1959).
51. P. J. Sides and Ch. W. Tobias, *J. Electrochem. Soc.*, **127**, No. 2: 288-291 (1980).
52. G. Kreysa and H. J. Kupls, *J. Electrochem. Soc.*, **128**, No. 5, 979-984 (1981).
53. P. J. Sides and Ch. W. Tobias, *J. Electrochem. Soc.*, **129**, No. 12: 2715-2720 (1982).
54. H. Vogt, *J. Appl. Electrochem.*, **13**, No. 1: 87-88 (1983).
55. O. Lanzi and R. F. Savinell, *J. Electrochem. Soc.*, **130**, No. 4: 799-802 (1983).
56. M. M. Leford-Quemere, F. Coeuret, and J. Legrand, *Electrochim. Acta*, **33**, No. 7: 881-890 (1988).
57. L. Coppola, O. N. Cavatorta, and U. Bohm, *J. Appl. Electrochem.*, **19**, No. 1: 100-104 (1989).
58. S. J. D. Van Stralen and W. M. Stuyter, *J. Appl. Electrochem.*, **15**, No. 4: 527-536 (1985).
59. H. Vogt, in: *A Comprehensive Treatise on Electrochemistry*, vol. 6 (Eds. E. Yeager et al.), Plenum Press, New York, 1983.
60. S. Hiraoka, I. Yamada, H. Mori, H. Sugimoto, N. Hakushi, A. Matsuura, and H. Nakamura, *Electrochim. Acta*, **31**, No. 3: 349-354 (1986).
61. H. Vogt, *Electrochim. Acta*, **32**, No. 4: 633-636 (1987).
62. H. Vogt, *Electrochim. Acta*, **29**, No. 2: 175-180 (1984).
63. H. Vogt, *Electrochim. Acta*, **30**, No. 2: 265-270 (1985).
64. A. Wein, *Elektrokhimiya*, **23**, No. 5: 658-661 (1987).
65. A. V. Sokirko and Yu. I. Kharkats, *Elektrokhimiya*, **25**, No. 10: 1306-1312 (1989).
66. P. Meares, J. F. Thain, and D. G. Dawson, in: *Membranes. A Series of Advances* (Ed. G. Eisenman), vol. 1. *Macroscopic Systems and Models*, M. Decker Inc., New York, 1972, pp. 55-124.
67. N. I. Nikolaev, *Diffuziya v membranakh* (Diffusion in Membranes), Khimiya, Moscow, 1980 (in Russian).
68. M. A. Spitsyn, V. N. Andreev, and L. I. Krishtalik, in: *ISE 37-th Meeting*, Vilnius, Aug. 24-31, 1986. Ext. Abstr., vol. 2, 397-399.
69. A. Narebska, S. Koter, and W. Kujawski, *J. Membr. Sci.*, **25**, No. 2: 153-170 (1985).

3 Modelling of Electrochemical Processes 285

70. J. Dukovic and Ch. W. Tobias, *J. Electrochem. Soc.*, **134**, No. 2: 331-343 (1987).
71. L. I. Kheifets and A. B. Goldberg, *Elektrokhimiya*, **10**, No. 8: 1204-1208 (1974).
72. I. Rousar, *J. Appl. Electrochem.*, **17**, No. 1: 134-146 (1987).
73. H. Vogt, *Physicochim. Hydrodyn.*, **8**, No. 4: 373-382 (1987).
74. A. Shah and J. Jorne, *J. Electrochem. Soc.*, **136**, No. 1: 144-158 (1989).
75. D. Ziegler and J. W. Evans, *J. Electrochem. Soc.*, **133**, No. 3: 567-576 (1986).
76. P. A. Aleksandrov, I. S. Gramberg, V. E. Kazarinov, V. S. Krylov, and L. M. Marmorshtein, *Elektrokhimiya*, **22**, No. 7: 929-932 (1986).
77. Yu. Ya. Gurevich and Yu. I. Kharkats, *Elektrokhimiya*, **25**, No. 3: 383-386 (1989).
78. L. I. Sedov, *Mekhanika sploshnoi sredy* (Mechanics of Continuous Media), Nauka, Moscow, vol. 1, 1983 (in Russian).
79. V. Yu. Golitsyn, O. V. Bobreshova, and S. F. Timashev, *Teor. Osn. Khim. Tekhnol.*, **23**, No. 3: 399-403 (1989).
80. *Magnitnoe pole i elektrodinamicheskie sily v zone rasplava moshchnykh elektrolizerov alyuminiya* (Magnetic Field and Electrodynamic Forces in the Melt Zone of Heavy-Duty Aluminium Electrolyzers) (Ed. E. A. Meerovich), *Izv. AN SSSR*, Moscow, 1962 (in Russian).
81. V. I. Eberil, A. A. Kostin, and V. P. Archakov, *Elektrokhimiya*, **18**, No. 8: 1016-1023 (1982).
82. T. Z. Fahidy, *J. Appl. Electrochem.*, **13**, No. 5: 553-563 (1983).
83. E. Z. Gak and V. S. Krylov, *Elektrokhimiya*, **22**, No. 6: 829-834 (1986).
84. A. A. Korchagin, *Elektronnaya obrabotka materialov* (Electron-Beam Machining of Materials), No. 5: 9-12 (1982).
85. Yu. G. Mikhalev, L. A. Isaeva, and P. V. Polyakov, *Elektrokhimiya*, **21**, No. 4: 519-523 (1985).
86. V. V. Skripachev, Yu. G. Mikhalev, L. A. Isaeva, and P. V. Polyakov, *Elektrokhimiya*, **19**, No. 1: 30-35 (1983).
87. A. P. Grigin, V. A. Petrov, and N. P. Pet'kin, *Elektrokhimiya*, **20**, No. 9: 1197-1201 (1984).
88. G. A. Ostroumov, *Vzaimodeistvie elektricheskikh i gidrodinamicheskikh polei. Fizicheskie osnovy elektrogidrodinamiki* (Interaction of Electric and Hydrodynamic Fields. Physical Fundamentals of Electrohydrodynamics), Nauka, Moscow, 1979 (in Russian).
89. S. S. Dukhin and B. V. Deryagin, *Elektroforez* (Electrophoresis), Nauka, Moscow, 1976 (in Russian).
90. E. V. Filippov and V. S. Krylov, *Elektrokhimiya*, **20**, No. 7: 917-920 (1984).
91. E. V. Filippov and V. S. Krylov, *Elektrokhimiya*, **21**, No. 4: 440-443 (1985).
92. V. S. Krylov and A. D. Davydov, *Khim. promyshlennost'*, No. 11: 676-679 (1981).
93. A. I. Dikusar, A. D. Davydov, A. N. Molin, and G. R. Engelhardt, *Elektrokhimiya*, **33**, No. 7: 963-965 (1987).
94. T. Z. Fahidy, *J. Appl. Electrochem.*, **14**, No. 2: 231-240 (1984).
95. T. Z. Fahidy, *J. Electrochem. Soc.*, **131**, No. 5: 1054-1059 (1984).
96. J. T. Mulvale and T. Z. Fahidy, *Electrochim. Acta*, **31**, No. 2: 173-180 (1986).
97. T. Z. Fahidy, *Electrochim. Acta*, **29**, No. 10: 1321-1326 (1984).
98. T. Z. Fahidy, *J. Electrochem. Soc.*, **132**, No. 7: 1575-1578 (1985).
99. T. Z. Fahidy, *J. Appl. Electrochem.*, **17**, No. 1: 57-66 (1987).
100. T. Z. Fahidy, *J. Appl. Electrochem.*, **17**, No. 4: 841-848 (1987).
101. G. R. Engelhardt and A. D. Davydov, *Elektrokhimiya*, **24**, No. 11: 1511-1517 (1988).
102. N. S. Demidova, *Elektrokhimiya*, **25**, No. 2: 160-166 (1989).
103. S. A. Martem'yanov and B. M. Grafov, *Elektrokhimiya*, **24**, No. 3: 373-376 (1988).
104. A. Steinchen and A. Sanfeld, in: *Sovremennaya teoriya kapillyarnosti* (Modern Theory of Capillarity), Khimiya, Leningrad, 1980, pp. 301-315 (in Russian).
105. O. Lev and L. M. Pismen, *Electrochim. Acta*, **31**, No. 4: 451-455 (1986).
106. V. V. Nechiporuk and I. L. Elgurt, *Elektrokhimiya*, **24**, No. 1: 122 (1988).
107. V. V. Nechiporuk and I. L. Elgurt, *Elektrokhimiya*, **24**, No. 11: 1566-1568 (1988).

108. Yu. Ya. Gurevich and Yu. I. Kharkats, *Dokl. AN SSSR*, **303**, No. 4: 890-893 (1988).
109. T. Teorell, in: *Voprosy biofiziki* (Problems in Biophysics), Nauka, Moscow, 1964, pp. 29-48 (in Russian).
110. R. H. Aranow, *Proc. Natl. Acad. Sci.*, USA, **50**, No. 6: 1066-1070 (1963).
111. V. N. Mataruev, *Elektrokhimiya*, **17**, No. 2: 258-261 (1981).
112. J. Arisawa and K. Misawa, *J. Membr. Sci.*, **42**, No. 1-2: 57-67 (1989).
113. A. Ya. Hochstein and A. N. Frumkin, *Dokl. AN SSSR*, **132**, No. 2: 388-391 (1960).
114. P. Russel and J. Newman, *J. Electrochem. Soc.*, **134**, No. 5: 1051-1059 (1987).
115. C. B. Diem and J. L. Hudson, *AIChE Journal*, **33**, No. 2: 218-224 (1987).
116. Yu. G. Mikhalev, P. V. Polyakov, and L. A. Isaeva, in: *Termodinamika neobratimykh protsessov* (Thermodynamics of Irreversible Processes) (Ed. A. I. Lopushanskaya), Nauka, Moscow, 1987, pp. 138-145 (in Russian).
117. R. Aogaki, K. Kitazawa, K. Fueki, and T. Mukaibo. *Electrochim. Acta*, **23**, No. 9: 867-874 (1978).
118. P. V. Mityushev and V. S. Krylov, *Elektrokhimiya*, **22**, No. 4: 552-555 (1986).
119. E. Nakache, M. Dupeyrat, and M. Vignes-Adler, *Faraday Disc. Chem. Soc.*, No. 77: 189-196 (1984).
120. I. Kristex and M. Nikolova. *ISE. 37-th Meet.*, Vilnius, Aug. 24-31, 1986. *Ext. Abstr.*, vol. 2, pp. 160-162.
121. S. F. Timashev. *Zhurn. Vsesoyuzn. khimich. ob-va im. D. I. Mendeleeva*, **32**, No. 6: 619-627 (1987).
122. A. P. Grigin, *Elektrokhimiya*, **22**, No. 11: 1458-1462 (1986).
123. A. P. Grigin and A. P. Shapovalov, *Mekhanika zhidkosti i gaza*, No. 5: 8-12 (1987).
124. B. Baranowski, *J. Nonequilibrium Thermodyn.*, **5**, No. 2: 67-72 (1980).
125. A. I. Lopushanskaya, V. V. Nechiporuk, V. V. Negrich, M. G. Bazovy, and P. M. Grigorishin, *Elektrokhimiya*, **17**, No. 2: 1782-1789 (1981).
126. V. V. Nechiporuk and I. L. Elgurt, in: *Termodinamika neobratimykh protsessov* (Thermodynamics of Irreversible Processes) (Ed. A. I. Lopushanskaya), Nauka, Moscow, 1987, pp. 125-137 (in Russian).
127. Ch. Iwakura, T. Edamoto, and H. Tamura, *Denki kagaku*, **52**, No. 9: 654-658 (1984).
128. Ya. M. Kolotyrkin, B. Sh. Galyamov, Yu. E. Roginskaya, R. R. Shifrina, and V. I. Bystrov, *Dokl. AN SSSR*, **241**, No. 1: 137-140 (1978).
129. M. Kramer and M. Tomkiewicz, *J. Electrochem. Soc.*, **131**, No. 6: 1283-1288 (1984).
130. D. C. Wright and D. Stroud, *J. Electrochem. Soc.*, **132**, No. 7: 1507-1511 (1985).
131. M. Nedyalkov and C. Gavach, *J. Electroanal. Chem.*, **234**, Nos. 1-2: 341-346 (1987).
132. A. V. Neimark and L. I. Kheifets, *Dokl. AN SSSR*, **301**, No. 3: 646-651 (1988).
133. E. Leiva, P. Meyer, and W. Schmickler, *J. Electrochem. Soc.*, **135**, No. 8: 1993-1996 (1988).
134. A. Katchalsky, and O. M. Kedem, in: *Voprosy Biofiziki* (Problems in Biophysics), Nauka, Moscow, pp. 49-69, 1964 (in Russian).
135. L. F. del Castillo and E. A. Mason, *Biophys. Chem.*, **9**, No. 2: 111-120 (1979).
136. E. A. Mason and A. P. Malinauskas, *Gas Transport in Porous Media: The Dusty-Gas Model*, Elsevier, Amsterdam, 1983, VIII, Vol. 7, p. 194.
137. J. W. Lorimer, *J. Membr. Sci.*, **25**, No. 2: 211-221 (1985).
138. E. A. Mason and L. F. del Castillo, *J. Membr. Sci.*, **23**, No. 2: 199-220 (1985).
139. J. L. Anderson and J. A. Quinn, *J. Biophys.*, **14**, No. 2: 130-150 (1974).
140. A. B. Goldberg and L. I. Kheifets, *Teor. Osn. Khim. Tekhnol.*, **24**, No. 3: 325-338 (1990).
141. Yu. M. Vol'fkovich, *Elektrokhimiya*, **19**, No. 3: 335-340 (1983).
142. Recent Developments in Nonequilibrium Thermodynamics. *Lecture Notes in Physics* (Eds. J. Casas-Vazquez, D. Jou, G. Lebon), Springer, Berlin, 1984, Vol. 199.
143. K. S. Pitzer, *J. Phys. Chem.*, **77**, No. 2: 268-277 (1973).
144. K. S. Pitzer and G. Mayorga. *J. Phys. Chem.*, **77**, No. 19: 2300-2308 (1973).

145. K. S. Pitzer and J. J. Kim. *J. Amer. Chem. Soc.*, **96**, No. 18: 5701-5707 (1974).
146. H. S. Harned and G. E. Mannweiler, *J. Amer. Chem. Soc.*, **57**, No. 10: 1873-1876 (1935).
147. H. S. Harned and M. A. Cook, *J. Amer. Chem. Soc.*, **59**, No. 10: 1890-1895 (1937).
148. H. S. Harned and J. M. Harris, *J. Amer. Chem. Soc.*, **50**, No. 10: 2633-2637 (1928).
149. V. V. Stender, *Elektroliticheskoe proizvodstvo khlora i shchelochei* (Electrolytic Production of Chlorine and Alkalis), ONTI-KHIMTEORET, Leningrad, 1935 (in Russian).
150. J. A. Rard and D. G. Miller, *J. Sol. Chem.*, **8**, No. 10: 701-716 (1979).
151. T. Mussini and A. Pagella, *Chim. Ind.*, **52**, No. 12: 1187-1191 (1970).
152. *Spravochnik khimika* (The Chemist's Handbook), Khimiya, Moscow, Leningrad, 1964, Vol. III, p. 1005 (in Russian).
153. R. N. Bhatia, K. E. Gubbins, and R. D. Walker, *Trans. Faraday Soc.*, **64**, Part 8, No. 548: 2091-2099 (1968).
154. P. T. Merenkov, *Uzb. Khimich. Zhurn.*, No. 5: 35-46 (1962).
155. S. Lengyel, J. Giber, Gy. Beke, and A. Vertes, *Acta Chim. Hung.*, **39**, No. 3: 357-363 (1963).
156. E. A. Kaimakov and N. L. Varshavskaya, *Uspekhi khimii*, **35**, No. 2: 201-228 (1966).
157. V. M. Serebritsky, *Dissertation*. Dnepropetrovsk, 1969 (in Russian).
158. L. M. Yakimenko and M. I. Pasmanik, *Spravochnik po proizvodstvu khlora, kausticheskoi sody i osnovnykh produktov* (Production of Chlorine, Caustic Soda and Chlorinated Products. Handbook), Khimiya, Moscow, 1976 (in Russian).
159. G. Angel, *Elektroliz khloristykh solei shchelochnykh metallov v vannakh s diafragmoi* (Electrolysis of Alkali Metal Chlorides in Diaphragm-Type Baths), ONTI-KHIMTEORET, Leningrad, 1935 (in Russian).
160. J. Balej, M. Kohoutkova, I. Paseka, and J. Vondrak, *Chem. prum.*, **14**, No. 1: 9-11 (1964).
161. A. B. Goldberg and L. I. Kheifets, *Elektrokhimiya*, **21**, No. 11: 1470-1474 (1985).
162. A. B. Goldberg and L. I. Kheifets, *Zhurn. prikl. khimii*, **62**, No. 10: 2263-2267 (1989).
163. N. Yokota, *Kagaku kogaku*, **22**, No. 8: 476-481 (1958).
164. M. Takahashi, *Soda Chlorine*, **29**, No. 8: 379-393 (1978).
165. D. V. Kokoulina and L. I. Krishtalik, *Elektrokhimiya*, **7**, No. 3: 336-352 (1971).
166. I. L. Kheifets, A. B. Goldberg, and A. F. Mazanko, in: *Itogi nauki i tekhniki* (Advances in Science and Technology), *Elektrokhimiya*, VINITI, Moscow, 1983, Vol. 19, pp. 244-276.
167. A. B. Goldberg. *Dissertation*, Moscow, 1983 (in Russian).
168. T. Mukaibo, *Denki Kagaku*, **20**, No. 10: 482-486 (1952).
169. A. K. Gorbachev, F. K. Andryushchenko, E. F. Maksimchuk, and V. N. Potapov, *Zhurn. prikl. khimii*, **58**, No. 6: 1275-1279 (1985).
170. A. Tasaka and T. Tojo, *J. Electrochem. Soc.*, **132**, No. 8: 1855-1859 (1985).
171. A. Tasaka and T. Tojo, *Soda Chlorine*, **39**, No. 2: 1-6 (1988).
172. A. Tasaka, A. Kimura, T. Yamahara, I. Hirosue, and T. Tojo, *Soda Chlorine*, **39**, No. 4: 22-28 (1988).
173. F. Hine, M. Yasuda, and K. Fujita, *J. Electrochem. Soc.*, **128**, No. 11: 2314-2321 (1981).
174. V. A. El'tsov, G. A. Vorob'ev, L. I. Yurkov, V. L. Kubasov, and V. B. Vorob'eva, *Zhurn. Prikl. Khimii*, **53**, No. 1: 128-132 (1980).
175. I. Rousar, V. Cezner, and J. Hostomsky, *Collect. Czechosl. Chem. Commun.*, **36**, No. 1: 1-17 (1971).
176. L. I. Kheifets and A. B. Goldberg, *Khim. Prom-st'*, No. 8: 459-466 (1985).
177. F. Hine, *Soda Chlorine*, **31**, No. 9: 1-16 (1980).
178. M. A. Lavrent'ev and B. V. Shabat, *Metody teorii funktsii kompleksnogo peremennogo* (Methods in Complex Variable Theory), Nauka, Moscow, 1973 (in Russian).
179. A. B. Goldberg and L. I. Kheifets, *Teor. Osnovy Khim. Tekhnologii*, **21**, No. 6: 805-810 (1987).
180. A. B. Goldberg and L. I. Kheifets, *Elektrokhimiya*, **23**, No. 3: 339-343 (1987).

181. A. Nidola, in: *Membranes and Membrane Processes* (Eds. E. Drioli, M. Nakagaki), Plenum Press, New York, 1986 pp. 281-298.
182. C. W. Walton and R. E. White, *J. Electrochem. Soc.*, **134**, No. 9: 565C-574C (1987).
183. L. I. Kheifets and A. B. Goldberg, *Teor. Osnovy Khim. Tekhnologii*, **16**, No. 5: 627-635 (1982).
184. L. I. Krishtalik, *Elektrokhimiya*, **15**, No. 4: 462-466 (1979).
185. A. B. Goldberg, O. P. Romashin, L. I. Kheifets, and V. M. Zimin, *Khim. Prom-st'*, No. 2: 106-108 (1987).
186. M. Takahashi and M. Noboru, *Soda Chlorine*, **29**, No. 3: 99-111 (1978).
187. E. S. Venttsel. *Teoriya veroyatnostei* (Theory of Probability), Nauka, Moscow, 1969 (in Russian).
188. A. B. Goldberg, A. G. Vaganov, S. G. Ogryz'ko-Zhukovskaya, L. I. Kheifets, and A. V. Shabalin, in: *Fundamental'nye i prikladnye aspekty elektrokataliza* (Fundamental and Applied Aspects of Electrocatalysis), 3rd All-Union Conference, Chernogolovka, Sept. 10-14, 1991, Abstract: Macrokinetics and porous-structure effects of the active layer in water electrolysis with a solid polymer electrolyte (in Russian).
189. A. B. Goldberg, L. I. Kheifets, Yu. M. Volfkovich, I. A. Yablokova, A. V. Shabashin, V. E. Sosenkin, and T. A. Safronova, *ibid.*, 1991, Abstract: Macrokinetics of a silver-hydrogen storage cell (in Russian).
190. A. B. Goldberg and L. I. Kheifets, *ibid.*, 1991 Abstract: Electrochemical production of chlorine and alkali using a solid polymer electrolyte (in Russian).
191. A. B. Goldberg, L. I. Kheifets, V. V. Bannikov, and F. I. L'vovich, *Khim. promyshlennost'*, No. 4: 227-229 (1990).
192. A. B. Goldberg, L. I. Kheifets, and V. I. Dyumulen, *Khim. promyshlennost'*, No. 5: 306-308 (1990).
193. R. Haase, *Electrochim. Acta*, **35**, No. 4: 749-751 (1990).
194. Yu. Ya. Gurevich, A. V. Noskov, and Yu. I. Kharkats, *Elektrokhimiya*, **27**, No. 2: 161-165 (1991).
195. E. Zhong, and H. L. Friedman, *J. Phys. Chem.*, **94**, No. 20: 7868-7872.
196. R. E. White, F. Jagush, and H. S. Burney, *J. Electrochem. Soc.*, **137**, No. 6: 1846-1848.
197. S. A. Shcherbinin, *Elektrokhimiya*, **27**, No. 5: 672-677 (1991).
198. T. Z. Fahidy. *Electrochim. Acta*, **35**, No. 6: 929-932 (1990).
199. H. Malchow. *Z. Phys. Chem.* (DDR), **271**, No. 4: 751-758 (1990).
200. J. Garrido and V. Compan, *Electrochim. Acta*, **35**, No. 4: 711-714 (1990).

4 Problems in Modelling and Intensification of Mass Transfer with Interfacial Instability and Self-Organization

L. M. Rabinovich

Karpov Institute of Physical Chemistry, Obukha 10, Moscow, 103064, USSR

4.1 Introduction

In chemical, petrochemical, metallurgical, energetic, biological and other branches of industry of common use are mass-transfer apparatuses and multiphase chemical reactors whose performance and efficiency are determined by the nature and intensity of transport processes occurring at interphase surfaces. The technological progress and innovative improvements in the design of such apparatuses and reactors necessitate in-depth studies of the interface and effect of its dynamics on interphase exchange processes under different hydrodynamic conditions.

A factor of prime importance determining the convective transport in two-phase liquid-liquid and gas-liquid systems is the interfacial instability that leads to the loss of hydrodynamic stability of the interfacial surface and to a spontaneous development of periodic convection patterns or chaotic pulsations within the interface and its adjacent regions. In the case when the subsurface convective motions exhibit a high degree of periodicity and give rise to a coherent structure, such as "rolls" or "cells" with the liquid circulating therein, one refers to the occurrence of an ordered interfacial convection. In other instances when these motions manifest a marked randomization reminiscent of the chaotic pulsations in a turbulent liquid, the term "interfacial turbulence" is commonly used. Such phenomena originate in the interplay of hydrodynamic, diffusional, thermal and chemical processes and are capable of sharply enhancing the interphase exchange and intensify a chemical process, for example, liquid-liquid extraction, metal extraction, absorption and chemisorption of gases, distillation, condensation and evaporation, membrane separation, phase-transfer catalysis, ion-exchange and electrochemical reactions, and so forth.

At present, the resurgence of scientific interest to the interfacial instability phenomena is stimulated by the achievements accomplished in the last decades in such rapidly expanding fields of knowledge as nonequilibrium thermodynamics, theory of self-organization, synergetics, dynamics of nonlinear systems. Multiphase systems exhibiting such effects are of common use in chemical engineering; they display all the properties requisite for the development of self-organizing processes in them. These systems are thermodynamically open, that is, capable of exchanging matter and energy with the surroundings, which provides for a permanent influx of negative entropy into them; the processes occurring therein are strongly shifted from thermodynamic equilibrium and can be described by nonlinear dynamic equations. They possess strong enough forces operated by both feedforward and feedback mechanisms; these systems have unstable stationary states, whereas their constituent subsystems exhibit a cooperative character, with the tendency to a spontaneous self-organization at a macromolecular level. The nonlinearity of the medium itself and its dissipative properties are of basic importance for the emergence of the aforementioned structures. The ordered convective surface flows and their characteristic velocity, concentration and temperature fields that arise spontaneously in interphase-transfer systems should be regarded as space-time dissipative structures formed in an initially uniform medium; in this aspect, the occurrence of an interfacial turbulence should be viewed as an order-to-chaos transition in deterministic systems. A key role in such self-organization should be assigned to the irreversibility originating in mass transport, heat transfer, and chemical reactions and to nonlinearities—both purely hydrodynamic and those associated with a variety of physicochemical factors—inherent in the systems.

The study of interface hydrodynamics is of particular importance in establishing the true macrokinetics of gas-liquid and liquid-phase reactions and in modelling and design of chemical engineering processes. In the mass and heat transfer producing interfacial instability and attended by the mentioned above effects, the interphase-exchange rate and the driving force (due to concentration or temperature) are related to each other in a qualitatively different manner vis-à-vis conventional systems with the hydrodynamic conditions playing the role of an external factor for transport processes. For this reason, theoretical and experimental studies of transport phenomena under interfacial instability conditions have developed into a self-sustained subdiscipline of physicochemical hydrodynamics.

The occurrence of feedback-like interactions between mass (heat) transfer and momentum transfer in such systems is manifest in that the diffusion (heat) flux at the interface, while exerting influence upon the velocity field, becomes, for its part, exposed to a response reaction and undergoes enhancement under certain conditions. Such an "autocatalytic" effect in mass transport provides for a number of important practical applications.

4 Modelling of Mass Transport Processes

Characteristic of the actual state-of-art of chemical engineering is the search for new ways to intensify the interphase heat and mass transfer at the macrokinetic level of an elementary act. Quite promising in this respect are regimes with advanced interfacial instability that develop under strongly nonequilibrium conditions and exhibit a nonlinearity between the mass and heat fluxes and their respective driving forces. It should be emphasized that the enhanced mass transport in systems with interfacial instability is achieved via more intense renewal of the interface surface, rather than via increasing the phase contact surface or via augmenting the driving forces (which leads to increased exergic losses). There have been reported situations (these will be dealt with below in Section 4.5) when the decrease in driving force elicited a spontaneous interfacial convection, with ensuing sharp increase in mass transport rate. In addition, this route to intensification is more "economical" than that effected via increase in the relative velocity of phase motion, which requires a significant energy expenditure to overcome the increasing hydraulic resistance. Thus, the use of interfacial instability phenomena appears to be a versatile and promising route to process intensification enabling to reduce the diffusional (or thermal) retardation near the interface, that is, precisely in the region where under normal conditions (in the absence of such effects) the mass or heat transfer is inhibited.

The development of mathematical problems in the theory of interfacial self-organization and the construction of physically meaningful mathematical models for mass and heat transfer with interfacial instability followed by the workout of reliable process design methods come at present to the foreground of practical interests. The development of effective methods for process control and optimization under interfacial instability conditions requires further quantitative studies that would allow to adequately describe the hydrodynamics, heat and mass transport as well as the conditions for the occurrence of self-organized dissipation structures and their evolution. Likewise, one cannot ignore the importance of experimental studies aimed at a detailed elucidation and interpretation of interfacial instability mechanisms and intervenient external factors.

Studies of interfacial self-organization phenomena and their effect on hydrodynamics and transport processes in multiphase systems are essential for the progress of a new interdisciplinary field, interfacial microstructural technology, one of the high-priority research lines, as emphasized in the famous report "Chemical Engineering Frontiers", referred to as the "Amundson Report" [1, 2]. To be noted, the discipline we are concerned with in this paper and which we conventionally name "transport processes with interfacial self-organization" has developed from the following theoretical and applied fields of knowledge: physicochemical hydrodynamics, macrokinetics of interphase processes and reactions, theory of self-organization on the one hand, and separation techniques and heat and mass transfer processes, on the other hand.

The major objective of this paper is to consider, from a single standpoint, the physicochemical hydrodynamics of various phenomena of interfacial instability and self-organization interrelated by their physical nature and primarily connected to the instability of interfacial tension (the Marangoni effect). The presented material is essentially based on the results obtained by the author and his group for physicochemical hydrodynamics studies at the Karpov Institute of Physical Chemistry in Moscow. The main issues concerned with are: construction of mathematical models for description of mass transport under interfacial instability conditions; studies on dynamics of perturbations and dissipation structures at the interface; design of mass transfer models for their application to chemical engineering; determination of the physicochemical system parameters corresponding to high-intensity mass transfer regimes.

The second Section of the paper is devoted to the interfacial instability and self-organization in liquid-liquid and gas-liquid systems and considers a variety of situations in which this instability is exhibited; discussed here are also the effects produced by interfacial instability on interface hydrodynamics and mass transfer kinetics. A classification of physical effects producing or affecting interfacial instability is proposed. The third Section is concerned with theoretical studies of the Marangoni interfacial instability mechanism in different systems. Instability criteria for a reactive two-layer system, a jet, a liquid drop, and a falling liquid film have been derived by linear analysis. By making use of nonlinear finite amplitude stability method and numerical finite difference methods, the perturbation dynamics has been studied and the structure of velocity and concentration fields computed. The fourth Section reviews various models for mass transfer under interfacial convection conditions. A novel semiempirical model containing a single empirical parameter has been proposed and tested on laboratory and industrial-scale extraction units. The fifth Section is concerned with certain applied aspects of the interfacial instability. Methods for mass transfer intensification as effected through the agency of interfacial hydrodynamic instability regimes have been discussed and the choice of high-efficient absorbents, extractants and solvents assisting in the interfacial self-organization has been considered.

It has not been our intention to give a comprehensive review of numerous papers on hydrodynamic aspects of interfacial instability; the interested reader is referred to review papers in [3-16]. Various mass transfer models for the interfacial convection governed by the Marangoni effect were considered in [17, 18].

4.2 Interfacial Instability Phenomena and Their Physical Nature

Numerous examples of convective flows spontaneously formed at the interface of liquid-liquid and gas-liquid systems with interphase heat or mass transfer have been reported in the literature [19-34]. In many instances, ordered convection patterns could be observed using a variety of optical visualization methods. The topological properties of convection patterns originating in the Marangoni effect have been studied in greater detail starting from the middle 1950s till the present time. In experiments with a system diethyl ether-0.16% aqueous butyl alcohol [23], convection cells are observed to form in the surface of a diethyl ether drop as n-butyl alcohol is transported from aqueous phase into the drop (photograph in Fig. 4.1). In a system 8% aqueous isobutyl alcohol-diethyl ether, the transport of isobutyl alcohol into the surrounding diethyl ether is even more vigorous, and eruptions are observed to occur in the surface of an alcohol-water drop (photograph in Fig. 4.2). Shown in Fig. 4.3 are the photographs of the convection patterns resembling rolls and cells that develop in the surface of an aqueous monoethanolamine (MEA) solution on chemisorption of carbon dioxide (initial MEA concentration 0.5 mol/l) [31]. Visualization was carried out by optical polarization method using an optically active liquid.

The temporal development of interfacial turbulence on chemisorption of carbon dioxide by a $0.5N$ aqueous potassium hydroxide solution is shown in Fig. 4.4. The pictures of convective motion were obtained by Karlov [32] using a Mach-Zehnder interferometer. A $45 \times 15 \times 15$ mm sample cell was filled with a liquid whose upper surface was in contact with a gas. The vertical bands have been created artificially by means of an optical wedge, and their arrangement in photograph 4.4a corresponds to the quiescent liquid at zero-of-time of phase contact. The succeeding pictures were taken at a second time interval between. Shown in photograph 4.4d is a fully formed undisturbed diffusion subsurface layer which subsequently becomes disturbed by emerging convective vortices (photograph 4.4e). In further course, the boundary layer suffers a complete destruction by the advancing interfacial turbulence reaching deep into the liquid phase (photograph 4.4f).

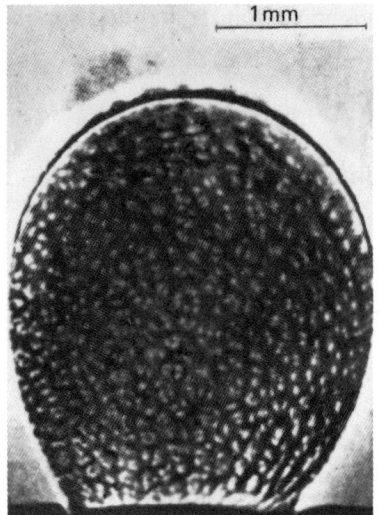 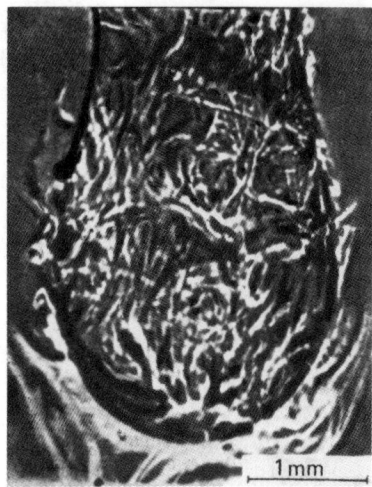

Fig. 4.1 Convection cells in the surface of a liquid drop

Fig. 4.2 Eruptions attendant to the extraction of a component from a liquid drop

The development of surface convection in absorption of carbon dioxide by different aqueous solutions was observed by Rabinovich and Ambartsumyan [33] by means of a shadow microscope using a Foucault knife-edge as a shadow stop aligned parallel to the gas-liquid interface. Shown in Fig. 4.5 is the measurement cell containing a liquid absorbent and filled with a gas to be absorbed. It is a quartz-glass cylinder 5 cm in diameter with wall thickness of 1 cm; in performing measurements, the cell symmetry axis is aligned horizontally. Typical stages of convection development on uptake of carbon dioxide by 20% aqueous monopropanolamine (MPA) solution are shown in Fig. 4.6a, and by 25% aqueous MEA solution, in Fig. 4.6b. On addition of a sparingly soluble surfactant isoamyl alcohol to the CO-MPA system, the convection becomes to a significant extent inhibited, whereas the intentional disrupture of the surfactant film has led to convective streams arising in a close vicinity of the disrupture region (Fig. 4.6c).

In Fig. 4.7 successive stages of the convection development in the organic phase (Fig. 4.7a) and the aqueous phase (Fig. 4.7b) on transfer of propionic acid in a dibutyl phthalate-water system are shown. The visualization has been performed by microinterferometric and Mach-Zehnder interferometric methods [34].

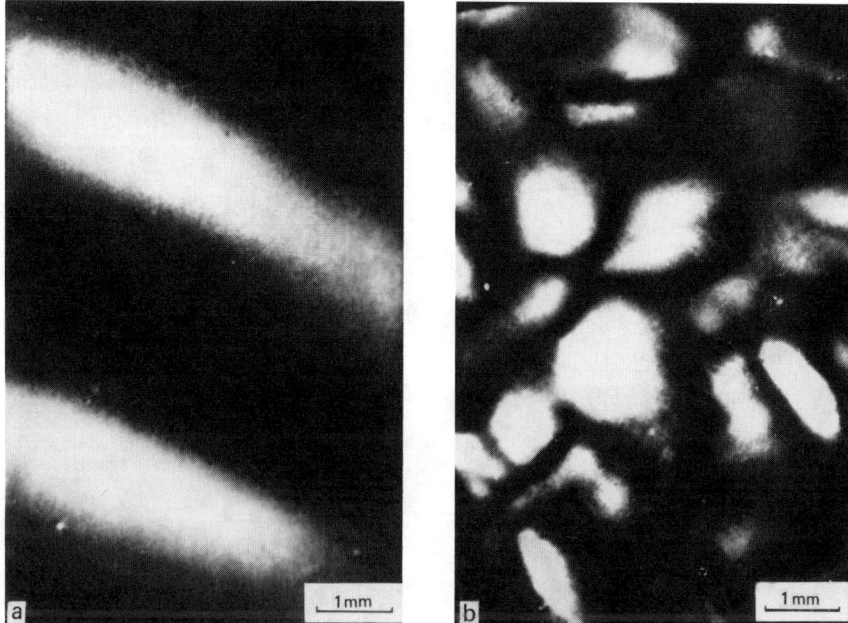

Fig. 4.3 Interfacial convection patterns of roll type (a) and cellular type (b) in a CO_2-MEA system (top view)

Alongside the convective motion visualization technique, the interfacial instability phenomena are widely studied using methods based on mass transfer kinetics measurements. For a typical example of such studies in extraction, absorption, and distillation one may refer to [35-58]. Sherwood and Wei [35] in their study of acetynic acid extracted from solutions have measured a mass transfer coefficient that was several times larger than that predicted by the penetration theory. These authors interpreted the observed discrepancy as due to a surface activity effect.

Shatokhin [45] in his kinetic studies of mass transfer from an isolated liquid drop in extraction attended by spontaneous interfacial convection has measured the instantaneous mass transfer coefficient as a function of the concentrational driving force (Fig. 4.8). The plateaus in the experimental curves correspond to diffusional mass transfer, in the absence of interfacial convection. The inflection in the experimental curve signifies the change from a steady-state mass transfer to an intensified mass-transfer when the interfacial

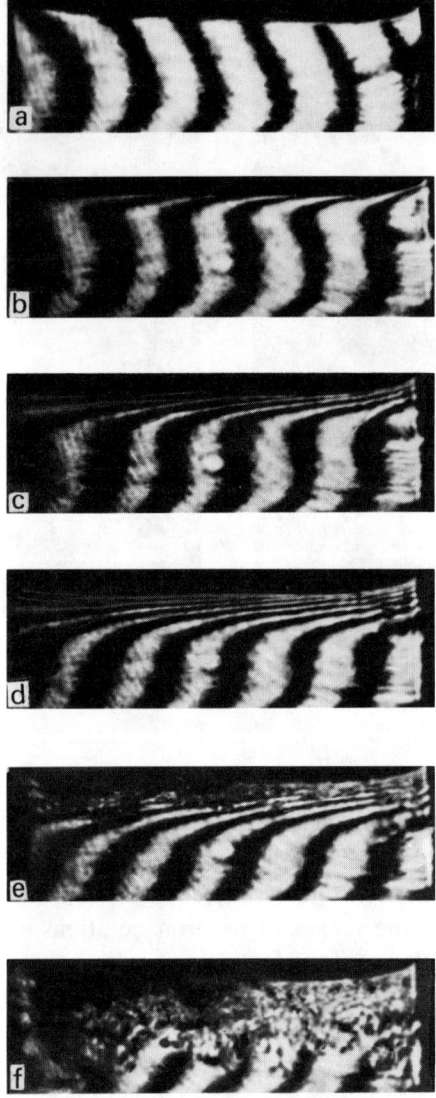

Fig. 4.4 Inerfacial turbulence development in chemisorption of CO_2 with aqueous potassium hydroxide solution

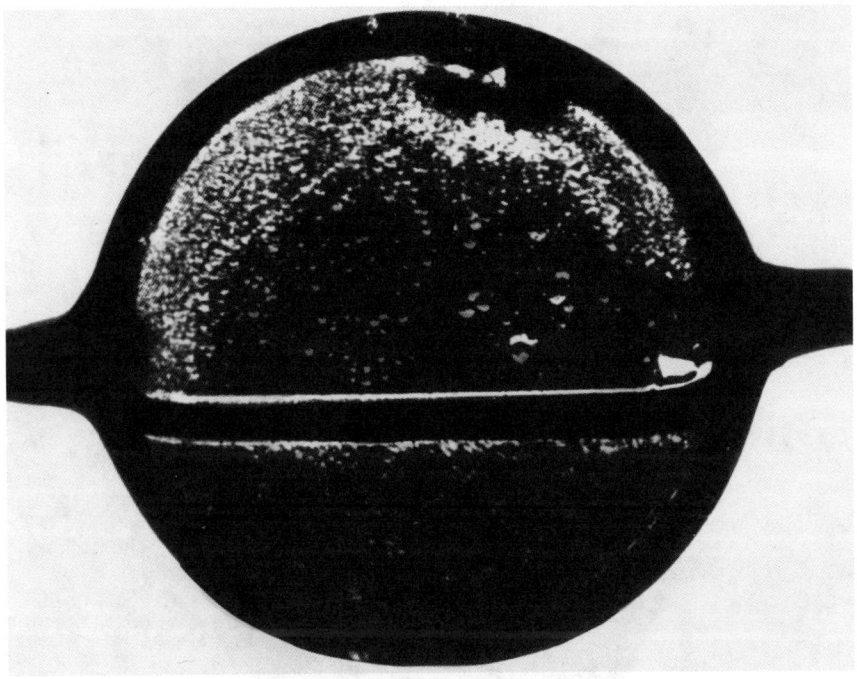

Fig. 4.5 Measurement cell filled with a liquid absorbent and an absorbed gas

convection sets in which becomes stronger with the rise in driving force.

Similar relationships were obtained by Ermakov et al. [46] in their studies of methylethyl ketone and acetic acid extracted from CCl_4 into water using a cell in which two quiescent liquid layers were separated by a plane interface (Fig. 4.9).

Of the experimental works concerned with the influence of surface convection on mass transfer rate in a chemically nonreacting gas-liquid system, a study by Brian and coworkers [47] should be mentioned. These authors studied the desorption of diethyl ether, triethylamine, acetone (agents capable of reducing the surface tension) from their dilute aqueous solutions in a nitrogen-filled liquid-film column. The mass transfer rate was measured by making use of inert gas tracers which were propylene for the liquid and water for the gaseous phase. The results have shown that at certain initial concentration of those agents the liquid-phase mass transfer coefficient becomes increased by a factor of 3.6 (maximum) as compared to that for propylene desorption

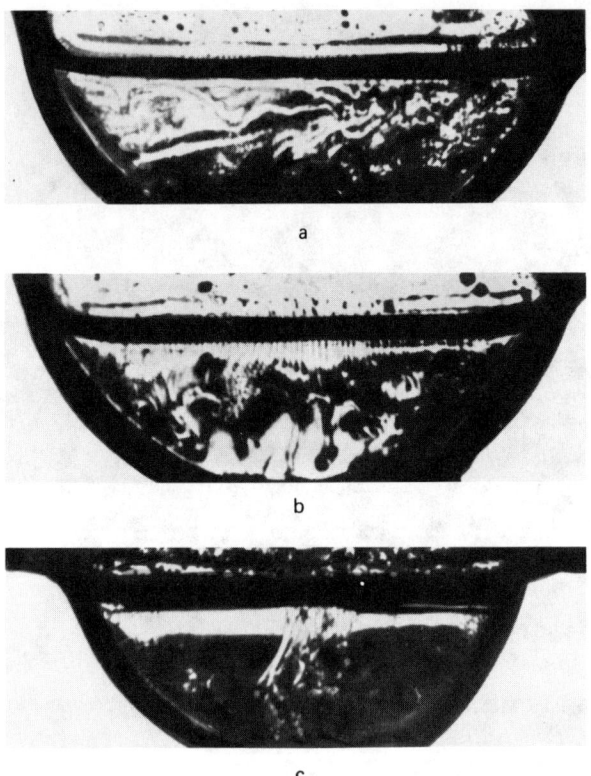

Fig. 4.6 Convective streams formed in the layer of a liquid absorbent in CO_2 chemisorption

from water. This provides an indirect evidence for the occurrence of surface convection in liquid phase; as distinct, the gas-phase mass transfer coefficient remained constant.

Hozawa et al. [49] in their studies of carbon dioxide desorption in nonaqueous solvents have measured a mass transfer coefficient noticeably higher than that predicted by the penetration theory. The curves 1 and 2 in Fig. 4.10 show time dependences for mass transfer coefficient calculated by Higbie model for methanol and toluene as absorbents.

Apparently, Brian and coworkers [50] were the first to obtain results that provided evidence for a role of surface convection in the absorption involving a chemical reaction. These authors studied the commercially important ab-

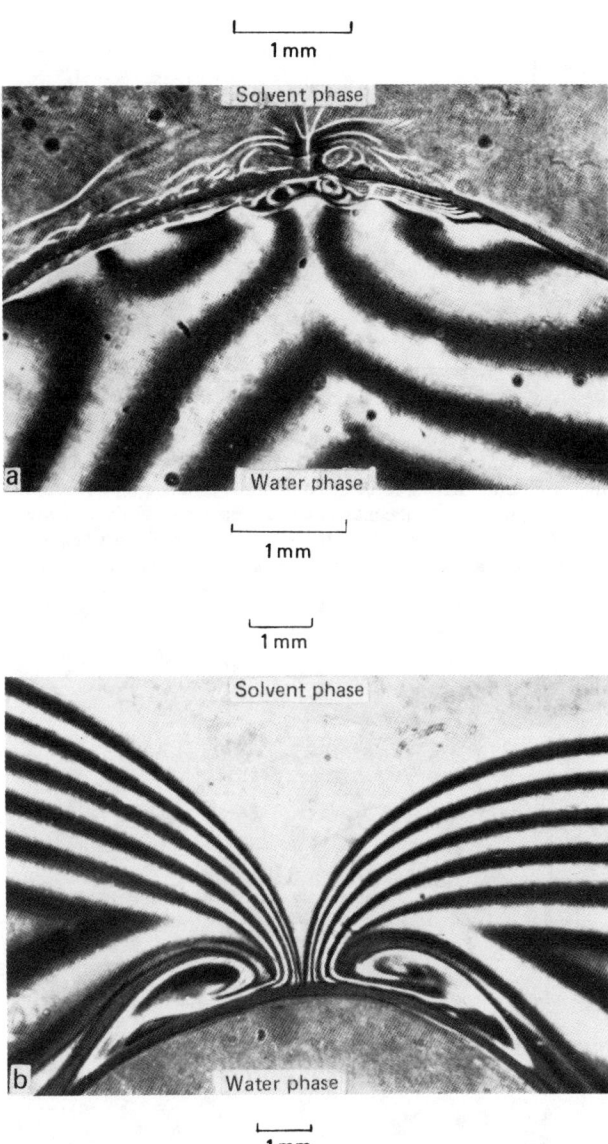

Fig. 4.7 Convective motions arising at the interface in extraction of propionic acid from dibutyl phthalate into water

300   Mathematical Modelling of Chemical Processes

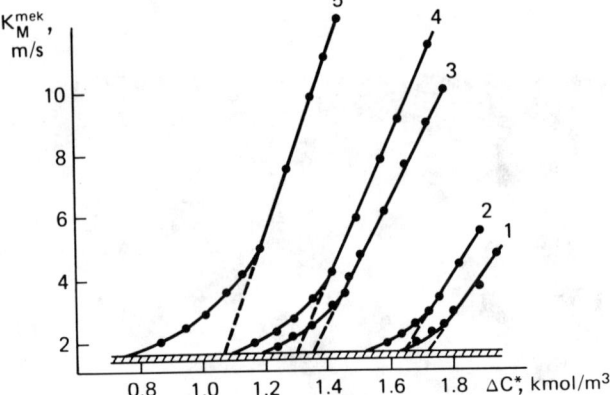

Fig. 4.8 Instantaneous mass transfer coefficient as a function of the concentrational driving force in extraction: •, transfer of acetic acid from carbon tetrachloride into water; △, transfer of acetic acid from dichloroethane into water; +, transfer of phenol from water into toluene; □, transfer of acetic acid from chloroform into water

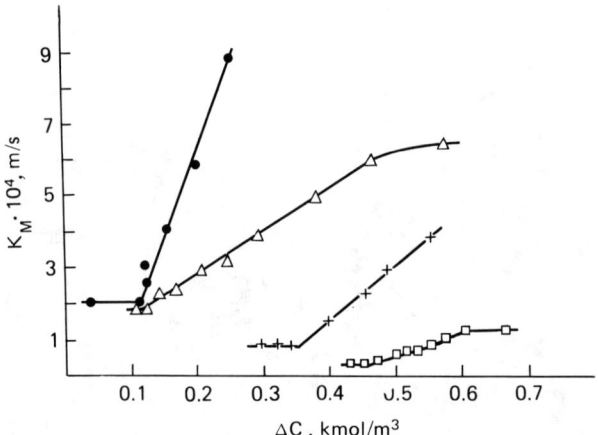

Fig. 4.9 Mass transfer coefficient vs. driving force for methylethyl ketone (MEK). Initial concentrations (in kmol/m^3); for MEK, 2,0; for acetic acid, 0 (curve 1), 0.55 (curve 2), 1.14 (curve 3), 1.55 (curve 4), 2.33 (curve 5)

sorption of carbon dioxide by aqueous MEA solution, with the simultaneous desorption of an inert tracer (propylene) in a wetted-wall tubular column of short length. The propylene desorption data (without CO_2 chemisorption) agree well with the results calculated by short-term phase-contact theory. However, the physical mass transfer coefficient was shown to increase significantly for the simultaneously occurring two mass transfer processes. Brian's hypothesis on surface instability in the CO_2-MEA system has been confirmed by numerous experiments, in most of which the liquid chemisorbent was either at rest, or its flow rate was low, with the Reynolds number varying within a narrow range. In this connection, the available experimental data that lend support to the possibility of mass transfer enhancement due to interfacial instability merit a special attention [51, 52]. Shown in Fig. 4.11 are the results that were obtained by Akselrod and Dilman in their studies of carbon dioxide chemisorption by aqueous MEA solution in a wetted-wall column 1 m long [51]. Nitrous oxide desorbed from a $1M$ MEA solution was used as a gas tracer. With the CO_2 gas-phase concentration increasing, a marked rise in mass transfer coefficient β_{N_2O} was observed which was explained by a surface convection effect. The desorption enhancement arising from this effect is much higher over that caused by the increase in the Reynolds number for the main liquid flow to a maximum of about 2100. Similar qualitative results were obtained by Akselrod and Dilman in their studies of N_2O desorption from $2.3M$ and $4M$ MEA solutions and helium desorption from a $1M$ MEA solution. The data demonstrating the mass transfer enhancement in a laminar jet were reported in [53].

Quite similar results on surface convection intensity in a laminar jet were obtained by Sada and coworkers [54] who studied the absorption of carbon dioxide by MEA and monoisopropyl alcohol solutions using ethylene as a gas tracer.

An analysis of experimental and theoretical studies of the interfacial phenomena conducive to hydrodynamic instability of the interface and its adjoining regions has allowed one to specify major physical effects responsible for the interfacial instability and exerting a sizeable influence on its development and characteristics (see Table 4.1). The classification of these effects as suggested in Table 4.1 is somewhat arbitrary considering that the mechanisms underlying the interfacial instability are most commonly quite complicated in character. Under real conditions, the above effects often manifest themselves simultaneously; therefore, it is important to specify the predominant role of one or another effect. In most cases, one would hesitate to draw a border

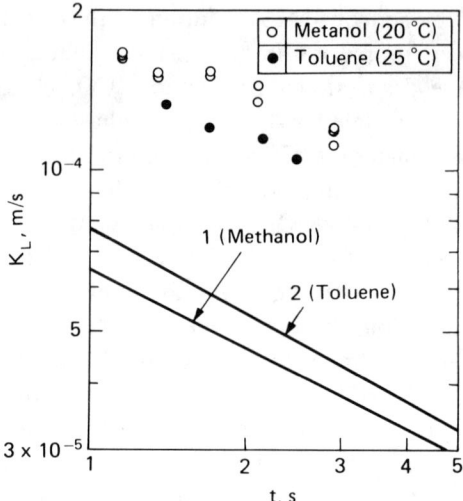

Fig. 4.10 Time dependence for mass transfer coefficient as calculated by Higbie model for methanol and toluene as absorbents

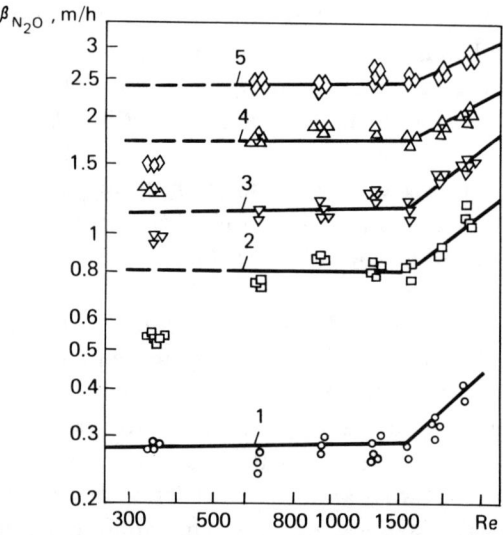

Fig. 4.11 The β_{N_2O} vs. Re relationship for an aqueous MEA solution (1 kmol/m³) at different CO_2 gas content (wetted-wall column, 1 m long): curve 1, 0%; curve 2, 10%; curve 3, 20%, curve 4, 50%; curve 5, 100% CO_2

4 Modelling of Mass Transport Processes

Table 4.1 Physical Effects Generating the Interfacial Instability or Playing a Role in Its Development

INTERFACIAL INSTABILITY			
Effects associated with surface tension	Thermal and concentrational effects	Hydrodynamic effects	External field effects
Marangoni effects (concentrational-capillary, thermocapillary, chemocapillary, electrocapillary, conformational-capillary). Laplace and Gibbs capillary effects.	Benard effects (thermal-and concentrational-gravitational). Thermal and concentrational-kinetic effects). Thermo- and diffusion phoresis. Internal and surface sources of heat and radiation. Phase transitions.	Forced convection and turbulence. Surface deformation. Shear stress. Dispersion. Coalescence and crushing Rayleigh-Taylor instability. Vibration.	Electroconvective instability. Magnetohydrodynamic effect. Thermolaser effect. Reduced gravitation and inponderability. Ultrasonic effect.
Kinetic effects		"Cross" and "double" effects	
Chemical reactions (homogeneous and heterogeneous). Rheokinetic effects. SAS adsorption Electrokinetic effects.		Soret and Dufour effects (thermodiffusive, diffusive-thermal). Stefan fluxes. Mass-transfer-induced fluxes. "Double" diffusion.	

line between them, and treating the experimental results in terms of a single "pure" effect is practically never possible.

Major effects capable of producing interfacial instability are the Marangoni effects [3, 59-62], which can lead to local tangential stresses at the interface due to surface tension gradient. The major source of Marangoni instability is the dependence of surface tension on concentration (solutal-capillary effect), temperature (thermocapillary effect), density of subsurface electric charges or dipoles (electrocapillary effect), electromagnetic field strength, chemical composition of the interface (chemicapillary effect), conformational structure of the surface layer. Besides, the surface tension instability originating either from composition inhomogeneity, or surface temperature leads to local changes of Laplace pressure and elastic properties of the surface film (Gibbs effect). These capillary effects exert influence on mechanical deformation and curvature of the interface.

Thermal and concentrational effects responsible for a nonuniform density distribution in gravitational field are also non-negligeable contributors to the convective interfacial flows. A classical example is the Bénard effect causing a thermogravitational convection [63, 64]. The dependence of the medium density on concentration may produce a concentrational-gravitational convection similar to the thermal Bénard convection in nature and properties.

Exo- and endothermic reactions in a system are also capable, under definite conditions, of producing a nonuniform density distribution and eliciting thereby convective flow (thermokinetic instability). Analogously, a chemical reaction can lead to a change in the density of the medium (concentrational-kinetic instability).

Noteworthy, the Bénard and Marangoni effects often manifest themselves simultaneously, for example, in liquid layers a few centimeters thick, heated from below. This results in the development of a mixed Bénard-Marangoni convection [10, 12, 13, 65, 66]. In thinner layers, the Marangoni effect is predominant, whereas in thicker layers, the Bénard effect. Most commonly, both effects are simultaneously operative in chemical reactions causing to change the density and surface tension of the liquid. For instance, in chemisorption of CO_2 by aqueous MPA solution (Fig. 4.6a), the development of convection in liquid phase is due to four factors: solutal and thermal Marangoni effects, concentrational- and thermokinetic effects. In the case of CO_2 absorbtion by an aqueous MEA solution (Fig. 4.6b), the solutal Marangoni effect is predominant.

The heat released by inner and outer sources other than a chemical reaction is also capable of producing a sizeable effect on interfacial convective stability. For example, the radiative processes in the medium can contribute significantly to the total heat and mass transfer and thus exert influence on the structure and stability of convective surface flows. Phase transitions can also give rise to an additional heat source (or heat efflux) at the interface.

The study of external hydrodynamic factors capable of affecting the interfacial stability is of great interest from both theoretical and practical standpoints. This is primarily connected with the role the interfacial stability plays in chemical engineering processes, for which the Reynolds number of the main flow is large reaching several thousands or even more. According to the theory and available experimental evidence, it is to be anticipated that the convective instability nascent within the interface can destabilize thin boundary layers (viscous, concentrational, or thermal) adjacent to the free surface and undisturbed by external turbulent pulsations. Still, the question about the existence of interfacial turbulence and its eventual contribution to the overall heat and mass transfer under the conditions of external turbulence (in turbulent flows, in a medium agitated by high-speed stirrers) remains open.

The effect of external fields on the interfacial instability is in many respects dependent on the properties of the liquid. If the liquid is an electric conductor (electrolyte, liquid metal), its flow characteristics are significantly influenced by magnetic field. This influence is due to a magnetohydrodynamic (MHD) interaction between the magnetic field and the electric currents induced in the moving fluid. The MHD forces exert action also on the convective flow in the surface of a conducting liquid and on the flow stability. The imposition of an external magnetic field leads, in general, to a flow stabilization. If the liquid is a dielectric, the imposed electric field can substantially affect its stability characteristics. Ion-exchange processes and ion-exchange reactions involving surfactant ions exert a marked electrohydrodynamic action on the convective interfacial stability.

Laser radiation locally focused on the interfacial surface is a method enabling one to study mass and heat transfer processes in a disturbed interface. The laser beam energy, on its uptake by the thin subsurface layer, creates a local overheating and initiates convective circulation motions enhancing the transfer processes. The laser technique has been used to intensify gas absorption in a number of chemical and biochemical systems [67].

Reduced gravitation or even imponderability play a special role in the interfacial instability phenomena. In the absence of gravitational forces and associated therewith convective motions, the Marangoni effect becomes a major factor in liquid dynamics and other processes of space technology.

A variety of kinetic effects contribute significantly to the interfacial instability. On the one hand, kinetic instabilities and transport instabilities are independent sources of both nonstationarity and nonuniformity of physical fields [68]. On the other hand, by virtue of numerous feedback interactions, kinetic instabilities can produce influence on momentum, mass and heat transfer and electric conduction.

The chemical bulk reactions on both sides of the interface surface affect the concentration and temperature gradients and their respective fluxes. Surface reactions affect the mass and heat balance and, in general, the entire situation in the interface, which results in a change of the effective surface tension. For this reason, reactions of both types are capable of influencing appreciably the conditions for the onset of Marangoni instability causing thereby changes in the characteristics of the resultant convective surface motion. The probability of composition and temperature fluctuations occurring in a liquid surface increases, if the mass transfer across it is accompanied by chemical reactions, since in this case the surface layer is made up of several components whose number may be greater over that in a conventional mass transfer. Considering the high rate of mass transfer processes in a chemically reacting system, the surface convection is expected to be more probable in chemisorption, rather than in physical absorption of gases. This is especially true of a mass transfer processes with irreversible chemical reaction. As a rule, such processes are substantially nonequilibrated, and the respective systems are quite often shifted from equilibrium. Experimental [28, 69] and theoretical [70-73] studies show that if a heterogeneous chemical reaction takes place at the interface, an interfacial instability and subsequent development of dissipative structures may occur also in the absence of a transit mass flux across the interface. In this case, the irreversibility as a prerequisite for self-organization is provided for by a nonlinear reaction and by diffusion of the reactants to the interface from the surrounding liquid bulk, rather than by transit mass flux. This fact is of crucial importance, since it casts light on the mechanism of interaction of chemical and hydrodynamic processes in the initiation of hydrochemical interface instability.

The influence of rheokinetic factors on interfacial instability is attributable to the fact that chemical reactions, in certain instances, produce changes in physicochemical characteristics of the medium and its rheological properties.

Adsorption/desorption plays an important role in transport processes, when the velocity of migration of the solute toward the interface is determined by the adsorption/desorption rate, or when the kinetics is mixed in character, that is, with the rates of adsorption/desorption and diffusion being close to one another.

The solutes exhibiting a tendency to accumulate within the liquid surface layer or to adsorb on the liquid surface are known to reduce the surface tension. This behaviour is especially displayed in strong surfactants which are practically insoluble in the liquid bulk and become tightly adsorbed on the liquid surface. Such a behaviour is exemplified, for instance, by fatty acids in water, or by alkali metals in mercury. If the surface concentration of a surfactant is small in comparison to its maximum concentration (that of a continuous monomolecular layer), that is, if the adsorbed particles form an ideal two-dimensional gas or a dilute "solution", the surface film retains all of its liquid characteristics; however, its capillary properties become modified. At high concentration, the adsorption layer changes to a condensed state, and the surface film acquires elastic properties. The occurrence of a layer of adsorbed surface-active substance (SAS) is especially manifest in the effect of damped capillary waves; a theory for this effect has been developed by Levich in [74]. The SAS concentration gradient normal to surface and the respective diffusion flux lead to Marangoni instability.

At the interface of two phases (of which one is an electrolyte solution), a narrow region of charged solution emerges, the so-called electric double layer. The double layer gives rise to electrokinetic phenomena, specific electrohydrodynamic effects originating from the relative motion of constituent phases with respect to each other. Convective motion in the subsurface regions can arise due to the imposition of an electric field (electrophoresis) both in a liquid-liquid system (for example, in water-in-oil or oil-in-water emulsions) and in a liquid-solid system (colloid particles in an electrolyte solution). Concurrent with electrophoresis, an opposite effect is also observed, i.e. in the motion of particles elicited by nonelectric (for example, gravitational) forces an electric field sets in that modifies the structure and dynamics of the subsurface region. According to Ruckenstein [75], the phoretic motion may be regarded as an effect due to interfacial tension gradient (electrocapillary Marangoni effect).

In currently used heat and mass transfer apparatuses, the common transmitting media are liquids with specific rheological properties that modify the behaviour of a free liquid surface. The nonlinear-viscous and viscoelastic properties of a non-Newtonian fluid define the pattern of the convective surface motion and its stability.

In certain situations, the specific surface viscosity plays a significant role in a modified dependence of interfacial tension (consequently its dynamic properties) on the rate of interfacial surface deformation; the notion of surface viscosity was first introduced by Boussinesq [76] and later worked out theoretically by Scriven [77] and Levich [78]. The surface viscosity is an essential factor, for example, in slow adsorption of surfactants that decelerate the gravitational fall of drops in a liquid.

The Van der Waals forces and disjoining pressure [79, 80] determine in many respects the character of convective flows in polymolecular wetting films widely used in flotation, drying, and impregnation of porous materials.

In certain multicomponent systems, a notable contribution to interfacial instability is due to the so-called "cross" effects arising from the energy, mass and momentum fluxes in which several driving forces are simultaneously operative. Among such effects, one should mention thermodiffusion (Soret effect) and diffusional heat conduction (Dufour effect) whose respective coefficients satisfy, as a rule, the Onsager reciprocal relations. Apart from these phenomena, under definite conditions (commonly, at high concentration and temperature gradients) effects due to the transfer of an additional momentum across the interface at the expense of mass or heat fluxes are observed.

In the absence of "cross" effects, when a simulataneous transfer of two components, or a transfer of heat and of one component take place (the so-called "bidiffusional" processes), the interfacial stability picture may be quite different from that involving a single component.

To summarize this brief review of the physical effects influencing the interfacial behaviour, we wish to emphasize that the major effects contributing to interfacial instability are the Marangoni effects. In what follows, we focus our attention on the theoretical consideration of phenomena concerned precisely with these effects. Despite the dissimilarity of physical causes conducive to Marangoni instability—be these mass transfer, chemical reactions, or charge transfer—the qualitative picture of interfacial convections originating from these effects has much in common.

A scenario for the development of self-organization processes in such systems may be demonstrated by referring to the interphase flux and its dimensionless characteristics, the Sherwood and Nusselt numbers, as a function of the Marangoni number characterizing the surface forces responsible for interfacial convection. A strict definition of the Marangoni number is given below in Section 4.3. At sufficiently small surface tension gradients and, accordingly, weak surface forces, the situation is close to equilibrium, and the steady state in the absence of convective flows exhibits an asymptotic stability. In such a state, the mass transport in a quiescent medium is effected by molecular diffusion only. Under such conditions, a diffusional mass transfer regime sets in; portion 1 of the curve in Fig. 4.12 corresponds to this regime. In a moving

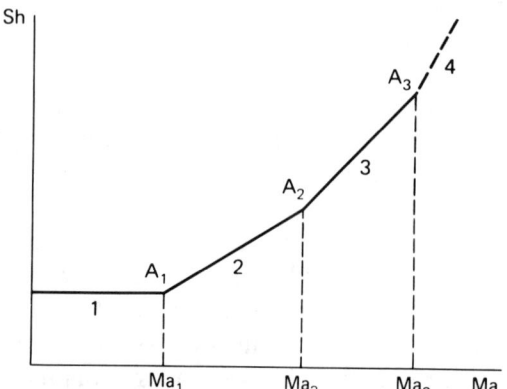

Fig. 4.12 A qualitative Sherwood number vs. Marangoni number relationship on transition from interfacial convection to interfacial turbulence

liquid flow, even a small rise in the Marangoni number may cause a change in the longitudinal velocity component, as, for example, in the surface of a laminar film without loss of hydrodynamic stability. As the surface tension gradient is allowed to increase and the Marangoni number attains a definite critical value (the first critical Marangoni number Ma_1), the hydrodynamic pattern undergoes a sharp change: convective circulation flows are observed to emerge at the surface (portion 2 of the curve in Fig. 4.12) leading to a substantial increase in interfacial flux. Further increase in both surface tension gradient and Marangoni number can lead to a disturbed stability pattern of the convective cells and induce the transition to cells of different type, for example, aggregated into clusters (portion 3 on the curve in Fig. 4.12). Finally, at the next critical Marangoni number, the surface is completely destabilized, and the system enters into a state of interfacial turbulence with a rather high mass transport rate, or a discontinuity of the surface occurs, which shows up, in particular, as eruptions. As the critical Marangoni numbers are reached in succession at points A_1, A_2 and A_3, the stability of the given branch of mathematical solution of the problem breaks down to bifurcate into new branches, under a changed balance of external forces (in this particular case, surface forces) and dissipative forces. These successive bifurcations can ultimately lead to a chaotic state (interfacial turbulence).

The relationship in Fig. 4.13 that has been obtained by Shatokhin [45] in his studies of the extraction of propionic acid at initial concentration C_0 from a carbon tetrachloride drop into water lends an experimental support to the scheme outlined above. Portion 1 of the curve corresponds to a diffusional mass transport regime; portions 2 and 3 correspond to an intensified mass transport regime: this is the stage where the formation and development of interfacial convection takes place with the transition at the points ΔC_1 and ΔC_2 from one type of convection to another. Portion 4 of the curve corresponds to the region of limited development of interfacial convection where the interfacial tension gradient growth either decelerates, or stops altogether. Within this region of concentrational driving force, the mass transport rate in the drop sharply increases to become commensurate with, or even greater than, the rate of efflux of the solute from the surface of the drop into the bulk of the continuous accepting phase, and the disperse phase thus no longer limits mass transport. Further intensification of interfacial convection and mass transfer rate is infeasible under such conditions, since this would have led to a limitation of transport process by the continuous phase and to the concentrational levelling of the transported solute within the drop along the

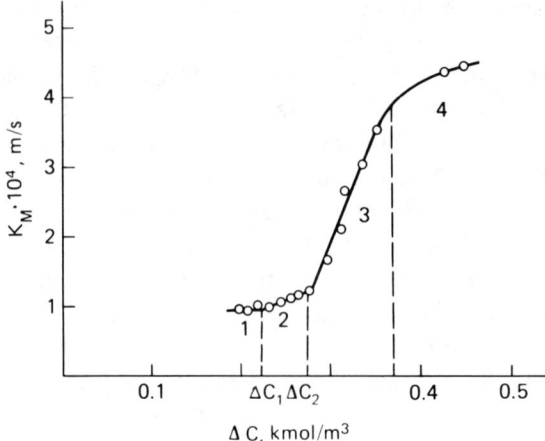

Fig. 4.13 Instantaneous mass transfer coefficient vs. extraction driving force in the transfer of propionic acid from a carbon tetrachloride drop into water (initial propionic acid concentration $C_0 = 0.7$ kmol/m^3)

radial coordinate and, possibly, to a reduced interfacial convection. Thus, the buildup of a limiting region, its extension, also the value of K_M are determined by the properties of the extraction system; these properties determine also the relation between diffusional phase resistances.

A generalization of interfacial effect data that manifest themselves in the hydrodynamic interfacial instability, modified mass transfer and physical factors underlying these effects allows one to define, in general terms, the notions of interfacial instability and self-organization.

The interfacial instability in liquid-liquid, gas-liquid and liquid-solid systems includes, as a general notion, the sum total of space-time instability effects relevant to velocity, concentration, temperature, and charge fields that arise in the interface or its adjoining regions either by interphase interaction, or by the action of external forces. In a narrower sense, the interfacial instability bears upon the effects of hydrodynamic instability of the interface due to interphase mass and heat transfer, ion exchange and chemical reactions. A major factor defining these effects is the surface tension gradient at the interface.

Among numerous interfacial instability phenomena, one can single out those accompanied by the spontaneous formation of ordered structures of the appropriate physical fields. In this case, we refer to **an interfacial self-organization** and interfacial dissipation structures. A major feature that distinguishes the interfacial self-organization from the self-organization in various homogeneous systems is the occurrence in the former of a feedback as effected via varying properties and characteristics of the interface. We wish to emphasize that the interfacial self-organization is capable of initiating a self-organization of transport processes not only at the level of elementary acts at the interface, but also on the scale of the entire apparatus, with the eventual change of its performance characteristics.

4.3 Theoretical Studies of Interfacial Instability and Self-Organization

The current state-of-art of the theory of interfacial instability and self-organization outlines the following major trends:

—definition of conditions for the occurrence of interfacial instability and determination of appropriate quantitative criteria;

—linear and nonlinear analysis of the transition from a steady to an unsteady interface state;

—study of the dynamics of interfacial dissipation structures and computation of their major parameters;

—nonlinear analysis of dissipation structure instability and transition to chaotic state (interfacial turbulence);

—performance of computer experiments in all stages of the interfacial instability study;

—finding of strict solutions to convective mass and heat transfer equations;

—development of mass transfer models (primarily semiempirical ones) for interfacial convection;

—computation of diffusion and heat fluxes across the interface and appropriate heat and mass transfer coefficients;

—determination of optimal mass and heat transfer parameters for separators, purifiers and chemical reactors operating under high-performance conditions.

Experimental studies on interfacial instability—both its physical mechanisms and practical applications—focus on the following major issues:

—studies of interfacial convection and turbulence by optical visualization methods;

—kinetic studies of mass and heat transfer in systems with interfacial instability both at the level of elementary acts and on the scale of an apparatus or a reactor;

—selection of high-effective absorbents, extractants, reagents and solvents whose kinetic properties would provide for the buildup of interfacial convection;

—determination of conditions for high-intense mass and heat transfer regimes; development of methods for interphase transfer enhancement using interfacial instability effects.

A point to be noted is that accurate measurement methods have not yet gained a wide acceptance in studies of varying (both in time and space) velocity, concentration and temperature fields at the interface. This deficiency presents limitation to a comprehensive theoretical analysis of small-scale structures under interfacial instability conditions.

4.3.1 Basic Equations for Mathematical Models

Mathematical models that are currently used for the description of interfacial instability and self-organization in two-phase systems are chiefly based on the equations of continuous-medium mechanics, heat and mass transfer, nonequilibrium thermodynamics, macroscopic electrodynamics, physicochemical and chemical kinetics; in addition, other equations describing perturbative effects as those mentioned in Section 4.2 are used. These models include also balance equations for momentum, matter and energy, as well as boundary conditions for description of the interface behaviour and feedback relations between the interface and the forces operative in the bulk phase. The feedback is the major contributor to the nonlinearity of mathematical models, and plays a crucial role in the development of interfacial self-organization.

It has not been our intention to go into mathematical details of interfacial surface dynamics and thermodynamics; the interested reader is referred to excellent works [77, 81-87] in this field.

We write now a system of equations commonly used in most models on interfacial instability and self-organization. This system includes equations of motion for a viscous incompressible Newtonian fluid, convective diffusion and heat conduction, with allowance made for the sources within each of the two contact phases:

$$\varrho^{(i)} \left[\frac{\partial \mathbf{v}^{(i)}}{\partial t} + (\mathbf{v}^{(i)} \nabla) \mathbf{v}^{(i)} \right] = - \nabla p^{(i)} + \mu^{(i)} \nabla^2 \mathbf{v}^{(i)} + \mathbf{G}^{(i)} \qquad (3.1)$$

4 Modelling of Mass Transport Processes

$$\nabla \mathbf{v}^{(i)} = 0 \tag{3.2}$$

$$\frac{\partial C_j^{(i)}}{\partial t} + (\mathbf{v}^{(i)} \nabla) C_j^{(i)} = D_j^{(i)} \nabla^2 C_j^{(i)} + R_j^{(i)} \tag{3.3}$$

$$\frac{\partial T^{(i)}}{\partial t} + (\mathbf{v}^{(i)} \nabla) T^{(i)} = \chi^{(i)} \nabla^2 T^{(i)} + Q_j^{(i)} \tag{3.4}$$

Here the superscripts $i = 1, 2$ refer to phases 1 and 2, and the subscript j denotes the ordinal number of a component in a mixture; v is the velocity vector; ϱ is the density; p is the pressure; μ is the dynamic viscosity; C is the solute concentration; D is the solute diffusivity; T is the absolute temperature; χ is the thermal diffusivity. The term G which describes the mass force in Navier-Stokes equation (3.1) takes into account the dependence of density on temperature and concentration. Commonly, the Newton-Boussinesq model is used to account for such gravitational effects [88]. The net source terms R and Q in Eqs. (3.3) and (3.4) take into account chemical reactions occurring in the bulk phase with heat release other than zero, as well as the heat or mass sources of other types. The right-hand sides of Eqs. (3.3) and (3.4) may also contain terms that take into account the Soret and Dufour effects providing that these contribute substantially to the Fick and Fourier fluxes. For gas-phase processes, it should be kept in mind that the density in all the equations is a variable quantity.

The boundary conditions include balance equations for normal and tangential forces at the interface of two immiscible continuous media:

$$p^{(2)} - p^{(1)} + \sigma \left(\frac{1}{r_1} + \frac{1}{r_2} \right) = 2\mu^{(2)} \frac{\partial v_n^{(2)}}{\partial x_n} - 2\mu^{(1)} \frac{\partial v_n^{(1)}}{\partial x_n} \tag{3.5}$$

$$\gamma \left(\tau \cdot \left[\frac{\partial \mathbf{v}^{(s)}}{\partial t} + (\mathbf{v}^{(s)} \nabla) \mathbf{v}^{(s)} \right] \right) - (\nabla \sigma \cdot \tau) + \mu^{(s)} \nabla_s^2 \mathbf{v}^{(s)} =$$

$$\mu^{(1)} \left(\frac{\partial v_n^{(1)}}{\partial x_\tau} + \frac{\partial v_\tau^{(1)}}{\partial x_n} \right) - \mu^{(2)} \left(\frac{\partial v_n^{(2)}}{\partial x_\tau} + \frac{\partial v_\tau^{(2)}}{\partial x_n} \right) \tag{3.6}$$

Here the index s refers to the interface surface; n and τ denote the normal and tangential unit vectors to the surface; r_1 and r_2 are the principal curvature radii of the surface; σ is the surface tension; $\mu^{(s)}$ is a phenomenologic constant denoting the surface viscosity; γ is the surface density of the transported solute at the interface; t is the time; x_n and x_τ are the normal and tangential coordinates.

The condition of mass balance must be fulfilled at the interface. For each of the surface-active substances transported across the deforming interface [89], we have:

$$\frac{\partial \Gamma_k}{\partial t} + \nabla_s(\Gamma_k v^{(s)}) + \Gamma_k(\nabla_s \mathbf{n})(v \cdot \mathbf{n}) = D_k^{(s)} \nabla_s^2 \Gamma_k + J_k^{(s)} + R_k^{(s)} \quad (3.7)$$

Here Γ_k is the surface concentration for the k-th adsorbed species; $\mathbf{n}$ is the local unit normal; $D^{(s)}$ is the surface diffusion coefficient for SAS molecules; $J_k^{(s)}$ is the net flux of molecules of the k-th SAS from the solution bulk toward the interface surface; $R_k^{(s)}$ is the net flux from chemical sources at the interface.

In the case of a fast adsorption-desorption, the transport toward the interface is diffusion-controlled, and $J_k^{(s)} = \sum_{i=1}^{2} (D^{(i)} \nabla_n C_k^{(i)})_s$. A relation between Γ and the volume concentration at the interface, $C^{(s)}$, is established by means of the adsorption isotherm within the hypothesis of a local thermodynamic equilibrium between the interface and the adjoining solution layers.

In case if the transport toward the interface is controlled by adsorption-desorption processes, the quantity $J^{(s)}$ represents an overall adsorption flux. Equation (3.7) should be supplemented by a condition for equality of the adsorption and diffusion fluxes from the bulk.

The equation of state at the interface gives the surface tension vs. temperature and SAS concentration relationship:

$$\sigma = \sigma_0 + \frac{\partial \sigma}{\partial T}(T - T_0) + \sum_{k=1}^{N} \frac{\partial \sigma}{\partial \Gamma_k}(\Gamma_k - \Gamma_{k0}) \quad (3.8)$$

To the system of equations (3.1)-(3.8), a number of dimensionless parameters are brought in correspondence, such as Reynolds, Peclet, Weber, Schmidt, Prandtl, Lewis, Nusselt, Sherwood, Rayleigh, Marangoni numbers, reaction parameters (analogues of Thele and Damköhler numbers) and others. Now, we consider the Marangoni numbers and reaction parameters. By general definition, the Marangoni number is represented by the ratio of the surface forces associated with the surface tension gradient to the viscous dissipative forces,

$$\text{Ma} = \frac{\Delta \sigma l}{\mu \nu}$$

where $\Delta \sigma$ is the surface tension change per characteristic length of the convection cell l. This ratio characterizes the relative contribution of the respective

4 Modelling of Mass Transport Processes

terms to Eq. (3.6). If diffusion and mass transport are considered, the concentrational Marangoni number is introduced,

$$\mathrm{Ma}_C = \frac{(\partial\sigma/\partial\Gamma)\Delta\Gamma l}{\mu D},$$

where $\Delta\Gamma$ is the surface concentration change per characteristic length of the convection cell. For thermocapillary effects, the temperature Marangoni number is introduced,

$$\mathrm{Ma}_T = \frac{(\partial\sigma/\partial T)\Delta T l}{\mu \chi},$$

where ΔT is the temperature change per length l along the surface. The Marangoni numbers are interrelated as

$$\mathrm{Ma}_C = \mathrm{Ma}\,\mathrm{Sc}, \quad \mathrm{Ma}_T = \mathrm{Ma}\,\mathrm{Pr},$$

$$\mathrm{Sc} = \frac{\nu}{D}, \quad \mathrm{Pr} = \frac{\nu}{\chi}$$

where Sc is the Schmidt number and Pr is the Prandtl number.

In solving specific problems as well as in modelling chemical processes, it is convenient, using the Marangoni numbers, to replace the characteristic concentration and temperature change along the surface by the respective driving forces for mass and heat transfer. In an analogous manner, the characteristic dimensions of convection structures (which can be determined only through solving the problem in question) may be replaced by a characteristic linear scale of the system, for example, the phase depth for limiting the interphase transport, or the thickness of a layer exhibiting concentration or temperature gradients.

In what follows, we understand by the reaction parameter R a dimensionless quantity defined as the ratio of the chemical reaction rate to the diffusional transport rate of reactants involved. The formula for this parameter is dependent on the type of kinetic equation for chemical reaction rate. By way of example, for a monomolecular surface reaction, the parameter takes the form

$$R = \frac{K_s l}{D}$$

where K_s is the surface reaction rate constant; D is the reactant diffusivity, l is the characteristic length scale of a system. For a bimolecular reaction carried out in the bulk, we have

$$R = \frac{KC_0 l^2}{D}$$

where K is the bulk reaction rate constant, C_0 is the characteristic concentration of one of the reactants.

An Approach to Constructing Mathematical Models for Interfacial Instability in a Liquid-Solid System

The problems in mathematical modelling of phase-transfer processes as outlined above are commonly referred to liquid-liquid or gas-liquid systems. However, there is available experimental evidence for the occurrence of a convective surface instability at the liquid-solid interface [90-93]. A hydrodynamic model for interfacial instability in such a system is suggested below.

Let us consider, by way of example, a system "electrolyte solution-solid dielectric surface". Suppose, an external electric field having a tangential component to the interface surface is imposed on the system. Such a situation may occur, for example, in electrophoresis, in electrode processes, or in the transport of ions from solution to the membrane surface (membrane electrodialysis). As is known, near the solid surface in contact with a solution, an electric double layer is formed, whose liquid diffuse junction develops a noncompensated electric charge. The motion of the liquid within this layer is described, in the general case, by the Navier-Stokes equation with the term $\varrho_e \mathbf{E}$ representing the mass force. Here $\mathbf{E}$ is the electric field strength and ϱ_e is the electric charge density related to the solution potential φ through Poisson's equation, $\nabla^2 \varphi = -4\pi \varrho_e/\varepsilon$, ε being the dielectric permittivity. If ions are transported to the surface, the potential distribution within the layer is dependent on the ion concentration distribution; in the case of interfacial instability, the potential distribution can vary throughout the surface. The relation between the potential and the ion concentration is established via convective diffusion equation, with allowance made for the ion migration

$$\frac{\partial C}{\partial t} + (v\nabla)C = D\nabla^2 C + zuF \, \text{div} \, (C \cdot \nabla \varphi) \tag{3.9}$$

where C is the concentration of a transported solute; D is the solute diffusivity; z is the ion charge expressed in protonic charge units; u is the ion mobility;

F is the Faraday constant. The existence of a relation between the potential and the concentration may provide for an interfacial instability and the development of self-organized convection structures. A potential φ_s is assigned to the solid surface or, to be more exact, to the inner boundary of double layer where the flow velocity tends to zero. In our particular case of interfacial instability, the outer diffuse layer potential and the zeta potential (also called the electrokinetic potential) are quantities variable throughout the interface; they are determined by solving the problem under specified conditions.

Outside the double layer with the electroneutrality condition holding for any elementary volume of the solution, the liquid motion is likewise described by a Navier-Stokes equation which, however, lacks the mass force associated with an electric field. At the double diffuse layer boundary, the condition for velocity continuity as well as the balance of normal and tangential stresses must hold.

Such an approach in which the liquid phase is assumed to be partitioned into two subregions (double layer and external solution), each described by a separate equation, provides a means to solve the problem for the boundary condition of stress balance at the solid surface in the absence of a liquid-solid slip. The attempts of traditional approach to the description of interfacial instability in a liquid-solid system, with the hydrodynamics and mass transfer related to each other through an interface boundary condition, have proved to be little effective, since it is precisely the condition for the sticking of the liquid that "freezes" this relation.

To conclude, we wish to note that the occurrence of a surface convection in the system of interest may be interpreted as a display of the Marangoni effect. A similar conclusion was made by Ruckenstein [75] who considered phoretic motions. To make the picture descriptive, we use a boundary layer approximation to specify the flow within the double layer in neglect of the dynamic pressure gradient. Then, through substituting Poisson's equation into the equation of motion and integrating, we obtain

$$\mu \frac{\partial v_x}{\partial y} = \varepsilon E_x \frac{\partial \varphi(x)}{\partial y} \tag{3.10}$$

where x and y are, respectively, the coordinates directed tangentially and normally to the surface into the bulk of the solution. Setting $\varepsilon E_x \partial \varphi / \partial y = \partial \sigma / \partial x$, Eq. (3.10) can be rewritten in the form

$$\mu \frac{\partial v_x}{\partial y} = \frac{\partial \sigma}{\partial x} \tag{3.11}$$

where σ is the effective surface tension dependent on the ion concentration distribution within the double layer. Equation (3.11) in this particular case may be regarded as the boundary condition for an external problem on convective ion diffusion outside the double layer. Thus, the surface convection in a system "liquid electrolyte-solid surface" is viewed as a motion driven by surface tension gradient.

The results for interfacial instability studies in a variety of liquid-liquid and gas-liquid systems are dealt with in the following Sections.

4.3.2 Conditions and Criteria for Interfacial Instability

An important stage in studying the interfacial instability is to specify the conditions under which a stationary uniform state of the interface becomes unstable with respect to disturbances of velocity, pressure, concentration or temperature, as well as other state variables. Reliable data on the stability of a system toward small perturbations constitute the basis for the architecture of an interfacial instability theory. At present, the number of works concerned with the linear analysis of Marangoni instability in gas-liquid and liquid-liquid systems totals several hundreds.

Pioneering studies in the field were the works of Pirson [94], Scriven and Sternling [95, 96]. In [94], the heat transfer in an initially stagnant liquid layer of finite thickness, with its free surface in contact with the surrounding gas phase, was considered. The paper [95] was concerned with the mass transfer between two quiescent liquid phases, infinite in depth and separated by a strictly plane interface. A major result of the Sternling-Scriven model was that two parameters were specified enabling to define the stability of a system, namely, the ratio of the diffusion coefficients for a solute transferred between the phases, and the ratio of the kinematic viscosity coefficients of these phases.

Numerous papers have been reported in the literature concerned with the analysis, chiefly a linear one, of the conditions for occurrence of Marangoni convection in a variety of systems, for instance, in stagnant and moving layers [97-115, 226, 227], liquid films [116-126], drops [127-131], and jets [125]. However, the situations considered do not by any means exhaust the eventual conditions for interfacial instability subject to various complicating factors.

Given below are recent results of a linear analysis of the concentrational Marangoni instability (solute-capillary effect) that we have obtained for the following systems: a bilayer system in the presence of mass transfer and a

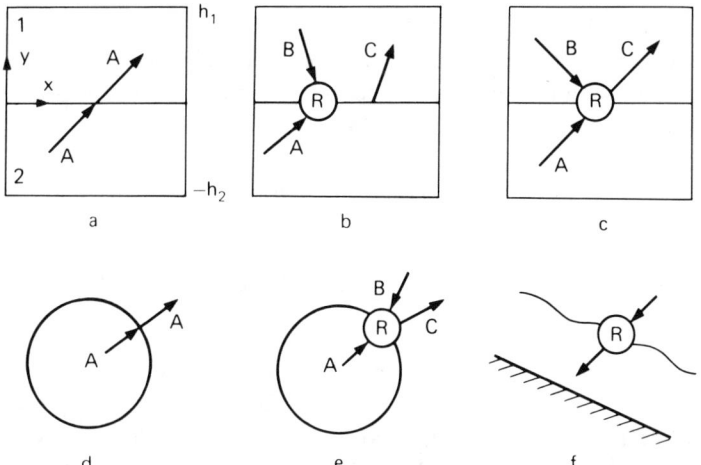

Fig. 4.14 Models for two-phase systems with a planar and a spherical interface

chemical surface reaction (Fig. 4.14a, b, c); a two-phase system with a spherical interface (Fig. 14d, e), also in the presence of a surface reaction or a laminar liquid film into which a component is chemisorbed from the gas phase (Fig. 14f).

For each of the above systems, dispersion equations relating a perturbation wave number k with the complex frequency ω to all physicochemical parameters of the system in question have been derived using the conventional formalism of the linear perturbation theory,

$$F\left(\omega, k, \frac{D_1}{D_2}, \frac{\mu_1}{\mu_2}, \frac{h_1}{h_2}, \text{Ma}, \text{R}, \text{Sc}_1, \text{Sc}_2, \ldots\right) = 0 \qquad (3.12)$$

From (3.12), an equation for neutral (marginal) stability is obtained by setting $\omega = 0$.

In greater detail, the transfer from one phase into another of a solute affecting the surface tension (Fig. 4.14a) has been studied in [106, 110-112].

Here we propose a novel instability criterion for a system with a given sign of surface activity ($\partial\sigma/\partial c > 0$ or $\partial\sigma/\partial c < 0$)

$$\text{Sgn}\,\frac{h^2 - d}{1 - d} = \begin{cases} 1, & \text{if } \uparrow \text{ or } \downarrow \quad \text{(a)} \\ -1, & \text{if } \uparrow \text{ and } \downarrow \quad \text{(b)} \end{cases} \qquad (3.13)$$

Here $h = h_1/h_2$, $d = D_1/D_2$; the signs ↑ and ↓ denote the mass transfer direction for the occurrence of an interfacial instability.

For a system with the specified direction of mass transfer and with variable surface activity, the instability criterion takes the form

$$\text{Sgn} \frac{h^2 - d}{1 - d} = \begin{cases} 1, & \text{if } \frac{\partial \sigma}{\partial c} > 0 \text{ or } \frac{\partial \sigma}{\partial c} < 0 & \text{(a)} \\ -1, & \text{if } \frac{\partial \sigma}{\partial c} > 0 \text{ and } \frac{\partial \sigma}{\partial c} < 0 & \text{(b)} \end{cases} \quad (3.14)$$

The physical meaning of the criteria derived implies the following. Given the surface activity of a transferred solute, there exist three regions for the diffusion coefficient ratio d which correspond to two different instability regimes. One of these is when the instability arises only in one of the two directions (3.13a); another one is when the system loses its stability due to mass transfer in either direction (3.13b). The boundaries for these three regions of d values are demarcated by $d = 1$ and $d = h^2$. Analogous regions can be defined for a given direction of transport of a solute with the surface activity distinct in sign [Eqs. (3.14a) and (3.14b)].

The existence of an instability as produced by transfer of the same solute in both directions has been confirmed by numerous experiments on extraction in liquid-liquid systems. For example, according to Ermakov and Shatokhin [43], a spontaneous interfacial convection occurs on transfer of acetic acid from butyl acetate into water and, conversely, from water into butyl acetate. A similar effect is observed for propionic acid or acetone transferred between carbon tetrachloride and water. While quantitatively comparing criteria (3.13) and (3.14) to experimental results, one should keep in mind that the thickness of the layers within which concentration gradients (acting as driving forces for each phase) are operative play a role of parameters h_1 and h_2. In the absence of forced stirring, these may be the depths of the contact phases. If each phase is stirred thoroughly, the thickness of the subsurface diffusion layer within which the entire concentration drop occurs should be taken for parameters $h_{1,2}$.

Further presented are some linear analysis results on convective instability of bilayer systems at whose interface a chemical reaction takes place between reactants delivered via diffusion from the bulk of both phases (Fig. 4.14b, c). In the case of a solute partly involved in reaction at the interface on its transport from phase 2 to phase 1 (Fig. 4.14b), the mass balance in the surface takes the form

$$D_1 \frac{\partial C_1}{\partial y} - D_2 \frac{\partial C_2}{\partial y} = D_3 \frac{\partial C_3}{\partial y} = KC_2 C_3, \quad C_1 = \beta C_2, \quad \text{at} \quad y = 0 \qquad (3.15)$$

Here C_1, C_2 are the concentrations of a transferred solute A in phases 1 and 2, respectively; C_3 is the concentration of a reactant B dissolved in phase 1; K is the reaction rate constant; β is the interphase equilibrium constant. The surface tension as a function of the solute A concentration is approximated by a linear function

$$\sigma = \sigma_0 - \alpha_1 C_1, \quad \text{at} \quad y = 0 \qquad (3.16)$$

In case if the reactants A and B dissolved in phases 1 and 2 enter into an irreversible chemical reaction at the interface to become completely converted to a product C which is subsequently lost into the bulk of phase 1, the mass balance in the surface takes the form

$$D_1 \frac{\partial C_1}{\partial y} = D_2 \frac{\partial C_2}{\partial y} = -D_3 \frac{\partial C_3}{\partial y} = -KC_2 C_3, \quad \text{at} \quad y = 0 \qquad (3.17)$$

Here C_1, C_2, and C_3 are the concentrations of species C, A, and B, respectively. The surface tension in this case is dependent on the concentrations of reactant A and product C:

$$\sigma = \sigma_0 - \alpha_1 C_1 - \alpha_2 C_2 \qquad (3.18)$$

The system in question have trivial stationary states

$$\psi^{(i)} = 0, \quad c_j = a_j y + b_j \quad (j = 1, 2, 3) \qquad (3.19)$$

Here ψ_1, ψ_2 are the stream functions in phases 1 and 2,

$$v_x^{(i)} = \partial \psi^{(i)} / \partial y; \quad v_y^{(i)} = -\partial \psi^{(i)} / \partial x$$

By virtue of the nonlinearity of boundary conditions (3.15) and (3.17), one arrives at multiple stationary states (3.19); for this reason, in performing computations the values for constants a_j and b_j are chosen in accordance with the condition for positive concentrations at the interface.

Next, we consider the stability of stationary states (3.19) with respect to small perturbations of the stream function $\tilde{\psi}^{(i)}$ and concentration $\tilde{c}_j$, which are represented by separate Fourier components with a dimensionless real wave number k and a complex frequency ω

$$\tilde{\psi}^{(i)} = \Psi^{(i)}(y) e^{ikx + \omega t}, \quad \tilde{c}_j = A_j(y) e^{ikx + \omega t} \qquad (3.20)$$

Here Ψ_y^i and $A_j(y)$ are the perturbation amplitudes, i is the imaginary unit.

By substituting (3.20) into Eqs. (3.1), (3.3) and linearizing them with respect to $\tilde{\psi}^{(i)}$ and C_j, we obtain a system of ordinary differential equations for perturbation amplitudes. A nontrivial solution to these equations exists only at definite values of the spectral parameter ω. The decrements Re ω are found as the eigenvalues of a boundary problem, and the respective eigenfunctions $\Psi^{(i)}$ and A_j determine the structure of characteristic perturbations for velocity and concentration. The solutions to equations of (3.12) type allow to construct neutral stability curves that demarcate stability regions (Re $\omega < 0$) from instability regions (Re $\omega > 0$). These curves provide a means for exploring the stability boundary as a function of both physicochemical and regime parameters; besides, they allow to determine critical Marangoni numbers and critical reaction number R. The perturbation wavelengths corresponding to extremum points in neutrality curves define thus the dominant wave modes which are the first to experience growth as the Marangoni numbers surpass respective critical values; these modes determine to a significant extent the dissipation structure produced by convective instability and perturbation stabilization.

Shown in Fig. 4.15 are neutral stability curves for the case of a solvent extraction involving an interfacial chemical reaction (Fig. 4.14b) [73]. The chosen Marangoni number Ma and reaction parameter R were Ma $= \alpha_1 a_0 h_1 / \mu_1 D_3$, $R = K a_0 h_1 / D_3$; a_0 is the component A concentration at $y = -h_2$. The computations have been carried out at $h = h_1/h_2 = 1$, $\mu = \mu_1/\mu_2 = 0.5$, $d_1 = D_2/D_3 = 5/6$, $\beta = 5$, and at different values of $d = D_1/D_2$. An analysis of the curves in Fig. 4.15 shows that the system stability is to a large extent dependent on the ratio d of the diffusion coefficients of a solute affecting the surface tension. Depending on the values of d, the system can lose its stability upon transport of substances acting to reduce the surface tension (SAS), or to increase it (surface-tension increasing substances, for short SIS). The critical Marangoni numbers corresponding to the extrema in the curves can grow in absolute value with both the chemical reaction rate constant and the reaction number R rising. This points to a possible stabilizing effect of a chemical reaction on the system stability.

At definite values of d, the growing reaction number R leads to drastic changes in system stability. As exemplified by Fig. 4.16, given $d = 0.7$, the system instability at $R = 0 - 1$ is produced only by the transfer of a SAS; at $R = 1.6 - 3$, only by the transfer of a SIS. However, at $R = 1.45 - 1.5$, the system instability is initiated by the transfer of either SAS, or SIS. It is to be inferred therefore that as the reaction number R and, in particular, the

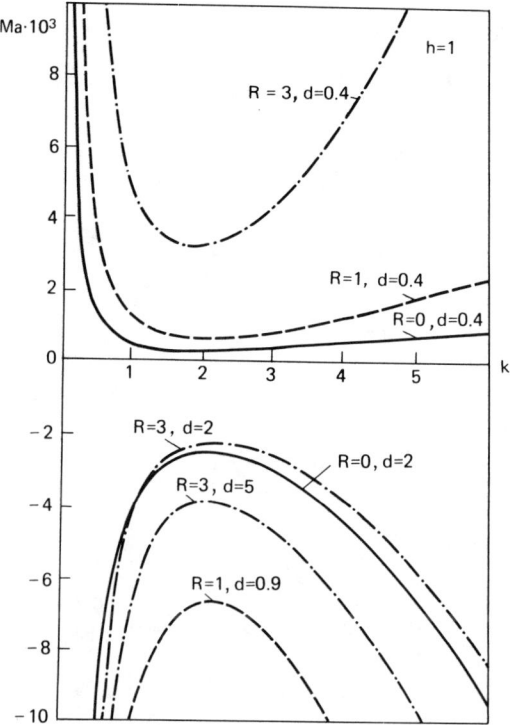

Fig. 4.15 See the text for explanation

reaction rate constant are made to vary, the system containing a solute of specified surface activity suffers a change in stability.

Shown in Fig. 4.17 are the curves that characterize a relationship between the numbers Ma and R corresponding to the stability boundary ($\omega = 0$). The regions above the curves (at Ma > 0) and below the curves (at Ma < 0) correspond to unstable states. The vertical asymptotes at $d = 0.2$ and 0.3 provide evidence for the occurrence of an instability in the same system on transfer of either SAS, or SIS.

Presented in Fig. 4.18 are the neutral stability curves for an interface reaction in which one of the reactants and the product affect the surface tension (Fig. 4.14c) [73, 132]. Here $\text{Ma}_A = \alpha_2 a_0 h_2 / \mu_2 D_2$, $\text{Ma}_C = \alpha_1 b_0 h_1 / \mu_1 D_3$, b_0 is the reactant B concentration at $y = h_1$. The course of the curves in Fig. 4.18 shows that, at definite ratios of the phase depths, the system is unstable in

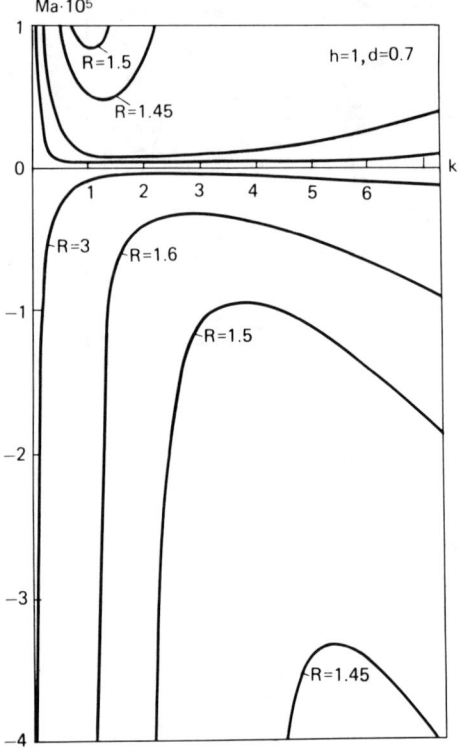

Fig. 4.16

the presence of both a surface-active and surface-inactive reactant. The conclusion as to a substantial influence of the parameter h on the conditions for the occurrence of an interfacial convection and its intensity agrees with the experimental results in [69, 133]. Shown in Fig. 4.19 are the curves for Ma_A as a function of R that separate the stability regions (below the curves) from instability regions (above the curves) at different Ma_C and d. The nonmonotonic course of these curves, each exhibiting a minimum, attests to a stabilizing influence of the chemical reaction on convective stability of the system at both large and small reaction numbers. This conclusion bears a clear physical meaning, since at $K \to 0$, the surface diffusion fluxes also tend to zero according to (3.17), which results in a short supply of the reactants to the reaction surface. At $K \to \infty$, the concentrations throughout the interface surface become levelled out, and the surface tension gradients decrease; therefore, in order

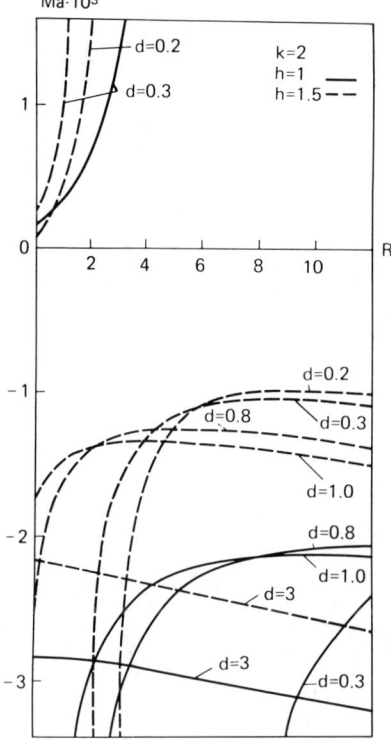

Fig. 4.17

to maintain the convective instability regime, one must augment the surface activity and the respective Marangoni numbers. The curves depicting the relationships between the critical Marangoni numbers Ma_A^{cr} and Ma_C^{cr} at different values of d and h are shown in Fig. 4.20. The course of these curves provides evidence for a variety of unstable regimes in a system of two components exhibiting surface-active properties and participating in a chemical reaction.

In studying the hydrodynamic stability of a liquid drop [134, 135] the cases of a transit flux of a component (Fig. 4.14d) and an interfacial chemical reaction (Fig. 4.14e) have been considered. A specific feature of the formulated problem vis-à-vis the known analysis by Sorensen and Hennenberg [129] was that we have taken into account the mixed adsorption-diffusion kinetics of interphase transfer and undisturbed volume concentration profiles in the general form—either exponential, or raised to a power. The exponential pro-

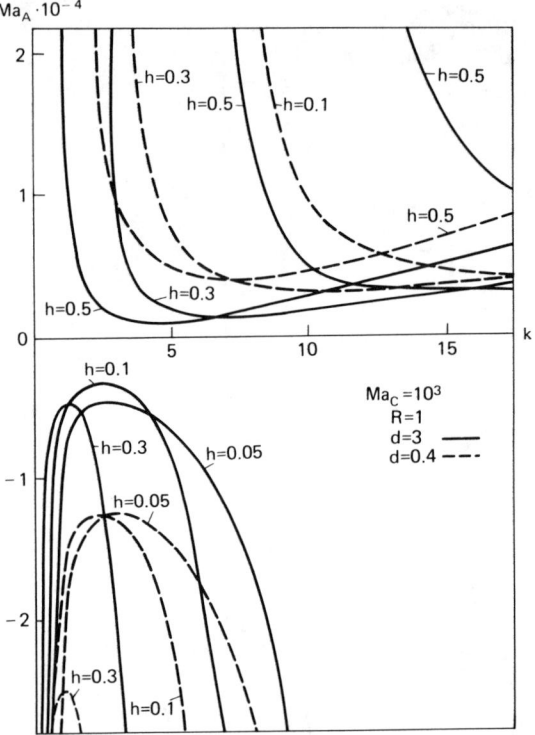

Fig. 4.18

files are of special interest, since they give a good approximation of the volume concentration distribution at the initial stage of a diffusion process when, according to the available experimental evidence, an interfacial instability develops.

In considering the extraction of a surface-active agent (Fig. 4.14d), a general equation for neutral instability (in the absence of oscillations) has been obtained. In critical situations, when the SAS transport is diffusion-controlled ($D_i \ll aK_i^c$) or when the limiting stage of the transfer process is adsorption ($D_i \gg aK_i^c$), this equation takes the form:

$$\mathrm{Ma} = H \frac{K_a(l+1) + D_* El}{l(l+1)(2l+1)(\varphi_{12} - D_*\varphi_{11})K_a}, \quad D_i \ll aK_i^c$$

4 Modelling of Mass Transport Processes

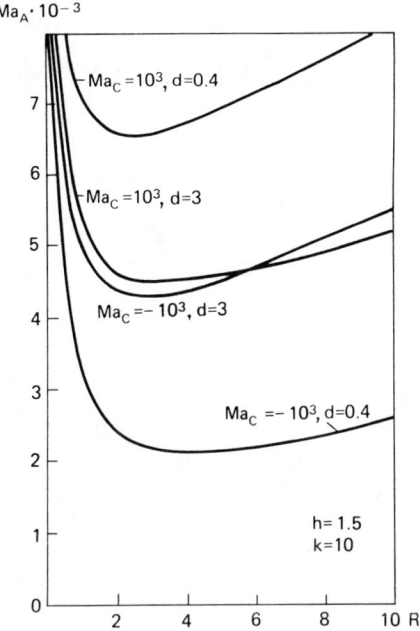

Fig. 4.19

$$\text{Ma} = H \frac{D_*(1 + E)}{l(l + 1)(2l + 1)(l^{-1}\varphi_{l2}K_a - D_*^2(l + 1)^{-1}\varphi_{l1})}, \quad D_i \gg aK_i^c \quad (3.21)$$

$$H = (1 + \mu_*)(2l + 1) + \varepsilon_1 l(l + 1) - 2\varepsilon_2, \quad \mu_* = \mu_2/\mu_1, \quad D_* = D_2/D_1$$

$$\varepsilon_1 = \frac{\xi + \varepsilon}{a\mu_1}, \quad \varepsilon_2 = \frac{\varepsilon}{a\mu_1}, \quad K_a = \frac{K_2^c}{K_1^c}, \quad E = \frac{K_2^\Gamma}{K_1^\Gamma}$$

$$\Gamma_{0l} = \frac{\Gamma_0 l(l + 1)}{aa_2 n_2 C_{20}}, \quad K_i^c = \left[\frac{\partial J_i(C_i, \Gamma)}{\partial C_i}\right]_0, \quad K_i^\Gamma = \left[\frac{\partial J_i(C_i, \Gamma)}{\partial \Gamma}\right]_0,$$

$$\varphi_{l1} = [(2l + 1)(2l + n_1 + 1)(2l + n_1 - 1)]^{-1},$$
$$\varphi_{l2} = [(2l + 1)(2l + n_2 + 1)(2l + n_2 + 3)]^{-1}$$

$$\text{Ma} = \frac{d\sigma}{d\Gamma} \frac{dC_{20}}{dr} \frac{a^2}{\mu_1 D_1} \frac{K_1^c}{K_1^\Gamma}$$

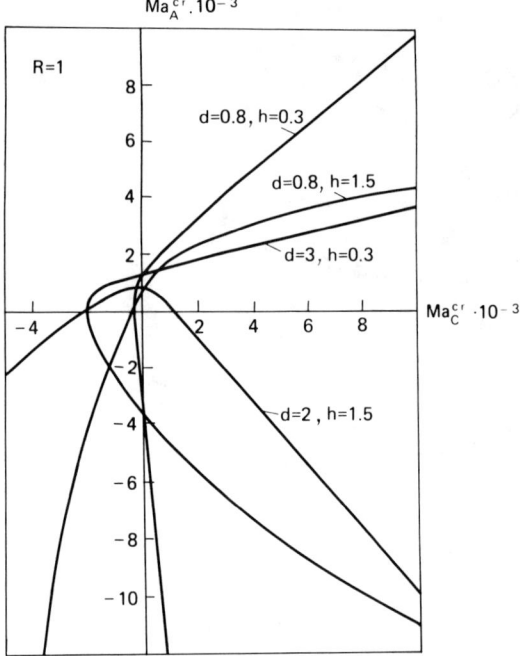

Fig. 4.20

Here the subscript $i = 1$ refers to the exterior liquid, and $i = 2$, to the liquid within the drop; a is the drop radius; K_i^c and K_i^Γ are the adsorption and desorption constants; ξ and ε are the surface dilatational and shear viscosities; l is the number of a perturbation harmonic; n_i are the constants that characterize exponential spherically symmetric stationary profiles of unperturbed concentrations; Γ is the surface concentration of a component; C is the solute concentration; C_{20} is the unperturbed concentration at the center of the liquid drop; r is the radial coordinate.

Suppose, the SAS is transported from a liquid drop into the surroundings; then $dc_{20}/dr < 0$, $d\sigma/d\Gamma < 0$, and $Ma > 0$. In this case, an instability may arise, if

$$D_* < F(l) = \begin{cases} \varphi_{12}/\varphi_{11}, & D_i \ll aK_i^c \\ K_a\varphi_{12}(l+1)/(l\varphi_{11})]^{1/2}, & D_i \gg aK_i^c \end{cases} \quad (3.22)$$

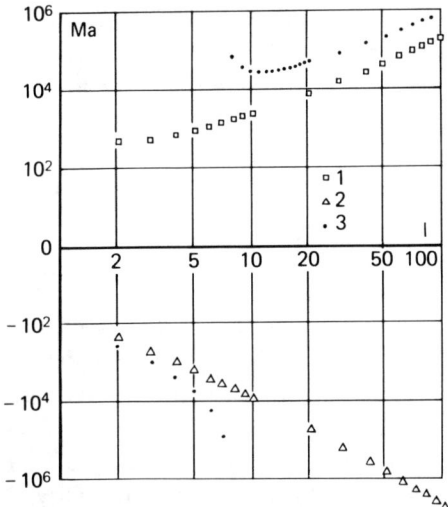

Fig. 4.21

In case if the mass transport is directed into the drop, the inequality sign in (3.22) must be reversed.

The discrete function $F(l)$ has a limit and is, therefore, bounded. We denote the minimal and maximal values of $F(l)$ by F_{min} and F_{max}. It follows then from (3.21) that, given $D_* < F_{min}$ or $D_* > F_{max}$, the interfacial instability can arise only if the SAS is transferred in one direction. By contrast, if $F_{min} < D_* < F_{max}$, instability may arise with the SAS transferred in either direction. Thus, whereas there is a single critical value of the ratio of the diffusion coefficients $D_* = 1$ for a plane interface between two phases of infinite depth [105], in the case of a liquid drop there are two critical values that divide the entire range of D_* into three subregions, each being characterized by a distinct direction of SAS transfer corresponding to the unstable regime. This conclusion is in a qualitative accord with the result that follows from formula (3.13) and holds for a bilayer system of finite phase depth. In Fig. 4.21, plotted on logarithmic scale are the neutral-stability Marangoni numbers as a function of the perturbation harmonic number l at $\varepsilon_1 = 0.01$, $\varepsilon_2 = 0.02$, $E = 1.5$, $K_a = 5.0$, $n_1 = 1$, $n_2 = 2$ for different values of D_* in the case of a slow adsorption. Given these parameters, one has $F_{min} = 1.51$ and $F_{max} = 2.24$. The set of points 1 correspond to $D_* = 1.5 < D_{min}$; the set of points 2, to

$D_* = 2.25 > F_{max}$; and the set of points 3, to $D_* = 2.0$, $F_{min} < D_* < F_{max}$.

In examining conditions that provide for interfacial instability in a system with a chemical surface reaction, a situation has been considered when surface-active agents are transported to the interface to be adsorbed thereon and to enter into reaction (Fig. 4.14e). The linear analysis of hydrochemical stability for a liquid drop in the presence of one or two chemically reacting species has revealed the following features [135]. The chemical instability, however low may the surface activity of the reacting species be, leads to a hydrodynamic instability. The adsorption of reactants at the interface, when proceeding at a finite rate, favors the occurrence of a hydrochemical instability and affects in a marked manner the stability of higher modes. Depending on the ratio of the reaction rate to the adsorption rate, the Marangoni effect can either stabilize a chemically unstable system, or destabilize a chemically stable system. Under definite conditions, the dissipative processes affected by surface and volume viscosities can promote a hydrochemical instability. This seemingly paradoxical conclusion turns out, generally speaking, to be quite natural for oscillationally coupled systems and has analogies, for example, in the theories of the flutter and Poiseuille flow stability. The above effect arises as a result of the phase relations between perturbation modes being such that these perturbations are sustained by the energy from an external source. In our case, such a source is the material flux across the interface. The rise in viscosity may render these relations more favourable from the standpoint of energy extraction from a source and thus augment the instability.

Given below are the results for a single reacting species that have been obtained in the approximation of a low reaction involving small-size drops, viz., $a^2/D_i\tau \ll 1$, $a^2/\nu_i\tau \ll 1$, where τ is the characteristic time for the chemical surface reaction. Shown in Fig. 4.22 are the regions corresponding to different types of system stability plotted on the $f - |Ma|$ coordinates, where f is a parameter characterizing the reaction rate constant, Ma is the Marangoni number, $Ma = (d\sigma/d\Gamma)(\tau^2/a^2)$. The broken line I is a demarcation line between the instability region (on the left) and the stability region (on the right). Each of these two regions is divided by the parabola II into two subregions, respectively, 1, 2 and 3, 4. The subregions 2 and 3 correspond to unstable and stable oscillatory regimes; by contrast, no oscillations occur in subregions 1 and 4. As is seen in Fig. 4.22, a specific feature of the given system is the possibility for instability to occur only in the case of $f < f_{cr} < 0$. The f_{cr} must be greater over a critical value associated parametrically with adsorption-desorption and

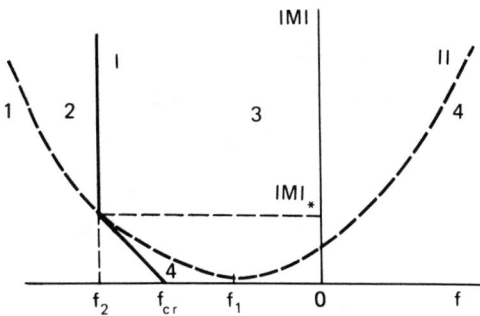

Fig. 4.22

diffusion as well as with the volume and surface viscosities. Noteworthy, the instability can manifest itself at however small Marangoni number $|Ma|$. Otherwise stated, the purely chemical instability, with a chemical reaction however weakly coupled to hydrodynamics ($|Ma| \neq 0$), leads to a hydrodynamic instability. Within a range of $f_2 < f < f_{cr}$, the increase in $|Ma|$ leads to stabilization of the system and to suppression of instability. The system subject to $f < f_2$ is unstable at any $|Ma|$.

Another fact is worth mentioning. In the absence of hydrodynamic coupling ($|Ma| = 0$), no interface movement is possible, and chemical instability is the only one to be anticipated. However, if the hydrodynamic coupling is operative ($|Ma| \neq 0$), instability occurs also at $f < f_{cr}$, and the instability threshold grows in absolute value. The above-said holds only for the modes with $l \geqslant 1$, such that a hydrodynamic mode with $l = 0$ is non-existent and, consequently, hydrodynamics does not affect chemical stability.

Further, we wish to note the effect of a finite adsorption rate on stability of the system in question. With growing l, the boundary of the instability region I shifts toward higher absolute values of f; this shift is larger for a fast, rather than slow, adsorption. It is to be inferred therefore that the slow adsorption destabilizes the system.

In [136], a model for interfacial instability of a liquid plane layer moving in a gravitational field under the condition of a mass exchange with the surrounding gas phase accompanied by chemical reaction was suggested (Fig. 14f). The velocity profile within the subsurface layer was approximated by a constant U. A linear analysis of this model has yielded the dispersion equation below:

$$F_1(\omega, k)\,\mathrm{Ma} - F_2(\omega, k) = 0 \tag{3.23}$$

$$F_1(\omega, k) = k_* + F(W)[k_*^2 + i\,\mathrm{Sc}^{-1}(\mathrm{Fo} + \mathrm{Pe}\,k_*)]^{1/2} + \\ + (2/\sqrt{\pi})(\varkappa\beta/\Gamma_0)(l^2/\beta d_0)[N_1 + F(W)N_2];$$

$$F_2(\omega, k) = \{2 + F(W)[2 + i\,\mathrm{Sc}^{-1}(\mathrm{Fo} + \mathrm{Pe}\,k_*)k_*^{-2}]\} \times \\ \times \{i(\mathrm{Fo} + \mathrm{Pe}\,k_*) + (D_s/D)k_*^2 + \mathrm{Bi} + R + (l/\beta)[k_*^2 + i(\mathrm{Fo} + \mathrm{Pe}\,k_*)]^{1/2}\}$$

$$F(W) = \frac{(\mathrm{Fo} + \mathrm{Pe}\,k_*)^2 - \mathrm{Sc}\,G\,(1 + W^{-1}k_*^2) - 2i\,\mathrm{Sc}(\mathrm{Fo} + \mathrm{Pe}\,k_*)k_*^2}{\mathrm{Sc}\,G(1 + W^{-1}k_*^2)k_* + 2i\,\mathrm{Sc}(\mathrm{Fo} + \mathrm{Pe}\,k_*)[k_*^2 + i\,\mathrm{Sc}^{-1}(\mathrm{Fo} + \mathrm{Pe}\,k_*)]^{1/2}k_*}$$

$$N_1 = \int_0^\infty \exp\{-(l/d_0)^2 s^2 - [k_* + [k_*^2 + i(\mathrm{Fo} + \mathrm{Pe}\,k_*)]^{1/2}]s\}\,ds$$

$$N_2 = \int_0^\infty \exp\{-(l/d_0)^2 s^2 - [\gamma + [k_*^2 + i(\mathrm{Fo} + \mathrm{Pe}\,k_*)]^{1/2}]s\}\,ds;$$

$$\mathrm{Ma} = \frac{(\partial\sigma/\partial\Gamma)\Gamma_0 l}{\mu D}, \quad \mathrm{Fo} = \frac{\omega l^2}{D}, \quad \mathrm{Pe} = \frac{Ul}{D},$$

$$\mathrm{Sc} = \frac{\nu}{D}, \quad \mathrm{Bi} = \frac{ql^2}{\beta_g D}, \quad R = \frac{K_s l}{D}, \quad W = \frac{\varrho g l^2}{\sigma_0}$$

$$G = \frac{gl^3}{\nu D}, \quad k_* = kl, \quad d_0 = (4DL/U)^{1/2},$$

$$\gamma = [k_*^2 + i\,\mathrm{Sc}^{-1}(\mathrm{Fo} + \mathrm{Pe}\,k_*)]^{1/2}$$

Here ω and k are the complex frequency and the real perturbation wave number; Fo, Ma, Pe, Sc, W, Bi denote, respectively, the Fourier, Marangoni, Peclet, Schmidt, Weber, and Biot numbers; G and R are the gravitation and reaction numbers; Γ_0 is the limiting surface concentration; ϱ is the liquid density; l is a constant with the dimension of a length (in subsequent computations taken equal to unity); L is the characteristic scale of the layer length; D and D_s are the volume and surface diffusion coefficients in the liquid phase; K is the surface reaction rate constant; q is the gas-phase mass transfer coefficient; g is the gravitational constant; β and β_g are the constants for local equilibrium between the surface and the subsurface layer in the liquid and the gas phases, respectively.

The marginal stability curves (at $\omega = 0$) reflecting the influence of Pe, Re, Sc, W and R numbers on Marangoni number as a function of perturbation wave number are shown in Figs. 4.23 and 4.24. An analysis of the curves in Figs. 4.23 and 4.24 shows that, with the Peclet and Reynolds numbers increasing (Re = Pe Sc^{-1}), the system stability threshold rises. This conclusion is in a qualitative agreement with the result in [113], where a linear analysis of Marangoni instability in a bilayer system showed the critical Marangoni numbers to grow with the velocity of the relative motion of contact layers. The course of the curves in Figs. 4.23 and 4.24 is also illustrative of the fact that, with increasing Weber number, the critical Marangoni numbers tend to decrease. The curves in Fig. 4.25 show that the increase in Schmidt number produces a stabilizing effect on the system, that is, the critical Marangoni numbers increase with growing Sc. The curves in Fig. 4.26 illustrate both the stabilizing and destabilizing effects of a surface reaction on the stability of a system with respect to perturbations of different wavelength.

To conclude this Section, we wish to emphasize that, despite the exceptional importance of linear analysis results for understanding the qualitative characteristics of stability of a multiphase system, they, most commonly, fail to account for many specific features of the transition to an unstable state and do not provide full information on the dissipation structures as produced by nonlinear interactions of perturbations. For this reason, the use of nonlinear analysis methods, including numerical methods for solving equations of macrokinetics and continuum mechanics, is necessary both for an in-depth study of the interfacial structure dynamics and for interfacial flux computations.

4.3.3 Dynamics of Interfacial Structures and Computation of Convection Pattern Parameters

By applying the linear analysis to interfacial instability, one can obtain an expression for the incremental growth of infinitesimal perturbations, formulate the conditions for neutral stability, and determine critical values of major dimensionless parameters. However, the linear analysis fails to specify both the nature of disturbed stability of a system and the properties of resultant stabilized convection patterns. To get a deeper insight one resort to a nonlinear analysis of the finite amplitude stability. Various aspects of the nonlinear Marangoni instability analysis were dealt with in [137-150]. Application of this analysis to hydrochemical Marangoni convection generated by a heterogeneous reaction [151, 152] is given below. The approach we have used

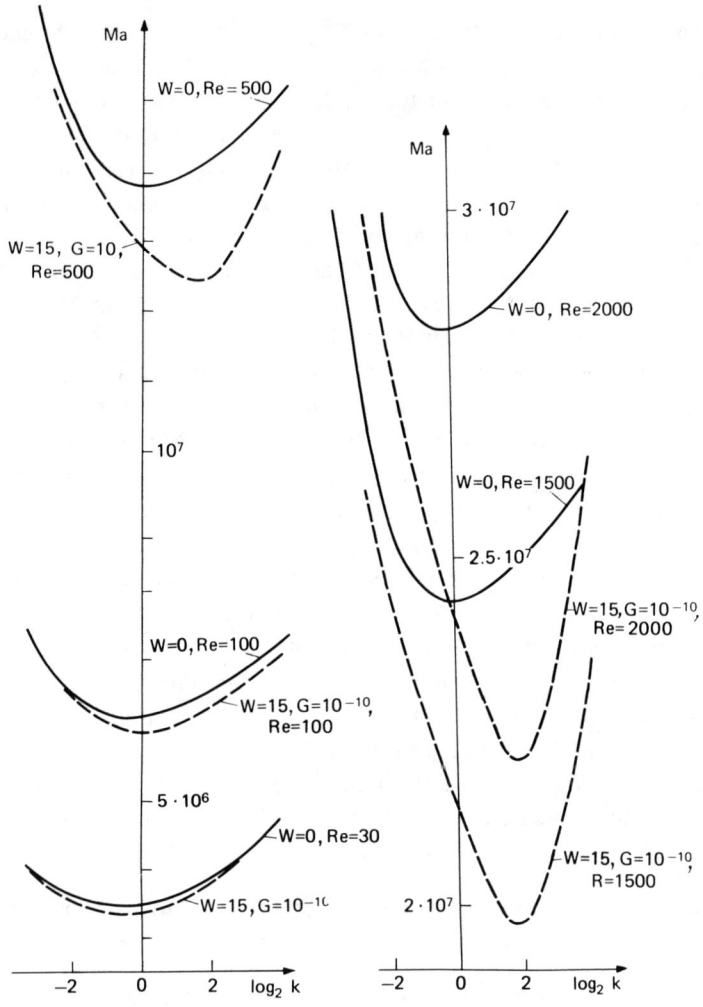

Fig. 4.23

is based on a hypothesis relating the hydrochemical instability to Landau-Hopf bifurcation [153-155]. According to this hypothesis, the transition to a turbulent motion is regarded as a sequence of post-critical bifurcations of the sets of unstable periodic or quasiperiodic Navier-Stokes equation solutions into analogous sets of higher dimensionality. It is assumed also that a stabilization of such solutions is possible under slightly supercritical conditions due to non-

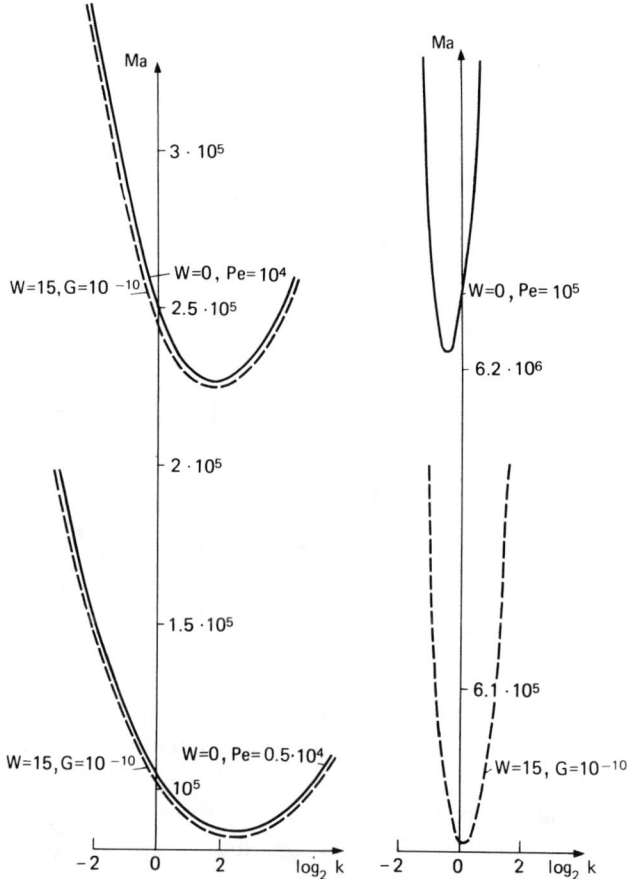

Fig. 4.24

linear interactions. This hypothesis, although untenable in treating the development of turbulence in a liquid flow, holds nonetheless in many other instances, in particular, when applied to interfacial convection. This circumstance justifies the assumption of a smallness of the stabilized structures and their closeness to harmonic modes. The method, commonly used in the study of weakly nonlinear structures, has evolved from a small-parameter method known in the nonlinear theory of hydrodynamic stability; conceptually, it is close to the method of Stuart [156] and Watson [157] that was applied to a treatment of nonlinear stability of Poiseuille and Couette flows.

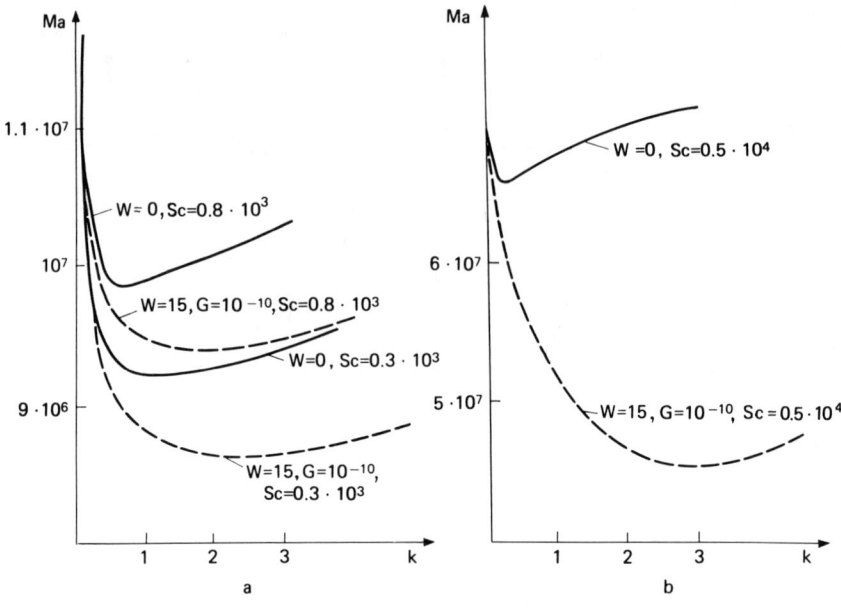

Fig. 4.25

The statement of a problem and the results of a linear stability analysis are given below. Further, the results for the nonlinear stage of hydrochemical instability development and the computed parameters of dissipative structures formed are presented. The nonlinearity arising from a heterogeneous reaction is shown to play a crucial role in the buildup of a convective flow. As has been established, there occur two types of transition to instability in the liquid-liquid system of interest: the soft regime conducive to interfacial convection exhibiting an ordered pattern of convective flows, and the hard regime conducive to interfacial turbulence, that is, chaotic flow pulsations with a sufficiently broad wave spectrum.

In real physicochemical systems, the post-critical instability stage can be significantly affected by nonlinearities of different type, for example, those specific of Navier-Stokes equation or convective diffusion, and by the nonlinearities arising from surface deformations, concentration and temperature effects. In chemical engineering processes, of special interest are the effects associated with the nonlinear dependence of chemical reaction rate on reactant concentration. The model proposed takes into account only the nonlinearity connected with a two-component chemical reaction of second order.

4 Modelling of Mass Transport Processes

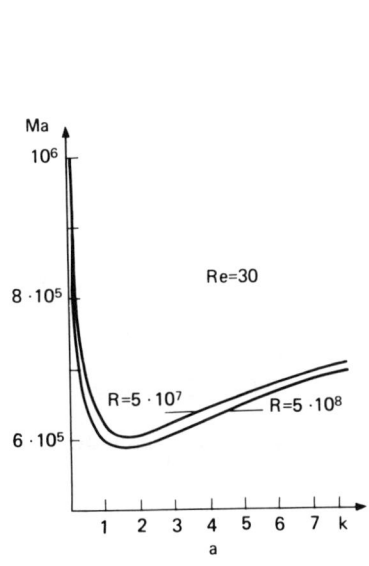
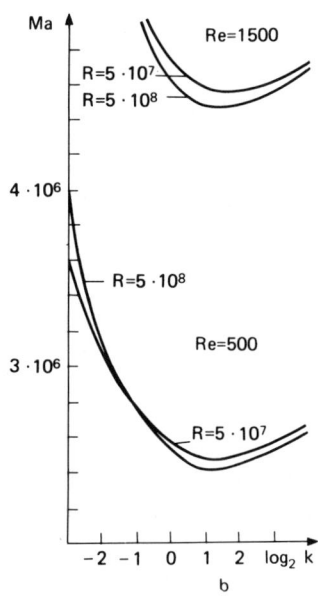

Fig. 4.26

The model considers the development of hydrochemical instability near a plane interface ($y = 0$) separating two immiscible quiescent liquids of infinite depth. Each liquid contains solute reactants that diffuse toward the interface to enter into a chemical reaction whose rate is given by the expression $KC_{s1}C_{s2}$, where C_{sj} ($j = 1, 2$) is the surface concentration of a reactant, and K is the reaction rate constant. The unperturbed concentration fields are linear in both liquids: $C_j^0 = C_{sj}^0 + (-1)^{j-1}\beta_j y$, β_j being constant concentration gradients. The surface tension coefficient is dependent on the reaction concentration and can be approximated by a linear function $\sigma = \sigma_0 - \alpha_1 C_{s1} - \alpha_2 C_{s2}$. The value of σ_0 is assumed to be large enough to render the interface practically undeformable.

The space-time evolution of the perturbations of stream function ψ_j and concentration field C_j is described by linearized nonstationary Navier-Stokes (3.1) and convective diffusion (3.3) equations. The interface boundary conditions for these equations are continuity of the tangential components of the velocity vector and equality to zero of the normal components. Hydrodynamic and chemical interactions are accounted for by the boundary conditions at the interface that describe the balance of the tangential stress and mass balance; in this particular case, these take the form:

$$\left(\frac{\partial^2}{\partial y^2} - \frac{\partial^2}{\partial x^2}\right)(\mu_2\psi_2 - \mu_1\psi_1) = \frac{\partial\sigma}{\partial x} \quad (3.24)$$

$$D_1\frac{\partial C_1}{\partial y} = -D_2\frac{\partial C_2}{\partial y} = KC_1C_2$$

The linear analysis has yielded a neutral stability equation relating the dimensionless wave number k (for perturbations with zero complex frequency) to major dimensionless parameters of the system:

$$8k^3 + 8(R_1 + R_2)k^2 - (\text{Ma}_1 + \text{Ma}_2)k -$$
$$- \text{Ma}_1 R_2(1 - D_1/D_2) + \text{Ma}_2 R_1(1 - D_2/D_1) = 0 \quad (3.25)$$

Here each of the contact phases has been assigned a Marangoni number Ma_j and a reaction number R_j,

$$\text{Ma}_j = \frac{(\partial\sigma/\partial C_{sj})\beta_j l^2}{(\mu_1 + \mu_2)D_j}, \quad R_j = \frac{KC_{sj}^0 l}{D_j}$$

It follows from the neutral stability equation that perturbations with wave numbers $k < k_1$ (where k_1 is a root to Eq. (3.25)) are unstable. An analysis of Eq. (3.25) has shown the system stability to be dependent in a complicated manner on the value the physicochemical parameters take. In some instances, the system remains stable at any Marangoni number; in others, this number has a threshold critical value beyond which the instability develops; this critical value is determined by the extent the ratio $D_* = D_1/D_2$ differs from unity. By and large, the enhanced surface activity of reactants lowers the system stability threshold, whereas the increased reaction rate makes it rise. For small wave number perturbations, the above relationships may fail to hold, which is suggestive of an influence of the space scale on hydrochemical interaction.

Shown in Fig. 4.27 are the neutral stability curves for the cases of one and two surface-active reactants in a system at $D_* > 1$. The curves 1 and 2 correspond to the case when the second reactant is surface-neutral ($\text{Ma}_2 = 0$). Each curve is composed of two branches separated by an asymptote $k_* = R_2(D_* - 1)$; therefore, the system is unstable at any negative activity of the first reactant ($\text{Ma}_1 < 0$). With the first reactant exhibiting a positive activity ($\text{Ma}_1 > 0$), the instability is possible only if $\text{Ma}_1 > \text{Ma}_1^{cr} > 0$. The curves 3 and 4 refer to the case when the second reactant displays surface activity, and $\text{Ma}_2 = 500$. If the second reactant is surface-inactive ($\text{Ma}_2 = -500$), the

4 Modelling of Mass Transport Processes

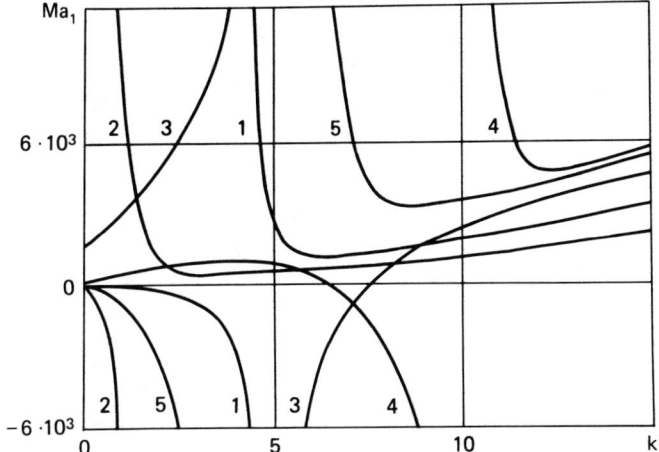

Fig. 4.27 Neutral stability curves for systems containing one or two surface-active reactants at $D_* > 1$: $Ma_2 = 0$ (curves 1, 2); $Ma_2 = 500$ (curves 3, 4); $Ma_2 = -500$ (curve 5)

behaviour of the system is described by the curve 5 bearing resemblance to curves 1 and 2. However, at $Ma_1 > 0$, the system becomes more stable, and the Ma_1^{cr} grows in magnitude.

Shown in Fig. 4.28 are parametric neutral stability curves illustrative of the Ma_1 vs. Ma_2 relationship. Neutral curves resembling the curve of type 1 in Fig. 4.28 are observed both at $D_* > 1$, $k > R_2(D_* - 1)$ and at $D_* < 1$, $k > R_1(D_*^{-1} - 1)$. Simultaneously, the system stability tends to decrease with the increasing surface activity of the second component. At $D_* > 1$, $k < R_2(D_* - 1)$ or at $D_* < 1$, $k < R_1(D_*^{-1} - 1)$, the parametric curves take a form similar to that of curve 2 in Fig. 4.28, and the system becomes more stable with increasing Ma_2.

The parametric Ma_1 vs. R_1 curves for neutral stability are shown in Fig. 4.29. They resemble the curve of type 1 for perturbations at wave numbers subject to $k^2 > 1/8\, Ma_2(1 - D_*^{-1})$, $k > R_2(D_* - 1)$ or $k^2 < 1/8\, Ma_2(1 - D_*^{-1})$, $k < R_2(D_* - 1)$. With R_1 growing, the system becomes more stable toward such perturbations. Otherwise, neutral stability curves tend to be of the type 2, and the system stability decreases with increasing R_1. The parametric Ma_1 vs. R_2 curves for neutral stability are shown in Fig. 4.30. At $D_* < 1$, they resemble the type 1, and at $D_* > 1$, the type 2. The curve 2 has an asymptote

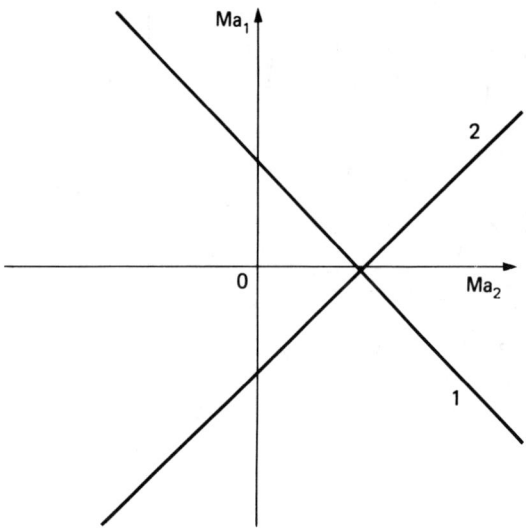

Fig. 4.28 Typical Ma_1 vs. Ma_2 curves for a neutral stability state

$R_2^* = k(D_* - 1)^{-1}$; therefore, at $R_2 < R_2^*$, the system stability increases sharply with R_2 growing. The same is true of $R_2 > R_2^*$, providing that the first reactant is inactive. In case if this reactant is surface-active, the system becomes unstable.

Dispersion curves showing the perturbation growth increment $\alpha = \mathrm{Re}\,\omega$ as a function of wave number k at $D_* = 1$ are represented in Fig. 4.31a. At $\alpha > 0$, the perturbations are unstable; by contrast, at $\alpha < 0$, they are stable. The curve 1 in Fig. 4.31a corresponds to a stable state of the system at $Ma_1 < Ma_1^{cr}$; the curve 2 corresponds to an unstable state at $Ma_1 > Ma_1^{cr}$. The only growing perturbations are those with $k < k_*$. The dispersion curves in Fig. 4.31b correspond to a case of $D_* < 1$. Curve 1 corresponds to a stable state at $0 < Ma_1 < Ma_1^{cr}$; curve 2, to a neutral stability state at $Ma_1 = Ma_1^{cr}$; curve 3, to an unstable state at $Ma_1 > Ma_1^{cr}$. The only growing perturbations are those with $k'_* < k < k''_*$, where k'_* and k''_* are roots to equation (3.25).

The above results characterize the stability of a system toward perturbations with an infinitesimal amplitude. Within a more general context of dissipation structure dynamics, it is necessary to explore the system stability with respect to finite amplitude perturbations. To this effect, we have considered weakly-

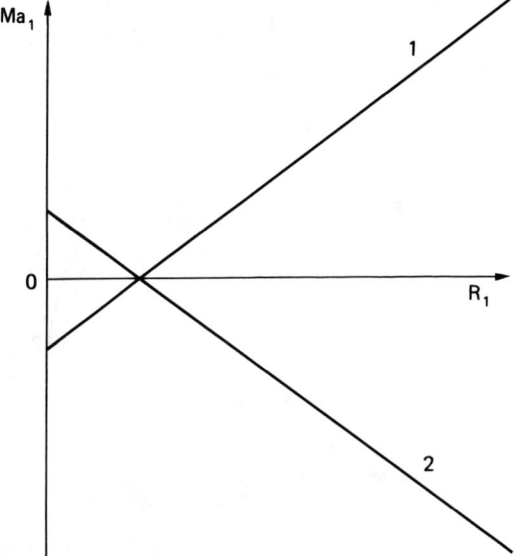

Fig. 4.29 Typical Ma_1 vs. R_1 curves for a neutral stability state

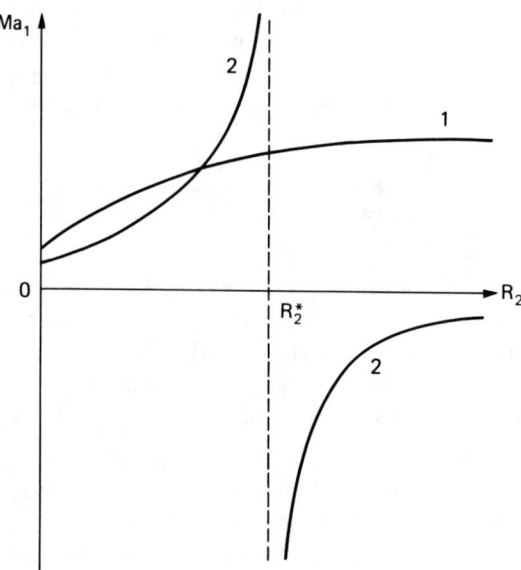

Fig. 4.30 Typical Ma_1 vs. R_2 curves for a neutral stability state

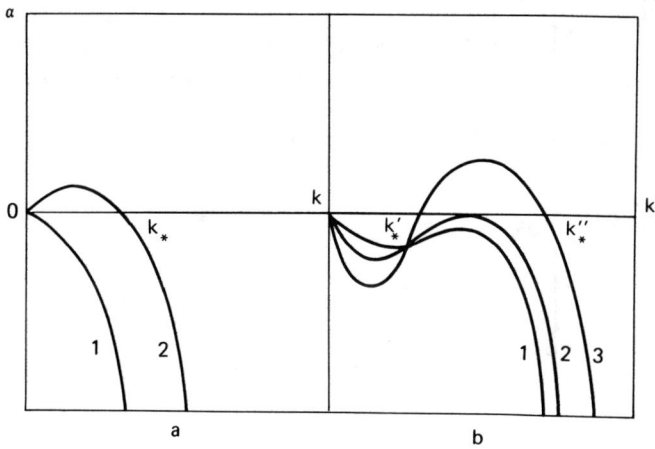

Fig. 4.31 Typical dispersion relationships: a, $D_* = 1$; b, $D_* < 1$

nonlinear almost periodic convective flows generated by interfacial instability. The functions that describe such flows may be regarded as almost harmonic ones, that is, representable by the sum of the main terms corresponding to the fundamental harmonic and the even-numbered harmonics with coefficients whose order of smallness increases with the harmonic number. Thus, the unknown functions in the system of equations (3.1)-(3.3) are represented in the form of a Fourier expansion

$$a(t, x, y) = a_0(y) + \sum_{n=1}^{\infty} [A_n(y)e^{inkx} + A_n^*(y)e^{-inkx}]e^{n\omega t} \tag{3.26}$$

where an asterisk denotes a conjugate complex number; a is any of the quantities ψ_j and c_j; n is the number of a harmonic; $\omega = \Omega + i\omega_0$. Solutions of type (3.26) describe plane (two-dimensional) disturbances.

Strictly speaking, expressions of type (3.26) can be solutions to the formulated problem only at $\omega = 0$. However, as shown in [158], the computations can be carried out for small values of $|\omega| \neq 0$, passing subsequently to the limit $|\omega| \to 0$. If the square of the fundamental mode (first mode) amplitude $\varepsilon = a_1 a_1^*$ is small, then one may accept $a_n a_n^* \sim \varepsilon^n$ and $a_0 a_0^* \sim \varepsilon^2$. Therefore, assuming ε to be small, but nonetheless finite in magnitude, we have confined ourselves to the first modes of expansion (3.26) including the zeroth mode. For simplicity, $|\omega|$ is assumed to have an order of the small parameter ε.

4 Modelling of Mass Transport Processes

The suggested weakly-nonlinear approach has yielded a dispersion equation distinct from (3.12) in that the former contains parameters which describe the disturbance amplitudes for velocity and concentration. For a nonlinear growth increment and an oscillatory frequency we have

$$\Omega = \Omega_0(k) + \Omega'(k)A_{11}^2 + \Omega''(k)A_{21}^2, \quad \omega_0 = 0 \qquad (3.27)$$

where A_{11}, A_{21} are the concentrational disturbance amplitudes in phases 1 and 2, respectively; Ω_0 is the disturbance growth increment by linear analysis; $\Omega'(k)$ and $\Omega''(k)$ are known functions of parameters Ma_j, R_j, D_*.

It follows from (3.27) that in the case of small, but finite perturbation amplitudes, the system in question develops a monotonic instability ($\omega_0 = 0$), also identified by linear analysis. However, for certain values of the parameters involved, the sign of the increment Ω can be reversed, and Ω may become arbitrarily large in absolute value, which contradicts its postulated smallness in terms of (3.27). It should be inferred, therefore, that along with the monotonic, there exists an instability of another type that emerges in the nonlinear stage of disturbance development and is poorly tractable by the method proposed.

The qualitative behaviour of a dispersion curve as the system parameters are allowed to vary is shown in Fig. 4.32. A typical dispersion curve for infinitesimal perturbations $A_{j1} \to 0$ is shown in Fig. 4.32a. The curve exhibits a single maximum at $\partial \sigma / \partial C_{s2} = 0$ corresponding to the maximal growth waves. Shown in Fig. 4.32b is a dispersion curve at $A_{j1}^2 \ll 1$ in the case of $k_1 < k_0 < k_2$, where k_0 is the wave number such that $|\Omega| \to \infty$; k_1 and k_2 are the roots to equation $\Omega'(k) = 0$. The dispersion curve exhibits two maxima lying on both sides of the vertical asymptote $k = k_0$. In the vicinity of these maxima, the disturbances are unstable. The maximum value of disturbance growth increment is markedly smaller than the respective value for maximal growth waves according to the linear theory. The dispersion curve plotted at the same value of A_{11} as that in Fig. 4.32b, but in this case subject to $k_1 < k_2 < k_0$ is shown in Fig. 4.32c. The curve has two maxima lying on the left side of the vertical asymptote $k = k_0$. The respective values of increment Ω are also smaller than the analogous value for maximal growth wave by linear theory. In the vicinity of $k = k_0$, the value of Ω tends to $\pm \infty$, and the proposed theory fails to hold, which should be attributed to the initial assumption of $|\omega|$ being a small value. The tendency of $|\omega|$ to ∞ at $k \to k_0$ appears to be an artifact; for this reason, the most probable Ω vs. k dependence is that shown in dashed line in Fig. 4.32b, c. This assumption leads to an additional local maximum in the vicinity of $k = k_0$. In this particular case, the disturbances

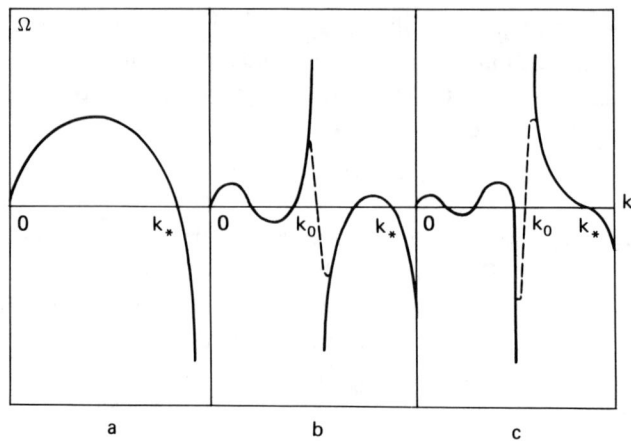

Fig. 4.32 Typical dispersion curves at different values of the square of the concentrational perturbation amplitude: a, infinitesimal perturbations ($A_{j1}^2 \to 0$); b, small finite perturbations ($A_{j1}^2 \ll 1$) at $k_1 < k_0 < k_2$; c, small finite perturbations ($A_{j1}^2 \ll 1$) at $k_1 < k_2 < k_0$

can generate, along with the monotonic instability, an instability of different type. A comparison of the curves in Fig. 4.32a, b, c shows that, under the conditions of a slightly postcritical state ($\text{Ma} - \text{Ma}^{\text{cr}} \ll \text{Ma}^{\text{cr}}$), the characteristic shape of dispersion curves becomes changed. This is manifested by the occurrence of new local incremental maxima which are smaller in magnitude than the disturbance growth increment in the linear stability theory. This signifies that the nonlinear effects as produced by a chemical heterogeneous reaction can lead to the stabilization of although small, but finite-amplitude disturbances and to the occurrence of secondary convective flows with an ordered structure.

The possible types of transition to instability can be characterized from the standpoint of the disturbance growth increment as a function of the square of the concentrational perturbation amplitude A_{11}^2 at different fixed wave numbers (Fig. 4.33). It follows from (3.27) for $k < k_*$ that the disturbance amplitude can either stay at a definite level (curve 1 in Fig. 4.33), or further increase to produce a stronger disturbance destabilization (curve 2 in Fig. 4.33).

In the short-wave region of $k > k_*$, the system is stable to infinitesimal perturbations. The stability is also retained with respect to finite magnitude per-

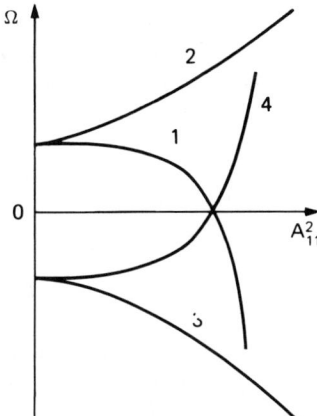

Fig. 4.33 Typical curves for nonlinear perturbation growth increment as a function of the square of its amplitude at different instability regimes (see text for explanation)

turbations, if $\Omega'(k) < 0$ and $\Omega''(k) < 0$. However, at $\Omega'(k) > 0$ and $\Omega''(k) > 0$, the system develops instability toward finite perturbations. Relevant situations are examplified by curves 3 and 4 in Fig. 4.33.

Stabilized periodic structures arise only in the former case which corresponds to the so-called "soft" type of instability development [154]. The condition of stationarity requires the increment $\Omega(k)$ to become zero. The condition of periodicity (decay of all types of disturbances except the only one specified by a definite k) requires that Ω attain a maximum as a function of k at fixed squares of the amplitudes [159]. These conditions having been taken into account, the following equations for estimating the parameters k and A_{11}^2 of a stabilized convection pattern are obtained:

$$\Omega = 0, \quad \frac{d\Omega}{dk} = 0, \quad \frac{d^2\Omega}{dk^2} < 0 \qquad (3.28)$$

The solutions for k and A_{11}^2 must be real and positive. Otherwise, no stabilized convection pattern exists, and the transition to instability is effected via "hard" mechanism. In this case, the heterogeneous nonlinear reaction fails to stabilize augmenting disturbances, which results in a simultaneous generation of waves of different periods [154] and finally leads to interfacial turbu-

Table 4.2 The nonlinear increment of disturbance growth as a function of the square of its amplitude in different regimes of instability development (see text for explanation)

k, C_j, ψ	$r \approx 1$, $p \ll 1$	$r \ll 1$, $p \gg 1$	$r \ll 1$, $p \gg 1$
kl	$0.357\sqrt{Ma}$	$0.362\sqrt{Ma}$	$0.221\sqrt{Ma}$
$\dfrac{A_{11}^2}{(C_{s1}^0)^2}$	$0.248 \dfrac{Ma}{Ti_2^2}$	0.886	$0.538 \left(\dfrac{Ti_1}{Ti_2}\right)^2$
$\dfrac{A_{21}^2}{(C_{s2}^0)^2}$	$1.08 \cdot 10^{-4} \dfrac{Ma}{Ti_1^2}$	$0.0107 \left(\dfrac{Ti_2}{Ti_1}\right)^2$	1.31
$\dfrac{B_1^2 l^2}{D_1^2}$	$-0.015 \dfrac{Ma^3}{(Ti_1 Ti_2)^2}$	-0.196	$-0.072 \dfrac{Ma^2}{Ti_2^2}$

lence. Under the conditions of a small postcriticality ($Ma_1 \ll 1$) and the diffusion coefficients of both reactants being close, the instability has been shown to develop by "hard" mechanism. As distinct, the instability sets in by "soft" mechanism, if the diffusion coefficient of the surface-active reactant is much smaller than that of the second reactant. In this case, concentration patterns with different amplitudes may arise in the contact phases.

Conditions (3.28) are essential for the formation of dissipation structures and constitute thus the basis for computing major hydrodynamic characteristics of interfacial convection. By solving the system of equations (3.28), the characteristic size k^{-1} of convection patterns and the respective concentrational disturbance amplitudes and convective flow velocities can be determined. The estimates of characteristic scales for generated convection patterns in different limiting cases are listed in Table 4.2.

A numerical solution of system (3.28) has shown that the interfacial convection manifests itself through buildup of two-dimensional rolls whose size is scaled inversely proportional to the square root of the Marangoni number. The intensity of rotational motion of the liquid diminishes with growing chemical reaction rate constant; this effect is more pronounced under the conditions of augmented postcriticality.

4.3.4 Computer Simulation of Interfacial Convection

An effective method for studying the dynamics of dissipation structures under mass and heat exchange conditions is computer simulation of interfacial convection using Navier-Stokes and convective diffusion and heat conduction equations [160-172]. Computer experiment results, although less generalizing than those obtained by a rigorous mathematical analysis, proffer a number of essential advantages. Whereas most analytical studies involving nonlinear perturbation methods are, as a rule, restricted to weakly nonlinear problems and to small postcritical parametric regions, the numerical modelling allows to reveal many essential features of interfacial self-organization in a far postcritical region.

Given below are the results of a computer interfacial convection modelling for quiescent gas-liquid and liquid-liquid systems [171, 172].

The first model system was a two-dimensional horizontal liquid layer spread over a solid surface; the free nondeforming liquid surface was in contact with the surrounding gas phase [171]. One of the gas components capable of affecting the surface tension could be absorbed into the liquid or, being dissolved in the liquid, could be desorbed from it.

Nonlinear analysis was applied to study the interfacial instability featuring a solute-capillary Marangoni convection and dissipation. The disturbance was assumed to develop from a point perturbation of the solute concentration field at a central point of the surface layer and was monitored numerically on a computer. A system of equations including the Navier-Stokes and convective diffusion equations were solved numerically using the modified program complex NEPTUN. The computational algorithm was based on a conservative difference scheme with directed differences; details concerning the computation of concentration fields are given in [173].

The computed results are represented in Fig. 4.34 in the form of level lines of the stream function and concentration field of the transported surface-active solute. The computations were carried out using parameters $Ma = 10^3$, $Sc = 10^3$, $Bi = 10^{-2}$, $l/h = 6$, where Bi denotes the Biot number characterizing the mass-transport rate in the gas phase; l and h are the length and the thickness of the liquid layer. The numerical solution has yielded a steady convective motion whose pattern features four pairs of rolls implying thereby that in this particular case, the harmonic $n = 4$ predominates, in accordance with the results of linear stability analysis for the system in question. This convective motion was shown to set in within a relatively short period of time

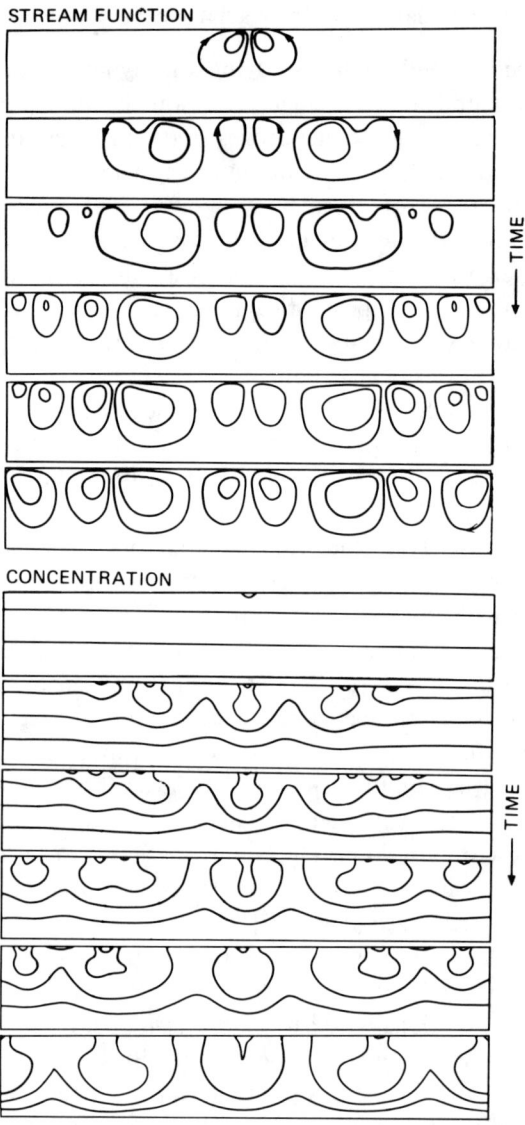

Fig. 4.34 Velocity field (top) and concentration field (bottom) in interfacial liquid-layer convection (computer simulation results)

$4.4 \cdot 10^{-2}$ as expressed in dimensionless time units tD/h^2; in further course, the pattern of the convective motion remained unaltered. The concentration field, as distinct from the convective motion, took a longer time, $tD/h^2 \approx 1$, to establish. This delay in concentration field stabilization may be attributed to the characteristic diffusion time in the liquid, $t_D = h^2/D$, being much greater over the characteristic time of the viscosity-mediated momentum transfer, $t_\nu = h^2/\nu$.

The chosen positive parameters Ma and Bi correspond to the adsorption of compounds acting to increase the surface tension. Computer experiments with these parameters, varied in value and sign, have shown that the interfacial convection can be produced by desorption of solutes capable of reducing the surface tension; however, it never occurs in absorption of compounds acting to reduce the surface tension, as well as in desorption of inactive solutes. The stream function and concentration fields for an interfacial convection with Ma < 0 and Bi < 0 exhibit patterns similar to those shown in Fig. 4.34.

The second model was a system of two stagnant horizontal liquid layers separated by a plane interface ($y = 0$) and sandwiched between two solid surfaces $y = h_1$ and $y = -h_2$ [172]. A surface-active solute was transported from the lower ($-h_2 \leqslant y < 0$) into the upper ($0 < y < h_1$) layer. By analogy with the first model, we have analyzed changes in the concentration field locally perturbed at a central point of the interface surface; it was assumed that the bilayer system persisted in an unperturbed state at zero-of-time:

$$\psi_j = \omega_j = 0, \quad a_j = \frac{m - D_*^{j-1}(y/h_1)}{m + D_* h} \qquad (3.29)$$

where the indices $j = 1, 2$ denote two contacting phases 1 and 2; ψ is the stream function, ω is the vortex; a is the dimensionless concentration; m is the local interphase equilibrium constant; $D_* = D_1/D_2$ is the ratio of the solute diffusivities of the two respective phases; $h = h_2/h_1$.

Computations have been performed at postcritical (according to the linear theory) Marangoni numbers in a computational grid nonuniformly condensing toward the interface. The velocity vortex at the interface was determined from the condition of tangential velocity component continuity. In formulating the solid surface boundary condition for ω, a difference formula of second-order accuracy was used [173]. The results of a numerical solutions are presented in Fig. 4.35, in which the equidistant stream lines are shown on the left, and the equidistant concentration level lines, on the right side in the region of $0 \leqslant x \leqslant l/2$ (l being the integration path). The patterns shown in Fig. 4.35

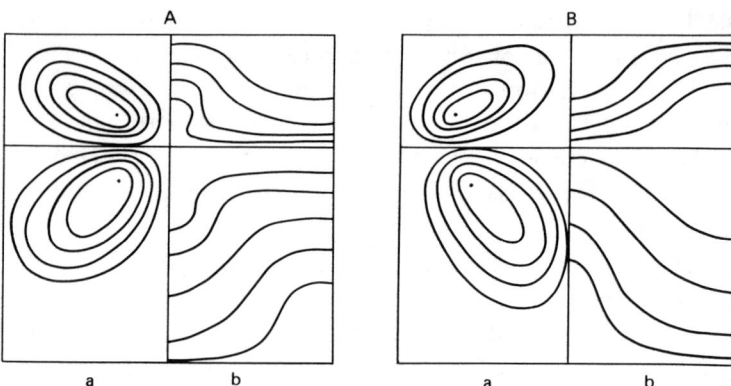

Fig. 4.35 Stream lines (a) and lines of constant concentration (b) for interphase mass transfer in a two-layer system: A, $h = 1.5$, $Ma = 10^5$; B, $h = 1.5$, $Ma = -10^5$

correspond to fields stabilized within the period of time $t \approx h_1^2/D_2$ for the cases of $D_* = 0.7$, $m = 1$, $Sc_1 = 4 \cdot 10^3$, $Sc_2 = 10^3$, at $h = 1.5$, $Ma = 10^5$ and $Ma = -10^5$.

The results obtained provide evidence that a regime of advanced convection becomes materialized in the above situations. Noteworthy, the computer experiments we have performed lend support to the results of linear stability analysis for the cases of $Ma > 0$ and $Ma < 0$. They also agree with the instability criteria (3.13) and (3.14); in particular, they confirm the eventual development of convective instability for transported solutes differing in surface activity sign, when $D_* = 0.7$ and $h = h_1/h_2 = 2/3$ ($h^2 < D_* < 1$).

The numerical experiments have shown that at $|Ma| < 10^3$, a weak convection develops, if the liquid velocity is small enough in both phases and does not affect the concentration distribution; if so, the mass transport regime that sets in is close to a purely diffusional regime.

In greater detail, the effects of convective motion on the rate of interphase mass transfer and the computational results for local and average material fluxes across the interface will be dealt with in Section 4.4.

To briefly summarize this Section, we wish to point out that the transition from ordered convection patterns to interfacial turbulence is of special interest for interface fluid dynamics and is one of the central problems in synergetics—"order-to-chaos" transition in determinate systems. The methods of nonlinear system dynamics effectively used in the study of hydrodynamic systems appear to be also promising for a theoretical treatment of this phenomenon. To exemplify pioneering endeavours in this, one may refer to the work by Lorenz [174] who studied numerically a finite-dimensional discrete system of equations which was a simplification of the system used by Saltzman [175] for the description of Bénard finite-amplitude convection. The model thus constructed was that of a rotating liquid layer heated from below which developed a number of different periodic circulation regimes, with the subsequent transition into a region of aperiodic motions.

This model of Bénard convection, despite its failure to provide a fully realistic picture for the generation of turbulence in the liquid, has nonetheless proved to be quite instructive and useful for studying the thermal convection and highlighted in many respects the trend later outlined by the works of Ruelle and Takens [176, 177]. These researchers introduced the notion of a strange attractor which was a volume-confined topological set in the phase space and which acted to attract all, or almost all external integral curves which, once inside the confined volume, became unstable. They have shown that the jump-like transition to an aperiodic (stochastic) motion through a number of periodic regimes is a common feature of nonlinear fluid dynamic equations; this new state of the system is characterized by a complete disorder with a fastly decaying autocorrelation function. The occurrence of a strange attractor serves as a criterion for generation of a stochastic regime within the system under study.

An attempt was made in [178] to study the transition from an ordered Marangoni convection to interfacial turbulence as generated by the mass transport of SAS in the liquid layer whose free surface was in contact with the gas phase. By making use of the Galerkin-type method for a finite-dimension approximation to the Navier-Stokes and convective diffusion equations, dynamic models of third and fourth order were constructed. At sufficiently large Marangoni numbers, the solutions to these models resembled a Lorenz attractor by some of their properties. One may also cite the works [179, 180] which pursued a similar trend in exploring the Marangoni convection.

4.4 Mass Transfer Models Involving Spontaneous Interfacial Convection

An effective practical application of interfacial instability effects may be expected only if one has a good understanding of the nature of convection patterns and has a means to control the conditions for their buildup;

besides, one must know effective methods enabling to quantitate the mass transfer accompanied by interfacial convection. Because of the nonlinearity of partial differential equations used for describing mass transport under the conditions of spontaneous interfacial convection (SIC), it is impossible in the general case to derive explicit expressions for the mass transfer rate. Industrial processes exhibiting interfacial instability are commonly run under conditions remote from a critical state and are complicated by a variety of interfering factors; for this reason, the rigorous quantitative estimation of interphase fluxes is an exceptionally arduous task.

This section is devoted to the kinetics and mathematical modelling of a mass transfer under SIC conditions. An analysis of the available experimental evidence on extraction, absorption, distillation and other processes has allowed to single out the following major features of a SIC mass transfer:

—the SIC mass transfer coefficients are markedly higher than the respective coefficients of a diffusional mass transfer regime (in the absence of spontaneous interfacial convection);

—the mass transfer coefficients are time-dependent in systems with induced phase motion (stirred diffusion cells, forced-flow cells, wetted-wall columns, moving liquid drops and films), whereas in a diffusion regime they are constant;

—the time dependence of mass transfer coefficients in systems with quiescent phases is intractable by the penetration theory;

—the mass transfer coefficients are dependent on the driving force and the properties of the interface, primarily, interfacial tension;

—the driving force of a SIC process has a critical value which, once surpassed, renders the mass transfer coefficient intractable by a diffusion theory;

—under SIC conditions, several transport regimes can be operative differing among themselves in the dependence of their mass-transfer coefficients on physicochemical parameters of the system.

Three major trends emerge in the studies of mass transfer kinetics under SIC conditions;

—establishment of empirical relationships and correlations for mass transfer coefficients as a function of the governing physico-chemical parameters of a system under study;

—construction of semiempirical mass transfer models enabling to describe the major experimental features with the use of a minimal set of empirical parameters;

—computation of mass transfer characteristics within the framework of an adopted model system.

4 Modelling of Mass Transport Processes

We now proceed with a brief survey of the aforementioned trends; for greater details, the reader is referred to the review papers [17, 18].

4.4.1 Empirical Correlations

The emergence of a spontaneous interfacial convection (SIC) is primarily due to the interphase mass transfer and dependence of the interfacial tension on concentration of the transported solute relating the hydrodynamic, diffusional and chemical processes. Therefore, major physical quantities that characterize the conditions for occurrence and development of spontaneous interfacial convection and mass transfer kinetics are the driving force of the process, ΔC, and the change in interfacial tension, $\Delta\sigma$, which is proportional to ΔC, that is, $\Delta\sigma = (\partial\sigma/\partial C)\Delta C$; here $\partial\sigma/\partial C$ is the interfacial activity of the transported component.

The currently known empirical correlations relate the major interphase transfer characteristics under SIC conditions, viz., the mass transfer coefficients K_M, the mass transfer enhancement factor $\varkappa = K_M/K$, or the dimensionless Sherwood number $Sh = K_M L/D$, to the driving force of interphase transfer, to the interfacial tension change $\Delta\sigma$, or to a dimensionless parameter, the Marangoni number $Ma = L(\mu D)^{-1}(\partial\sigma/\partial C)\Delta C$. This quantity is the major parameter for SIC characterization; it is derived as the ratio of the characteristic diffusion time $\tau_d \sim L^2/D$ to the characteristic time of liquid motion driven by the interfacial tension gradient $\tau_s \sim L\mu[(\partial\sigma/\partial C)\Delta C]^{-1}$; in other terms, it may be defined as the ratio of the force actuated by the interfacial tension gradient and concentration gradient to the viscous force.

Correlations as defined above have been described by linear functions in a number of works concerned with experimental studies of extraction [36-39, 181, 182], distillation [55-57] and absorption [58]. In [36], a relationship for the mass transfer coefficient under SIC conditions as a function of the interfacial pressure difference:

$$K_M = K_D[1 + \alpha(\Delta P - \Delta P_{cr})], \quad \Delta P \geqslant \Delta P_{cr},$$
$$K_M = K_D, \quad \Delta P < \Delta P_{cr} \tag{4.1}$$

where ΔP_{cr} is the critical value of interfacial pressure difference at which the SIC is initiated. The quantities α and ΔP_{cr} are determined by the physicochemical properties of a system; besides, ΔP_{cr} is also dependent on the direction of mass transport. In this particular case, the mass transfer enhancement coefficient is also a linear function of the difference $\Delta P - \Delta P_{cr}$.

In [37-39], a linear mass transfer coefficient K_M vs. driving force relationship was obtained for an extraction carried out under SIC conditions. As shown in [39], the slope of a straight line plotted on the K_M vs. ΔC coordinates is dependent on the initial strength of interfacial tension.

In [181], an empirical equation for the mass transfer kinetics under SIC conditions in a stirred diffusion cell was proposed:

$$-\frac{1}{S}\frac{dC}{dt} = K_D\Delta C + K_D K \Delta C(\Delta C - \Delta C_{cr}) \qquad (4.2)$$

where S is the specific phase-contact surface; K_D is the diffusion mass transfer coefficient; ΔC_{cr} is the critical driving force for SIC initiation; K is a coefficient which depends on the system parameters and which characterizes both the intensity of spontaneous interfacial convection and the mass transfer enhancement. From (4.2), the expressions below are derived for the mass transfer coefficient under SIC conditions and for the mass transfer enhancement coefficient:

$$K_M = K_D[1 + K(\Delta C - \Delta C_{cr})]$$
$$\varkappa = 1 + K(\Delta C - \Delta C_{cr}) \qquad (4.3)$$

The parameters of Eq. (4.3) and the partition coefficient k_e for a number of liquid extraction systems with SIC-assisted mass transfer are summarized in Table 4.3 [182].

Apparently, formulas (4.2) and (4.3) hold at $\Delta C > \Delta C_{cr}$. In [183-187], the coefficient K was studied as a function of the physicochemical parameters of liquid extraction systems, primarily, the interfacial activity $\partial\sigma/\partial C$ [183] and the phase viscosities [184-186]. A proportionality $K \sim \partial\sigma/\partial C$ was derived for a number of systems in experiments run in a stirred diffusion cell.

As shown in [185, 188] relations (4.2) and (4.3) hold for the mass transfer from moving liquid drops in a limiting disperse phase under SIC conditions. The extractive mass transfer in moving drops under SIC conditions was also studied in [40, 189, 190]; a similar process in nascent drops was dealt with in [23, 27, 41-43, 191, 192]. Semiempirical methods for calculating the end effect under SIC conditions were proposed in [41, 43, 188, 192].

A linear correlation between the Sherwood and Marangoni numbers was proposed in [182]:

$$\text{Sh} = \begin{cases} \text{Sh}_0 + \alpha(\text{Ma} - \text{Ma}^{cr}), & \text{Ma} \geq \text{Ma}^{cr} \\ \text{Sh}_0, & \text{Ma} < \text{Ma}^{cr} \end{cases} \qquad (4.4)$$

Table 4.3 Physicochemical parameters of extraction systems studied

Extraction system	No	ΔC_{cr} kmol/m^3	$\Delta\sigma/\Delta C \cdot 10^3$ kg·m^3/kmol·s^2	$\eta_1 \cdot 10^3$ kg/m·s	$\eta_2 \cdot 10^3$ kg/m·s	$D \cdot 10^9$ m^2/s	$\varrho_1 \cdot 10^{-3}$ kg/m^3	$l_1 \cdot 10^2$ m	$Ma_1^* \cdot 10^{-7}$	$l_2 \cdot 10^4$ m	$Ma_2^* \cdot 10^{-5}$
Dichloroethane + acetic acid → water	1	0.18	12.6	0.832	1.00	1.58	1.252	4	6.9	2.3	3.9
Benzene + acetic acid → water	2	0.42	4.49	0.65	1.00	2.04	0.879	4	5.7	1.4	2.0
Carbon tetrachloride + acetic acid → water	3	0.31	11.27	0.97	1.00	1.37	1.59	4	10.5	1.8	4.7
Carbon tetrachloride + propionic acid → water	4	0.7	5.8	0.97	1.00	1.25	1.59	4	14	4.1	14

where α is an empirical parameter dependent on the physicochemical properties of a liquid extraction system; $Sh_0 = K_D L/D$ is the Sherwood number corresponding to a diffusion transfer regime.

Experimental data treated in terms in this correlation for a number of liquid extraction systems are shown in Figs. 4.36. 4.37 [181, 182]. The experiments were carried out in a stirred diffusion cell containing two immiscible liquids of equal volume separated by a plane interface. The parameters for calculating the critical Marangoni number Ma^{cr} are included in Table 4.3; here ΔC_{cr} is the critical driving force; η_1 and η_2 are the dynamic viscosities of the donating and the accepting phases at the initial extractant concentration C_0; ϱ and D are the density and the solute diffusivity in the donating phase (mass-transfer limiting phase) at concentration C_0; l_1 is the layer depth of the limiting phase; l_2 is the thickness of the subsurface diffusion layer estimated as the ratio of the molecular diffusion coefficient of the donating phase to the mass transfer coefficient in the absence of spontaneous interfacial convection; Ma_1^* is the critical Marangoni number with $l = l_1$ in its formula; Ma_2^* is the critical Marangoni number with $l = l_2$. The curves for Marangoni number $Ma = (\partial\sigma/\partial C)(\Delta C l/\eta D)$ were calculated using $l = l_1$ (Fig. 4.36) and $l = l_2$ (Fig. 4.37). Physicochemical parameters of the extraction systems are listed in Table 4.3. The first kink in the curves corresponds to the transition from

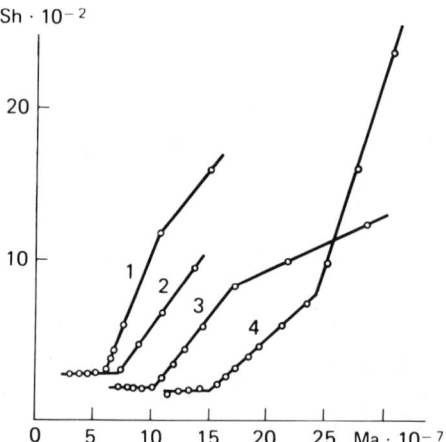

Fig. 4.36 Sherwood number Sh vs. Marangoni number Ma relationship (computed at $l = l_1$)

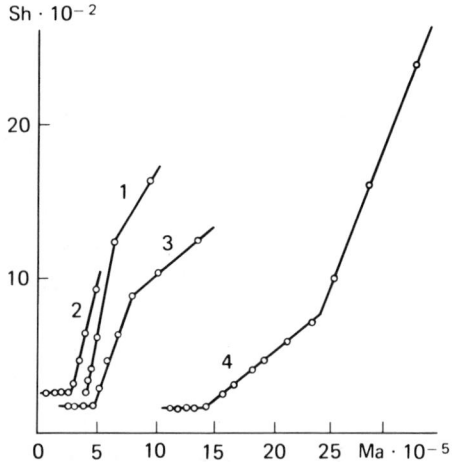

Fig. 4.37 Sherwood number Sh vs. Marangoni number Ma relationship (computed at $l = l_2$)

a diffusional mass transfer regime to a SIC regime, whereas the second kink indicates the breakdown of stable dissipation patterns produced by the first bifurcation and corresponds therefore to the transition into a new convective-diffusion regime of spontaneous interfacial convection.

Linear correlations, analogous to (4.1), (4.3), and (4.4), were obtained in a number of experiments concerned with the study of interfacial instability effect on distillation kinetics [55-58]. As is known, the distillation systems are classified, by the way the interfacial tension change affects the mass transfer rate, into "positive" systems (with the interfacial tension σ increasing during distillation, "negative" (with σ decreasing), and "neutral" (with σ remaining unchanged). The mass transfer rate in positive systems is notably higher over that in negative systems; in particular, the number of transfer units in the former systems is markedly large [55, 193]. In [55], a linear relationship between the number of transfer units and the quantity $M = -(\partial\sigma/\partial a)(a - a^*)$, the so-called "stabilization factor" characterizing a maximum interfacial tension change was established (here a is the concentration of the transported solute in the liquid phase, and a^* is the equilibrium concentration). In [56], a linear relationship for the liquid-phase mass transfer coefficient β_1 as a function of M was reported for a number of positive systems. In [57], this linear relationship was expressed via Marangoni number $Ma = (\sigma' - \sigma)\theta/D_1\mu_1$ and took the form

$$\beta_l = \beta_{l0}(1 + B\,\mathrm{Ma}) \tag{4.5}$$

where σ' is the interfacial tension corresponding to the liquid-phase composition at the interface; θ is the liquid film thickness; D_l and μ_l are the liquid-phase diffusion and dynamic viscosity coefficients; B is an empirical factor dependent on the ratio of the phase resistances.

Nonlinear relationships for mass-transfer coefficients in solvent extraction under SIC conditions were derived in [39, 194-197]. The major idea advanced in [39, 194-196] was a hypothesis that the mass transfer kinetics for SIC-assisted solvent extraction is determined by the ratio of the "specific mass-transfer energy" $F = C^{-1}(C - C_0)RT$ to the "interfacial surface energy" $H = \sigma S_0$ (here C_0 is the equilibrium concentration of the solute in the donating, mass transfer limiting phase; S_0 is the interface area per g-mole of the transported solute within the monomolecular surface layer; R is the universal gas constant, T is the absolute temperature). The correlation for the mass transfer coefficient under different SIC conditions was formulated in [196] in the final form:

$$\log\left(\frac{K_M}{K_D}\right) = \alpha + \beta\frac{H}{F} \tag{4.6}$$

where α and β are coefficients dependent on the phase viscosity.

Despite the predictive value of correlation (4.6) that gives a fairly good empirical description of the mass transfer kinetics for a large number of solvent extraction systems, the hypothesis underlying this correlation is not quite satisfactory from a physical standpoint, since it says nothing of the nature of the spontaneous interfacial convection arising from the interfacial tension gradient. The main parameter H/F includes the interfacial tension rather than the interfacial activity of the agent primarily responsible for the SIC effect [198].

A nonlinear dependence of the mass transfer coefficient on both the interfacial tension gradient and driving force under SIC conditions is typical of absorption and chemisorption processes [48, 52, 53, 199-201]. The experimental results as reported in [52, 199, 200] were analyzed in [53] to yield a correlation

$$\mathrm{Sh} = 0.033\,\mathrm{Ma}^{0.515}\,\mathrm{Sc}^{0.74} \tag{4.7}$$

which holds true when subject to the condition of $0.3\,\mathrm{Ma}^{0.515}\mathrm{Sc}^{0.24} \gg 1$. It is to be noted that correlation (4.7) presupposes a proportionality between interfacial tension gradient in the circulation cell, $\partial\sigma/\partial x$, and the interface flux J_0 of the solute responsible for spontaneous interfacial convection.

A similar exponential dependence was obtained in [201] for the mass transport of a number of surfactants desorbed from aqueous solutions. For the mass transfer enhancement factor, a correlation was reported,

$$\varkappa = (Ma/Ma^{cr})^{0.4 \pm 0.1} \tag{4.8}$$

where Ma^{cr} is the critical Marangoni number for each given system. The inclusion of the critical Marangoni number Ma^{cr} in correlation (4.8) agrees with a concept advanced in [202] in studying the desorption of various surface-active components from liquid mixtures.

It should be emphasized that in the experiments in [201], the mass transfer enhancement factor did not tend to infinity, but was rather confined to a certain limit with the increasing Marangoni number. This is due to the fact that as both the liquid circulation velocity in the convection cell and the surface renewal rate increase, the mass transfer becomes limited by the desorption of surfactants from solution.

Different mass transfer regimes under SIC conditions have been studied experimentally [39, 40, 45, 52, 182, 194-196, 203-207]. In most cases, two types of regimes may be singled out: the regime of advanced interfacial convection in which the major contribution to mass transport is due to the surface renewal by vortices produced in spontaneous interfacial convection, and the transient regime in which both the SIC and the forced convection (stirring in a diffusion cell, wave formation in a falling liquid film, etc.) play a significant role. The transition from one regime to another is determined by the driving force of transfer process, phase viscosity, and the hydrodynamics of the system.

4.4.2 Semiempirical Models

One of the first semiempirical models was that proposed by Linde [25, 205]. In this experiments with a stirred diffusion cell, the mass transfer kinetics was shown to exhibit a nonexponential behaviour. For its description, the author suggested to introduce a nonlinear term $(\partial C/\partial z)^n$ into the diffusion flux formula

$$J = -K_1 \left(\frac{\partial C}{\partial z} \right) - K_n \left(\frac{\partial C}{\partial z} \right)^n \tag{4.9}$$

where K_1, K_n and n are empirical constants, and z is the coordinate pointed into the interior from the surface.

It was shown in [205], that the kinetic mass transfer curves could be described by Eq. (4.9) at $n = 2$ or $n = 3$ depending on the conditions of a process. The empirical constants K_1 and K_n were determined for a number of liquid extraction systems.

However, such an approach seems to be somewhat formal. Linde and collaborators in [25] made an attempt to substantiate the nonlinear equation (4.9) for a diffusion flux and considered the emergence of an interfacial instability as a step-to-step feedback process of perturbative self-enhancement, with the self-enhancement coefficient proportional to the "driving force of interfacial instability", $(\partial \sigma/\partial C)(\partial C/\partial z)$. While acknowledging the originality of such an approach it must be emphasized that the assumptions as to the dependence of the enhancement and self-enhancement coefficients on the driving force of interfacial instability are rather speculative and lacking in substantiation.

The model that has been suggested by Bakker et al. [20] appears to be physically more meaningful. These authors assume that the SIC pattern arises as an aggregation of convection cells, the characteristic size of the cell l being comparable to the diffusion layer thickness $\delta_D \sim \sqrt{\pi Dt}$. Assuming further that the liquid phases within each cell attain a state of equilibrium during the period of a single circulation, the authors have obtained the following expression for the mass transfer enhancement factor

$$\varkappa = \frac{\pi}{2} \frac{l}{\delta_D} \tag{4.10}$$

The estimates of this factor fall within a range of 1.5-3.5.

The model as outlined in [20], while treating satisfactorily certain experimental results, fails, however, to provide a plausible interpretation of the experimentally observed ten-fold or even several ten-fold mass transfer amplification [182, 184, 189]. To be noted, Eq. (4.10) features the cell size as a major parameter characterizing the spontaneous interfacial convection; this size is dependent on both the interfacial activity of the transported solute and the driving force of mass transfer and is subject to variation during this process.

Ruckenstein proposed a semiempirical model within the framework of a penetration theory in which the residence time θ' of a fluid element in the interface of the circulation cell was the major model parameter. This author has considered two extreme situations: (a) the fluid element fails to become completely renewed during the period of circulation; (b) the fluid element becomes completely renewed during the period of circulation; accordingly, two formulas for mass transfer coefficient have been derived:

4 Modelling of Mass Transport Processes

$$\text{(a)}\ K = \sqrt{\frac{4D}{\pi t(\theta'/\theta)}}, \quad \text{(b)}\ K = \sqrt{\frac{4D}{\pi \theta'}}$$

where θ is the complete circulation time for the fluid within the cell, at $\theta'/\theta \approx 0.25$. As has been emphasized in [208], either situation may occur in practice. In the model of interest, the quantity θ' is related to the cell size l and the fluid circulation velocity v_s; with allowance made for the tangential stress balance, the interfacial tension drop $\Delta\sigma$ across the cell length is expressed through these two quantities as

$$|\Delta\sigma| \sim (\varrho_1 v_1^{1/2} + \varrho_2 v_2^{1/2}) l^{1/2} v_s^{3/2} \tag{4.11}$$

where ϱ_i and v_i ($i = 1, 2$) are the density and kinematic viscosity of the respective phase.

The model as suggested in [208], received further development in [209]. With a view of establishing a relationship between the Sherwood and Marangoni numbers, the mass transport of a surface- and chemically inert solute from one phase into the other was considered similar to [208, 210]. It was assumed that the process was controlled by the convective diffusion of this solute from the interface into the bulk of the accepting phase. Also, it was assumed that the Peclet number corresponding to the established convective flow was large and the diffusional penetration depth was small, which thus made it possible to use the approach of a subsurface diffusional boundary layer. Within this layer, the tangential velocity component could be regarded independent of the transverse coordinate y and equal to U at the surface. These assumptions allowed one to formulate a problem for the convective diffusion of a solute at constant concentrations C_s and C_0, respectively, at the interface and in the bulk of a phase in the form

$$U(x)\frac{\partial C}{\partial x} - y\frac{dU(x)}{dx}\frac{\partial C}{\partial y} = D\frac{\partial^2 C}{\partial y^2}$$

$C = C_s$, at $y = 0$; $C = C_0$, at $y \to \infty$; $C = C_0$, at $x = 0$.

The problem having been solved, a formula for the Sherwood number is obtained that defines a dimensionless diffusion flux for the transported solute across the interface:

$$\text{Sh} = \frac{1}{\sqrt{\pi D}} \int_0^l U(x) \left[\int_0^x U(\xi)\, d\xi \right]^{-1/2} dx \tag{4.12}$$

where

$$\text{Sh} = \frac{1}{C_0 - C_s} \int_0^l (\partial C/\partial y)_{y=0}\, dx = \frac{1}{D(C_0 - C_s)} \int_0^l J(x)\, dx,$$

l stands for the convection cell length and $J(x)$, for the flux of the solute across the interface. By substituting the expression for $U(x)$ from the tangential stress balance at the interface, we obtain the desired relationship

$$\text{Sh} = b\left[1 + \frac{\varrho_2}{\varrho_1}\left(\frac{\nu_2}{\nu_1}\right)^{1/2}\right] \text{Sc}^{1/6} \text{Ma}^{1/3} \tag{4.13}$$

where b is an empirical factor, $\text{Sc} = \nu/D$ is the Schmidt number, and $\text{Ma} = \Delta\sigma l/\varrho_1 \nu_1 D$ is the Marangoni number.

In [211], a relationship between the Sherwood and Marangoni numbers was derived on the basis of an energy balance between the surface forces and viscous forces

$$\int_S \sigma S\, dS = 2\mu \int_V E^2\, dV, \tag{4.14}$$

where S and V are the free surface area and the volume of a convection cell adjacent to the interface; E is the fluid strain rate tensor; σ and μ are the surface tension and dynamic viscosity. The left-hand side of Eq. (4.14) is determined solely by the surface and the right-hand side, by the volume distribution of strain rate. For a unique determination of the quantity E, the harmonic law for velocity distribution in the circulation cell was assumed. The pattern of the flow was represented by parallel arrays of stationary two-dimensional vortices. The strain rate tensor within a circulation cell was represented in the diagonal form as

$$E = \|\varepsilon_{nn}\|, \quad \varepsilon_{nn} = \partial u_n/\partial n, \quad n = x, y$$

where ε_{nn} refers to the strain rate in the direction of the principal axes; u_n is the velocity component; x and y are the coordinates, tangential and normal to the interface.

The surface tension distribution is related to the material flux in the deforming surface,

$$\sigma(x) = \sigma_0 - \alpha^2 (\Delta\sigma/D)\beta(x), \tag{4.15}$$

$$\beta(x) = \left(\frac{D}{\pi}\right)^{1/2}\left[\exp 2\int_0^x \varepsilon_S\, dx \bigg/ \int_0^x \exp\left(2\int_0^x \varepsilon_S\, dx\right) dx\right]^{1/2} \tag{4.16}$$

where $\Delta\sigma$ is the maximum change of the surface tension within the confines of a single convection cell; β is the surface mass transfer coefficient; U is the surface velocity; α is an empirical factor; $\sigma_0 = \sigma\,(x = x_0)$; x_0 is the cell size.

From relations (4.14)-(4.16) the desired relationship is derived

$$\text{Sh} = \alpha \text{Ma}^{1/2}$$

$$\text{Sh} = \frac{1}{D} \int_0^{x_0} \beta(x)\,dx, \quad \text{Ma} = \frac{\Delta\sigma x_0}{\mu D} \tag{4.17}$$

The raising of Marangoni number to different powers in (4.13) and (4.17) is explained as follows. It was assumed, in deriving (4.17), that the work of the surface forces associated with the surface tension gradient was spent on overcoming the viscous dissipative forces operative within the entire depth of the convection cell. In deriving (4.13) to estimate the tangential component of the viscous stress tensor $\mu(\partial u_x/\partial y)_{y=0}$, it was assumed that $y \sim \delta \sim (\nu x_0/U_0)^{1/2}$, where U_0 is the characteristic velocity at the interface surface. Therefore, if the depth of viscosity-assisted penetration for the momentum is set equal to x_0, we obtain $\text{Sh} \sim \text{Ma}^{1/2}$, whereas at $\delta \ll x_0$, we have $\text{Sh} \sim \text{Ma}^{1/3}$.

The models as developed in [208, 209, 211] give a physically clear and meaningful picture of a mass transfer under SIC conditions; still, they all share a common drawback: the formulas for mass transfer coefficient and Sherwood number include characteristic parameters of the circulation cell, viz., the cell size and the interfacial tension drop which are subject to variation during the mass transfer and which are very difficult (if possible altogether) to measure for a real system.

A model that was developed in [212] on the basis of the penetration theory as applied to a system with forced phase mixing is free of the aforementioned limitation. According to this model, the spontaneous interfacial convection leads to a change in the mean residence time of a fluid element at the surface, θ, as compared to the respective time θ_0 in the absence of spontaneous interfacial convection:

$$\frac{1}{\theta} = \frac{1}{\theta_0} + \frac{1}{\theta'} \tag{4.18}$$

where θ' is the residence time for the fluid element in the surface of a circulation cell. Using (4.18) and the mass transfer coefficient $K_M = \sqrt{4D/\pi\theta}$ with allowance for the tangential stress balance at the interface and assuming the concentration gradient along the surface to be proportional to the normal surface stress, these authors have obtained an expression for the mass transfer coefficient,

$$K_M^2 = K_D^2 + \gamma K_M \Delta C, \quad \gamma = (\beta/\mu)\partial\sigma/\partial C \tag{4.19}$$

where β is an empirical coefficient. Formula (4.19) gives a good description of the experimental results. It follows from (4.19) that at $K_M \gg K_D$, K_M is proportional to ΔC, the fact, as has already been noted, born out by a wealth of experimental evidence.

The semiempirical model for mass transfer under SIC conditions as proposed in [212] has a number of limitations. First, formula (4.19) contains no critical driving force ΔC_{cr} responsible for the SIC effect; for this reason, (4.19) holds, generally speaking, only if the deviation of the driving force from its critical value is large enough, when the system is no longer "reminiscent" of its transition through the critical state. Second, relationship (4.18) is valid only if the characteristic vortex scales for both forced convection and spontaneous interfacial convection are of the same order of magnitude. In the general case, θ must in a certain manner be dependent also on the ratio of these scales.

The approach as based on the surface renewal theory was further developed in [213] assuming the partition function for the fluid particle residence time at the interface to be a superposition of two partition functions corresponding to the bulk and interfacial turbulences. Formulas for the mass transfer enhancement factor for different types of surface renewal were obtained. The mass transfer enhancement factor has been shown to be a function of a single parameter which is the ratio of the surface renewal rates in the presence and in the absence of spontaneous interfacial convection. However, as the authors themselves caution, the practical use of the derived formulas requires a knowledge of the hydrodynamics of a system and its interfacial convection characteristics.

Somewhat different approach to the mass transfer modelling under SIC conditions was suggested in [214, 215] assuming that the regions adjoining the interface in a liquid-liquid system are the layers of uniform isotropic turbulence tractable by the Kolmogorov theory [216]. Assuming also, in accordance with [216], that (i) the vortices with a characteristic scale λ_0 corresponding to the inertial limit, that is, with $\lambda_0 = \nu/\bar{\varepsilon}_0$ (ν being the kinematic viscosity and $\bar{\varepsilon}_0$, the turbulent pulsation velocity) are major contributors to the turbulent mass transfer; (ii) the overall energy dissipation ε is the sum of a bulk component ε' (due to mixing) and a surface component ε^s (due to spontaneous interfacial convection), $\varepsilon = \varepsilon' + \varepsilon^s$; (iii) the proportionality $\varepsilon_1^s/\varepsilon_2^s = \varepsilon_1'/\varepsilon_2'$ holds for phases 1 and 2, the authors have obtained the following formula for the mass transfer coefficient under interfacial turbulence conditions [215]:

$$K_M = K_M^0[1 + (K_M^s/K_M^0)^4]^{1/4} \tag{4.20}$$

Relation (4.20) contains only one empirical parameter K_M^s. Also, the hypothesis on proportionality of the ratios of the bulk and surface dissipation energies in both phases was supported experimentally in that work; however, no attempt was made to correlate K_M^s to the major parameters, viz., the interfacial activity of the solute and the mass transport driving force. The approach as outlined in [214, 215] appears to be intriguing enough; still, it applies only to systems exhibiting a very intense spontaneous interfacial convection (chaotic motions of subsurface liquid, or interfacial turbulence) and can hardly be expected to hold for systems with ordered or oscillatory cellular convection. To be noted, both the isotropic nature of the turbulence produced by phase boundary movements and the applicability of Kolmogorov theory remain a controversial issue.

The above review suggests thus the need of a novel semiempirical SIC mass transfer model that would reflect, on the one hand, the main physical features of a process and, on the other hand, would not contain SIC characteristics for empirical parameters. Such a semiempirical model meeting these requirements has been advanced in [217, 218]. The following physical concepts focusing on the pattern and parameters of spontaneous interfacial convection constitute the basis of this model. It has been assumed that the spontaneous interfacial convection can be represented by a regular pattern of convection cells accomodated near the interface. A general conceptual formula relating the diffusion flux J_0 across unit area of convection cell surface to the characteristic velocity v_s of the fluid circulating in the cell is derived from the balance equations for the tangential interface stresses and for the transported solute. It is assumed also that the characteristic time for fluid circulation in the cell, $\tau_s \sim l/v_s$ (l being the cell size) has the same order of magnitude as the characteristic time of hydrodynamic response, $\tau_\nu \sim l^2/\nu$. The concentration change across the diffusional boundary layer depth in the cell, $\Delta_\delta C$, is presumed to be proportional to the mass transfer driving force through the phase depth ΔC,

$$\Delta_\delta C = b\Delta C, \quad \Delta C = (C_{1o} - C_{2o}/m) \tag{4.21}$$

where b is an empirical parameter; c_{1o} and c_{2o} are the solute concentrations in the bulk of phases 1 and 2; m is the interphase equilibrium constant. Relation (4.21) with the parameter $b \ll 1$ bears both upon the conditions for occurrence of circulation cells under interfacial instability and upon the nonideality of surface renewal. A reason for this nonideal renewal is that the solute-rich liquid on its transport toward the cell center from the phase bulk becomes mixed to the solute-lean liquid which is carried away from the periphery and in part brought back toward the cell surface by circulation. This having been assumed, a formula was derived for interface mass flux J_0

$$J_0 = \frac{b^2}{\eta_1 + \eta_2} \left| \frac{\partial \sigma}{\partial C} \right| \text{Sc}^{-1/2} (\Delta C)^2 \tag{4.22}$$

where η_1, η_2 are dynamic viscosities for the respective phases 1 and 2; Sc is the Schmidt number of the phase limiting mass transfer. For a single cell, the mass transfer coefficient $K_M^0 = J_0/\Delta C$ is proportional to the driving force

$$K_M^0 = K_{SIC}^0 \Delta C, \quad K_{SIC}^0 = b^2 (\eta_1 + \eta_2)^{-1} |\partial \sigma / \partial C| \text{Sc}^{-1/2} \tag{4.23}$$

Formula (4.23) can be applied to estimating the mass transfer across the phase contact surface providing that the entire surface is occupied by circulation cells. However, this takes place only under sufficiently intense spontaneous interfacial circulation, if $\Delta C \gg \Delta C_{cr}$, for example, in the buildup of a disperse phase. In the general case, it is assumed that only a portion ε of the surface is occupied by convection cells, the ε being dependent on the deviation of the driving force from its critical value ΔC_{cr} at which the spontaneous interfacial convection arises. This fact having been taken into consideration, the following formulas were obtained for the total solute flux per unit interface area J and for the mass transfer coefficient $K = J/\Delta C$ at $\Delta C - \Delta C_{cr} \ll \Delta C_{cr}$

$$J = K_D \Delta C + K_{SIC} \Delta C (\Delta C - \Delta C_{cr}) \tag{4.24}$$

$$K_M = K_D + K_{SIC} (\Delta C - \Delta C_{cr}), \quad K_{SIC} = B(\eta_1 + \eta_2)^{-1} |\partial \sigma / \partial C| \text{Sc}^{-1/2}$$

Here K_D is the diffusion-regime mass transfer coefficient, $B = b^2 f$ is an empirical parameter that takes account of the external hydrodynamic conditions. We have $J = K_D \Delta C$, $K_M = K_D$ for $\Delta C < \Delta C_{cr}$.

We introduce the Sherwood and Marangoni numbers into (4.24) to obtain, in a dimensionless form,

$$\text{Sh} = \begin{cases} \text{Sh}_0 + B \, \text{Sc}^{-1/2} (\text{Ma} - \text{Ma}^{cr}), & \text{if Ma} \geq \text{Ma}^{cr} \\ \text{Sh}_0, & \text{if Ma} < \text{Ma}^{cr} \end{cases} \tag{4.25}$$

$$\text{Ma} = \frac{L}{(\eta_1 + \eta_2) D} \left| \frac{\partial \sigma}{\partial C} \right| \Delta C$$

4 Modelling of Mass Transport Processes

For the mass transfer enhancement factor $\varkappa = \text{Sh}/\text{Sh}_0$, we have:

$$\varkappa = 1 + B\,\text{Sc}^{-1/2}(\text{Ma} - \text{Ma}^{\text{cr}}), \quad \text{if Ma} > \text{Ma}^{\text{cr}} \tag{4.26}$$

Formulas (4.24)-(4.26) confirm the empirical relationships as established in [57, 58, 181, 182], whereas the linear K_{SIC} vs. $\partial\sigma/\partial C$ relationship agrees with the experimental results in [183] lending thereby support to the adequacy of the model suggested. The more so, relations (4.24)-(4.26), as referred to the evidence in [181, 182], hold also when $\Delta C - \Delta C_{\text{cr}} \geqslant C_{\text{cr}}$ and $\text{Ma} - \text{Ma}^{\text{cr}} \geqslant \text{Ma}^{\text{cr}}$.

Formulas (4.24)-(4.26) contain two empirical parameters $B = b^2 f$ and Ma^{cr}. These parameters must be determined experimentally for each system with a specified geometry for the interface and the external hydrodynamic conditions defined. Besides, these parameters must be correlated to physicochemical and hydrodynamic characteristics. An essential point is that the factor B is independent of the interfacial activity of the solute; in the general case, its functional correlation must take the form:

$$B = B(\text{Re}, \text{Pe}, m) \tag{4.27}$$

Numerous studies of hydrodynamic stability of the interface exposed to SAS mass transport (see Section 3.2) have shown that the general correlation for Ma^{cr} must be

$$\text{Ma}^{\text{cr}} = F(\eta_1/\eta_2, D_1/D_2, m, \text{Re}, \text{Pe}, \text{Sc}) \tag{4.28}$$

The proposed model has allowed to draw certain important conclusions concerning both the mass transfer rate and the characteristics of convection patterns. For example, one can derive a time dependence for the characteristic cell size l. Indeed, suppose the entire interface surface has a cellular structure. Then, for the driving force changing in time we have $d(\Delta C)/dt \sim -(\Delta C)^2$ from (4.22); the integration yields $(\Delta C)^{-1} - (\Delta C)_0^{-1} \sim t$ and, since $l \sim (\Delta C)^{-1}$, we obtain $l = l_0 + at$, a being a positive constant. Thus, the cell size grows linearly with time. Precisely, such a behaviour for circulation cells under interfacial convection has been reported by a number of authors [19, 20, 25].

The parameters of surface convection cells are intimately related not only to the mass transport driving force, but also to the interfacial activity of the solute. Within the adopted model, the following relations for the SIC parameters v_s and l have been derived:

$$v_s = b(\eta_1 + \eta_2)^{-1}|\partial\sigma/\partial C|\Delta C, \quad l \sim \nu/v_s \tag{4.29}$$

It ensues therefrom that the fluid circulation velocity is directly proportional, and the cell size is inversely proportional to the interfacial activity $|\partial\sigma/\partial C|$. For the interfacial tension gradient $\partial\sigma/\partial x$ along the cell we have

$$\frac{\partial\sigma}{\partial x} \sim \frac{\partial\sigma}{\partial C} J_0(\nu D)^{-1/2} \qquad (4.30)$$

Relation (4.30) shows this gradient to be proportional to the interfacial activity of the solute, $\partial\sigma/\partial C$, and to its flux J_0 across the interface. This result agrees with the hypothesis in [52, 53] on a proportionality of the solute flux to the interfacial tension fluctuations. The SIC parameters, responsive to the solute interfacial activity, are also dependent on the viscosity of the contact phases. The convection cell size grows with increasing viscosity, whereas the fluid circulation velocity tends to decrease. Therefore, the more intense is the mass transfer, the smaller is the convection cell size and the faster is the fluid circulation in the cell.

The proposed model has been successfully applied to the mass transfer computation for a number of solvent extraction systems [192, 217-220]. Listed in Table 4.4 for a number of the solvent extraction systems studied are physicochemical parameters and the computed coefficient K_{SIC} as a measure of enhanced mass transfer in a stirred extractive diffusion cell containing two immiscible liquids of equal volume separated by a plane interface. The empirical coefficient B has been determined by making use of formula (4.24) and of the experimental mass transfer enhancement factor K_{SIC}. For all of the systems studied, the deviations for the factor B from its average did not exceed 3%.

In [192], the proposed model has been used to calculate the mass transfer in the formative period of a liquid drop (the so-called "end effect"). For the degree of extraction within the period of drop formation t_*, the following expression was obtained:

$$A = 1 - \alpha^{-2}(\alpha\coth\alpha - 1)$$
$$\alpha = 3\sqrt{3}b[(\eta_1 + \eta_2)^{-1}|\partial\sigma/\partial C|(\Delta C)_0 m^{-1}(a_*/t_*)]^{1/2}Sc^{-1/4} \qquad (4.31)$$

where a_* is the radius of the drop formed. For the empirical parameter b, the following correlation with the hydrodynamic conditions for formation of the drop was established:

$$b^2 = 3.2 \cdot 10^{-6} Pe^{0.5} \qquad (4.32)$$

where $Pe = ud/D$ is the Peclet number, u is the liquid outflow velocity from a capillary of diameter d; D is the solute diffusivity inside the drop. Relation (4.31), with reference to (4.32), treats nicely the end-effect experimental data

4 Modelling of Mass Transport Processes

Table 4.4 Physicochemical parameters of extraction systems used for calculating K_{SIC} and B

Extraction system	$\eta_1 \cdot 10^3$ kg·m^{-1}×s^{-1}	$\eta_2 \cdot 10^3$ kg·m^{-1}×s^{-1}	$\varrho \cdot 10^{-3}$ kg·m^{-3}	$d\sigma/dC \cdot 10^3$ kg·mol^{-1}×s^{-2}·m^3	$D \cdot 10^9$ m^2·s^{-1}	$K_{SIC}^2 \cdot 10^5$ m^2·kmol^{-1}×	$B \cdot 10^3$ s^{-1}
Carbon tetrachloride + acetic acid → water	0.97	1.00	1.59	16.3	1.37	13.29	0.34
Benzene + acetic acid → water	0.65	1.00	0.879	3.7	2.04	4.66	0.38
Dichloroethane + acetic acid → water	0.83	1.00	1.252	8.5	1.58	7.39	0.32
Carbon tetrachloride + propionic acid → water	0.97	1.00	1.59	12.0	1.25	7.69	0.28
Chloroform + propionic acid → water	0.57	1.00	1.49	5.2	2.12	7.15	0.29
Carbon tetrachloride + acetone → water	0.97	1.00	1.59	9.8	1.33	7.00	0.30
Water + propionic acid → water	1.00	0.97	1.00	20.0	1.13	10.47	0.31
Benzene + propionic acid → water	0.65	1.00	0.88	5.1	1.80	4.83	0.31
Kerosene + propionic acid → water	0.97	1.00	0.70	6.2	1.25	3.53	0.32
Toluene + propionic acid → water	0.58	1.00	0.86	8.3	2.09	8.1	0.28
Heptane + propionic acid → water	0.41	1.00	0.68	12.5	2.99	15.6	0.25
Water + butyric acid → benzene	1.00	0.65	1.00	11.0	1.13	9.1	0.31

for a large number of solvent extraction systems [192]. It also provides a plausible explanation of the significant discrepancy observed between the experimental end effect (under SIC solvent extraction conditions) and the end effect as predicted by the Higbie and Ilkovič theories [221].

In [219, 220], this semiempirical model has been applied to describe the mass transfer kinetics of solvent extraction in a moving drop. The driving mass transfer force of the moving drop is smaller as compared to that of the drop *in statu nascendi*, and for this reason it would be erroneous to assume that in spontaneous interfacial convection, the entire phase contact surface is occupied by convection cells. Therefore, to estimate the degree of extraction, the general formula (4.24) was used, with due account of the fraction of the surface area allotted to the convection cells. It has been shown that under SIC conditions, the degree of extraction grows with time much faster as compared to the diffusional mass transfer regime.

Golovin has extended the semiempirical SIC model to a multicomponent mass transfer and to the mass transfer in a chemically reacting system [222]. For the flux of a surface-inert solute across the interface in the presence of convective flows arising from the transport of a surface-active agent, the following formula was derived:

$$J^{in} = K_D^{in} \Delta C^{in} + K_{SIC}^* \Delta C^{in}(\Delta C^{in} - \Delta C_{cr}^{in}) \qquad (4.33)$$
$$K_{SIC}^* = B(\eta_1 + \eta_2)^{-1} |\partial \sigma / \partial C|(\Delta C_0 / \Delta C_0^{in})$$

Here, the superscript "in" refers to the inert component. A comparison of formulas (4.33) and (4.24) suggests that the transport of the surface-inert solute under the SIC conditions due to the surface-active agent is effected in such a manner as if the surface-inert component itself was responsible for the spontaneous interfacial convection. However, the coefficient K_{SIC}^* is dependent on the ratio of the initial driving forces for the surface-active and surface-inert components $\Delta C_0 / \Delta C_0^{in}$. This signifies that the enhancement of the surface-inert component mass transfer assisted by spontaneous interfacial convection due to the transport of the surface-active component is dependent on the relative concentration of these two components.

For the case of SIC-assisted mass transfer accompanied by a chemical reaction at the interface, a relation for the Sherwood number was obtained:

$$Sh = \frac{bR}{Ma} \left[\frac{R\,Sc^{1/2}}{2} - \left(\frac{1}{4} R^2 Sc + R\,Ma\,Sc^{1/2} \right)^{1/2} + 1 \right] \qquad (4.34)$$

Here R is the dimensionless reaction rate constant; b is an empirical parameter; the Schmidt number Sc refers to the phase that limits the mass transfer. It follows from (4.34) that the mass transfer rate in the presence of a chemical reaction cannot grow too large with increasing spontaneous interfacial convection intensity characterized by Marangoni number. A plausible explanation is that as the convective streams make the surface renewal rate increase, the mass transport becomes inhibited by the chemical reaction; this fact has found support in numerous experiments and theoretical model computations.

4.4.3 Analytical Relations

Exact formulas for the mass fluxes under interfacial convection conditions can be obtained either by making use of a simplified model, or by imposing strict constraints on the problem's conditions.

A common approach used in the theoretical study of mass transport is to specify a hydrodynamic structure for the interfacial convection (or surface convection) and for the respective distribution of convective flow velocity. In [223-225], a kinetics has been considered for stationary mass transfer across the plane interface between two stagnant liquid layers exhibiting a hydrodynamic circulation pattern in the form of vortex regions on both sides of the interface

$$(\mathbf{v}_j \cdot \nabla C_j) = D_j \nabla^2 c_j \quad (4.35)$$

$$\psi_j = \frac{1}{\pi^2} A_0 b a_j \sin\left(\frac{\pi x}{a_j}\right) \sin\left(\frac{\pi y}{b}\right) \quad (4.36)$$

Here the subscripts $j = 1, 2$ refer to phases 1 and 2, $\mathbf{v}$ is the velocity vector, ψ is the stream function; C is the concentration of the transported solute; A_0 is a constant characterizing the intensity of a circulating liquid; a_j, b are constants characterizing the size of a vortex region. Equation (4.35) has been solved in the approximation of small and large Peclet numbers, also using a numerical method that was a version of the projection-iteration method. The following formula for mass transfer coefficient using a large Peclet number approximation for the diffusion boundary layer has been obtained [224]:

$$K = \sqrt{2A_0/\pi^3} m\sqrt{D_1/D_2}/(m\sqrt{D_1} + \sqrt{D_2}) \quad (4.37)$$

Here m is the Henry coefficient. Formula (4.37) agrees with formula (4.13) derived on the basis of a hydrodynamic model assuming the occurrence of Marangoni convective vortices. In fact, we can, following the line of reasoning in [208], calculate A_0 as a function of the interfacial tension drop $\Delta\sigma$ through a length b and substitute it into (4.37) to obtain

$$K = \text{Ma}^{1/3} F(\varrho_j, \nu_j, D_j) \quad (4.38)$$

where F is a known function of the physical and transport bulk parameters of the contact phases. Relation (4.38) is exactly coincident with (4.13) for the case of limiting resistance of one of the phases.

For small Peclet numbers, the following formula for the SAS mass transfer coefficient was derived in [223]:

$$K = \beta \text{Ma} \quad (4.39)$$

where the coefficient β is determined by the parameters that specify diffusional resistance of the mass transfer limiting phase.

The above approach enabling one to define the flow velocity profile has been used in a number of studies concerned with the mass transfer rate in falling liquid films with circulatory convection cells [228-231]. In [228, 229], a nonstationary velocity profile was given by formulas

$$v_x = U\{1 + A[\sin k(x - wt)]\cos(k_1 y)\}$$
$$v_y = -AU(k/k_1)[\cos k(x - wt)]\sin(k_1 y)$$
(4.40)

where v_x and v_y are the tangential and normal film velocity components. In solving the nonstationary equation for convective diffusion, the approach of a boundary diffusion layer was used assuming the longitudinal component of the velocity vector to be independent of, and the transverse component, by contrast, linearly dependent on the transverse coordinate y [232, 233]. The analysis in [232, 233] has allowed to obtain a solution to the problem only for the phase wave velocity w differing significantly from the velocity of the main liquid flow. Further development of the mass transport theory in wave-like films using the self-modelled solutions for convective diffusion equations have been suggested in [230, 234]. These solutions provide a generalization of the results in [232] for travelling waves at any point of the film surface for the phase velocity markedly different from the moving film velocity. A reverse situation, when the phase velocity of circulation cells is equal to the average velocity of the moving film has been considered in [228, 229]. Here, the problem was solved by means of an integral transformation for small-amplitude circulations. The inverse integral transformation has allowed to derive an expression for the interfacial mass flux in the form of a series expanded to the powers of the circulatory flow amplitude. The mass flux increases as both the amplitude and the linear size of circulation cells in the transverse direction become large, and decreases with decrease in the longitudinal cell size.

In [231], an exact analytical solution to the problem of nonstationary mass transfer for rotational convective flow (4.40) at an arbitrary ratio of the linear velocity and the phase velocity of convection cells has been obtained in the approximation of free-surface boundary layer. The thickness of the boundary diffusion layer was assumed to be essentially smaller than the characteristic cell size along the y-axis. If so, the velocity vector components in the convective diffusion equation can be approximated by their respective values at the interface surface ($y = 0$), and the problem takes the form

$$\frac{\partial C}{\partial t} + U\{1 + A\sin[k(x - wt)]\}\frac{\partial C}{\partial x} - UAky\cos[k(x - wt)]$$
$$\times \frac{\partial C}{\partial y} = D\frac{\partial^2 C}{\partial y^2}$$
(4.41)

$$C = C_0 \ (t = 0), \ C = C_0 \ (x = 0), \ C = C_s \ (y = 0), \ C = C_0 \ (y \to \infty)$$

Changing to the dimensionless quantities $\theta = (C - C_0)/(C_s - C_0)$, $\xi = kx$, $\zeta = ky$, $z = k(x - wt)$ and introducing the self-modelled variable $\gamma = \zeta/\delta(z, \xi)$, we obtain the solution to problem (4.41) in the form

$$\theta = 1 - \frac{2}{\sqrt{\pi}} \int_0^\gamma \exp(-\tau^2) \, d\tau \tag{4.42}$$

$$\delta = 2 \left\{ \frac{Dk}{w} \int_z^\alpha \left[\left(1 - \frac{U}{w}\right) - \frac{AU}{w} \sin \tau \right] d\tau \right\}^{1/2} \left[\left(1 - \frac{U}{w}\right) - \frac{AU}{w} \sin z \right]^{-1}$$

$$\xi = \int_z^\alpha \frac{(U/w)(1 + A \sin \tau)}{(1 - U/w) - (AU/w) \sin \tau} \, d\tau$$

This solution allows one to calculate both the time-averaged local mass flux across the interface and the enhancement factor for mass transport accompanied by surface convection,

$$\varkappa = \frac{\sqrt{\xi}}{2\pi} \int_0^{2\pi} \left\{ \frac{(w/U)^{1/2}[(1 - U/w) - (AU/w) \sin z]}{\left[\int_z^\alpha [(1 - U/w) - (AU/w) \sin \tau] \, d\tau \right]^{1/2}} - 1 \right\} dz \tag{4.43}$$

The condition for the so-called concentrational independence of convection cells [230] is defined in terms of parameters which make the integrand denominator in (4.43) become zero. In this case, at any $\xi \neq 0$ and $z \neq z_* = \arcsin[(w - U)/AU]$, a single value of parameter α can be chosen to satisfy the equation for ξ in (4.42). The limiting process $z \to z_*$ shows that the mass flux in this particular case is equal to the flux of a laminar flow at a constant rate w. Formula for $\varkappa$ can be obtained in an explicit form when the amplitude of circulation streams is small ($A \ll 1$) in the absence of the "concentrational dependence" for convection cells and the reverse convective flow ($|1 - U/w| > AU/w$),

$$\varkappa = \frac{A^2 U}{4w(1 - U/w)^2} \left\{ 1 - \left(1 - \frac{U}{w}\right) \frac{1}{\xi} \sin \left[\frac{w}{U} \left(1 - \frac{U}{w}\right) \xi \right] + \right.$$

$$\left. + 3 \frac{U}{w} \frac{1}{\xi^2} \sin^2 \left[\frac{w}{2U} \left(1 - \frac{U}{w}\right) \xi \right] \right\} \tag{4.44}$$

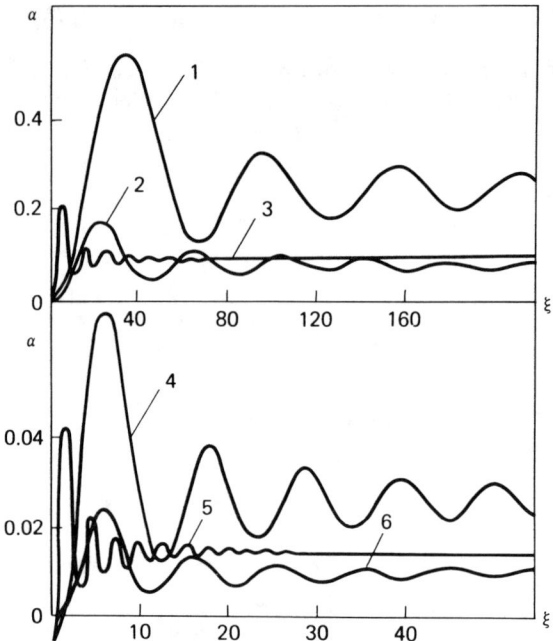

Fig. 4.38 Interfacial mass transfer enhancement factor as a function of dimensionless longitudinal coordinate ξ: 1, $A = 0.1$, $U/w = 0.9$; 2, $A = 0.1$, $U/w = 1.2$; 3, $A = 0.3$, $U/w = 0.6$; 4, $A = 0.3$, $U/w = 2.5$; 5, $A = 0.3$, $U/w = 0.3$; 6, $A = 0.1$, $U/w = 0.6$

It follows therefrom that the time-averaged local mass flux across the interface exhibits an oscillatory behaviour, and in the limit of $\xi \to \infty$, the mass transport rate is always greater in the presence of convection. The mass transport can be enhanced by virtue of a surface convection with a large amplitude and with a phase velocity tending to the flow velocity. The curves in Fig. 4.38 depicting numerical solutions of Eqs. (4.42) and (4.43) show that the highest mass transport rate is achieved in a regime of "concentrational independence" for convection cells.

The outlined above approach to mass transport modelling under SIC or surface convection conditions has a substantial constraint arising from the arbitrary choice of convective motion characteristics, viz., the velocity field and the convection cell dimensions. These characteristics enter as independent parameters into the formulas for mass transport coefficients and must be determined experimentally, which entails considerable difficulties. Besides, a strong feedback is operative between the mass transport and the spontaneous interfacial convection pattern, resulting in a continual change of the convection patterns during the mass transfer. For this reason, the results as reported in the cited above papers hold in situations where the convective pattern characteristics may be assumed invariable.

In a number of papers [235-242], situations have been considered where the Marangoni surface convection in falling liquid films does not lead to interfacial hydrodynamic instability, but is, nonetheless, capable to alter in a marked manner the longitudinal velocity component of a laminar flow and to exert thereby a non-negligible effect on the mass transport characteristics. A variety of approaches are used to model the mass transport for such situations: (i) specifying the velocity profile [235]; (ii) constructing asymptotic solutions [236-238]; (iii) finding the exact solutions to simplest nonlinear problems [239-242].

In [235], the mass transfer of a surface-active agent in a falling liquid film with the velocity profile

$$u = U\left[\left(2 - \frac{h}{\mu U}\frac{d\sigma}{d\Gamma}\frac{d\Gamma}{dx}\right)\frac{y}{h} - \left(1 - \frac{h}{\mu U}\frac{d\sigma}{d\Gamma}\frac{d\Gamma}{dx}\right)\left(\frac{y}{h}\right)^2\right]$$

$$U = \frac{3}{2}\frac{Q}{h} + \frac{h}{4\mu}\frac{d\sigma}{d\Gamma}\frac{d\Gamma}{dx}, \quad Q = \int_0^h u\,dy = \text{const} \tag{4.45}$$

has been considered. Here u is the tangential velocity component (along the x-axis), μ is the dynamic viscosity, h is the film thickness, Γ is the SAS surface concentration, $d\sigma/d\Gamma = \text{const}$ is the surface activity, Q is the flow rate. The relation between the SAS concentration gradient and the film thickness $h(x)$ is defined as $d\Gamma/dx = \alpha\mu Q/(d\sigma/d\Gamma)h^2$, where α is a numerical multiplier, which is determined from the problem solution. The stationary equation for convective diffusion in a thin-layer approximation is solved by a Karman-type method, which enables this equation to be reduced to an ordinary differential equation with respect to $h(x)$ possessing an exact solution. For the mass transfer coefficient of a surface active solute in a liquid film, according to (4.16), the following formula has been obtained:

$$K = \left(\frac{3}{4} + \frac{\alpha}{4}\right)^{1/2}\left(\frac{DQ}{\pi}\right)^{1/2}\left(h^2\int_0^x \frac{dx}{h}\right)^{-1/2} \tag{4.46}$$

The numercial multiplier α can be varied within a range of $-6 < \alpha < 2$, which is determined by the conditions for limiting thickness of the film at $x \to \infty$ and for non-retardation of its surface ($U > 0$); this complies with the requirements for a real process, for example, as that operative in a falling-film absorber. It follows from (4.46) that the agents capable of reducing the surface tension ($d\sigma/d\Gamma < 0$) and exhibiting a negative concentration gradient along the surface ($d\Gamma/dx < 0$) make the flux of a neutral solute increase. Agents that act to increase the surface tension ($d\sigma/d\Gamma > 0$) are also capable of enhancing the mass transfer provided that their surface concentration gradient is positive, which implies that they accumulate at the surface.

Three characteristic regions for the flow of a SAS-containing laminar film have been specified in [236]: the entry region ($0 < x < l_1$); the main flow region ($l_1 \leqslant x \leqslant l_2$); deceleration, or exit, region ($l_2 < x < l_3$). With a surfactant dissolved in the liquid, a significant extension of the entry region is observed in comparison with the pure liquid film. It has been assumed that in the asymptotic profile region ($l_1 < x < l_2$), the surfactant produces no effect on the flow regime, whereas near the exit cross-section, it makes the film layer become thicker and the deceleration region more extended. The solutions to stationary two-dimensional Navier-Stokes and convective diffusion equations in both the entry and exit regions are found by perturbation method using $\varepsilon = h_0/L$ as a small parameter, where $h_0 = (3\nu^2 \text{Re}/g)^{1/3}$ is the film thickness throughout the flow region (l_1, l_2), L is the chosen film length scale; ν is the kinematic viscosity, Re is the Reynolds number, and g is the gravitational constant. For the Sherwood number corresponding to the diffusion gas flow absorbed by the film with a SAS dissolved therein, the following formulas have been obtained in the region of $0 < x < l_1$ [236]:

$$\text{Sh} = \text{Sh}_0 - \frac{1}{3}\sqrt{\frac{6\text{Pe}}{\pi\varepsilon}}\left[\int_0^1 F(X)\,dX - \lim_{x \to 0}\left\{\left(\int_0^1 F(X)\,dX\right)\bigg/\sqrt{X}\right\}\right] \quad (4.47)$$

in the region of $l_2 < x < l_3$ [237]:

$$\text{Sh} = \text{Sh}_0 - \frac{4}{\sqrt{\pi}}(\sqrt[3]{4} - 1)^{3/2}\theta_0^{1/2}\left[A_0\int_0^{A_0^{-1/2}} e^{-\tau^2}\,d\tau - A_0 - \frac{1}{3}\right] \quad (4.48)$$

where $A_0 = \varepsilon A$, $\theta_0 = h_0/(l_3 - l_2)$.

Here Sh_0 is the Sherwood number for a SAS-free film; the function $F(X)$ and the quantity A are obtained by solving the respective hydrodynamic problems.

4 Modelling of Mass Transport Processes

The application of asymptotic methods to the problem of mass transport in the film containing an insoluble SAS in its surface has been given in [238]; the quantity $\varepsilon = \Delta\sigma/\mu\bar{u}$ is used for a small parameter (here $\Delta\sigma$ is the characteristic surface tension drop through the film length l; $\bar{u}$ is the cross-section-averaged fluid velocity. For the mass transport coefficient enhanced by Marangoni convection, the following expression was obtained:

$$\varkappa = 1 + \frac{3gh_0^4 D}{16\nu l D_s^2} \text{Ma} \tag{4.49}$$

where g is the gravitational acceleration; h_0 is the initial film thickness; l is the characteristic film length scale; D_s is the surface diffusion coefficient for the insoluble SAS; D is the diffusion coefficient for the transported solute; Ma is the Marangoni number.

Macroscopic measurements of the flow velocity profile and of the mass transport parameters subject to Marangoni effect and chemical reaction have been considered in [241, 242], exemplified by a falling laminar plane film containing insoluble surfactants in its free surface. Such plane films exhibiting sizeable surface tension gradients were observed in experiments on chemisorption of carbon dioxide by aqueous monoethanolamine solutions [51, 52]. The stationary laminar film flowing down an oblique wall in the presence of insoluble surfactants involved in a chemical reaction has been treated in the quasi-one-dimensional approximation of a thin liquid layer,

$$\nu \frac{\partial^2 u}{\partial y^2} + g \sin\theta = 0 \tag{4.50}$$

Here θ is the tilt angle of the wetted wall relative to the horizontal; ν is the kinematic viscosity; other notations are those specified in (4.45). At the film surface ($y = h(x)$), the balance equations for tangential stress and mass take the form

$$\mu \frac{\partial u}{\partial y} = \delta \frac{d\Gamma}{dx}, \quad \delta = \frac{d\sigma}{d\Gamma} = \text{const}$$

$$\frac{d(U\Gamma)}{dx} = D_s \frac{d^2\Gamma}{dx^2} + R(\Gamma) \tag{4.51}$$

where $R(\Gamma)$ defines the rate of a chemical surface reaction as a function of concentration and describes thus a large class of surface reactions according to the scheme adsorption + chemical conversion as well as a variety of self-catalyzed reactions.

Problem (4.50), (4.51), supplemented with appropriate boundary conditions, reduces to a set of two nonlinear ordinary differential equations:

$$\frac{g\sin\theta h^3}{3\nu} + \frac{\delta h^2}{2\mu}\frac{d\Gamma}{dx} = Q$$

$$\frac{d}{dx}\left[\Gamma\left(\frac{g\sin\theta h^2}{2\nu} + \frac{\delta h}{\mu}\frac{d\Gamma}{dx}\right)\right] = D_s\frac{d^2\Gamma}{dx^2} + R(\Gamma) \qquad (4.52)$$

In the absence of a chemical reaction ($R(\Gamma) \equiv 0$), system (4.52) has an exact solution

$$\frac{L_1}{L}\xi = d_s\left[\frac{5}{2}\zeta - \frac{13}{18}\frac{\zeta}{1-\zeta^3} + \frac{17}{4^{1/3}18}\ln\frac{(1-4^{1/3}\zeta)^2}{1+4^{1/3}\zeta+4^{2/3}\zeta^2} - \right.$$

$$-\frac{55}{108}\ln\frac{(1-\zeta)^2}{1+\zeta+\zeta^2} - \frac{17}{3^{3/2}4^{1/3}}\arctan\frac{4^{1/3}2\zeta+1}{3^{1/2}} + \qquad (4.53)$$

$$\left. + \frac{55}{3^{1/2}18}\arctan\frac{2\zeta+1}{3^{1/2}}\right] + \frac{L'}{L}\left[\ln\frac{1-\zeta^3}{1-4\zeta^3} - \frac{1}{2(1-\zeta^3)}\right]\text{Sgn }\delta + \text{const}$$

where

$$\xi = \frac{x}{L'}, \quad \zeta = \frac{h}{L}, \quad L = \left(\frac{12\nu Q}{g}\right)^{1/3}, \quad L_1 = \left|\frac{2\delta q}{3\varrho g Q}\right|, \quad d_s = \frac{D_s}{Q}$$

q denotes a constant flux of the surfactant at the surface and is the first integral of the second equation in (4.52); the integration constant in (4.53) is expressed by $\zeta_0 = h_0/L$ at $h(x=0) = h_0$; the quantities L and L_1 may be regarded as scale factors for h and for a characteristic distance over which the value of h is liable to a significant change. The ζ vs. ξ curves subject to $\zeta_0 < 1$ and $d_s = 0$ are shown in Fig. 4.39. For inactive solutes, a uniform state is observed to set in at $x \to \infty$ ($\xi \to \infty$),

$$h = h_* = \left(\frac{3\nu Q}{g}\right)^{1/3}, \quad \Gamma = \Gamma_* = 2q\left(\frac{\nu}{9gQ^2}\right)^{1/3} \qquad (4.54)$$

For agents capable of reducing the surface tension, a stationary regime of the above type is feasible only with films of finite extension. The inclusion of the surface diffusion effect may predict a basically different behaviour of the film [241].

With a view to explore the effect which a surfactant can produce, through changing the flow rate, on the mass exchange between the film and the surrounding gas, we have calculated the diffusion flux $J(\xi)$ of a surface-neutral

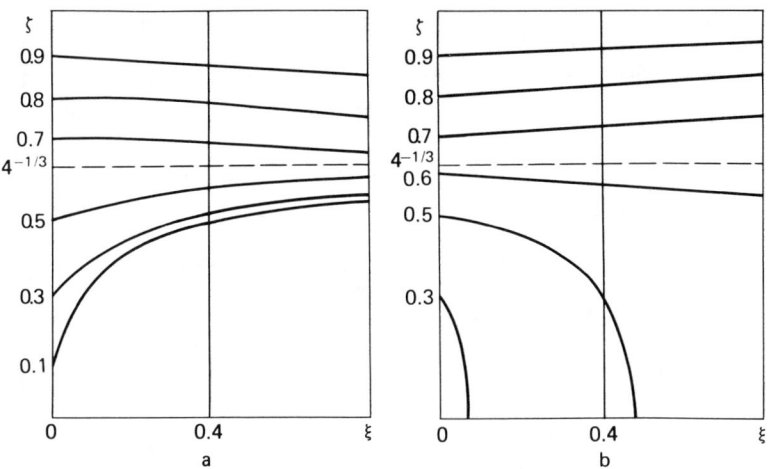

Fig. 4.39 Dimensionless film thickness ζ vs. dimensionless longitudinal coordinate ξ relationship for (a) inactive and (b) active solutes

solute in the film surface (according to formula (4.12)) and the mass transfer enhancement factor $\varkappa = J(\xi)/J_*$. Here J_* is diffusional flux of the solute across the film of constant thickness h_* corresponding to a given value of Q in accordance with (4.54); the surfactant concentration in the film surface is presumed to be independent of x. The $\varkappa$ vs. ξ relationships are shown in Fig. 4.40; they correspond to the curves in Fig. 4.39. As is seen, both active and inactive solutes enhance the mass transport only if the initial dimensionless film thickness is $\zeta_0 < \zeta_* = 4^{-1/3}$; otherwise, the mass transport is retarded.

In the presence of a chemical reaction ($R \neq 0$) relation, (4.52) can be reduced to a system of dimensionless equations which, at $D_s = 0$, takes the form [242]:

$$\frac{dH}{dX} = \frac{(1-H^3)(4-H^3)}{\varepsilon H(2H^3+4)\gamma} - \frac{3H^2 f(\gamma)}{\gamma(2H^3+4)} \quad (4.55)$$

$$\frac{d\gamma}{dX} = \frac{1-H^3}{\varepsilon H^2}$$

where

$X = x/L_2$, $\quad H = h/h_*$, $\quad \gamma = \Gamma/\Gamma_*$, $\quad L_2 = Q/h_* K$,
$f(\gamma) = 2R(\Gamma_*\gamma)/3K\Gamma_*$, $\quad h_* = (3\nu Q/g)^{1/3}$,
$\varepsilon = (9/\nu Q g^2)^{1/3}(\delta\Gamma_*/2L_2\varrho)$

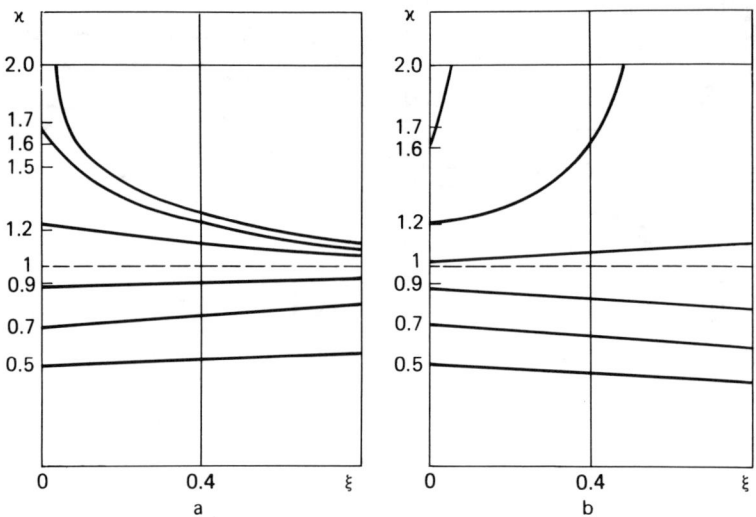

Fig. 4.40 Mass transfer enhancement factor $\varkappa$ vs. ξ relationships corresponding to curves in Fig. 4.39

Here K is the chemical reaction rate constant, of dimension t^{-1} (t is the time).

The state of equilibrium of system (4.55) at $H = 1$, $\gamma = 1$ ($f(1) = 0$) corresponds to a film flow established along the longitudinal coordinate. The linear stability analysis reveals that this state for a SAS-containing film is unstable, that is, the surface SAS concentration and the film thickness do not tend to a state of uniformity; quite contrary, the system "picks up speed" and enters into a regime intractable by the quasi-one-dimensional scheme. By contrast, in a film containing an inactive solute, the state of equilibrium at $f'(1) < 0$ is stable, and both the film thickness and the solute concentration tend to their constant values $H = 1$ and $\gamma = 1$. A point of interest is the occurrence of a stable state of equilibrium of a "focus" type. This signifies that at $f'(1) < -3/(8\varepsilon)$, the film thickness and the surface concentration experience damping oscillations along the longitudinal coordinate (Fig. 4.41). Thus, a wavy ripple-like pattern, decaying downstream, is formed in the film surface.

In the case of "heavy films" at high flow rate, or films containing a weakly-active SAS, the quantity ε becomes a small parameter, and we can apply the perturbation method to solve set (4.55) by imposing the boundary conditions:

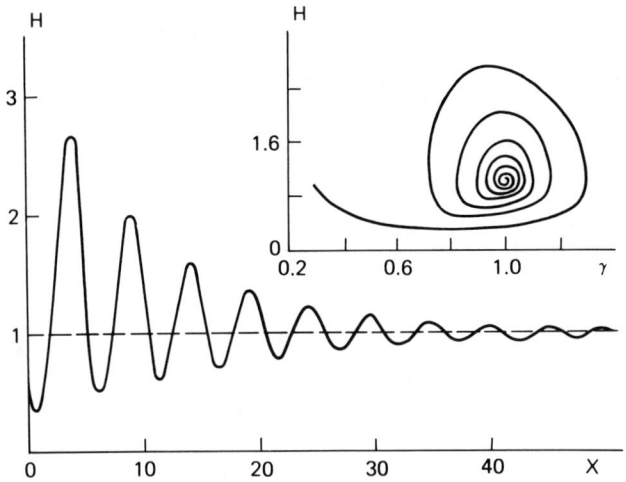

Fig. 4.41 Dimensionless film thickness H as a function of dimensionless longitudinal coordinate X; dependence of H on dimensionless concentration γ (inset)

$h = h_0$, $\Gamma = \Gamma_0$ ($x = 0$). Since in set (4.55) the small parameter ε makes part of the derivative of higher order, the problem thus becomes a singularly perturbed one. A solution when sought in the form of a power series in ε using the regular perturbation theory would fail to satisfy the two boundary conditions above. For this reason, the method of matched asymptotic expansions was used in seeking a solution to the problem. Composite expansions have been thus obtained:

$$H = 1 - \frac{\varepsilon}{3}\left[f(\gamma_0) - f(\alpha)\exp\left(-\frac{3X}{2\varepsilon\alpha}\right)\right] + 0(\varepsilon) \tag{4.56}$$

$$\gamma = \gamma_0 - \frac{2}{3}\varepsilon f(\gamma_0)\left(\gamma_0 + \int_\alpha^{\gamma_0} \frac{yf'}{f}\,dy\right) + \frac{2}{3}\varepsilon\alpha f(\alpha)\exp\left(-\frac{3X}{2\varepsilon\alpha}\right) + 0(\varepsilon)$$

where $\alpha = \Gamma_0/\Gamma_*$; γ_0 implies a zero approximation and is given as an implicit function of X,

$$\int_\alpha^{\gamma_0} \frac{d\gamma}{f(\gamma)} = X$$

The integrated flux of a compound absorbed by the film, $I = \int_0^x J(x)\,dx$, takes, in accordance with (4.12) the form

$$I = I_0\left(1 + \frac{1}{2\varepsilon}\langle U_1/U_0\rangle\right), \qquad (4.57)$$

$$I_0 = 2C_0(DU_0x/\pi)^{1/2}, \qquad \langle U_1/U_0\rangle = \frac{1}{x}\int_0^x \frac{U_1}{U_0}\,dx$$

Here $U = U_0 + \varepsilon U_1$ is the flow velocity at the film surface; C_0 is the gas component concentration near the film surface; D is the diffusion coefficient for the gas dissolved in film. It ensues from (4.56) that

$$\langle U_1/U_0\rangle = \frac{2}{3}(\gamma_0 - \alpha)/X - (4/9)\varepsilon \alpha f(\alpha)\exp(-3X/2\varepsilon)$$

Neglecting the terms of the order of $0(\varepsilon)$, formulas are obtained for the integrated flux and its relative change,

$$I = I_0\left(1 + \frac{1}{3}\varepsilon\frac{\gamma_0 - \alpha}{X}\right)$$

$$\Delta(X) = \frac{I - I_0}{I_0} = \frac{\varepsilon}{3}\frac{\gamma_0(X) - \alpha}{X} \qquad (4.58)$$

Thus, the relative change $\Delta(X)$ as produced by chemically reacting surfactants diminishes as a function of $1/X$ with growing film length, and the film tends to an undisturbed state with $H = 1$.

In modelling the mass transport accompanied by an interfacial convection, the most rigorous of the above approaches is the one based on the nonlinear analysis of interfacial instability and on the computation of major parameters of the emergent convection patterns [139, 145, 148, 151, 152]. Such an analysis applied to weakly nonlinear convection patterns has shown that at Marangoni numbers deviating but slightly from the critical value, $Ma - Ma^{cr} \ll Ma^{cr}$, the disturbance amplitudes for velocity and concentration are proportional to the square root of the postcritical parameter, $\sqrt{Ma - Ma^{cr}}$. This generates an additional convective mass flux proportional to the postcritical increment: $J_{conv} \sim Ma - Ma^{cr}$. Thus, the mass transfer enhancement factor becomes linearly related to the Marangoni number, $\varkappa = 1 + \alpha(Ma - Ma^{cr})$, where α is a function of the viscosity and solute diffusivity [145, 148].

The results of a nonlinear analysis as applied to hydrochemical instability generated by an interfacial chemical reaction in a liquid-liquid system have been presented above in Section 4.3.3. An ordered interfacial convection has been considered when the fluid flow is shaped into two-dimensional rolls, and the longitudinal component of the velocity vector can be represented in the form

$$V_x = U e^{i\frac{2\pi x}{L}} + U^* e^{-i\frac{2\pi x}{L}} \qquad (4.59)$$

where U is the amplitude of convective motion velocity, L is the characteristic linear size; x is a coordinate directed along the interface; the asterisk denotes a conjugate complex number. In the case of $D_1 \ll D_2$ and a small postcriticality $Ma \ll 1$ ($Ma^{cr} = 0$), the following estimates for the characteristic size of convective patterns and for the convective motion amplitude at the interface have been derived [243]:

$$L \sim \frac{l}{\sqrt{Ma}}, \quad |U| \sim \frac{D_1(R_1 + R_2)Ma}{lR_1 R_2}, \quad \text{at } M \ll 1$$

$$L \sim \frac{l}{\sqrt{Ma}}, \quad |U| \sim \frac{D_1 Ma^{3/2}}{lR_1 R_2}, \quad \text{at } M \gg 1 \qquad (4.60)$$

where $M = \sqrt{Ma}/(R_1 + R_2)$, $Ma = Ma_1$; other notations are the same as in (3.25).

By substituting the velocity component (4.59) and the quantities L and $|U|$ in (4.60) into (4.12), we obtain a relation connecting the transfer rate of the target component (which is in all respects a neutral species) to the physicochemical parameters of the system. Extending this relation to any value of M (providing that the computed interfacial convection parameters hold for each particular value of M), the following formula is obtained for the Sherwood number:

$$Sh \sim \left[\frac{D_1(R_1 + R_2)^2}{DR_1 R_2}\right]^{1/2} M^n \qquad (4.61)$$

Here D is the diffusion coefficient for a neutral species; exponent n alters with growing M from 1/2 at $M \ll 1$ to 1 at $M \gg 1$. The Sherwood number increases monotonically from zero with growing M, which is due to the interfacial mass transport rate growth. Simultaneously, the convective motion intensifies with rise in postcriticality. The most efficient mass transport regimes are those which are implemented at highest postcriticality, that is, at the highest Marangoni numbers possible, providing that the specified above convective-

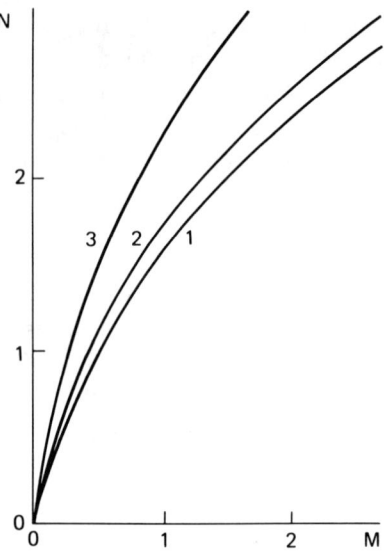

Fig. 4.42 Reduced Sherwood number N vs. postcriticality parameter M for different computed values of parameter r: 1, $r = 0$; 2, $r = 0.2$; 3, $r = 1$

flow pattern is retained. The Sherwood number is dependent on the reaction parameters R_j in a more complicated manner. However, it follows from (4.61) that as the R_j's increase, the interphase transport rate of a neutral species decreases, which should be attributed to the increased rate of dissipation processes associated with a chemical reaction, and to a decrease in postcriticality. The analytical asymptotic behaviour of the Sherwood number as expressed by (4.61) is compared in Fig. 4.42 to the results of numerical computations. The computations have been carried out on the basis of (4.12) using available data on the linear size of convection cells and on the intensity of rotational motion of the fluid inside them as reported in [152]. The dependence for the quantity $N = \text{Sh}\,(D/D_1)^{1/2}(R_1 R_2)^{1/2}(R_1 + R_2)^{-1}$ as a function of M is in a qualitative agreement with formula (4.61). However, curves 1-3 corresponding to different ratios $r = R_2/(R_1 + R_2)$ differ among themselves, which is in accord with formula (4.61) where the exponent n is a function of r.

With a view to explore the possibility of a control of the interphase transport rate of a target compound, formula (4.61) can be conveniently represented by

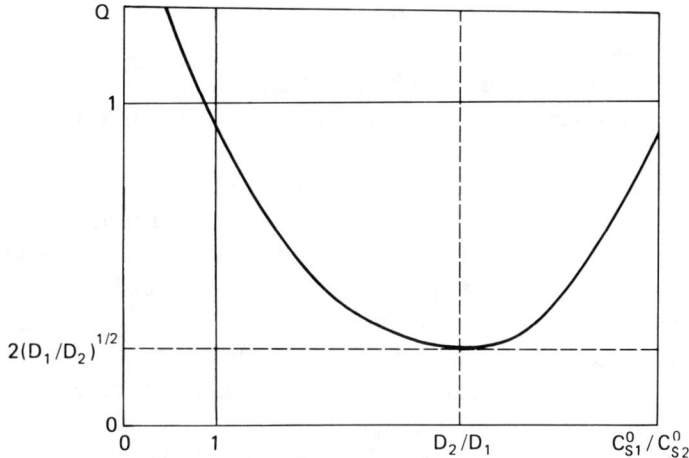

Fig. 4.43 Qualitative dependence of reduced Sherwood number Q on the ratio of initial surface solute concentration C_{s1}^0/C_{s2}^0

making use of (3.24) as a function of the system and regime parameters:

$$\text{Sh} \sim \left(\frac{D_2}{D}\right)^{1/2} \left[\frac{|\partial\sigma/\partial C_{s1}|}{K(\mu_1 + \mu_2)}\right]^{n/2} \left[\left(\frac{C_{s1}}{C_{s2}}\right)^{1/2} \bigg/ \left(1 + \frac{D_1 C_{s1}^0}{D_2 C_{s2}^0}\right)\right]^{n-1} \quad (4.62)$$

It follows from (4.62) that intense mass transport regimes are achieved with an appropriately chosen set of system parameters. The solvents must have low viscosity; the reactant controlling the reaction rate must exhibit a sufficiently high surface activity, and the reaction rate constant must be preferably small. The interfacial transport rate grows if the diffusion coefficient for the target species in the limiting phase becomes smaller in comparison to the diffusion coefficient of a reactant exhibiting no surface activity.

For practical purposes, the transfer rate is convenient to control via proper choice of the initial surface concentration for reactants. Shown in Fig. 4.43 is a general relationship for $Q = \text{Sh}\,(D_2/D)^{-1/2}[|\partial\sigma/\partial C_{s1}|/K(\mu_1 + \mu_2)]^{-n/2}$ as a function of C_{s1}^0/C_{s2}^0. It is to be inferred from this relationship that a maximum mass transport rate is achieved in excess of a reactant lacking in surface activity. By contrast, with a surface-active reactant taken in excess, the mass transport rate decreases by a factor of $(D_1/D_2)^{1/2}$. The mass transport rate is at its minimum if the condition $C_{s1}^0/C_{s2}^0 = D_2/D_1$ holds for the ratio of the reactant concentrations. A physical explanation of the fact is that if the system has one of the reactants in excess, conditions are thereby provided for a large postcriticality.

The results obtained may be regarded as a direct evidence for the strong effect of hydrochemical convection on the interfacial transport rate of chemically inert compounds. Noteworthy, the development of interfacial instability and the respective increase in the convective mass flux toward the interface leads to an acceleration of the diffusion-controlled reaction responsible for this instability. This is due to the fact that the rate of a heterogeneous reaction is determined by the rate at which the reactants are delivered to the interface. The interfacial convection facilitates this delivery and thereby accelerates the reaction. Thus, the optimal regimes that provide for a high mass-transport rate under hydrochemical convection conditions are in fact regimes that ensure an intense course of the reaction itself.

To conclude this Section, we wish to note that the results obtained by weakly nonlinear analysis hold chiefly for small deviations of the system parameters from their critical values. A relation between the Sherwood and Marangoni numbers for an advanced interfacial convection falling into a postcritical parametric region can be established by making use of finite-dimension approximation methods, for example, the Galerkin-type method [178-180], or methods of computer mass transfer modelling; the latter are dealt with in the following Section.

4.4.4 Computer Mass Transfer Models

Numerical computer-aided methods provide a most effective approach to the mathematical mass transport modelling when one is confronted with such handicaps as spontaneous interfacial convection or a substantial nonlinearity of the governing equations. Computer experiments allow one to perform a detailed parametric examination of a process as well as to obtain an information that might be inaccessible by physical experiment. At the same time, the effectiveness of a computer modelling as carried out on the basis of Navier-Stokes and convective diffusion equations can be improved significantly if one is in possession of reliable data including the results that can be obtained by linear and nonlinear analyses of the hydrodynamic stability of the system of interest, in particular, concerning the conditions for the occurrence of interfacial convection in it.

Computer mass transfer modelling under interfacial convection conditions was carried out in [160-162, 171, 172]. Liquid-liquid extraction was considered

4 Modelling of Mass Transport Processes

in [160, 172], desorption of surface-active agents in [161, 162], and absorption and chemisorption, in [171]. Computer experiments using complete nonstationary nonlinear equations have yielded the velocity and concentration distributions in circulation cells and the estimates of interfacial material fluxes.

The results of a numerical analysis of interfacial instability and the development of convection patterns in gas-liquid and liquid-liquid systems have been given in Section 4.3.4. The derived distributions for convection velocity and transported solute concentration enable calculating the interfacial mass transfer rate and estimating the mass transfer enhancement factor due to interfacial convection.

With a view of studying the effect of convective motion on the mass transport between a liquid layer and the surrounding gas phase, local and surface-averaged Sherwood numbers, Sh and Sh_*, were computed in [171], as well as the respective enhancement factors f and f_* defined as the relative change in the Sherwood number during transition from a diffusion regime (in the absence of convection) to convective diffusion regime,

$$f = \frac{Sh - Sh_0}{Sh_0}, \quad f_* = \frac{Sh_* - Sh_0}{Sh_0}$$

$$Sh_* = \frac{1}{N} \sum_{i=1}^{N} Sh_i$$

(4.63)

Here Sh_0 is the Sherwood number corresponding to an undisturbed state of the system in the absence of convection; N is the number of the nodes of the computation grid at the interface. The curves that characterize distribution

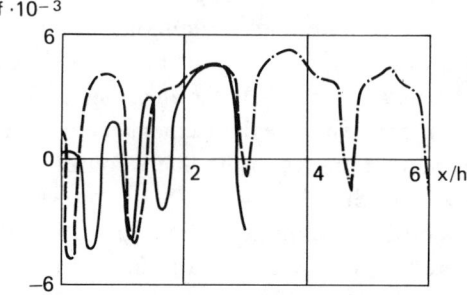

Fig. 4.44 Local mass transfer enhancement factor f along the longitudinal layer coordinate at different times

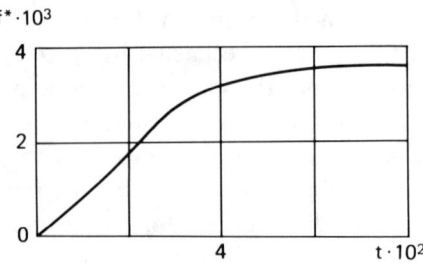

Fig. 4.45 Average mass transfer enhancement factor f_* as a function of dimensionless time

of the quantity f along the layer surface at different times are shown in Fig. 4.44 (the physical system parameters were those reported above in Section 4.3.4). The curves in Fig. 4.44 correspond to different dimensionless times $t_* = (D/h^2)t$; the solid line corresponds to $t_* = 1.2 \cdot 10^{-2}$; the dashed line, to $t_* = 2.92 \cdot 10^{-2}$; the dashed-dotted line, to $t_* = 4.42 \cdot 10^{-2}$. The distribution curves are symmetric with respect to the axis perpendicular to the surface at its central point. The f_* vs. t_* relationship is shown in Fig. 4.45. A maximum of f_* is attained at the instant of a full development of the flow pattern ($t_* = 4.4 \cdot 10^{-2}$) and subsequently remains practically constant. Computations have shown that even at a small Marangoni number (Ma = 10^3), the interfacial convection regime allows to reach a 30% increase of the mass transfer enhancement factor relative to its maximum value $(f_*)_{max} = B = 10^{-2}$. Numerical computations have shown that the growth of Marangoni number is concomitant with a substantial enhancement of the mass transport rate in liquid layer. The gas-phase mass transfer rate as determined by the Biot number B becomes a limiting factor to the interphase transport at sufficiently intense liquid-layer convection, which is reached at a definite Marangoni number. To study the effect of interfacial convection on mass transport rate between two liquid layers, the surface-averaged Sherwood numbers have been computed in [172] for each phase as well as the mass transfer enhancement factor $\varkappa$ defined as the ratio of the average Sherwood numbers to the Sherwood numbers in the stationary distribution (3.29), $\varkappa = Sh_1/Sh_1^{(s)} = Sh_2/Sh_2^{(s)}$. Shown in Fig. 4.46 are the curves for $\varkappa$ as a function of the dimensionless time $t_* = (D_2/h_1^2)t$ at the formative stage of the interfacial convection pattern. Starting from the time $t_* \approx 0.15$, the value of $\varkappa$ undergoes a change only in the 3rd decimal place, this change tending to diminish with time. Thus, after the convection pattern has established, the value of $\varkappa$ becomes practically constant. As demonstrated

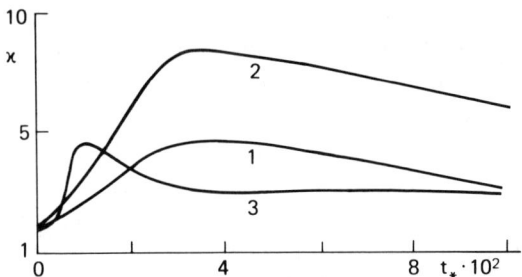

Fig. 4.46 Average mass transfer enhancement factor x of a two-layer system as a function of dimensionless time: 1, $h = 1.5$, $Ma = 10^5$; 2, $Ma = -10^5$; 3, $h = 1$, $Ma = 10^5$

by the graphs in Fig. 4.46, in each case considered, the interfacial convection is capable to intensify substantially the mass transfer in a two-layer system.

We wish to point out that the computations dealt with in this Section focus on mass fluxes capable of affecting the surface tension. Simultaneously, the transport rate for surface-neutral species under interfacial convection conditions may be markedly higher over that for the surfactants that generate interfacial instability.

It is to be emphasized, in concluding Chapter 4.4, that the computations as based on nonlinear and numerical analyses of interfacial instability fail to provide, despite their high informativeness, a method for determining the mass transport characteristics that could be used in chemical engineering design. The nonlinear analysis and based thereupon analytical models hold only for a limited number of idealized systems, for example, stagnant liquid layers, or systems with the interfacial convection patterned into vortex rolls. Both the computer experiment and the laboratory experiment stand in need of additional correlations between the physicochemical parameters and mass transport characteristics for a system under study. Therefore, today the most promising route in an aspect of chemical engineering design appears to be the construction of semiempirical models that might describe the major physical features of interfacial convective mass transfer using a minimal number of empirical parameters.

4.5 Applied Aspects of Interfacial Self-Organization

The study and active use of interfacial self-organization phenomena are gaining an ever-increasing importance for multiphase processes in chemical engineering since they provide basically new possibilities for the construction of self-organizing systems of high level of sophistication and for a purposeful design of self-regulatory technologies. Novel methods have emerged enabling an active intervention into intrinsic mechanisms operative both on the microscopic level and on the global scale of a transport process in a chemical apparatus. A clear understanding of such mechanisms and of the effects they produce in a complex system is an essential prerequisite for the design of high-effective technologies based on self-organization principles. Of particular emphasis is the resonant character of self-organized interactions, since a weak, but highly tuned action on a system capable of self-organization can produce a much greater effect than a powerful, but misapplied "off-target" impulse.

The concept of interfacial self-organization is the guideline in a search for effective means of chemical process control. Parametric studies of the interfacial instability and self-organization as performed by general mathematical models concerned with major hydrodynamic and physicochemical effects as well as by specialized models of a particular process provide a basis for the development of methods aimed at further improvement of commercial technologies, primarily, those involving the intensification of an interphase transfer. The implementation of these methods does not require large capital investments and, in the long run, allows to update the industrial plant equipment and to reduce energy expenditures. A number of means to control the mass transfer through the agency of interfacial instability and self-organization may be suggested.

1. Regime control, that is, a control as effected through variable regime parameters. This type of control envisages the change of the initially stable state of interface to a state of interfacial instability and self-organized mass transport. For example, in a quiescent liquid medium this signifies the transition from a molecular diffusion regime to a convective diffusion regime. In a moving liquid medium, this is the transition to a regime exhibiting both an enforced and an interfacial convection. In certain systems one may, through making vary the regime parameters, attain a state of resonant excitation, most commonly realizable only within a narrow parametric region.

2. Physicochemical composition control. This mode of control bears upon the determination of optimal physicochemical parameters responsible for an intense mass transport under spontaneous interfacial convection conditions. It also implies the use of high-effective media (absorbents, chemisorbents, extractants, reagents, solvents, and so forth) capable of affecting the kinetics

of a process and eliciting thereby the interfacial instability. To be noted, a variation in the composition of a system with the resultant change in density, viscosity, diffusivity, heat conductivity, surface tension and other characteristics can initiate the transition from a subcritical state (with reference to interfacial stability) to a postcritical state.

3. External field control. This mode of control is based on the physical mechanisms of a coherent interaction between the internal processes and external fields of different nature. An external field imposed on a system can initiate self-organization and cause thereby a drastic change in the transport processes operative within the system.

4. System geometry control is effected through variation of the geometric parameters of a system modifying the conditions for emergence and development of interfacial convection.

5. Thermal and concentrational control as effected through variation of the temperature and concentration distributions at the outer boundary of a system.

6. Combined control, which implies a simultaneous use of mass exchange, heat and chemical processes as a governing factor of interfacial self-organization.

7. Control as effected through the agency of surfactants and microadditives. This control is based on the addition into a system of small quantities of surface-active agents capable of initiating interfacial instability.

Let us dwell in some detail on certain specific features of the aforementioned control methods.

To exemplify a successful application of the regime control to an industrial process, one may refer to the mass exchange intensification in a solvent extraction of residual chlorinated phenols in the commercial production of amine salt-based herbicides [45, 244]. A mere 5% reduction of the initial concentration of the reaction mass to be purified has provided for a near 3-fold increase in mass transfer rate. The reason for such a marked effect is the transition from a diffusion regime of mass transport to an interfacial convection regime which spontaneously sets in as the Marangoni number has surpassed its critical value. To be noted, the lowering of the initial amine salt concentration makes augment the surface activity coefficient $\partial\sigma/\partial C$, with the resultant increase in both Marangoni number and solvent extraction rate.

A similar mass transfer enhancement can also be achieved through changing a physicochemical parameter, for example, the solvent viscosity. The solvent viscosity change, even apparently small, can initiate the transition to a spontaneous interfacial convection regime by raising the Marangoni number to a critical value.

The proper choice of liquid composition may be instructively illustrated by carbon dioxide absorption in the synthetic ammonia production. The interfacial turbulence effect is manifested in the CO_2 absorption by an aqueous solution of monoethanolamine (or other appropriate amine); however, this effect is absent when a heated potash solution is used as a solvent. A high-priority task in thechnology is the search for such physical adsorbents of carbon dioxide which, while being readily accessible and of lower cost than conventional chemisorbents, could be capable of producing interfacial turbulence. Among organic solvents, methanol may be mentioned as a good candidate for this purpose [49].

Of the external fields capable of exercising a control function over the interfacial self-organization, magnetic and electric fields claim attention whose influence on the mass transport rate is primarily associated with the growth of kinetic coefficients due to the buildup of microstreams and steady convection patterns as well as to the modified properties of the medium by field action. For example, the imposition of an electric field on a liquid with a small relaxation time causes the occurrence of surface tension gradients and the formation of stable vortex cells at the interface. Magnetic field can also produce a similar effect; however, the practical implementation of its control function is disfavored by high energy and capital costs.

The control function through monitoring the absolute gravitational force constitutes the basis of space materials science. An exceptionally advantageous feature of any space technology is that the absolute gravitational force under extraterrestrial conditions can be decreased by a factor of 5 orders of magnitude, or even more. To be noted, the role of Bénard effects under reduced gravitation conditions becomes less important, in contrast to the predominant role of Marangoni effects in many heat and mass transfer processes.

Among the geometric parameters essentially affecting the interfacial instability and the development of convection patterns is the ratio of the depths of the interacting phases, or the ratio of the thickness of the layers within which the mass-transport driving force is operable. Other parameters intervenient in interfacial self-organization are the radii of liquid drops, of gas bubbles, of fluid streams and the film thickness. A change in the liquid layer thickness affects the relative contributions of Marangoni and Bénard effects to the natural convection development in heat and mass transfer. A point to be noted is that these effects differ essentially from each other in their response to the orientation of a system in gravitational field. The Bénard effects are notably dependent on the angle between the liquid layer plane and the gravitational force vector, whereas the Marangoni effects are much less susceptible to the spatial orientation.

Currently, an advanced technological procedure aimed at a higher efficiency appears to be the simultaneous realization of mass transfer, heat and chemical processes within a single apparatus. As an example, one may refer to chemirectification, chemidesorption, chemiextraction, extractive crystallization, or membrane catalysis. At the same time, it is precisely owing to a cooperative action of inherently distinct processes that the self-organization phenomena can arise and develop in a system. A special role in interfacial self-organization is assigned to resonant interactions between various heterogeneous processes carried out in a single multiphase system.

At present, the intensifying effect as produced by surface-active agents on mass transport and reaction processes in heterophase systems has gained support from the large enough experimental evidence acquired both in the laboratory and at the chemical plant. However, in most cases, a solid theoretical substantiation of such effects is lacking. Viewed from the standpoint of interfacial instability, additives apt to change the surface tension in interphase transport are the most promising for the intensification of multiphase processes. However, they must not be strong surface-active agents as to form a resilient film at the interface and to prevent thereby the development of convective interfacial motions. The addition of chemically reacting agents which, in the presence of hydrochemical interaction, generate interfacial instability may also be regarded as a possible means to control mass transport, in particular, its intensification. To be noted, the control over the mass transfer of a surface-neutral substance can be exercised through a simultaneous transport of a compound capable of eliciting interfacial instability. In [245], experimental data have been reported on a cooperative mass transport of cresol and acetic acid (acting as an interfacial convection initiator) in a system carbon tetrachloride-water. The results obtained provide evidence that in the course of a multicomponent transport—either one-directional or reversed—at high concentration gradients typical of spontaneous interfacial convection, the effects of diffusional and hydrodynamic interaction between the transported species become noticeable, in conformity with the basic postulates of Onsager theory [246] and solute Marangoni effect theory. Analogues results have been obtained in [247] for a one-directional cotransport of iodine and acetic acid from carbon tetrachloride into an aqueous potassium iodide solution under the conditions of spontaneous interfacial convection and dispersed organic phase.

To conclude this brief review of the means of technology control using the interfacial instability and self-organization effects, we wish to note that their effective implementation is feasible only on the basis of thorough theoretical and experimental studies of the mechanisms of physical and chemical interactions conducive to interfacial self-organization.

Further, we intend to consider a number of examples demonstrative of the practical utility of interfacial instability and self-organization in improving important chemical technologies such as absorptional and chemisorptional gas purification, solvent extraction, and some others.

Currently, there is an impressive body of experimental evidence on interfacial convection observed to occur in chemisorption processes. Most commonly, the laboratory experiments were carried out on systems with different types of the gas-liquid contact, such as a liquid film moving across a vertical or a horizontal plane surface, a liquid flow in a stacked-packing column, gas bubbling, also vertical or horizontal liquid jets in a gaseous medium [53]. Given in Table 4.5 [53] are surface convection characteristics for a $1 M$ monoethanolamine solution on its uptake of carbon dioxide in reactors of different type. Here, the intensity of interfacial convection is characterized by the mass transfer enhancement factor χ defined as the ratio $\chi = \beta/\beta_0$, where β and β_0 are the mass transfer coefficients of a tracer compound measured

Table 4.5 Comparative data on surface convection intensity for absorption of CO_2 by $1M$ aqueous monoethanolamine solution in apparatuses of different type (Re, liquid Reynolds number; C_g, CO_2 gas concentration; χ, mass transfer enhancement factor)

Apparatus	Re	C_g(%, v/v)	χ
Vertical wetted-wall column:			
$H = 1$ m	950	100	8.4
	300	100	8.2
$H = 0.12$ m	260	100	4.7
Vertical disk column, $H = 0.5$ m, solution feed rate 0.0205 m^3/h	—	100	3.8
Vertical laminar jet, $H = 0.115$ m	600	100	4.3
Bubble reactor, $H = 0.2$	—	60	1.4-1.5
Horizontal oscillating jet, $H = 0.14$ m	1725	100	1.6
Horizontal trough, $H = 0.4$ m	27	100	2.2
	54	100	1.8

4 Modelling of Mass Transport Processes

in the presence and in the absence of surface convection, respectively. The highest mass transfer enhancement is achieved in vertical wetted-wall columns. The value of χ in the liquid-film reactors rises with the film length. This result agrees with the conclusions in [26, 248, 249] stating that the duration of phase contact affects the surface convection intensity.

The factor χ is smaller for reactors with a horizontally moving liquid layer than that for wetted-wall reactors. Apparently, the horizontal liquid layer is a more stable system in comparison to a falling liquid film whose wave motion makes more probable the appearance of local velocity and composition nonuniformities at the surface.

The surface convection, as evidenced by the data in Table 4.5 is markedly weaker in bubble reactors. The experiments with a disk-type column [53] have shown that its factor χ takes a value intermediate between the respective values for a wetted-wall column and a bubble reactor. It is to be inferred therefore that as the turbulence of the main flow becomes more intense, the surface convection tends to decrease. A plausible reason of the latter effect may be the breakdown of the conditions requisite for the surface convection buildup because of the interference of turbulent vortices emerging to the surface from the bulk. In particular, under nonstationary mass transfer conditions, a local quasi-stationary distribution for transported solute concentration suffers breakdown, that is, the local distribution which is responsible for the generation of interfacial convection and which forms in a certain period of time, the so-called interfacial convection delay time. Another reason for interfacial convection retardation is a faster levelling of the interfacial tension because of turbulent pulsations near the interface.

To be noted, the available experimental data allow to suggest that the critical Marangoni number increases with Reynolds number. If the Reynolds number increases with the Marangoni number remaining constant, a situation may arise when Ma goes down below its critical value, with the resultant cessation of the surface convection. Although the available data on the surface convection intensity under turbulent liquid flow conditions are rather scarce, we would like to point out that the above concepts conform with the results presented in Fig. 4.11, these latter providing evidence that at Re > 1600, a tendency to a decay of the surface convection intensity is observed. However, with the Marangoni number sufficiently large, the surface (or interfacial) convection can exist also under advanced turbulence conditions and play an important role in interfacial mass transfer. An explanation is that, at large Reynolds numbers and small-scale turbulent pulsations, there exists near the free interface, a thin diffusion layer whose resistance to mass transfer cannot be removed by bulk turbulence. Simultaneously, both the interfacial convection and interfacial turbulence, originated at the interface are capable to destroy this diffusional surface layer and to reduce thereby its resistance to mass transfer.

Solvent extraction, commonly effected in countercurrent column extractors, is another example of a profitable use of interfacial instability for increasing the efficiency of industrial mass-exchange processes in disperse systems. The efficiency of a column is characterized by the mass transfer rate which in fact determines the column height that would suffice to achieve the desired degree of extraction of a solute. The increase of mass transfer coefficient through transition to a spontaneous interfacial convection regime allows to run a shorter column keeping the desired degree of extraction at the former level, or to rise the degree of extraction operating with the same column height. The SIC effect may be therefore regarded as a promising route to increasing the efficiency of extraction columns.

To estimate the action of spontaneous interfacial convection on the performance of a spray column, the transfer of iodine from carbon tetrachloride drops into water and the co-transport of iodine and an acid (acetic, propionic) in the same system were studied in [45] on a laboratory set-up. Surface-active iodine was used as a tracer, and its mass transfer coefficient was chosen as a measure for the intensity of spontaneous interfacial convection produced by the carboxylic acid transfer. The curves shown in Fig. 4.47 characterize the ratio A/A_0 as a function of the column height. Here A_0 is the degree of iodine extraction in the absence of spontaneous interfacial convection (diffusion regime) and A is the degree of iodine extraction in the presence thereof; C_0 is the initial acid concentration in the donating phase. The results in Fig. 4.47 show that the ratio A/A_0 is always larger than unity, which means that the desired degree of iodine extraction under interfacial convection is achieved at a smaller column height. Besides, the ratio A/A_0 tends to its asymptotic value dependent on the intensity and duration of the SIC regime.

The laboratory studies [45] have provided evidence for a substantial influence of spontaneous interfacial convection on the performance of perforated-plate extractors normally providing for a high intensity of the mass transfer at sufficiently high capacity (overall load, up to $60 \, m^3/m^2 \cdot h$). The high mass exchange rate in these extractors is explained by multiply repeated processes of coalescence and redispersion of liquid drops. At the instant of its being dispersed, the spherical liquid particle with a uniformly distributed

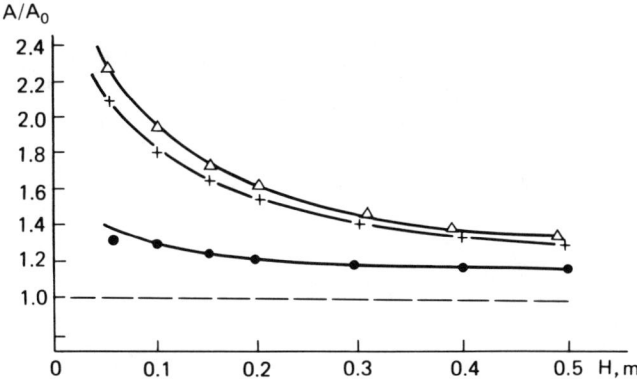

Fig. 4.47 Relative degree of extraction for iodine A/A_0 as a function of the column height; A is the degree of iodine extraction in cotransport: △, with acetic acid (initial concentration $C_0 = 0.7$ kmol/m); +, with acetic acid ($C_0 = 0.4$ kmol/m^3); •, with propionic acid ($C_0 = 0.5$ kmol/m^3)

solute exhibits a maximum of the mass transfer coefficient. As the particle migrates, the mass transfer coefficient and the SIC intensity both become decreased because of an ever-growing nonuniformity of the transported solute distribution in the direction of mass transfer. However, each next coalescence and redispersion restore a uniform distribution of the solute inside the drop, which helps to maintain both the SIC regime and a high mass transfer rate. Shown in Fig. 4.48 are the experimental curves for the overall degree of extraction of carboxylic acid as a function of the column height for a system carbon tetrachloride-water.

The results of laboratory experiments have been used in designing a commercial process for extractive purification of an aqueous solution of dimethylamine salt of 2,4-dichlorophenoxyacetic acid (2,4-DA) from the contaminant 2,4-dichlorophenol (DCP). The purified product is a very effective herbicide. Tetrachloroethylene and toluene were used as extractants in the experiments. The experimental data have provided evidence on the surface activity of chlorophenol and on the buildup of an intense spontaneous interfacial convection in the systems studied under definite regime and parametric conditions. They have provided a clue to the choice and substantiation of the most intense technologic regime for the transfer of DCP from the aqueous into the organic phase, which is an illustrative demonstration of the potential utility of SIC effects in industrial solvent extraction.

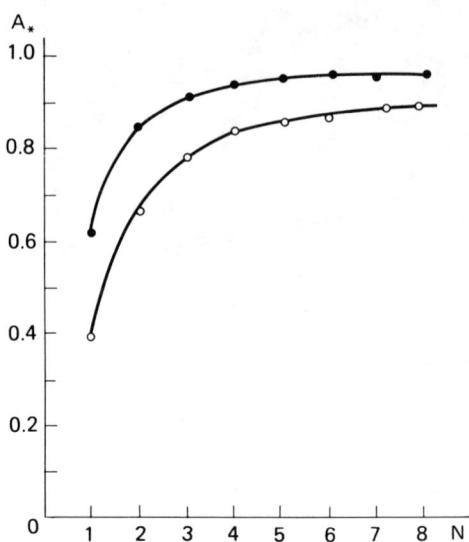

Fig. 4.48 Variation of the total degree of extraction along the column height (N is the number of sections) for a system CCl_4-H_2O in the transport of: •, acetic acid (initial concentration in disperse phase $C_0 = 0.7$ kmol/m^3); ○, propionic acid ($C_0 = 0.8$ kmol/m^3)

A concentrational variation of the transported solute in the phases along the extractor height [45] has been used to study the kinetics of chlorophenol transport from 2,4-DA solutions of different concentration in a laboratory column. The degree of extraction of chlorophenol at each separate plate (Fig. 4.49) could be determined from the kinetic curves. The results obtained have shown that owing to spontaneous interfacial convection, the degree of extraction in the first plates (following the course of the disperse phase) is nearly four-fold the degree of extraction at the end of the process, with the spontaneous interfacial convection subsided to zero.

The experimental data on the variation of degree of extraction of chlorophenols with column height from 2,4-DA solutions at different extractant concentration as obtained on a pilot plant are shown in Fig. 4.50 [45]. The amine salt and extractants were fed countercurrent into the column. Tetrachloroethylene used as extractant was fed into the column top, and the salt, into the bottom section of the column; using toluene, the extractant was fed

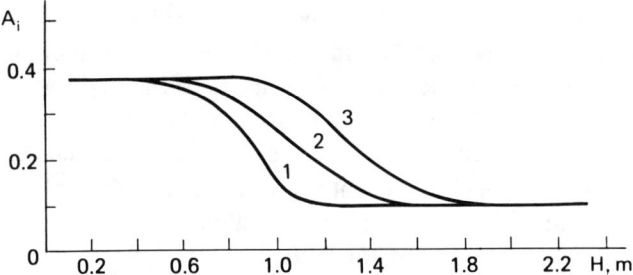

Fig. 4.49 Degree of extraction along the column height for a separate plate in extraction of chlorophenols with perchloroethylene. Chlorophenol concentration in the initial 2.4-DA solution: 1, 1.24% (by mass); 2, 2.55%; 3, 4.48%; 5, 1.0%

into the bottom section, and the mass to be purified, into the top. The results represented in Fig. 4.50 show that, in extracting chlorophenol from a 40% 2,4-DA solution under mild SIC regime, a larger extractor height is needed to go down to a given chlorophenol level in the raffinate. By contrast, the

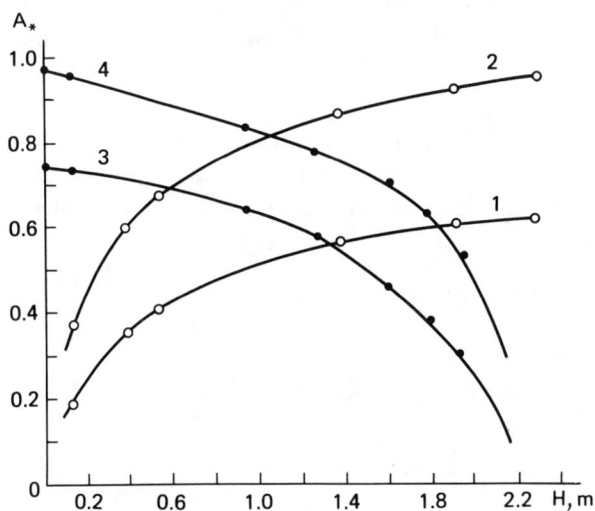

Fig. 4.50 Variation of the total degree of chlorophenol extraction along the column height for 2,4-DA solutions. Initial 2,4-DA concentration: 1, 38.90% (by mass), 2, 21.23% (○, extraction with perchloroethylene); 3, 39.70%, 4, 20.70% (●, extraction with toluene)

use of a salt solution diluted to 22% has led to a substantial increase in SIC intensity, which makes possible to run a shorter extraction column. This mode of action on the mass transfer intensity belongs to the regimen control type; its physical nature has been discussed earlier in this Section.

A point to be noted is that the occurrence of spontaneous interfacial convection in mass transfer must be taken into account in designing solvent extraction processes. Otherwise, the design height of a column extractor may be unnecessarily greater over that required for a given performance regime. By way of example, the actual height of the working section of an extractor operated in a SIC regime and used in the mentioned above experiments [45] was four meters. By way of comparison, the calculations of mass transfer efficiency in neglect of spontaneous interfacial convection by models of Kronig-Brink [250] and Handlos-Baron [251] predicted the column heights 5.6 and 2.6 m, respectively.

The data obtained have provided evidence that the enhancement of mass transfer due to spontaneous interfacial convection must be taken into account in estimating the efficiency of an industrial process. Given below is a method for calculating the major characteristics of a countercurrent extraction column operated in spontaneous interfacial convection regime [222] within the framework of the semiempirical model as outlined in Section 4.4.2.

Let, in a countercurrent extraction column, the disperse phase concentration be a, the continuous phase concentration be b, the partial mass transfer coefficient in disperse phase K_d, and in continuous phase, K_c. For the mass transfer effected in a SIC regime, the coefficients K_d and K_c are, in accordance with the model in [218], are linearly dependent of the mass transfer driving force:

$$K_d = K_d^0 + K_{SIC}^{(d)}[(a - a_s) - (a - a_s)_{cr}]$$
$$K_c = K_c^0 + K_{SIC}^{(c)}[(b_s - b) - (b_s - b)_{cr}]$$
$$K_{SIC}^{(d)} = B(\eta_1 + \eta_2)^{-1}|\partial\sigma/\partial a_s|Sc^{-1/2} \qquad (5.1)$$
$$K_{SIC}^{(c)} = B(\eta_1 + \eta_2)|\partial\sigma/\partial b_s|Sc^{-1/2}$$
$$K_{SIC}^{(c)} = \frac{1}{m} K_{SIC}^{(d)}$$

Here a_s and b_s are the surface concentrations in the disperse and continuous phases, $a_s = b_s/m$; m is the local equilibrium factor; $(a - a_s)_{cr}$ and $(b_s - b)_{cr}$ are the critical driving forces responsible for the occurrence of spontaneous interfacial convection; K_d^0 and K_c^0 are the partial diffusional mass transfer coefficients.

4 Modelling of Mass Transport Processes

We assume that the mass transfer is limited by the disperse phase, and that $a_s = b/m$, $K_d = \bar{K}_d$; $\bar{K}_d$ and $\bar{K}_c$ are the overall mass transfer coefficients. From the differential mass transfer equations, we have for the elementary column height dh:

$$dM = \bar{K}_d fS(a - a^*)dh = \bar{K}_c fS(b^* - b)dh$$

from the material balance equation $V_d da = V_c db$, with allowance made for (5.1), an equation for the degree of extraction A for the disperse phase is derived:

$$\frac{dA}{dZ} = p[1 + \varkappa(p - p_{cr})], \quad \text{at } p \geqslant p_{cr};$$

$$\frac{dA}{dZ} = p, \quad \text{at } p < p_{cr}$$

where

$$A = (a_1 - a)/a_1 - a_2^*, \quad p = 1 - nA_0 - (1-n)A,$$

$$p_{cr} = \frac{(a^* - a)_{cr}}{a_1 - a_2^*}, \quad Z = \frac{\bar{K}_d^0 fS}{V_d} h,$$

$$n = \frac{V_d}{V_c} m, \quad \varkappa = \frac{K_{SIC}^{(d)}(a_1 - a_2^*)}{\bar{K}_d^0}$$

Here a_1 is the concentration at the disperse phase inlet; a_2^* is the equilibrium concentration at the continuous phase inlet; Z is the dimensionless column height; f is specific phase contact area; S is the column cross-section; V_d and V_c are the feed rates for disperse and continuous phases, respectively; $\varkappa$ is a dimensionless parameter characterizing the mass transfer enhancement under SIC conditions; p_{cr} is the dimensionless mass transfer driving force.

Let us consider the case of $n < 1$, implying that the capacity of continuous phase is larger over that of disperse phase. The following formulas are derived for the dimensionless column height Z and the corresponding degree of extraction A_0 from (5.2):

(i) $Z = \dfrac{1}{(1 - \varkappa p_{cr})(1 - n)} \ln \dfrac{[1 + \varkappa(1 - A_0 - p_{cr})](1 - nA_0)}{[1 + \varkappa(1 - nA_0 - p_{cr})](1 - A_0)}$, \hfill (5.3)

at $1 - nA_0 > p_{cr}$, $1 - A_0 > p_{cr}$

(ii) $Z = \dfrac{1}{(1 - \varkappa p_{cr})(1 - n)} \ln \dfrac{1 - nA_0}{[1 - \varkappa(1 - nA_0 - p_{cr})]p_{cr}} +$

$+ \dfrac{1}{1 - n} \ln \dfrac{p_{cr}}{1 - A_0},$ (5.4)

at $1 - A_0 < p_{cr} < 1 - nA_0$

Formula (5.3) corresponds to the case when the spontaneous interfacial convection regime extends throughout the entire column height; formula (5.4) corresponds to the case when the spontaneous interfacial convection extends only to a portion of the column height, and the mass transfer in the remained portion is effected by diffusion regime. The height Z, defined by (5.3) or (5.4), is smaller as compared to the height Z_0 of the column run under SIC-free conditions,

$$Z_0 = \dfrac{1}{n - 1} \ln \dfrac{1 - nA_0}{1 - A_0} \quad (5.5)$$

In the absence of spontaneous interfacial convection, (at $\varkappa \to 0$), formulas (5.3) and (5.4) reduce to (5.5).

Shown in Fig. 4.51 are the curves for the ratio Z/Z_0 as a function of the parameter $\varkappa$ which characterizes the SIC-assisted mass transfer enhancement. In other terms, the quantity Z/Z_0 is the ratio of the height of transfer unit in the presence of spontaneous interfacial convection to the height of transfer unit in the absence thereof. Curve 1 corresponds to the case (i), curves 2-4 correspond to the case (ii). As is seen in Fig. 4.51, the column height needed for achieving a given degree of extraction and a given height of transfer unit can be significantly reduced if the process is made to run under spontaneous interfacial convection conditions.

Now, let us take a cursory look at the role the interfacial instability effects play in the ion transport across liquid membranes. Liquid membranes are gaining an ever-widening application in extraction, concentration and separation of various compounds [252-254]; besides, they can be used as models for special functions of cellular biomembranes [255]. A liquid membrane is a layer of organic phase separating two aqueous solutions. This intermediate layer can largely vary in thickness and can be either in a free state, or supported on a porous solid substrate providing for the mechanical strength of a heterogeneous system. Processes involving such composite emulsions are referred to as the membrane, or emulsion, extraction. Factors determinative of the interfacial liquid-membrane instability are the formation of oversaturation

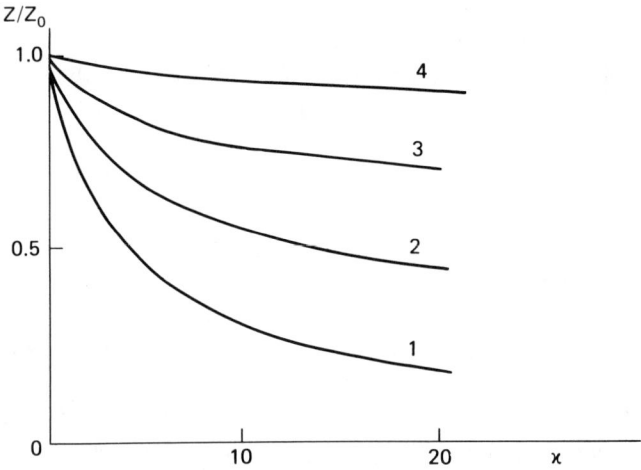

Fig. 4.51 Relative column height Z/Z_0 as a function of parameter x: 1, $n = 0.5$, $A_0 = 0.8$, $p_{cr} = 0.1$; 2-4, $n = 0.33$, $A_0 = 0.9$, 2, $p_{cr} = 0.15$, 3, $p_{cr} = 0.3$, 4, $p_{cr} = 0.5$

zones, microheterogeneity zones and condensed interfacial films, also the action by an electromagnetic field [256]. The occurrence of microheterogeneity zones at the liquid-liquid interface produces a strong effect on the mass transfer rate. On the one hand, this makes decrease the transit mass flux due to surface shielding; on the other hand, the extraction rate is observed to increase owing to tangential interfacial tension gradients and to the spontaneous interfacial convection produced by Marangoni effect. However, this mechanism is distinctly manifest only at a free interface provided that the microheterogeneity zones have suffered no phase transformation. The phase transformation occurs only at a definite concentration of the extract in aqueous solution [256, 257] and is accompanied by a jump-like change of the ion transport rate as sequent to sharp variations of rheologic interfacial properties. A study of adsorption at the toluene-water interface in extracting tetrachloropalladate with trioctylamide [258] revealed the formation of a condensed interface film of varying thickness accompanied by spontaneous interfacial convection arising from local variations of interfacial tension because of a nonuniform afflux of the solute to the interface.

An appreciable mass transfer enhancement under hydrodynamic interfacial instability conditions was recorded in [69, 133] concerned with SIC studies in metal extraction of a number of system used in hydrometallurgy. The ion transport rate was observed to increase by more than two orders of magnitude. The available experimental evidence have shown that the breakdown of hydrodynamic interfacial stability and the occurrence of a spontaneous motion of the liquid at the interface is invariably concomitant with the formation of a condensed interface film in extraction. A typical example is the extraction of lanthanides and certain actinides from nitric acid solutions with di-2-ethylhexylphosphoric acid dissolved in a nonpolar diluent such as octane, decane, or toluene [256, 259].

The spontaneous disrupture of a continuous interface film, the recommenced motion of the interface and its adjacent liquid layers are frequently observed effects. Occasionally, this process is periodic in character [260]. The concentration of reacting species exerts a marked effect on the development of hydrodynamic instability. The concentration of both metal and extractant has been found to exhibit a critical value which, once surpassed, initiates the transition of mass transfer to a SIC regime. The concentration range for occurrence of spontaneous interfacial convection in such systems is wide enough and falls into 10^{-1} throughout several hundred g/l in metal concentration and into 10^{-3} throughout $1M$ in extractant concentration.

Experimental data in the literature show (see Section 4.3.1) that the hydrodynamic instability as produced by mass transfer can arise at a non-free interface and can stimulate the ion transport [90-93]. Thus, the passage of an alternating electric current through impregnated membranes makes increase the ion transport rate [91], which should be attributed to the occurrence of convective motions in the membrane phase. The authors of the paper [90] believe that a 1 Hz/s a.c. current is optimal, since it is capable of producing tangential motions at the interface.

Along with their use in separation processes, the Marangoni instability effects can also be applied to electrochemical processes. A significant (over an order of magnitude) mass transfer coefficient enhancement due to the development of cellular convection patterns was reported in [261-265] in the electrodeposition of lead from a melt on the liquid electrode surface. The interfacial instability plays a role in electroanalytical chemistry in polarographic studies of the kinetics of processes involving the adsorption of organic surface-active substances. In a system mercury—electrolyte solution with a surfactant additive, polarographic maxima of the 3rd type were observed produced by the tangential motion of the interface because of the buildup of regions with locally increased interfacial tension [262, 263].

It is to be noted that the major concern of the present work has been to clarify the role the interfacial instability and interfacial self-organization phenomena play in the intensification of various interphase transfer processes. Still, these phenomena are also of special interest for the industrial processes where the interfacial hydrodynamic instability and the development of convection patterns and inhomogeneities in density, composition and temperature are objectionable. As an example, one may refer to the light-sensitive materials technology. The quality of cine and photographic films, also of magnetic tapes is to a large extent dependent on the degree of homogeneity of the polymer coatings used. A technologic derangement in the manufacture of such materials can create undesirable conditions for the development of interfacial instability in the stage of both application and curing of multilayer coatings. To avoid these effects, one must maintain the technology in a regime that would provide for hydrodynamic stability of the interface surface.

Similar complications may arise in many other technologic processes such as vaporization, solidification, drying, crystal growth, production of metals and alloys, polymerization. Therefore, in order to further improve the quality of products, the respective technology must be worked out on the basis of a knowledge about interfacial structures enabling to choose suitable physicochemical regime ane design parameters that would prevent the detrimental interference of interfacial instability.

To conclude this Section, we wish to point out that the phenomena of interfacial instability and interfacial surface self-organization play an important role, apart from chemical engineering, in many other industrial fields including metallurgy [266], electronic industry [267], space crystal growth techniques and production of semiconductors and high-quality optical glasses [268-270], photography [271, 272], and biological industry [67].

4.6 Conclusion

The issues in the theory of instability, self-organization and mass transport at the interface that have been considered above are of fundamental importance for physicochemical hydrodynamics, self-organization, nonlinear mass and heat transfer and microkinetics of heterophase reactions. These issues are also of practical significance for a correct design of multiphase technologies and for the development of highly-efficient energy-saving methods of their intensification. Theoretical studies with potential application to practice may be arbitrarily divided into two groups. The first group embraces results which allow to determine optimal conditions for the implementation of a process and to choose parameters corresponding to an intense mass transfer regime. The second group includes theoretical methods enabling to calculate the operating conditions for separation and reaction engineering processes under interfacial convection conditions.

A progress in the study of interfacial phenomena is to a large extent determined by the efficiency of theoretical methods chosen for solving specific problems. From the standpoint of a quantitative interpretation, a classical hydrodynamic approach based on a general solution of the balance equations for momentum, mass, energy, and state appears to be the preferable one. The role of one or another interfacial process responsible for the disturbance of hydrodynamic stability may be accounted for by choosing appropriate boundary conditions.

The data obtained by stability analysis allow to define the regions of physicochemical and regime parameters where the occurrence of interfacial instability and the mass transfer enhancement may be anticipated. Major factors affecting the interfacial instability and, consequently, the rate of a process are the interphase driving force, the interfacial activity of a transported solute (reactant), and the mass transfer direction. A stability analysis reveals the relation of these parameters to other physicochemical characteristics such as viscosity and solute diffusivities, characteristic phase dimensions (the diameter of liquid drops and gas bubbles, the liquid membrane thickness), the phase contact time, and others. The linear analysis results constituting a basis for subsequent computations are subject to certain limitations arising from adopted simplifications in the statement of a problem and the very essence of a linear approximation. A simplification, for example, is the assumption of a linear dependence of interfacial tension on the concentration of transported solute. The assumption about a constancy of a number of other physicochemical parameters also introduces certain limitations into the generality of the results obtained; however, it is felt that this does not affect substantially the basic conclusions and can cause alterations in critical parameters. Another limitation inherent in the linear analysis is that this analysis does not specify necessary and sufficient conditions for the occurrence of interfacial instability, since an instability is possible with respect to finite amplitude perturbations. Therefore, a nonlinear perturbation analysis is needed for a detailed study of the critical conditions for transition to an intense regime. This analysis allows to obtain data on the interfacial convection pattern and its evolutions and provides a way of computing, starting from the basic principles, the interphase fluxes and the mass transfer enhancement factor. Despite the fact that this purpose can be achieved via analytical methods only for a limited class of ideal systems, the nonlinear analysis results in a qualitative aspect appear to hold for real multiphase systems also. By way of example, one may refer to the linear relationship between the mass transfer enhancement factor and the postcriticality parameter characterizing the deviation of the driving force or the Marangoni number from a critical value. The nonlinear analysis results provide an insight into the physical nature of such a relationship and constitute a basis for technology design involving correlations between empricial parameters and physicochemical characteristics of a system.

The role of computer modelling in the studies of nonlinear mass transfer effects under interfacial instability as carried out through numerical solution of complete equations of hydrodynamics and convective diffusion can hardly be overestimated. However, in chemical engineering design and in choosing appropriate means of process control, the computer models become sufficiently informative only in conjunction with analytical correlations.

When speaking about different approaches to the nonlinear interface instability analysis, a special role of finite-dimension Galerkin-type approximations for the governing equations should be pointed out, with an emphasis on dynamic models (which in fact are systems of nonlinear, nonstationary ordinary differential equations) that can be constructed. Results thus obtained provide information about the initiation and evolution of self-organized dissipative structures at high postcriticality [274] and about the mechanisms of "order-to-chaos" transition [275-277]. It should be noted, however, that the results of such an analysis may, generally speaking, depend on the dimension of the approximation system; therefore, one must be cautious in predicting the behaviour of a system starting from small-mode approximation results. Even so, the small-mode approximation may happen to be a good aid in understanding the interfacial convection and mass transport effects.

In exploring the interfacial instability, one may turn to irreversible thermodynamics methods [278]. The postulates of irreversible continuum thermodynamics treating the excess entropy and local potential mainly in homogeneous media can be also applied to the instability of heterophase systems with heterogeneous reactions. The thermodynamic analysis allows, in the general form, to define criteria for hydrodynamic instability and to reveal the role of perturbation parameters [278, 279]. However, the use of these criteria for a quantitative interpretation meets with serious problems since this requires an experimental determination of phenomenologic coefficients, which, given our current knowledge of interfacial instability, is quaite and arduous task. Along with the thermodynamic analysis, an interesting attempt appears to be the use of hydrodynamic fluctuation theory methods for treating effects at the interface of surfactant solutions [280].

Now we wish to briefly discuss a number of basic problems whose urgent solution is important for an advantageous use of interfacial instability in optimizing operative technologies and in designing novel chemical engineering processes. We will also touch upon certain promising research trends of immediate practical interest.

An intriguing problem in interfacial instability theory is the interaction between forced convection (or main turbulent flow) and interfacial convection and their cooperative action on the mass transfer kinetics. The data on physical mechanisms of interfacial instability (Section 4.2) and the experimental results for chemisorption kinetics (Section 4.5) have provided evidence for an ambiguous character of such an interaction. On the one hand, developed momentum pulsations in the phase bulk favor the formation of nonuniformities at the interface and contribute thereby to the interfacial instability buildup. On the other hand, the bulk turbulence can inhibit surface-tension-driven disturbances and, besides, can reduce substantially the contribution of interfacial convection to mass transfer. Therefore to be able to solve this problem, one stands in need of further studies using advanced methods of fluid mechanics and turbulence theory.

Among other hydrodynamic problems, one may mention the effect of large-scale interfacial tension gradient (as arisen, for example, from a chemical reaction, heat release, or an external field effect) on the motion and mass transport in a liquid phase. This gradient, produced because of the nonuniformity of velocity, temperature and concentration fields, is capable of changing, in a marked manner, the velocity of gas bubble, liquid drop or film motion and, consequently, the mass transfer rate, even without a loss of hydrodynamic interface stability.

For a correct parametric design of mass and heat exchangers and chemical reactions operated in an interfacial instability regime, one must know in what a manner the Marangoni effect exerts influence on the motion and mass transfer in drops and in bubbles in the presence of hydrodynamic, diffusional or thermal interactions. To that end, one must also study the effects of the interfacial instability produces on coalescence, coagulation and disintegration in the disperse phase as well as on the behaviour of a moving assemblage of drops or bubbles.

A special attention in the interfacial instability theory is focused on the study of heat effects associated with exothermal chemical reactions, phase transitions, also with heat release in extraction, absorption and other separation processes. These effects are capable of producing alterations in the physicochemical characteristics and rheologic properties of a medium. In addition, they can initiate the thermocapillary convection, an independent contributor to the mass exchange process kinetics.

Studies of interfacial instability and self-organization arising from hydrochemical interactions and the mathematical modelling of gas-liquid reactors operated under appropriate regimes cover an important field of research concerned with the macrokinetics of diffusion-controlled reactions, phase-transfer catalysis and mass transfer accompanied by chemical reactions. The evergrowing use of reactors combining chemical and mass exchange processes makes such studies still more timely.

Among nonlinear effects that exert influence on the interfacial instability mechanism, we wish to turn attention to those which may be helpful in developing the methods of control over mass transfer. This, to be able to design controls as effected through the agency of physicochemical and regime parameters it is essential, in the first place, to study the effect of temperature and concentration on the surface activity, viscosity and diffusivity, that is, the parameters which define the Marangoni number. The study of nonlinear effects is of prime importance also when the interfacial instability originates in an interaction of forces differing in their physical nature, or when the interfacial convection is subject to action by a free convection of distinct nature.

Concerning the mass transfer control methods, it should be pointed out that the relevant procedure as it has been outlined above does not yield results that would suffice for providing an effective control of a real commercial process: to this effect, the computed data must be supplemented with an additional physical information about the process in question. The reason for this resides, on the one hand, in numerous simplifications to which the formulation of a mathematical model is necessarily confined; on the other hand, despite gross simplifications made, an accepted model is often too complicated, and its processing requires the employment of a high-speed computer. The truth is also that, in many instances, even simplest mathematical models are seldom attempted for chemical processes accompanied by interfacial instability. Therefore, a strategy of further studies must proceed from two major premises of which one is the inclusion of the most essential features of a real process, and the other is, in a broader sense, the improvement of mathematical implementation of the respective model. It is felt, however, that in the actual state-of-art, of major importance appears to be the development of a general methodology aimed at the study of this novel class of theoretical and practical problems in chemical engineering.

To summarize, we wish to stress once again that systematic studies in this particular field of chemical engineering science can be undoubtedly fruitful, for both theory and practice, results which will provide a deeper insight into interfacial processes and will prove to be a good aid in resolving problems concerned with technology intensification and process control as well as in the design of highly efficient mass-transfer apparatuses and gas-liquid reactors of new generation for their use in the interfacial engineering.

References

1. *The Amundson Report on the Future of Chemical Engineering*, Chem. Eng. Progress, **83**, No. 12, 62 (1987).
2. L. E. Scriven, *Chem. Eng. Progress*, **83**, No. 12, 65 (1987).
3. L. E. Scriven and C. V. Sternling, *Nature*, **187**, 186 (1960).
4. V. G. Levich, *Physicochemical Hydrodynamics*, Prentice-Hall, Englewood Cliffs, New Jersey, 1962.
5. D. B. R. Kenning, *Appl. Mech. Rev.*, **21**, 1101 (1968).
6. V. G. Levich and V. S. Krylov, *Annual Rev. Fluid Mech.*, **1**, 293, Annual Revs. Palo Alto, California (1969).
7. J. T. Davies and E. K. Rideal, *Interfacial Phenomena*, Academic Press, New-York, London, 1963.
8. J. C. Berg, A. Acrivos, and M. Boudart, *Adv. Chem. Eng.*, **6**, 61 (1966).
9. E. L. Koschmieder, *Adv. Chem. Phys.*, **26**, 177 (1974).
10. C. Normand, Y. Pomeau, and M. G. Velarde, *Rev. Modern. Phys.*, **49**, 581 (1977).
11. *Dynamics and Instability of Fluid Interfaces* (Ed. T. S. Sørensen), Lecture Notes in Physics, vol. 105, Springer-Verlag, Berlin, Heidelberg, New York, 1979.
12. M. G. Velarde and C. Normand, *Scientific American*, **243**, 78 (1980).
13. M. G. Velarde and J. L. Castillo, in: *Convective Transport and Instability Phenomena* (Eds J. Zieper and H. Oertel), Braun Verlag, Karlsruhe, pp. 235-264, 1981.
14. H. Linde, *ibid.*, pp. 265-296.
15. Y. A. Buyevich and L. M. Rabinovich, in: *Fluid Dynamics of Interfaces*, Mir Publishers, Moscow, pp. 5-18, 1984 (In Russian).
16. C. A. Miller and P. Neogi, *Interfacial Phenomena. Equilibrium and Dynamic Effects*, Dekker, New York, Basel, 1985.
17. A. A. Golovin and L. M. Rabinovich, *Teor. Osn. Khim. Tekhn.*, **24**, 592 (1990).
18. A. A. Golovin and L. M. Rabinovich, *A.I.Ch.E. Journal* (submitted for publication).
19. A. Orell and J. W. Westwater, *A.I.Ch.E. Journal*, **8**, 350 (1962).
20. C. A. Baker, P. M. Van Buytenen, and W. J. Beek, *Chem. Eng. Sci.*, **21**, 1039 (1966).
21. H. Sawistowski, in: *Recent Advances in Liquid-Liquid Extraction* (Ed. C. Hanson), Chap. 9, Pergamon Press, Oxford, 1971.
22. H. Linde and E. Schwartz, *Z. Phys. Chem.*, **224**, 331 (1963).
23. H. Linde and B. Sehrt, *Mber. dt. Akad. Wiss. Berl.*, **7**, 341 (1965).
24. H. Linde and L. Künke, *Wärme une Stoffübertragung*, **2**, 60 (1969).

25. H. Linde, P. Schwartz, and H. Wilke, in: *Dynamics and Instability of Fluid Interfaces* (Ed. T. S. Sørensen). Lecture Notes in Physics, Springer-Verlag, Berlin, Heidelberg, New York, vol. 105, pp. 75-119, 1979.
26. W. J. Thomas and E. McK. Nicholl *Trans. Instn. Chem. Engrs.*, **47**, 325 (1969).
27. J. D. Thornton and T. J. Anderson, *Int. J. Heat Mass Transfer*, **24**, 1847 (1981).
28. P. J. Thompson, W. Baten, and R. J. Warson, *Int. Chem. Eng. Symp. Ser.* No. 88, 231 (1984).
29. K. Warmuzinski, *Inz. Chem. Proc.* (*Poland*), **6**, 147 (1985).
30. G. Varquez, G. Antorrena, and J. M. Navaza, *Ann. Quim.*, **A83**, 647 (1987).
31. A. V. Vorob'ev, V. V. Dilman, V. V. Olevskii, L. M. Rabinovich, M. G. Slin'ko, and S. P. Timashev, *Theor. Found. Chem. Eng.*, **20**, 474 (1987) (Translated from Teor. Osn. Khim. Tekhn. **20**, 766 (1986)).
32. S. P. Karlov, *Personal communication*, December, (1985).
33. L. M. Rabinovich and E. N. Ambartsumyan, *Personal communication*, May, 1988.
34. Y. Nakaike, Y. Tadenuma, T. Sato, and K. Fujinawa, *Int. J. Heat Mass Transfer*, **14**, 1951 (1971).
35. T. K. Sherwood and J. C. Wei, *Ind. Eng. Chem.*, **49**, 1030 (1957).
36. N. G. Maroudas and H. Sawistowski, *Chem. Eng. Sci.*, **19**, 919 (1964).
37. H. Sawistowski and L. I. Austin, *Chem. Ing. Techn.*, **39**, 224 (1967).
38. D. R. Olander and L. B. Reddi, *Chem. Eng. Sci.*, **19**, 67 (1964).
39. M. V. Ostrovskii, A. A. Abramson, and I. I. Barsukov, *Teor. Osn. Khim. Tekhn.*, **7**, 512 (1973).
40. R. Marr and F. Moser, *Chem. Ing. Techn.*, **47**, 619 (1975).
41. J. D. Thornton and K. H. Javed, *Int. Chem. Symp. Ser.*, No. 88, 203 (1984).
42. J. D. Thornton, T. J. Anderson, K. H. Javed, and S. K. Achwal, *A. I. Ch. E. Journal*, **31**, 1069 (1985).
43. A. A. Ermakov and V. I. Shatokhin, *Zh. Prikl. Khim.*, **57**, 2244 (1984).
44. Yu. I. Konshin and A. A. Ermakov, *Zh. Prikl. Khim.*, **59**, 2222 (1986).
45. V. I. Shatokhin, *Ph.D. Thesis*, Karpov Institute of Physical Chemistry, Moscow, 1986.
46. A. A. Ermakov, I. G. Golovina, and V. A. Danilov, *Zh. Prikl. Khim.*, **53**, 2695 (1980).
47. P. L. T. Brian, J. E. Vivian, and S. T, Mayr, *Ind. Eng. Chem. Fundam.*, **10**, 75 (1971).
48. V. Linek, *Chem. Eng. Sci.*, **27**, 627 (1972).
49. M. Hozawa, N. Komatsu, N. Imaishi, and K. Fujinawa, *J. Chem. Eng. Jap.*, **17**, 173 (1984).
50. P. L. T. Brian, J. E. Vivian, and D. C. Matiatos, *A. I. Ch. E. Journal*, **13**, 28 (1967).
51. Yu. V. Akselrod and V. V. Dilman, and Yu. V. Furmer, *Teor. Osn. Khim. Tekhn.*, **5**, 676 (1971); **7**, 683 (1973).
52. Yu. V. Akselrod and V. V. Dilman, *Teor. Osn. Khim. Tekhn.*, **14**, 837 (1980).
53. Yu. A. Akselrod, *Gas-Liquid Chemisorption Processes. Kinetics and Simulation*, Khimiya, Moscow, 1989 (in Russian).
54. E. Sada, H. Kumazawa, and M. Butt, *J. Chem. Eng. Jap.*, **10**, 487 (1977).
55. F. P. Moens, *Chem. Eng. Sci.*, **27**, 275 (1972).
56. Yu. N. Grymzin, V. A. Lotkov, and V. A. Maljusov, *Teor. Osn. Khim. Tekhn.*, **13**, 811 (1979).
57. Yu. N. Grymzin, S. Y. Kvashnin, V. A. Lotkhov, and V. A. Maljusov, *Teor. Osn. Khim. Tekhn.*, **16**, 579 (1982).
58. I. G. Misko and Yu. N. Garber, *Zh. Prikl. Khim.*, **62** , 84 (1989).
59. J. Thomson, *Phil. Mag.*, Ser. 4, **10**, 330 (1955).
60. C. Marangoni, *Sull'espansione delle goccie di un liquido galleggiante sulla superficie di altro liquido*, Fusi, Pavia, 1865.
61. C. Marangoni, *Ann. Phys.* (*Poggendorff*), **143**, 337 (1871).
62. C. Marangoni, *Nuovo Cimento (3)*, **5/6**, 239 (1871); (3), **3**, 97 (1878).
63. H. Bénard, *Rev. Gen. Sci. Pures Appl.*, **11**, 1261 (1900).

64. H. Bénard, *Ann. Chim. Phys.*, **23**, 62 (1901).
65. J. Pantaloni, R. Bailleux, J. Salen, and M. G. Velarde, *J. Non-Equilib. Thermodyn.*, **4**, 201 (1979).
66. P. Cerisier, J. Pantaloni, and C. Peréz-Garcia, *Physicochem. Hydrodynamics*, **10**, 341 (1988).
67. A. I. Vinarov, V. V. Kafarov, V. A. Bykov, E. S. Shitikov, and S. P. Karlov, *Dokl. Acad. Nauk SSSR*, **284**, 986 (1985).
68. L. M. Pismen, *Chem. Eng. Sci.*, **35**, 1950 (1980).
69. A. A. Pichugin, V. V. Tarasov, V. A. Arutiunyan, and S. V. Goryachev, in: *Intern. Solvent Extraction Conference Papers*, Nauka, Moscow, Vol. 2, pp. 101-103, 1988.
70. Z. Dagan and L. M. Pismen, *J. Colloid Interface Sci.*, **99**, 215 (1984).
71. Yu. A. Buyevich, A. V. Vyaz'min, L. M. Rabinovich, and M. G. Slin'ko, *Dokl. Phys. Chem., Proc. Akad. Sci. USSR, August*, 170 (1987) (Translated from Dokl. Akad. Nauk SSSR, **292**, 1157 (1987)).
72. E. S. Perez de Ortiz and M. A. Mendes Tatsis, in: *Intern. Solvent Extraction Conference Papers*, Nauka, Moscow, Vol. 2, pp. 104-107 (1988).
73. L. M. Rabinovich and I. A. Balikova, *Int. Chem. Eng. Symp. Ser.* No. 119, 267 (1990).
74. V. G. Levich, *Zh. Eksp. Teor. Fiz.*, **10**, 1296 (1940).
75. E. Ruckenstein, *J. Colloid Interface Sci.*, **83**, 77 (1981).
76. J. Boussinesq, *Ann. Chim. Phys.*, **29**, 349 (1913).
77. L. E. Scriven, *Chem. Eng. Sci.*, **12**, 98 (1960).
78. V. Levich, *Physicochem. Hydrodynamics*, **2**, 85 (1981).
79. B. V. Derjaguin and N. V. Churaev, *J. Coll. Interface Sci.*, **66**, 389 (1978).
80. B. V. Derjaguin and N. V. Churaev, *Wetting Films*, Nauka, Moscow, 1984 (in Russian).
81. R. Aris, *Vectors, Tensors and the Basic Equations of Fluid Mechanics*, Prentice-Hall, Englewood Cliffs, New Jersey, 1962.
82. J. C. Slattery, *Chem. Eng. Sci.*, **19**, 379 (1964).
83. J. C. Slattery, *Ind. Eng. Chem. Fundam.*, **6**, 108 (1967).
84. J. C. Slattery, *Chem. Eng. Commun.*, **4**, 149 (1980).
85. H. T. Davis and L. E. Scriven, *Adv. Chem. Phys.*, **49**, 357 (1982).
86. R. Defay, I. Prigogine, and A. Sanfeld, *J. Coll. Interface Sci.*, **58**, 498 (1977).
87. J. Kovac, *Physica*, **86A**, 1 (1977).
88. S. Chandrasekhar, *Hydrodynamics and Hydromagnetic Stability*, Clarendon Press, Oxford, 1961.
89. H. A. Stone, *Phys. Fluids*, A2, 111 (1989).
90. P. Joos and Van der Bogaert, *J. Colloid Interface Sci.*, **56**, 213 (1976).
91. E. Cussler and D. Evans, *J. Membrane Sci.*, **6**, 113 (1980).
92. A. V. Vorob'ev, M. A. Krykin, Yu. M. Popkov, and S. F. Timashev, *Teor. Osn. Khim. Tekhn.*, **19**, 675 (1985).
93. S. F. Timashev, *Physical Chemistry of Membrane Processes*, Khimiya, Moscow, 1988 (in Russian).
94. J. R. A. Pearson, *J. Fluid Mech.*, **4**, 489 (1958).
95. C. V. Sternling and L. E. Scriven, *A. I. Ch. E. Journal*, **5**, 514 (1959).
96. L. E. Scriven and C. V. Sternling, *J. Fluid Mech.*, **19**, 321 (1964).
97. E. Ruckenstein and C. Berbente, *Chem. Eng. Sci.*, **19**, 329 (1964).
98. D. A. Nield, *J. Fluid Mech.*, **19**, 341 (1964).
99. K. A. Smith, *J. Fluid Mech.*, **24**, 401 (1966).
100. J. C. Berg and A. Acrivos, *Chem. Eng. Sci.*, **20**, 737 (1965).
101. P. L. T. Brian, *A. I. Ch. E. Journal*, **17**, 765 (1971).
102. P. L. T. Brian and K. A. Smith, *A. I. Ch. E. Journal*, **18**, 231 (1972).
103. H. J. Palmer and J. C. Berg, *J. Fluid Mech.*, **51**, 385 (1972).
104. T. S. Sørensen, M. Hennenberg, and A. Sanfeld, *J. Colloid Interface Sci.*, **61**, 62 (1977).

105. M. Hennenberg, A. Sanfeld, and P. M. Bisch, *J. I. Ch. E. Journal*, **27**, 1002 (1981).
106. J. Reichenbach and H. Linde, *J. Colloid Interface Sci.*, **84**, 433 (1981).
107. S. P. Lin and H. Brenner, *J. Colloid Interface Sci.*, **85**, 59 (1982).
108. J. L. Castillo and M. G. Velarde, *J. Colloid Interface Sci.*, **108**, 264 (1985).
109. H. A. Dijkstra and A. I. van de Vooren, *Int. J. Heat Mass Transfer*, **28**, 2315 (1985).
110. A. A. Nepomnyschii and I. B. Simanovskii, *Izv. Akad. Nauk SSSR, Mekh. Zhidk. Gaza*, No. 2, 3 (1986).
111. G. Frenzel and H. Linde, *Teor. Osn. Khim. Tekhn.*, **20**, 28 (1986).
112. M. G. Velarde and X.-L. Chu, *J. Colloid Interface Sci.*, **131**, 471 (1989).
113. W. Rong and A. Sanfeld, *Physicochem. Hydrodynamics*, **11**, 21 (1989).
114. K. Warmuzinski and J. Buzek, *Chem. Eng. Sci.*, **45**, 243 (1990).
115. H. A. Dijkstra, *J. Colloid Interface Sci.*, **136**, 151 (1990).
116. S. Whitaker, *Ind. Eng. Chem. Fundam.*, **3**, 132 (1964).
117. S. Whitaker and L. O. Jones, *A. I. Ch. E. Journal*, **12**, 421 (1966).
118. B. E. Anshus and S. L. Goren, *A. I. Ch. E. Journal*, **12**, 1004 (1966).
119. B. E. Anshus and A. Acrivos, *Chem. Eng. Sci.*, **22**, 389 (1967).
120. V. Ludviksson and E. N. Lightfoot, *A. I. Ch. E. Journal*, **14**, 620 (1968).
121. K. H. Wang, V. Ludviksson, and E. N. Lightfoot, *A. I. Ch. E. Journal*, **17**, 1402 (1971).
122. S. P. Lin, *A. I. Ch. E. Journal*, **16**, 375 (1970).
123. Ch. Boyadjiev and V. S. Krylov, *Chem. Eng. J.*, **6**, 225 (1973).
124. G. M. Homsy and R. J. Gumerman, *A. I. Ch. E. Journal*, **20**, 981, 1161 (1974).
125. L. M. Rabinovich, *Zh. Prikl. Mekh. Tekhn. Fiz.*, No. 2, 91 (1976).
126. V. V. Dilman and V. I. Naydenov, *Teor. Osn. Khim. Tekhn.*, **20**, 316 (1986); **23**, 575 (1989).
127. F. Deyhimi and A. Sanfeld, *C. R. Acad. Sci. Sér. D*, **279**, 437 (1974).
128. T. S. Sørensen, M. Hennenberg, A. Steinchen, and A. Sanfeld, *J. Colloid Interface Sci.*, **56**, 191 (1976).
129. T. S. Sørensen and M. Hennenberg, in: *Dynamics and Instability of Fluid Interfaces* (Ed. T. S. Sørensen), Lecture Notes in Physics, Springer-Verlag, Berlin, Heidelberg, New York, No. 105, pp. 276-315 (1979).
130. T. S. Sørensen, *J. Chem. Soc. Faraday Trans. II*, **76**, 1170 (1980).
131. A. Cloot and G. Lebon, *Physicochem. Hydrodynamics*, **6**, 453 (1985).
132. L. M. Rabinovich and I. A. Belikova, *Inzh. Fiz. Zh.*, **58**, 972 (1990).
133. A. Ya. Dupal, V. V. Tarasov, G. A. Yagodin, and V. A. Arutiunyan, *Koll. Zh.*, **50**, 355 (1988).
134. A. A. Golovin and L. M. Rabinovich, *Fluid Dynamics, September*, 272 (1988) (Translated from Izv. Akad. Nauk SSSR, Mekh. Zhidk. Gaza, No. 2, 137 (1988)).
135. A. A. Golovin and L. M. Rabinovich, *Zh. Prikl. Mekh. Tekhn. Fiz.*, No. 5, 101 (1988).
136. L. M. Rabinovich, *Fluid Dynamics*, **13**, 816 (1979) (Translated from Izv. Akad. Nauk SSSR, Mekh. Zhidk. Gaza, No. 6, 26 (1978)).
137. J. M. Scanlon and L. A. Segel, *J. Fluid Mech.*, **30**, 149 (1967).
138. S. P. Lin, *Progr. Heat and Mass Transfer Proc., Int. Symp. on Two-Phase Syst.*, **6**, 263 (1972).
139. S. S. Brodskii and A. M. Golovin, *Zh. Prikl. Mekh. Tekhn. Fiz.*, No. 2, 49 (1972).
140. S. H. Davis and G. M. Homsy, *J. Fluid Mech.*, **98**, 527 (1980).
141. J. P. Patzer and G. M. Homsy, *Phys. Fluids*, **24**, 567 (1981).
142. J. L. Castillo and M. G. Velarde, *J. Fluid Mech.*, **125**, 463 (1982).
143. S. Rozenblat, S. H. Davis, and G. M. Homsy, *J. Fluid Mech.*, **120**, 91 (1982).
144. S. Rozenblat, G. M. Homsy, and S. H. Davis, *J. Fluid Mech.*, **120**, 123 (1982).
145. G. Lebon and A. Cloot, *Acta Mechan.*, **43**, 141 (1982).
146. Yu. A. Buyevich, *Inzh. Fis. Zh.*, **48**, 230 (1985).
147. P. L. Garcia-Ybarra, J. L. Castillo, and M. G. Velarde, *Phys. Lett. A* **122**, 107 (1987).
148. K.-L. Ho and H.-C. Chang, *A. I. Ch. E. Journal*, **34**, 705 (1988).

149. X. L. Chu and M. G. Velarde, *Phys. Lett.* A **136**, 126 (1989).
150. A. Oron and P. Rosenau, *Phys. Rev.*, A **39**, 2063 (1989).
151. Yu. A. Buyevich, A. V. Vyaz'min and L. M. Rabinovich, *Dokl. Phys. Chem., Proc. Acad. Sci. USSR, January, 584 (1990)* (Translated from Dokl. Akad. Nauk SSSR, **307**, 629 (1989)).
152. Yu. A. Buyevich, A. V. Vyaz'min, and L. M. Rabinovich, *Dokl. Akad. Nauk SSSR*, **307**, 1135 (1989).
153. L. D. Landau, *Dokl. Akad. Nauk SSSR*, **44**, 311 (1944).
154. L. D. Landau and E. M. Lifshitz, *Fluid Mechanics*, Addison Wesley, Reading, Mass., 1959.
155. E. Hopf, *Comm. Pure Appl. Math.*, **2**, 303 (1948).
156. J. T. Stuart, *J. Fluid Mech.*, **9**, 353 (1960).
157. J. Watson, *J. Fluid Mech.*, **9**, 371 (1960).
158. Yu. A. Buyevich and S. V. Kudimov, *Inzh. Fiz. Zh.*, **54**, 406 (1988).
159. Yu. B. Ponomarenko, *Prikl. Math. Mekh.*, **28**, 688 (1964).
160. H. Wilke, *Z. Angew. Math. Mech.*, **9**, 437 (1980).
161. N. Imaishi, Y. Suzuki, M. Hozawa, and K. Fujinawa, *Kagaku Kogaku Ronbunshu*, **8**, 127 (1982).
162. N. Imaishi, M. Hozawa, K. Fujinawa, and Y. Suzuki, *Int. Chem. Eng.*, **23**, 466 (1983).
163. E. Chang and W. Wilcox, *J. Cryst. Growth*, **28**, 8 (1975).
164. P. A. Clark and W. Wilcox, *J. Cryst. Growth*, **50**, 461 (1980).
165. N. D. Kazarinoff and J. S. Wilkowski, *Phys. Fluids*, A **1**, 625 (1989).
166. A. A. Nepomnyschii and I. B. Simanovskii, *Izv. Akad. Nauk SSSR, Mekh. Zhidk. Gaza*, No. 3, 175 (1984).
167. D. Villers and J. K. Platten, *Physicochem. Hydrodynamics*, **6**, 435 (1985).
168. H. A. Dijkstra, *Physicochem. Hydrodynamics*, **10**, 493 (1988).
169. V. I. Polezhaev, A. V. Bune, N. A. Verezub et al., *Mathematical Modelling of Convective Heat and Mass Transfer on the Basis of Navier-Stokes Equations*, Nauka, Moscow, 1987 (in Russian).
170. S. V. Ermakov and A. I. Feonychev, in: *Hydromechanics, Heat and Mass Transfer in Material Processing* (Eds V. S. Avduevskii and V. I. Polezhaev), Nauka, Moscow, pp. 286-292, 1990.
171. G. G. Elenin, I. S. Kalachinskaya, and L. M. Rabinovich, *Fluid Dynamics, May, 827 (1988)* (Translated from Izv. Akad. Nauk SSSR, Mekh. Zhidk. Gaza, No. 6, 4 (1987)).
172. I. A. Belikova, B. P. Gerasimov, I. S. Kalachinskaya, and L. M. Rabinovich, *Teor. Osn. Khim. Tekhn.*, **24**, 30 (1990).
173. B. P. Gerasimov, T. G. Elizarova, I. S. Kalachinskaya et al., *Preprint No. 65*, Keldysh Institute of Applied Mathematics, USSR Academy of Sciences, 1985.
174. E. N. Lorenz, *J. Atmos. Sci.*, **20**, 130 (1963).
175. B. Saltzman, *J. Atmos. Sci.*, **19**, 329 (1962).
176. D. Ruelle and F. Takens, *Commun. Math. Phys.*, **20**, 167 (1971).
177. D. Ruelle, in: *Mathematical Problems in Theoretical Physics*, Lecture Notes in Physics, Springer-Verlag, Berlin, Heidelberg, New York, No. 80, p. 341, 1978.
178. A. N. Murovtsev, L. M. Rabinovich, and M. G. Slin'ko, in: *USSR Symp. on Macrokinetics and Chem. Gas Dynamics Proc.*, Chernogolovka, Vol. 1(2), p. 78, 1984 (in Russian).
179. G. G. Elenin, I. S. Kalachinskaya, and S. V. Solomatin, *Differ. Uravneniya*, **23**, 1169 (1987).
180. A. A. Golovin and L. M. Rabinovich, *Multiphase Science and Technology — An International Series* (submitted for publication).
181. A. A. Ermakov, Yu. A. Konshin, and V. I. Nazarov, *Zh. Fiz. Khim.*, **51**, 2151 (1977).
182. A. A. Ermakov, L. M. Rabinovich, and M. G. Slin'ko, *Dokl. Chem., Tekhn., Proc. Acad. Sci. USSR*, **301-303**, 100 (1988) (Translated from *Dokl. Akad. Nauk SSSR*, **303**, 429 (1988)).
183. Yu. A. Konshin, N. I. Parkhomenko, and A. A. Ermakov, *Zh. Prikl. Khim.*, **53**, 1975 (1980).
184. Yu. A. Konshin, A. A. Ermakov, and Z. P. Melezh, *Zh. Prikl. Khim.*, **53**, 1193 (1980).

4 Modelling of Mass Transport Processes

185. V. I. Shatokhin, A. A. Ermakov, and L. V. Kamneva, *Zh. Prikl. Khim.*, **54**, 1095 (1981).
186. V. I. Shatokhin, A. A. Ermakov, and V. A. Kirillov, *Zh. Prikl. Khim.*, **57**, 1556 (1984).
187. A. A. Ermakov and Yu. A. Konshin, *Zh. Prikl. Khim.*, **62**, 174 (1989).
188. V. I. Shatokhin, A. A. Ermakov, and M. Z. Maksimenko, *Zh. Prikl. Khim.*, **57**, 2512 (1984).
189. W. Fritz and T. Popova, *Chem. Ing. Tech.*, **42**, 1004 (1970).
190. H. Linde and B. Sehrt, *Z. Phys. Chem.*, **231**, 151 (1966).
191. H. Kroepellin, H. J. Neuman, and E. Prött, *Erdöl und Köhle*, **12**, 344 (1959).
192. A. A. Golovin, N. I. Polomarchuk, and A. A. Ermakov, *Khim. Prom.*, No. 12, 740 (1988).
193. G. Anderes, *Chem. Ing. Tech.*, **34**, 597 (1962).
194. M. V. Ostrovskii, G. T. Frumin, A. A. Abramson, *Zh. Prikl. Khim.*, **16**, 803 (1968).
195. S. K. Kalugina, M. V. Ostrovskii, and A. A. Abramson, *Zh. Prikl. Khim.*, **46**, 1378 (1973).
196. M. V. Ostrovskii and N. V. Golyakova, *Teor. Osn. Khim. Tekhn.*, **9**, 643 (1975).
197. H. Takeuchi and Y. Humata, *Int. Chem. Eng.*, **17**, 468 (1977).
198. I. I. Barsukov and S. K. Kalugina, *Teor. Osn. Khim. Tekhn.*, **17**, 448 (1988).
199. V. V. Dilman, Yu. V. Akselrod, and F. M. Khutoryansky, *Khim. Prom.*, No. 9, 693 (1976).
200. V. V. Dilman, Yu. V. Akselrod, and F. M. Khutoryansky, *Teor. Osh. Khim. Tekhn.*, **11**, 11 (1977).
201. N. Imaishi, Y. Suzuki, M. Hozawa, and K. Fujinawa, *Kagaku Kogaku Ronbunshu*, **6**, 585 (1980).
202. W. Clark and C. J. King, *A. I. Ch. E. Journal*, **16**, 64 (1970).
203. H. Sawistowski and G. E. Goltz, *Trans. Inst. Chem. Eng.*, **41**, 174 (1963).
204. H. Sawistowski and B. R. James, *Chem. Ing. Techn.*, **35**, 175 (1963).
205. H. Linde and D. Thiessen, *Z. Phys. Chem.*, **227**, 223 (1964).
206. H. Linde and K. Winkler, *Z. Phys. Chem.*, **225**, 223 (1964).
207. H. Linde and K. Winkler, *Z. Phys. Chem.*, **230**, 207 (1965).
208. E. Ruckenstein, *Int. J. Heat Mass Transfer*, **11**, 1753 (1968).
209. L. M. Rabinovich and M. G. Slin'ko, in: *Intern. Solvent Extraction Conference Papers*, Nauka, Moscow, Vol. 2, pp. 91-96, 1988.
210. F. K. Tsou, E. M. Sparrow, and R. J. Goldstein, *Int. J. Heat Mass Transfer*, **10**, 219 (1967).
211. L. M. Pikkov and L. M. Rabinovich, *Teor. Osn. Khim. Tekhn.*, **23**, 166 (1989).
212. A. Sethy and H. T. Gullnan, *A. I. Ch. E. Journal*, **21**, 575 (1975).
213. N. Imaishi, M. Hozawa, and K. Fujinawa, *J. Chem. Eng. Jap.*, **9**, 499 (1976).
214. M. Rechakova, V. Rod, and V. Hancil, *Coll. Czech. Chem. Commun.*, **43**, 582 (1978).
215. M. Rechakova and V. Rod, *Coll. Czech. Chem. Commun.*, **44**, 2780 (1979).
216. A. N. Kolmogorov, *Dokl. Akad. Nauk SSSR*, **30**, 299 (1941).
217. A. A. Golovin and L. M. Rabinovich, in: *Intern. Solvent Extraction Conference Papers*, Nauka, Moscow, Vol. 2, pp. 97-100, 1988.
218. A. A. Golovin, A. A. Ermakov, and L. M. Rabinovich, *Dokl. Chem. Techn.*, Proc. Acad. Sci. USSR, *January-June, 28 (1988)* (Translated from *Dokl. Akad. Nauk SSSR*, **305**, 921 (1989)).
219. A. A. Golovin and L. M. Rabinovich, in: *10th Int. Cong. Chem. Eng. (CHISA 90)*, Summaries, Praha, Vol. 5, D4.17, 1990.
220. A. A. Golovin, N. I. Polomarchuk, and A. A. Ermakov, *Teor. Osn. Khim. Tekhn.*, **24**, 450 (1990).
221. D. Ilkovič, *Coll. Czech. Chem. Commun.*, **6**, 498 (1934).
222. A. A. Golovin, *Ph. D. Thesis*, Karpov Institute of Physical Chemistry, Moscow, 1989.
223. W. Nitschke, P. Schwartz, V. S. Krylov, and H. Linde, *Teor. Osn. Khim. Tekhn.*, **19**, 311 (1985).
224. W. Nitschke, P. Schwartz, V. S. Krylov, and H. Linde, *ibid.*, 672 (1985).
225. W. Nitschke, P. Schwartz, V. S. Krylov, and H. Linde, *ibid.*, 729 (1985).
226. W. Dalle Vedove and A. Sanfeld, *J. Colloid Interface Sci.*, **84**, 318, 328 (1981).

227. W. Dalle Vedove, A. R. Marquez Garcia, and A. Sanfeld, *J. Colloid Interface Sci.*, **95**, 299 (1983).
228. E. Ruckenstein and C. Berbente, *Chem. Eng. Sci.*, **25**, 475 (1970).
229. E. Ruckenstein, *Canad. J. Chem. Eng.*, **49**, 62 (1971).
230. P. I. Geshev and A. M. Lapin, *Zh. Prikl. Mekh. Tekhn. Fiz.*, No. 6, 106 (1983).
231. A. V. Vyaz'min and L. M. Rabinovich, in: *Proc. USSR Symp. Heat Mass Equipment*, Moscow, p. 119, 1988.
232. E. Ruckenstein, *Chem Eng. Sci.*, **23**, 363 (1968).
233. E. Ruckenstein, *Chem. Eng. Sci.*, **25**, 1699 (1970).
234. P. I. Geshev, *Zh. Prikl. Mech. Tekhn. Fiz.*, No. 3, 104 (1987).
235. V. V. Dilman and L. M. Rabinovich, *Theor. Found. Chem. Eng.*, September, 44 *(1979)* (Translated from *Teor. Osn. Khim. Tekhn.*, **13**, 54 (1979)).
236. Ch. Boyadjiev and V. Beschkov, *Mass Transfer in Liquid Film Flows*, Publ. House Bulg. Acad. Sci., Sofia, 1984.
237. Ch. Boyadjiev and P. Mitev, *Chem. Eng. J.*, **14**, 225 (1977).
238. V. S. Krylov, V. A. Malusov, W. Nitschke, and V. A. Lotkhov, *Teor. Osn. Khim. Tekhn.*, **19**, 12 (1985).
239. E. Ruckenstein, *Inzh. Fiz. Zh.*, **7**, 116 (1964).
240. E. Ruckenstein, *Chem. Eng. Sci.*, **20**, 853 (1965).
241. Yu. A. Buyevich and L. M. Rabinovich, *Inzh. Fiz. Zh.*, **36**, 32 (1979).
242. A. A. Golovin and L. M. Rabinovich, *J. Eng. Phys. April*, 1156 (1988) (Translated from *Inzh. Fiz. Zh.*, **53**, 593 (1987)).
243. Yu. A. Buyevich, A. V. Vyaz'min, and L. M. Rabinovich, *Dokl. Akad. Nauk SSSR*, **312**, 904 (1990).
244. A. A. Ermakov, V. I. Shatokhin, A. A. Golovin, and L. M. Rabinovich, in: *Proc. Symp. Salavatn Scientists*, Ufa, p. 31, 1988.
245. A. A. Ermakov, V. I. Nazarov, and M. G. Slin'ko, *Dokl. Akad. Nauk SSSR*, **308**, 914 (1989).
246. L. Onsager, *Ann. N.Y. Acad. Sci.*, **46**, 241 (1945).
247. A. A. Ermakov, V. I. Nazarov, and V. S. Krylov, *Zh. Prikl. Khim.*, **59**, 2337 (1986).
248. W. J. Thomas and E. McK. Nicholl, *Chem. Eng. Sci.*, **22**, 1877 (1967).
249. P. V. Dankverts and A. Tawares da Silva, *Chem. Eng. Sci.*, **22**, 1513 (1967).
250. R. Kronig and I. Brink, *Appl. Sci. Res.*, **A2**, 142 (1950).
251. A. E. Handlos and T. Baron, *A. I. Ch. E. Journal*, **3**, 127 (1957).
252. L. Boyadzhiev, in: *Intern. Solvent Extraction Conference Papers*, Nauka, Moscow, Vol. 1, pp. 56-60, 1988.
253. S. Yu. Ivakhno, A. V. Afanasiev, and G. A. Yagodin, *Membrane Extraction of Inorganic Materials*, VINITI AN SSSR, Moscow, 1985 (in Russian).
254. V. V. Tarasov and A. A. Pichugin, *Uspekhi Khim.*, **57**, 990 (1988).
255. L. I. Boguslavskii, *Bioelectrochemical Phenomena and Interface*, Nauka, Moscow, 1978 (in Russian).
256. V. V. Tarasov, G. A. Yagodin, and A. A. Pichugin, *Kinetics of Inorganic Materials Extraction*, VINITI AN SSSR, Moscow, 1984 (in Russian).
257. G. A. Yagodin and V. V. Tarasov, *Solvent Extraction and Ion Exchange*, No. 2, 1 (1984).
258. V. E. Serga, L. D. Kulikova, and A. N. Popov, *Izv. Latvian Akad. Nauk. Ser. Khim.*, No. 2, 207 (1986).
259. V. A. Arutiunyan, *Ph. D. Thesis*, Mendeleev Institute of Chemical Engineering, Moscow, 1988.
260. A. A. Pichugin, V. V. Tarasov, and V. A. Arutiunyan, in: *Proc. USSR Symp. on Current Problems of Chemical Engineering*, Krasnoyarsk, Vol. 3, p. 182, 1986 (in Russian).
261. P. V. Polyakov, L. A. Isaeva, and Yu. G. Mikhalev, *Elektrokhimiya*, **16**, 1132 (1980).
262. A. N. Frumkin, E. V. Stenina, and N. V. Fedorovich, *Elektrokhimiya*, **6**, 1572 (1970).

4 Modelling of Mass Transport Processes

263. A. N. Frumkin, N. V. Fedorovich, and E. V. Stenina, in: *Elektrokhimiya*, VINITI AN SSSR, Vol. 13, p. 5, 1978 (in Russian).
264. R. Aogaki, K. Kitazawa, K. Fueki, and T. Mukaibo, *Electrochimica Acta*, **23**, 867 (1978).
265. V. V. Nechiporuk, N. N. Turash, and I. L. Elgurt, Electrokhimiya, **27**, 124 (1991).
266. P. Hammerschmid, *Stahl und Eisen*, **107**, 61 (1987).
267. S. P. Raspopov and A. G. Sukhodolskii, *Kvant. Elektronika*, **14**, 1709 (1987).
268. H. Wilke and W. Löser, *Crystal Res. Technol.*, **18**, 825 (1983).
269. V. I. Polezhaev, in: *Hydromechanics, Heat and Mass Transfer in Material Processing* (Eds Avduevskii and V. I. Polezhaev), Nauka, Moscow, pp. 9-26, 1990 (in Russian).
270. R. S. Subramanian, *Adv. Space Res.*, **3**, 145 (1983).
271. B. A. Bezuglyi, D. P. Krindach, and N. S. Mayorov, *Zh. Tekhn. Fiz.*, **52**, 2415 (1982).
272. B. A. Bezuglyi, *Zh. Tekhn. Fiz.*, **53**, 927 (1983).
273. N. K. Nelson and J. C. Berg, *Chem. Eng. Sci.*, **37**, 1067 (1982).
274. G. Nicolis and I. Prigogine, *Self-Organization in Non-Equilibrium Systems*, Wiley, New York, London, Sydney, Toronto, 1977.
275. H. Haken, *Synergetics. An Introduction*, Springer-Verlag, Berlin, Heidelberg, New York, 1983.
276. H. Haken, *Order and Chaos. Springer Series in Synergetics*, Springer-Verlag, Berlin, Heidelberg, New York, Vol. 17, 1982.
277. M. I. Rabinovich and M. M. Suschik, *Uspekhi Fiz. Nauk*, **160**, 3 (1990).
278. P. Glansdorff and I. Prigogine, *Thermodynamic Theory of Structure, Stability and Fluctuations*, Wiley-Interscience, London, New York, Sydney, Toronto, 1971.
279. *Modern Theory of Capillarity* (Eds. A. I. Rusanov and F. I. L. Gudrich), Khimiya, Moscow, 1980 (in Russian).
280. B. A. Noskov, *Koll. Zh.*, **52**, 84 (1990).

5 Hierarchical System Studies of Energy-Saving Schemes for Mixture Separation

V. M. Platonov, A. G. Kolokol'nikov, L. V. Baburina, and G. A. Meskhi

Karpov Institute of Physical Chemistry, Obukha 10, Moscow 103064, USSR

5.1 Introduction

Distillation is the most widespread chemical engineering process for mixture separation employed at temperatures of about 150°C (petroleum refining) to those of liquid helium and hydrogen. As reported [1], in the US national industry, distillation accounts for over 2% (or 2×10^{18} J) of the total of energy expenditure. For this reason, the methodology of approach to the design of energy-saving schemes for mixture separation comes to the foreground of urgent problems of chemical engineering. The solution of process design problems can be achieved in a large variety of ways; however, it is primarily defined by the physico-chemical properties of a mixture of interest. A convenient approach is to classify mixtures according to the degree of their nonideality: (i) ideal and pseudo-ideal mixtures; (ii) mixtures with a single α-line; (iii) azeotropic homogeneous mixtures; and (iv) azeotropic heterogeneous mixtures. In terms of intermolecular interaction type, the mixtures are divided into nonelectrolytes, electrolytes, and mixed solutions. In the general case, for the synthesis of an optimum separation scheme, a number of hierarchical levels must be considered; these are: (1) molecular level; (2) supramolecular level; (3) level of limiting distillation regimes: (4) level of operating regimes; (5) distillation complexes; (6) synthesis of an optimum scheme with heat recovery. We consider results that have been obtained from studying the above-mentioned levels; preliminarily, we focus on avoidable energy losses in distillation of solutions.

5.2 Analysis of Exergy Losses in Distillation

In analysing a distillation process, it is appropriate to distinguish its effective thermodynamic irreversibility which is due to the driving forces inherent in the process, and its lost irreversibility associated with mixing the process streams at different temperatures, concentrations, or pressures. The most rational solution is to provide for a uniform distribution of driving forces throughout the height of an apparatus; in doing so, the mixing effects must be reduced to a minimum.

The minimal thermodynamic work spent on separation of a solution by distillation carried out reversibly in an infinite nonadiabatic column is

$$A_{min} = \Delta H - T_f \Delta S \tag{1}$$

where
$$\Delta H = H(T_d) + H(T_w) - H(T_f)$$
$$\Delta S = DS_d + WS_w - FS_f$$
$$S_d = -R \sum X_d \ln \gamma X_d, \quad \text{etc.}$$

If relation (1) is considered at ambient temperature (T_0), then, as shown in [2], A_{min} includes also the work performed on reversible cooling of the mixture from T_0 down to T_f.

A minimum work spent on mixture separation by multicomponent distillation is obtained if only one component is drawn off from the column section. In the feed zone, the exergy loss is reduced to a minimum in mixing the streams inside the column, since their temperature and composition in the mixture to be separated and inside the column are the same (in the respective—vapour or liquid—phases).

This mode of mixture separation (as an alternative to separation of components contiguous by fugacity) allows the substantial reduction of the exergy consumption, and also suggests a novel approach to the design of innovatory schemes (complexes) based on reversible stream mixing.

The flow rates that provide for a reversible mixing at the feed zone are derived [3] from the material balance and phase equilibrium equations for the respective sections:

$$\frac{L^f}{F} \leqslant \frac{\alpha_p^f}{\sum_{i=1}^{p} \alpha_i^f x_i^f} \cdot \frac{V'^{(f)}}{F} \leqslant \frac{\sum_{i=1}^{p} \alpha_i^f x_i^f}{\alpha_1^f} \tag{2}$$

A simultaneous and complete depletion of the most volatile components is characterized by the equality sign in the above equation. The violation of the inequality leads to a mixing irreversibility in the feed zone. To be noted, relation (2) holds true for any mixtures, including those with $\alpha = \textit{Var}$, since the constant-composition zones are contiguous to the feed, and the operating line passes through the liquid-vapour node of the mixture to be separated.

By applying the material balance and phase equilibrium equations to the sections at the column ends (where one of the mixture components is completely depleted), the limiting flow rates are obtained:

$$\lim_{x_p \to 0} \frac{L}{V} = \frac{\alpha_p}{\sum_{i=1}^{p-1} \alpha_i x_i} \neq 0; \quad \lim_{x_1 \to 0} \frac{V'}{L'} = \frac{\sum_{i=2}^{p} \alpha_i x_i}{\alpha_1} \neq 0 \qquad (3)$$

Relations (3) determine the reflux and vapour flow rates at the column ends. These flow rates can be provided for either by large differential abstraction of heat from the infinite rectifying section (at the expense of the heat supplied to the stripping section), or by reflux draw-off from the sequent column.

Thus, in infinite reversible-distillation columns, the profile of heat abstraction (or heat supply) at the column ends is not reducible to zero.

An advantageous reflux draw-off from the next distillation column is possible only if the configuration of the scheme is characteristic of reversible distillation. Finally, in such distillation schemes, an additional supply of heat is not needed at all at the side draw-off of intermediate-volatile products—a feature disregarded by Grunberg [4]. All these facts—the avoided exergy losses in the feed zone, at the column ends under supply of external energy, in the intermediate draw-off, also the reduction of heat loss throughout the section length owing to a more uniform driving-force distribution—result in a substantial (30-70%) energy saving [5, 6, 7].

5.3 Molecular Level

From a methodical standpoint, the concept of a theoretical (equilibrium) separation stage in distillation modelling appears to be the most rational approach to an optimal structure of separation scheme. A knowledge of the equilibrium properties of a mixture to be separated allows one to specify all the thermodynamic constraints in obtaining the desired products (azeotropic separating manifolds) and to determine intimate physico-chemical characteristics of the mixture, such as α-manifolds and K-lines. In addition, knowing the mixture composition at the theoretical stage (theoretical plate), one can compare the energy consumption for separation schemes of different structure, since the vapour-liquid equilibrium at phase interface is a major factor preventing any large deviation of the distillation trajectory from that defined by the theoretical plate concept. This is explained by the principle of minimum entropy production under various mass-transfer conditions.

5.3.1 Binary Systems

a. Wilson Equation

In describing the phase behaviour of a homogeneous nonelectrolyte solution, of common use is the equation suggested [8] for calculating the activity coefficients. As has been shown [9], this equation provides an adequate characterization of all types of liquid-phase nonidealities (ln γ vs. X relations) for binary solutions and satisfies the Gibbs-Duhem equation

$$\sum_{i=1}^{N} x_i d \ln \gamma_i = 0 \qquad (4)$$

Shown in Fig. 5.1 are eight types of ln γ vs. X diagrams, each characterized by an appropriate system of inequalities.

1st type

$$\begin{cases} \left.\dfrac{\partial \ln \gamma_1}{\partial x_1}\right|_{x_1=0} < 0 \\ \left.\dfrac{\partial \ln \gamma_2}{\partial x_1}\right|_{x_1=1} > 0 \end{cases} \quad \text{or} \quad \begin{cases} \Lambda_{21}^2 - \dfrac{2}{\Lambda_{12}} + 1 < 0 \\ \Lambda_{12}^2 - \dfrac{2}{\Lambda_{21}} + 1 < 0 \end{cases}$$

2nd type

$$\begin{cases} \left.\dfrac{\partial \ln \gamma_1}{\partial x_1}\right|_{x_1=0} > 0 \\ \ln \gamma_1|_{x_1=0} > 0 \end{cases} \quad \text{or} \quad \begin{cases} \Lambda_{21}^2 - \dfrac{2}{\Lambda_{12}} + 1 > 0 \\ -\ln \Lambda_{12} - \Lambda_{21} + 1 > 0 \end{cases}$$

2'nd type

$$\begin{cases} \left.\dfrac{\partial \ln \gamma_2}{\partial x_1}\right|_{x_1=1} < 0 \\ \ln \gamma_2|_{x_1=1} > 0 \end{cases} \quad \text{or} \quad \begin{cases} \Lambda_{12}^2 - \dfrac{2}{\Lambda_{21}} + 1 > 0 \\ -\ln \Lambda_{21} - \Lambda_{12} + 1 > 0 \end{cases}$$

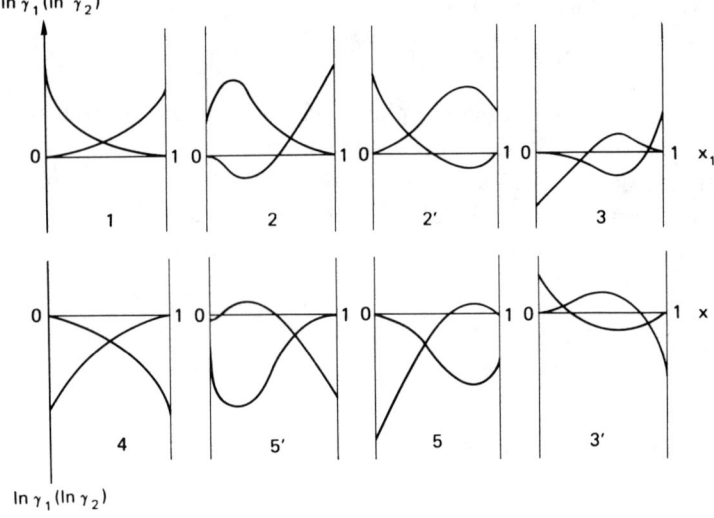

Fig. 5.1 Types of nonideality characteristic of binary mixtures

3rd type

$$\begin{cases} \ln \gamma_1|_{x_1=0} < 0 \\ \ln \gamma_2|_{x_1=1} > 0 \end{cases} \quad \text{or} \quad \begin{cases} -\ln \Lambda_{12} - \Lambda_{21} + 1 < 0 \\ -\ln \Lambda_{21} - \Lambda_{12} + 1 > 0 \end{cases}$$

3'rd type

$$\begin{cases} \ln \gamma_1|_{x_1=0} > 0 \\ \ln \gamma_2|_{x_1=1} < 0 \end{cases} \quad \text{or} \quad \begin{cases} -\ln \Lambda_{12} - \Lambda_{21} + 1 > 0 \\ -\ln \Lambda_{21} - \Lambda_{12} + 1 < 0 \end{cases} \quad (5)$$

4th type

$$\begin{cases} \dfrac{\partial \ln \gamma_1}{\partial x_1}\bigg|_{x_1=0} > 0 \\ \dfrac{\partial \ln \gamma_2}{\partial x_1}\bigg|_{x_1=1} < 0 \end{cases} \quad \text{or} \quad \begin{cases} \Lambda_{21}^2 - \dfrac{2}{\Lambda_{12}} + 1 > 0 \\ \Lambda_{12}^2 - \dfrac{2}{\Lambda_{21}} + 1 > 0 \end{cases}$$

5th type

$$\begin{cases} \dfrac{\partial \ln \gamma_2}{\partial x_1}\bigg|_{x_1=1} > 0 \\ \ln \gamma_2|_{x_1=1} < 0 \end{cases} \quad \text{or} \quad \begin{cases} \Lambda_{12}^2 - \dfrac{2}{\Lambda_{21}} + 1 < 0 \\ -\ln \Lambda_{21} - \Lambda_{12} + 1 < 0 \end{cases}$$

5'th type

$$\begin{cases} \dfrac{\partial \ln \gamma_1}{\partial x_1}\bigg|_{x_1=0} < 0 \\ \ln \gamma_1|_{x_1=0} < 0 \end{cases} \quad \text{or} \quad \begin{cases} \Lambda_{21}^2 - \dfrac{2}{\Lambda_{12}} + 1 < 0 \\ -\ln \Lambda_{12} - \Lambda_{21} + 1 < 0 \end{cases}$$

Type 1 is characterized by a positive deviation from Raoult's law, and type 4, by a negative deviation over the whole concentration range. The other types—2, 2', 3, 3', 5, 5'—exhibit a mixed $\ln \gamma$ vs. X behaviour. The domains corresponding to these $\ln \gamma$ vs. X relations (see Fig. 5.1), plotted in the $\Lambda_{12} - \Lambda_{21}$ space, are shown in Fig. 5.2.

As a global parameter (pressure or temperature) of a binary system is allowed to vary, the following changes, in accordance with the Vrevsky rule II [10], can be observed:

1. The system exhibiting a minimum-boiling azeotrope (maximum-pressure azeotrope) undergoes transition to a zeotropic state via the stage of a positive tangential azeotrope which thus becomes the high-volatile component.
2. The system exhibiting a maximum-boiling azeotrope (minimum-pressure azeotrope) undergoes transition to a zeotropic state via the stage of a negative tangential azeotrope which thus becomes the low-volatile component.
3. The system exhibiting two boundary azeotropes undergoes transition to a zeotropic state via the stage of formation of two tangential azeotropes corresponding to pure components of the solution. This case of a bifurcate transition takes place as the minimum-boiling azeotrope (maximum pressure) is located on the side of high-volatile components,

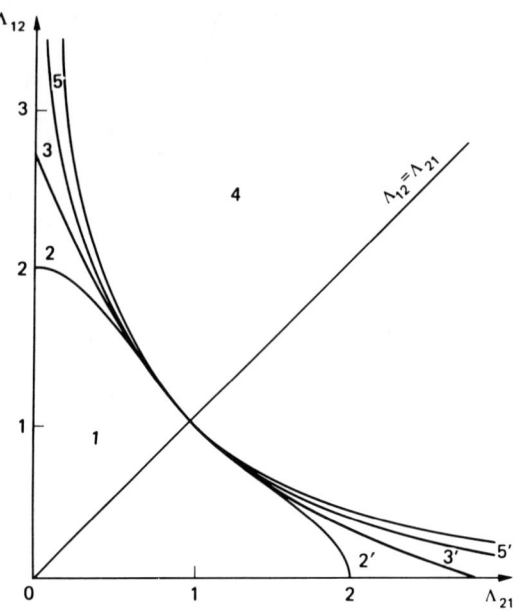

Fig. 5.2 Domains for existence of Wilson parameters corresponding to different types of nonideality in binary systems, 1, 2, 2′, 3, 3′, 4, 5, 5′ denote the nonideality types for binary systems

and the maximum-boiling azeotrope (minimum pressure) is located on the side of low-volatile component.

4. The system possessing two internal azeotropes undergoes transition to a zeotropic state through the stage of formation of an internal tangential azeotrope $\left(\text{a degenerate point at } \frac{\partial P}{\partial x_1} = \frac{\partial^2 P}{\partial x_2} = 0\right)$. Such a bifurcate transition takes place as the minimum-boiling azeotrope (maximum pressure) is located on the side of the low-volatile component, and the maximum-boiling azeotrope (minimum pressure) is located on the side of the high-volatile component.

In the parametric $\Lambda_{12} - \Lambda_{21}$ space of the Wilson equation, domains can be singled out which, at a given vapour pressure ratio of pure component (P_1^o/P_2^o) can give one azeotrope (positive or negative), two azeotropes (internal or boundary), or none altogether. The relative area of each domain characterizes the probability for formation of a binary azeotrope of respective type.

5 System Studies of Energy-Saving Schemes

For binary systems with an ideal vapour phase at $T = \text{const}$ and $P_2^o > P_1^o$, five parametric domains have been defined by graphical solution of the above inequalities:

(1) domain reproducing systems with one negative azeotrope,

$$\left(\frac{\partial P}{\partial x_2}\bigg|_{x_2=1} > 0, \quad \frac{\partial P}{\partial x_2}\bigg|_{x_2=0} < 0\right)$$

(2) domain reproducing systems with one positive azeotrope,

$$\left(\frac{\partial P}{\partial x_2}\bigg|_{x_2=1} < 0; \quad \frac{\partial P}{\partial x_2}\bigg|_{x_2=0} > 0\right)$$

(3) domain reproducing systems with two boundary azeotropes,

$$\left(\frac{\partial P}{\partial x_2}\bigg|_{x_2=1} < 0; \quad \frac{\partial P}{\partial x_2}\bigg|_{x_2=0} < 0\right)$$

(4) domain reproducing systems that contain two internal azeotropes

$$\left(\frac{\partial P}{\partial x_2}\bigg|_{x_2=1} > 0; \quad \frac{\partial P}{\partial x_2}\bigg|_{x_2=0} > 0\right)$$

$$\left(\frac{\partial^2 P}{\partial x_2^2}\bigg|_{0 \leqslant x_2 \leqslant 1} = 0; \quad \frac{\partial P}{\partial x_2}\bigg|_{0 \leqslant x_2 \leqslant 1} < 0\right)$$

(5) domain reproducing systems with no azeotrope

$$\left(\frac{\partial P}{\partial x_2}\bigg|_{x_2=1} > 0; \quad \frac{\partial P}{\partial x_2}\bigg|_{x_2=0} > 0\right)$$

$$\left(\frac{\partial^2 P}{\partial x_2^2}\bigg|_{0 \leqslant x_2 \leqslant 1} = 0; \quad \frac{\partial P}{\partial x_2}\bigg|_{0 \leqslant x_2 \leqslant 1} > 0\right)$$

A division of the parametric Λ_{12}-Λ_{21} space into respective domains at vapour pressures $P_1^o = 1.0$ and $P_2^o = 1.058$ is represented in Fig. 5.3. Curve a is defined by equation $\Lambda_{12} = \exp(1 - \Lambda_{21} - \ln \gamma_1)$, curve b, by equation $\Lambda_{21} = \exp(1 - \Lambda_{12} - \ln \gamma_2)$, and the curve c corresponds to a locus of points of the internal tangential azeotropes at different mixture concentrations such that the first and second derivatives of vapour pressure with respect to concentration become zero, or, which is the same, the system of equations below

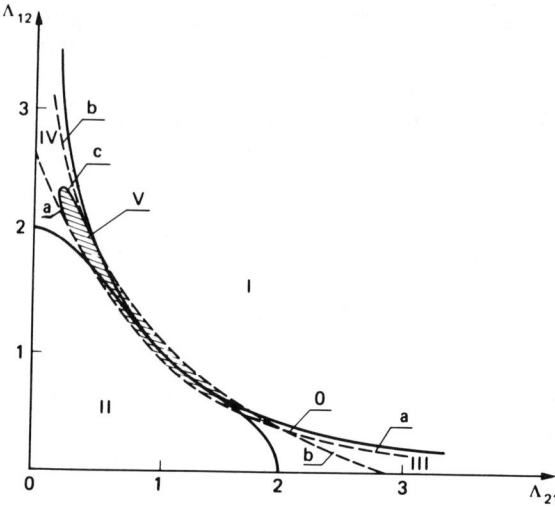

Fig. 5.3 Azeotropic regions within the Λ_{12}-Λ_{21} space:

I, with one negative azeotrope; II, with one positive azeotrope; III, with two boundary azeotropes; IV, with two internal azeotropes; V, with no azeotrope; C, single positive tangential azeotrope curve; a, positive tangential azeotrope curve; b, negative tangential azeotrope curve; O, the coexistence point for a positive and a negative tangential azeotrope

holds true:

$$\begin{cases} \gamma_1 P_1^0 - \gamma_2 P_2^0 = 0 \\ \dfrac{\partial \ln \gamma_1}{\partial x_1} - \dfrac{\partial \ln \gamma_2}{\partial x_1} = 0 \end{cases} \tag{6}$$

where

$$\begin{cases} \dfrac{\partial \ln \gamma_1}{\partial x_1} = \dfrac{\Lambda_{12} - 1}{x_1 + \Lambda_{12} x_2} - \dfrac{\Lambda_{12}}{(x_1 + \Lambda_{12} x_2)^2} + \dfrac{\Lambda_{21}^2}{(\Lambda_{21} x_1 + x_2)^2} \\ \\ \dfrac{\partial \ln \gamma_2}{\partial x_1} = \dfrac{1 - \Lambda_{21}}{\Lambda_{21} x_1 + x_2} + \dfrac{\Lambda_{21}}{(\Lambda_{21} x_1 + x_2)^2} - \dfrac{\Lambda_{12}^2}{(x_1 + \Lambda_{12} x_2)^2} \end{cases}$$

Given parameters in domain IV, a system of two internal azeotropes is obtained and, given parameters in domain III, a system of two boundary azeotropes is obtained. Shown in Fig. 5.4a, b are also systems at T = const and vapour pressure of pure components $P_1^0 = 1.0$, $P_2^0 = 1.058$ and the respective

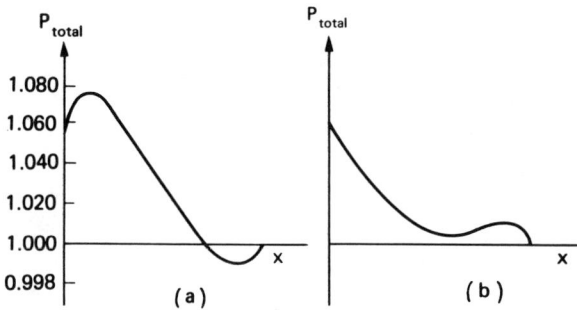

Fig. 5.4 Types of two-extremum P vs. x curves

a, at $\Lambda_{12} = 0.1$, $\Lambda_{21} = 2.7$ for region III;
b, at $\Lambda_{12} = 2.5$, $\Lambda_{21} = 0.2$ for region IV

parameters for (a) $\Lambda_{12} = 0.1$, $\Lambda_{21} = 2.7$ and for (b) $\Lambda_{12} = 2.5$, $\Lambda_{21} = 0.2$.

The above approach enables one to model binary systems with desired properties and to predict their behaviour as a global parameter of the system is allowed to vary.

In a three-component system whose two components form a binary system with a mixed deviation from ideality, the pattern of α_{ij}-lines characteristic of each diagram type is more complicated, with the eventual formation of two or even three internal azeotropes. We have derived, by computational simulation, a three-component diagram with two internal azeotropes with no azeotropic binary system present, and a three-component diagram with three internal azeotropes containing one azeotropic binary system.

These diagrams, each exhibiting a specific distillation curve and α_{ij}-pattern, are shown in Fig. 5.5a, b.

The parameters of binary (molecular) interaction can be calculated starting from one arbitrary experimental point of phase equilibrium, from activity coefficients at infinite dilution, and by using regression analysis of the available experimental vapour-liquid equilibrium data for a given binary system.

As has been shown by Aristovich et al. [11] and later by Silverman and Tassious [12], there can be envisaged several sets of parameters Λ_{12} and Λ_{21} for systems with negative deviation from Raoult's law ($\gamma_1 < 1$ and $\gamma_2 < 1$), that is, there are several roots in the Wilson equation solution satisfying the same activity coefficients at a given concentration of mixture components.

The location of domains corresponding to different roots of the Wilson equation in the parametric space Λ_{12}-Λ_{21} was considered in the work [13].

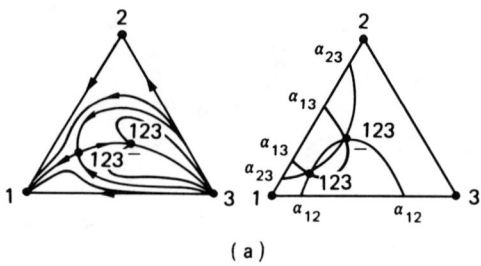

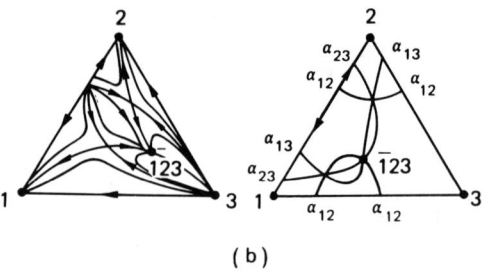

Fig. 5.5 Ternary systems possessing more than one internal azeotrope (shown are distillation curves and α-patterns)

a, system possessing two internal azeotropes and none of binary azeotropes;
b, system possessing three internal azeotropes and one azeotropic binary system

Conditions for the existence of each of the three possible sets of roots can be defined for negative deviations from ideality. Three domains for the existence of roots I, II, and III at $\gamma < 1$ are shown in Fig. 5.6. The domain for existence of roots II is defined by the inequality

$$1 - \frac{1}{t} - \ln(\gamma_1 t) > 0 \tag{7}$$

where

$$\ln(\gamma_1 t) = x_2 \left(\frac{\Lambda_{12}}{t} - \frac{\Lambda_{21}}{z} \right)$$

The expression (7) can be transformed to a simpler form,

$$\Lambda_{12} \Lambda_{21} > 1 \tag{8}$$

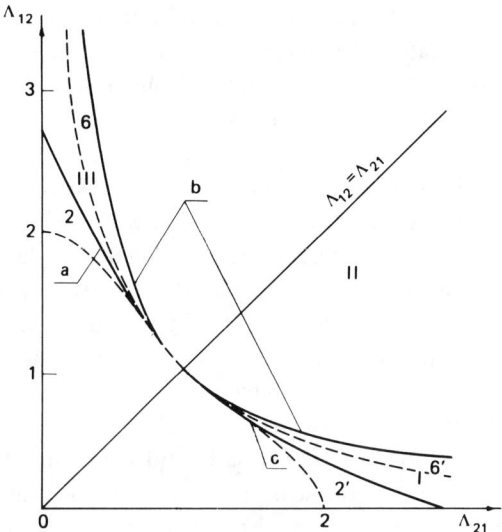

Fig. 5.6 Domains for existence of the Wilson equation roots:

a, curve $\ln \gamma_1^\infty = 0$; b, curve $\Lambda_{12}\Lambda_{21} = 1$; c, curve $\ln \gamma_2^\infty = 0$

The domain of Wilson parameters corresponding to the root II lies on the right side of the delimitative hyperbola $\Lambda_{12}\Lambda_{21} = 1$ (Fig. 5.6). The domain of Wilson parameters corresponding to the root I is defined by the curves $\ln \gamma_2^\infty = 0$ and $\Lambda_{12}\Lambda_{21} = 1$, whereas the domain corresponding to the root III is defined by the curves $\ln \gamma_1^\infty = 0$ and $\Lambda_{12}\Lambda_{21} = 1$. These domains are symmetric with respect to the straight line $\Lambda_{12} = \Lambda_{21}$. The course of the curve that delimits the parametric domains for mixtures with positive and negative deviations from ideality is dependent on the concentration of mixture components; however, it obeys the limiting condition $\ln \gamma_1(x) = 0$ and $\ln \gamma_2(x) = 0$.

The types 2 and 2′ exhibit no negative deviation from Raoult's law, and therefore for mixtures obeying this type of $\ln \gamma$ vs. X relationship, there can be found a single set of molecular interaction parameters at any mixture component concentration. For other types of nonideality (3, 3′, 5, 5′), characteristic is a negative deviation from Raoult's law within a limited range of concentrations, and if the solution is sought starting from one experimental point within this range (the overall state diagram is presumed to be largely unknown), then we run into the eventual finding of several solutions (several roots). Shown in Fig. 5.6 are also two special subdomains 6 and 6' that are characterized by purely negative deviation from ideality while belonging to different domains of root determination, viz. III and I.

We now consider the multiplicity of the Wilson equation for the case of a reliable knowledge of experimental data over a wide concentration range by making use of an idealized model system not subject to experimental errors.

Such a system can be derived starting from arbitrary molecular interaction (MI) parameters and their activity coefficients calculated over the whole concentration range. In [14], the search for possible solutions was carried out independently for each experimental point of the given concentration range.

By making use of the MI parameters chosen in the root III domain ($\Lambda_{12} = 2.66$, $\Lambda_{21} = 0.26$), we have calculated the activity coefficients corresponding to these MI parameters over the whole concentration range and found solutions at separate points in the root II and I domains for each set of activity coefficients. In Fig. 5.7, the selected parameter values are represented by point 5. The calculations have been performed for a concentration range of $x_1 = 0.05$ to $x_1 = 0.95$. In the root II domain, a solution of the Wilson

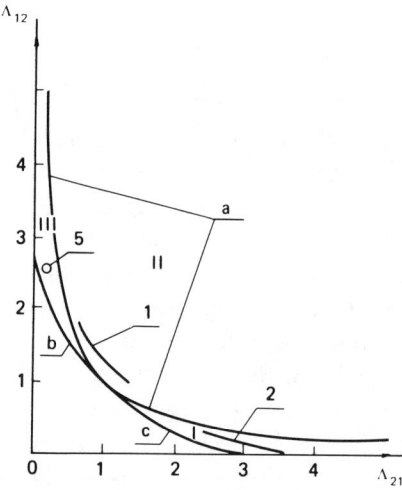

Fig. 5.7 Variation of molecular interaction parameters (MIP) within the concentration range of $x_1 = 0.05$ to $x_1 = 0.95$.

a, curve $\Lambda_{12}\Lambda_{21} = 1$; b, curve $\ln \Lambda_{12} - \Lambda_{21} + 1 = 0$; c, curve $\ln \Lambda_{21} - \Lambda_{12} + 1 = 0$. I, II, III are MIP domains corresponding to different roots

5 System Studies of Energy-Saving Schemes

equation can be found for each concentration. The calculations performed with the selected activity coefficients have defined sets of MI parameters that lie in the curve *1* within the root II domain.

The search for solution in the root I domain has also led to nonunique MI parameters (curve *2*); in this case, the solution has been obtained for a limited concentration range, from $x_1 = 0.05$ to $x_1 = 0.5$.

Thus, if the behaviour of activity coefficients within the whole concentration range is in accord with the MI parameter set chosen in the domain of existence of one of the roots, then no unique solution can be obtained in the domain of existence of two other roots over the whole concentration range.

Our study has provided evidence that the information about one experimental point of phase equilibrium is not always sufficient for deriving a true set of MI parameters to describe the actual behaviour of activity coefficients over the whole concentration range. In theory, two reliable experimental phase-equilibrium points at any concentration would have sufficed, and an identical solution for both concentrations in the domain of existence on one of the Wilson equation roots would have attested to the authenticity of MI parameter sets.

Different sets of MI parameters obtained for two reliable experimental points of vapour-liquid equilibrium, even if these sets belong to the same domain, testify to a non-authenticity of the MI parameters. Taking into account the experimental errors (especially for very dilute solutions) and the fact that, at close concentrations, accordingly close values of MI parameters can be obtained, it appears advantageous to use experimental phase-equilibrium data at concentrations $x_1 = 0.1$ and $x_1 = 0.9$.

When the vapour-liquid equilibrium data are available for a wide concentration range, the use of regression analysis is highly recommended. In principle, the regression analysis of the available body of experimental evidence can alone conduce to a nonoptimal solution, which may be due to three factors: (i) nonuniqueness of the Wilson equation, (ii) large experimental uncertainties, (iii) unfortunate guess of initial approximation. In this case, an optimal solution can be arrived at thorough examining the system for root multiplicity at the most reliable point of the vapour-liquid equilibrium and if, indeed, such a point is existent, the solution obtained for one experimental point can be used as an initial approximation in the regression analysis of all available information.

Such an approach allows one to exclude accidental errors arising from an inapt guess of the initial approximation.

In studying systems with a negative deviation from ideality by regression analysis, the mismatch function (that is, an appropriate function that provides a quantitative criterion for mismatch between the experimental and computational data) can yield as many as three local minima; however, preference should be given to the parameter sets that provide for the least mismatch. Precisely these MI-parameter sets should be regarded as the desired, that is, "authentic" ones.

For systems with the values of γ_i greater than unity, any initial approximation gives a unique set of the Wilson equation roots.

b. The NRTL and LEMF Equations

The Renon equation, or NRTL (Non-Random Two-Liquid) equation [15], similar to the Wilson equation, describes all types of nonideality for vapour-liquid binary systems. The studies of phase diagrams for binary systems at infinite dilution have led to the following inequality systems for each type of nonideality:

1st type
$$\begin{cases} r_{21} > -r_{12}g_{12}^2 g_{21} \\ r_{12} > -r_{21}g_{21}^2 g_{12} \end{cases}$$

2nd type
$$\begin{cases} r_{21} < -r_{12}g_{12}^2 g_{21} \\ r_{21} > -r_{12}g_{12} \end{cases}$$

2'-nd type
$$\begin{cases} r_{12} < -r_{21}g_{21}^2 g_{12} \\ r_{12} > -r_{21}g_{21} \end{cases}$$

3rd type
$$\begin{cases} r_{21} < -r_{12}g_{12} \\ r_{12} > -r_{21}g_{21} \end{cases}$$

3'-rd type
$$\begin{cases} r_{21} > -r_{12}g_{12} \\ r_{12} < -r_{21}g_{21} \end{cases}$$

4th type
$$\begin{cases} r_{21} < -r_{12}g_{12}^2 g_{21} \\ r_{12} < -r_{21}g_{21}^2 g_{12} \end{cases}$$

5th type
$$\begin{cases} r_{12} > -r_{21}g_{21}^2 g_{12} \\ r_{12} < -r_{21}g_{21} \end{cases}$$

5'-th type
$$\begin{cases} r_{21} > -r_{12}g_{12}^2 g_{21} \\ r_{21} < -r_{12}g_{12} \end{cases}$$

The domains for various types of nonideality of binary solutions at $\alpha_{12} = 0.4$ plotted in the parametric r_{12}-r_{21} space are presented in Fig. 5.8.

As has been established [16], the NRTL equation and the LEMF (Local Effective Molar Fractions) equation (suggested in [17]) have multiple solutions, that is, a set of roots satisfying the same activity coefficients.

The NRTL (LEMF) equation for a binary system of definite composition can be written in the following manner:

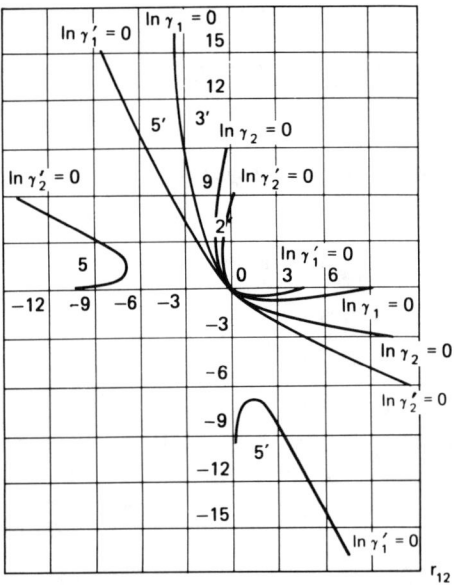

Fig. 5.8 Regions of nonideality for a binary solution according to the NRTL model at $\alpha_{12} = 0.4$

$$\ln \gamma_1 = x_2^2 \left\{ \frac{r_{21} \exp(-2\alpha_{12}r_{21})}{[x_1 + x_2 \exp(-\alpha_{12}r_{21})]^2} \right.$$

$$\left. + \frac{r_{12} \exp(-\alpha_{12}r_{12})}{[x_2 + x_1 \exp(-\alpha_{12}r_{12})]^2} \right\} \quad (10)$$

$$\ln \gamma_2 = x_1^2 \left\{ \frac{r_{12} \exp(-2\alpha_{12}r_{12})}{[x_2 + x_1 \exp(-\alpha_{12}r_{12})]^2} \right.$$

$$\left. + \frac{r_{21} \exp(-\alpha_{12}r_{21})}{[x_1 + x_2 \exp(-\alpha_{12}r_{21})]^2} \right\}$$

To determine the number of roots, Morozova and Platonov [18] have transformed the activity coefficient equations, based on equilibrium data at one experimental point, in such a manner as to make the activity coefficient of one of the mixture components a function of only one parameter. To that end, the following notations have been introduced

$$\begin{cases} F_1 = \ln \gamma_1 - \dfrac{x_2^2 r_{12} \exp(-\alpha_{12} r_{12})}{[x_2 + x_1 \exp(-\alpha_{12} r_{12})]^2} \\ \\ F_2 = \ln \gamma_2 - \dfrac{x_1^2 r_{12} \exp(-2\alpha_{12} r_{12})}{[x_2 + x_1 \exp(-\alpha_{12} r_{12})]^2} \end{cases} \quad (11)$$

Dividing the former expression by the latter, we obtain

$$r_{21} = \left(-\dfrac{1}{\alpha_{12}}\right) \ln \left[\dfrac{F_1 x_1^2}{F_2 x_2^2}\right] \quad (12)$$

where r_{21} is an explicit function of r_{12} and α_{12}.

By substituting r_{21} from Eq. (12) into Eq. (10), the following expression for the second activity coefficient is obtained:

$$\dfrac{\alpha_{12}}{x_1^2} \ln \gamma_2 = \alpha_{12} r_{12} \dfrac{\exp(-2\alpha_{12} r_{12})}{z^2} - \ln \dfrac{x_1^2 \xi_1}{x_2^2 \xi_2} \cdot \dfrac{\xi_1 \xi_2}{\theta^2} \quad (13)$$

where:
$z = x_2 + x_1 \exp(-\alpha_{12} r_{12})$

$\xi_1 = \ln \gamma_1 [x_2 + x_1 \exp(-\alpha_{12} r_{12})]^2 - x_2^2 r_{12} \exp(-\alpha_{12} r_{12})$

$\xi_2 = \ln \gamma_2 [x_2 + x_1 \exp(-\alpha_{12} r_{12})]^2 - x_1^2 r_{12} \exp(-2\alpha_{12} r_{12})$

$\theta = x_2 \xi_2 + x_1 \xi_1$

We now differentiate $\ln \gamma_2$ with respect to parameter r_{12} at fixed values of γ_1, x, and α_{12}.

$$\dfrac{\partial \ln \gamma_2}{\partial r_{12}} = \dfrac{\alpha_{12} \theta^2 \xi_2^2 x_2^2 [z - 2\alpha_{12} r_{12} x_2 + r_{12} \alpha_{12} z]}{[2\alpha_{12} x_2 \xi_2 \theta - x_1^2 \xi_1 z^2]}$$
$$+ [z - 2r_{12} \alpha_{12} x_2][-\xi_2 x_2^2 + \xi_1 x_1^2 \exp(-\alpha_{12} r_{12})]$$
$$- \xi_2 x_2^2 r_{12} \alpha_{12} z \quad (14)$$

By equating the expression for derivative from Eq. (14) to zero and making substitution of (12), we finally arrive at the equation of a curve in the parametric r_{12}-r_{21} space, with the partial derivative changing its sign:

$$\dfrac{\exp(-\alpha_{12} r_{12})(z - 2\alpha_{12} x_2 r_{12})}{[z - 2\alpha_{12} x_2 r_{12} + \alpha_{12} r_{12} z]} = \dfrac{[t - 2\alpha_{12} r_{21} x_1 + \alpha_{12} r_{21} t]}{\exp(-\alpha_{12} r_{21})[t - 2\alpha_{12} x_1 r_{21}]} \quad (15)$$

where $t = x_1 + x_2 \exp(-\alpha_{12} r_{21})$.

For each concentration, a system of curves can be plotted to divide the parametric space in Fig. 5.8 into domains each containing a set of parameters corresponding to one of the Renon equation roots.

In [18], the division of the parametric space for the NRTL equation at $x = 0.5$ and $\alpha_{12} = 0.4$ was exemplified.

In using the vapour-liquid equilibrium data at one experimental point, the NRTL equation as applied to systems with a positive deviation from Raoult's law can yield simultaneously as many as four roots; to systems with a negative deviation, five roots, and to systems with a mixed deviation, two roots.

Given the activity coefficients at a definite concentration, it has been suggested in [19], in seeking possible solutions, to plot a curve in the parametric space r_{12}-r_{21} that would correspond to the activity coefficient of the first component; then, a similar curve is plotted for the activity coefficient of the second component. The intersection points of these curves are precisely the equation roots.

Various curves described by Eqs. (10) at γ_1 and γ_2 = const for $\alpha_{12} = 0.4$, $x = 0.5$ are presented in Figs. 5.9 and 5.10. Two possible solutions (I and II) for $\ln \gamma_1 = 0.46$ (curve 1) and $\ln \gamma_2 = 0.58$ (curve 2) are shown in Fig. 5.9.

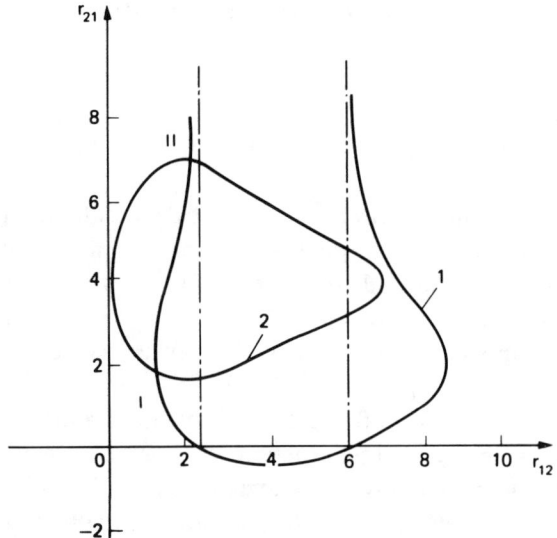

Fig. 5.9 Finding the solutions to an NRTL equation at $\alpha_{12} = 0.4$; $x = 0.5$; $\ln \gamma_1 = 0.46$ (curve 1); $\ln \gamma_2 = 0.58$ (curve 2). I, II, are the roots of the NRTL equation

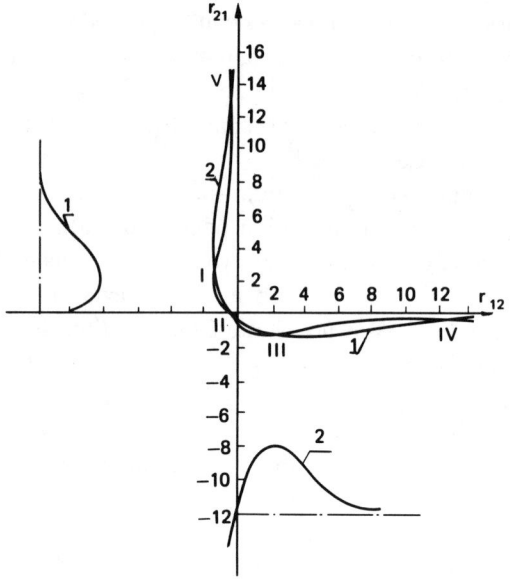

Fig. 5.10 Finding the solutions to a NRTL equation at $\ln \gamma_1$ (curve 1) = $\ln \gamma_2$ (curve 2) = -0.1; $x = 0.5$; $\alpha_{12} = 0.4$; I, II, III, IV, and V are the roots of the NRTL equation

Five possible solutions (I, II, III, IV, V) for $\ln \gamma_1$ (curve 1) = = $\ln \gamma_2$ (curve 2) = -0.1 are shown in Fig. 5.10.

Similar conclusions can be drawn about the LEMF equation ($\alpha_{12} = -1$). The study of equilibrium data at one experimental point [20] has revealed a picture opposite to that derived from the NRTL equation, that is, there is a possibility for the occurrence of up to five roots in the domain of positive deviations from ideality, four roots in negative deviations from ideality, and two roots in mixed deviations.

A mathematical study of NRTL and LEMF equations aimed at the search for possible solutions based on one experimental point has shown that these equations have multiple solutions for systems with any kind—positive, negative, or mixed—of deviation from Raoult's law; at infinite dilution, the NRTL equation has multiple solutions for systems with a negative deviation, and the LEMF equation gives a multiple solution for systems with a positive deviation. For this reason, the information provided by a scarce evidence on vapour-liquid equilibrium is not sufficient for deriving a "true" solution by NRTL and LEMF equations.

To obtain a unique solution, it suffices to have a reliable information on two experimental points of phase equilibrium, or, to be more concrete, one must make a rational use of the experimental evidence on phase equilibrium at concentrations $x_1 = 0.1$ and $x_2 = 0.9$. One can therefore draw a conclusion that the computational regression analysis of theoretically reliable data on phase equilibrium permit obtaining a unique solution which, as shown in Ref. [20], holds over the whole concentration range.

In order to exclude the ambiguity into which we run in applying regression analysis to experimental phase-equilibrium data, an essential preliminary condition is to find all feasible solutions for vapour-liquid equilibrium at one experimental point (the concentration having been selected with reference to the greatest reliability coefficient), which enables one to assess the mismatch function. The solution thus obtained should be used as initial approximation in regressing the available experimental data. This helps to avoid errors due to an unfortunate guess of the initial approximation.

The mismatch function can yield up to five local minima, each with its proper set of MI parameters; however, preferred are those sets that provide for the least sum of the squared differences between the experimental and computed points. It is these MI-parameter sets that are "true" ones, since they are presumed to be the most reliable in characterizing the whole concentration range of a vapour-liquid system.

5.4 Supramolecular Level

5.4.1 Three- and Four-Component Systems

The basic limitation in the structural analysis of a multidimensional azeotropic simplex is associated with specifying the type of singular points and major internal links between them both in the subsimplices (of different dimension) and, finally, the entire simplex, as well as with identifying boundary singular points that define the location of separating manifolds.

The solution of the problems in question and the development of a simple and efficient algorithm for structural analysis can be accomplished on the basis of the theory of conjugate singular points [21]. Studying the dynamic behaviour of a distillation system at a continuously varied equilibrium (for example, under differential distillation conditions, or due to a global parameter change of the system) enables one to reveal major links between singular points. For example, the continuous change of pressure in a system leads to either a fusion, or a breakdown, of conjugate singular points (azeotropes), which indicates the occurrence of a conjugate link between them. An analysis shows that the conjugate singular points are related to each other through a major link (in accordance with the definition suggested by Petlyuk [22], this is the link between singular points represented by a single joining C-line, or distillation line).

The conditions for singular point conjugation in designing a structural analysis algorithm are better defined by characteristic equations.

One may say that a singular point within an $(n - 1)$-dimensional simplex and a singular point in a simplex boundary element of dimension $(n - 2)$ are conjugate only if the root signs of characteristic equations

$$\begin{cases} \left| x_i^o \dfrac{\partial K_i}{\partial x_j} - \delta_{ij}\lambda \right|_{n-1} = 0 \\ (K_s^o - 1)\left| x_i^o \dfrac{\partial K_i}{\partial x_j} - \delta_{ij}\lambda \right|_{n-2} = 0; \quad i \neq s;\ j \neq s \end{cases} \quad (16)$$

are the same for all the mixture components, except a component "s" missing in the element of dimension $(n - 2)$; here $K_i = y_i/x_i$ is the vapour-liquid partition coefficient for the ith component; λ is the root of a characteristic equation; and δ_{ij} is the Kronecker delta which is 0 at $i \neq j$ and 1 at $i = j$.

To be more definite in the sequel, we introduce now the notions of a semiconjugate singular point and a semi-conjugate link. The semi-conjugate singular points are points that can be joined to each other through a semi-conjugate link (which is also the major link) and which in the limit are incapable of merging into a single point as the system parameters are made to vary. Such singular points are saddle points of the same order, but of opposite root sign for any two components of a mixture (irrespective of the system dimension).

For distillation subregions with the number of singular points equal to the number n of mixture components, the conditions for conjugation of singular points (Eq. (16)) forming a chain of $(n - 1)$ links must be fulfilled.

The distillation subregions with the number of singular points greater than the number of mixture components (excess subregions) possess semi-conjugate links; of these, characteristic are chains of links, or paths, of maximum length, $\geq (n - 1)$.

The structural analysis of a simplex of any dimension is carried out hierarchically, starting from the binary component links and ending in the desired

$(n - 1)$-dimensional structure. A set of pressures (boiling points) for all singular points of the system, with allowance made for the eventual bifurcate behaviour of the system, guarantees the occurrence of the only feasible links within the simplex.

The structural analysis of a vapour-liquid phase equilibrium diagram can be performed graphically (at each step, the $(n - 1)$-dimensional simplex is augmented with conjugate links characteristic of each subsimplex making part of the simplex of a given dimension); the internal singular points, starting from the quaternary ones, and, consequently, the links between them, are extended in succession beyond the contour of multidimensional composition simplex, as distinct from the structural distillation graph suggested in [23]. This allows preserving the pictorial demonstrativeness of graphical constructs with an increase in the number of system components.

Quite natural, a similar analysis can be carried out by making use of structural, or incidence, matrices with the aid of a computer. The structural analysis procedure as reported in [22] has been confined to systems with third-order azeotropes.

The algorithm that we propose in this work is based on the employment of the known procedure for constructing complex singular points as outlined in the paper [24], with testing the established links for conjugatedness at each step of the hierarchical analysis of the compositional simplex.

The links between singular points in an isobaric distillation process or a batch distillation are always pointed toward a higher boiling point (T_b), or, under isothermal conditions, toward lower vapour pressure.

The incoming links are typified by positive roots of the characteristic equation (16) and the outgoing, or exit, links, by negative roots. In order to establish a one-to-one correspondence between the number of positive (negative) roots of the characteristic equation and the number of incoming and exit links of a simplex, in filling the incidence matrix the excess incoming links (symbolized by "1") are enclosed in parentheses; by contrast, if the number of incoming links is smaller than the number of positive roots needed for characterization of a given singular point, extra symbols 1 are added. The number of symbols 1 not enclosed in parentheses in the incidence matrix columns corresponds to the number of positive roots of Eq. (16).

For the algorithm implementation, an initial information is needed; this information is supplied through enumerating the singular points, with an indication of their respective composition and T_b (pressure).

The links for binary components of a simplex are established in the following manner:

1. If a binary system has no azeotropes, the link is pointed from component i toward component j, providing $P_i > P_j$ ($T_i < T_j$), and vice versa.

2. If a binary system has an azeotrope, the link is pointed from component i to binary azeotrope ij, providing $P_i > P_{ij}$ ($T_i < T_{ij}$), and vice versa; a stable binary azeotrope is characterized by two incoming links and one positive root of Eq. (16), and in constructing the incidence matrix, one symbol 1 must be enclosed in parentheses.

In a three-component system, the type of unary singular points with respect to the Gibbs triangle is defined completely, and additional internal links that can be drawn are those that do not violate the previously established type of the unary singular point at the binary system stage. For unary saddle-type points, no additional links can be drawn within the Gibbs triangle; additional links can be drawn either from an unstable unary singular point, or toward a stable unary singular point. The type of a binary singular point within the three-component space is defined incompletely. The missing links in the Gibbs triangle are defined in the following manner:

1. If the system in question has no ternary azeotrope, the links are established between binary azeotropes with respect to the binary system; the direction of the links is defined by the difference in vapour pressure (T_b). If a three-component mixture lacks at least two binary singular points of one type, a conjugate link is established between the binary and unary singular points of the same type with respect to the Gibbs triangle.

If several structures are possible that satisfy the initial conditions, further experimental data are needed to specify the type of the binary singular point relative to the Gibbs triangle.

Schematically, the conjugation conditions for a three-component system with no internal singular points may be written as follows (in the parentheses is the singular point type relative to the binary system):

$$N_{13}, \ N_{23}(N_{22}) \rightarrow S_{23}(N_{22})$$
$$U_{13}, \ U_{23}(U_{22}) \leftarrow S_{23}(U_{22})$$
(17)

2. If a three-component subsimplex has a ternary azeotrope, its T_b is compared to the T_b of other singular points of the given Gibbs triangle. If the former T_b is higher over the T_b of any singular point in a side of the Gibbs triangle ($T_{ijk} > T_{ij}, T_i$), then the ternary azeotrope is a stable node; if $T_{ijk} < T_{ij}, T_i$, then it is an unstable node. The links are drawn toward binary azeotropes of the same type as that of the ternary azeotrope, but defined rela-

tive to the sides of the composition triangle, that is, relative to the binary systems.

If the T_b of a ternary azeotrope is intermediary within the T_b set of the singular points of the composition triangle (T_k, $T_{jk} > T_{ijk} > T_{ij}$, T_j), then the ternary singular point is a saddle, with two incoming links that enter into it from unstable triangle nodes, and two exit links extended to stable triangle nodes. If the system has more than two stable (unstable) unary singular points, then a semi-conjugate internal link arises between the ternary saddle-point azeotrope and the binary azeotrope.

Initially, links from the ternary singular saddle point toward binary singular points are established, and then, toward unary singular points.

The conjugation conditions for a three-component system with an internal azeotrope are written as follows (shown in the columns are the root signs of characteristic equation (16)):

$$
\begin{array}{c}
U_3 \\
\begin{vmatrix} + \\ + \end{vmatrix} \longleftarrow \begin{vmatrix} + \\ - \end{vmatrix} {}^{C^1_{23}(U_{22})}
\end{array}
$$

$$
\begin{array}{c}
N_3 \\
\begin{vmatrix} - \\ - \end{vmatrix} \longrightarrow \begin{vmatrix} - \\ + \end{vmatrix} {}^{C^1_{23}(N_{22})}
\end{array}
\qquad (18)
$$

$$
\begin{array}{c}
N_{13}, N_{23}(N_{22}) \qquad\qquad C^1_{33} \qquad\qquad U_{23}(U_{22}), U_{13} \\
\begin{vmatrix} - \\ - \end{vmatrix} \longrightarrow \begin{vmatrix} - \\ + \end{vmatrix} \longrightarrow \begin{vmatrix} + \\ + \end{vmatrix} \\
\\
C^1_{23}(U_{22}) \qquad\qquad\qquad C^1_{23}(N_{22}) \\
\begin{vmatrix} + \\ - \end{vmatrix} \qquad\qquad\qquad\qquad \begin{vmatrix} + \\ - \end{vmatrix}
\end{array}
$$

In constructing the incidence matrix for a stable ternary nodal singular point characterized by three incoming links and by two positive roots of the characteristic equation, one of the three symbols 1 must be enclosed in parentheses. For the ternary singular saddle point characterized by two incoming links and two exit links and by two roots of opposite signs, one 1 should be enclosed in parentheses.

In such a fashion, the structures for all three-component systems making part of a four-component system are established.

For a four-component system, the types of unary and binary singular points relative to the entire simplex can be defined starting only from the data available for the binary and ternary systems.

1. If a four-component system has no quaternary azeotropes, then the conjugate links are initially established between ternary azeotropes exhibiting the same type relative to the Gibbs triangle, that is, between the stable (unstable) nodes or between two saddle points. If a system lacks two ternary singular points of one type, then internal conjugate links between ternary and binary singular points are sequentially established. If the system possesses none of ternary azeotropes, then internal conjugate links are established between binary singular points.

The conjugation conditions for a four-component system with no internal azeotropes may be written in the following manner:

$$C_{34}^1(U_{33}) \to U_{34}(U_{33}), \quad U_{24}(U_{22}), \quad U_{14}$$
$$C_{34}^2(N_{33}) \leftarrow N_{34}(N_{33}), \quad N_{24}(N_{22}), \quad N_{14} \qquad (19)$$
$$S_{24}^1(N_{22}), \quad S_{34}^2(C_{33}^1) \leftarrow S_{34}^1(C_{33}^1), \quad S_{24}^2(U_{22})$$

For a four-component system with no internal azeotrope shown in Fig. 5.11b, only one conjugate link, 25-145, can be traced between the binary and ternary azeotropes (by analogy, a similar 25-135 conjugate link is shown in Fig. 5.11e).

2. If the system has a quaternary maximum-pressure (minimum-boiling) azeotrope, then the conjugate links are drawn from this azeotrope to ternary azeotropes of the same type. Shown in Fig. 5.11d is a system with internal maximum-pressure azeotrope 1345; for this system, the following internal conjugate links can be established: 1345-135, 1345-345, 1345-134, and 1345-145.

For an internal stable azeotrope characterized by three positive roots and four incoming links, one 1 must be enclosed in parentheses in constructing the incidence matrix.

For a four-component system possessing a first-order internal saddle-point azeotrope, typical is the occurrence of a separating nodal surface of stable type in which the quaternary azeotrope separatrices originate.

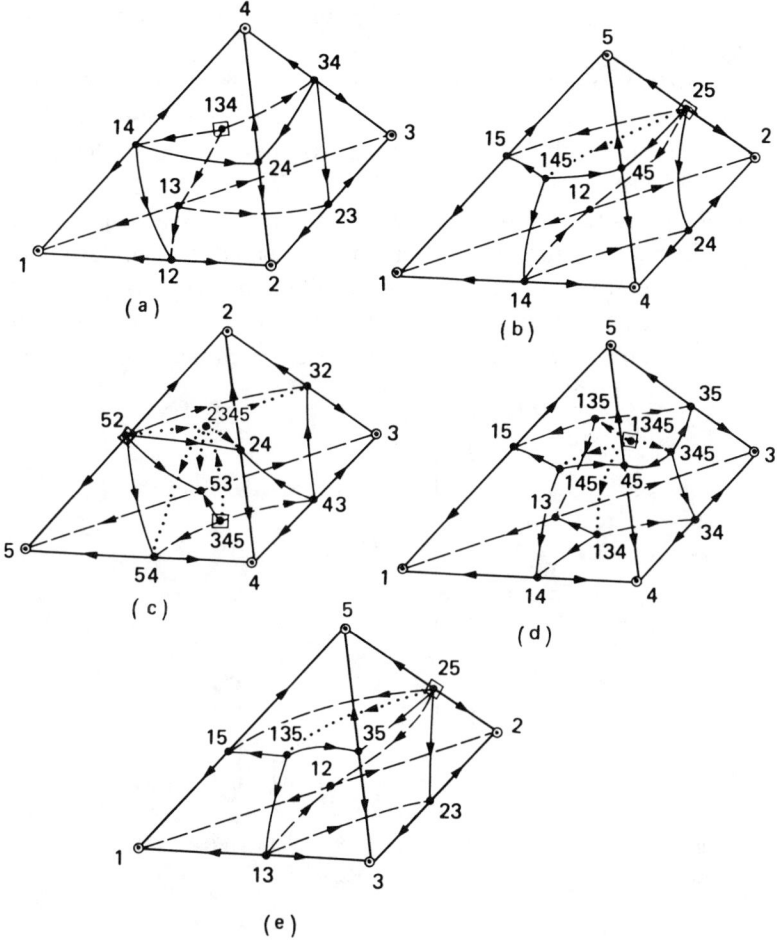

Fig. 5.11 Examples of structural diagrams within the concentration space of quaternary azeotropic systems

a, b, e, systems with no internal azeotrope;
c, a system with a quaternary saddle-point azeotrope; *d,* a system with a quaternary unstable saddle-point azeotrope

If the number of incoming links in a nodal stable-type surface is more than two, then the excess 1 symbols are enclosed in parentheses in constructing the incidence matrix, and a 1 is added to one incoming link.

In the case of a second-order saddle-point azeotrope, a separating nodal surface of unstable type occurs, with the respective separatrices making part of it. In constructing the incidence matrix, one of the two symbols 1 is taken in parentheses.

For a first-order quaternary saddle-point azeotrope, the conjugate singular points are stable nodes and saddles of second order relative to the four-component system (in the three-component system, they are saddles), whereas for a second-order quaternary saddle-point azeotrope, the conjugate points are unstable nodes and saddles of the first order. If a system has more than two stable (unstable) nodes, then a semi-conjugate link between the quaternary saddle-point azeotrope and the ternary saddle-point azeotrope arises.

The conditions for conjugation between the internal azeotropes and the boundary singular points of a four component system may conventionally be written in the form of a simple scheme:

$$
\begin{array}{c}
U_{44} \begin{vmatrix} + \\ + \\ + \end{vmatrix} \longleftarrow \quad C^1_{34}(U_{33}) \begin{vmatrix} - \\ + \\ + \end{vmatrix} \\
\\
N_{44} \begin{vmatrix} - \\ - \\ - \end{vmatrix} \longrightarrow \quad C^2_{34}(N_{33}) \begin{vmatrix} - \\ - \\ + \end{vmatrix} \\
\\
S^2_{24}(U_{22}), S^2_{34}(C^1_{33}) \begin{vmatrix} - \\ - \\ + \end{vmatrix} \longrightarrow \; S^1_{44} \begin{vmatrix} - \\ + \\ + \end{vmatrix} \longrightarrow \begin{vmatrix} + \\ + \\ + \end{vmatrix} \; U_{34}(U_{33}), U_{24}(U_{22}), U_{14} \nearrow \\
\\
S^1_{34}(C^1_{33}), S^1_{24}(N_{22}) \begin{vmatrix} + \\ - \\ + \end{vmatrix} \nearrow \\
\\
S^1_{24}(N_{22}), S^1_{34}(C^1_{33}) \begin{vmatrix} - \\ + \\ + \end{vmatrix} \longleftarrow \; S^2_{44} \begin{vmatrix} - \\ + \\ + \end{vmatrix} \longleftarrow \begin{vmatrix} - \\ - \\ - \end{vmatrix} \; N_{34}(N_{33}), N_{24}(N_{22}), N_{14} \nearrow \\
\\
S^2_{34}(S^1_{33}), S^2_{24}(U_{22}) \begin{vmatrix} + \\ - \\ - \end{vmatrix} \searrow
\end{array}
$$

(20)

For a four-component mixture, the links between an internal singular point and ternary azeotropes (whose type relative to the composition tetrahedron is undefined) are initially established. If the system has a quaternary saddle-point azeotrope, its type is defined by the azeotropy rule for four-component mixtures. Additional conjugate links are drawn toward binary singular points which have a "separating surface" in the four-component mixture. To be noted, it is only for such singular points that an additional link in the separating surface can be drawn without interfering with the preformed structure of a binary azeotrope at the three-component construct stage.

If the system has a quaternary saddle-point azeotrope and none of ternary azeotropes, then the order of the quaternary azeotrope is determined by the azeotropy rule for four-component mixtures free of ternary azeotropes, as recommended in Ref. [24]:

$$\sum_{k=1}^{2} 2^{k-1}(N_k^+ + C_k^+ - N_k^- - C_k^-) = -8(C_4^+ - C_4^-) \qquad (21)$$

where N_k^- (N_k^+) is the number of unstable (stable) k-nary nodes in the composition tetrahedron and C_k^- (C_k^+) is the number of k-nary saddles of the first (second) order.

A four-component mixture with the internal saddle-point azeotrope ($P_{25} > P_{345} > P_{34} > P_{2345} > P_{45} > P_{35} > P_5 > P_{23} > P_{24} > P_4 > P_3 > P_2$), shown in Fig. 5.11c, exemplifies the manner in which the structure of this system is determined after all of its conjugate links have been established.

The order of the quaternary saddle-point azeotrope is determined by the azeotropy rule after the conjugate link between this azeotrope and the sole ternary azeotrope 345 in the system has been established.

The type of the quaternary azeotrope 2345 is S_{44}^2, that is, it has an unstable nodal surface and two incoming links; one of these, linking 345-2345, has already been defined, whereas the other one links a second unstable node to an internal singular point, 25-2345.

Let us define the links dictated by a separating surface. Since the system in question has no more ternary azeotropes, additional links should be drawn from the singular point 2345 to binary singular points that have a separating surface; these are 23, 24, 35, and 45.

By making use of the procedure for defining links as specified by scheme (20), let us construct an incidence matrix for the system depicted in Fig. 5.11c. This matrix is represented by Table 5.1. The excess incoming links (that is, those redundant to the number of positive roots) are denoted in Table 5.1 by parenthetical symbols 1. The system has two unstable nodes: these are the

Table 5.1 Incidence Matrix for the System Shown in Fig. 5.11b
(SP signifies "singular point")

SP	[2]	[3]	[4]	[5]	23	24	{25}	34	35	45	{345}	2345
2	0	0	0	0	0	0	0	0	0	0	0	0
3	0	0	0	0	0	0	0	0	0	0	0	0
4	0	0	0	0	0	0	0	0	0	0	0	0
5	0	0	0	0	0	0	0	0	0	0	0	0
23	1	1	0	0	0	0	0	0	0	0	0	0
24	1	0	1	0	0	0	0	0	0	0	0	0
25	1	0	0	1	1	1	0	0	1	1	0	1
34	0	1	1	0	1	1	0	0	0	0	0	0
35	0	1	0	1	0	0	0	0	0	0	0	0
45	0	0	1	1	0	0	0	0	0	0	0	0
345	0	0	0	0	0	0	0	1	1	1	0	(1)
2345	0	0	0	0	(1)	(1)	0	0	(1)	(1)	0	0
SP type	U_{14}	U_{14}	U_{14}	U_{14}	S_{24}^1	S_{24}^1	N_{24}	C_{24}^2	S_{24}^1	S_{24}^1	N_{34}	S_4^2

binary azeotrope 25 and the ternary azeotrope 345 shown inside braces and four stable unary nodes 2, 3, 4, 5 shown inside square brackets.

Through examining the incidence matrix for singular points, we can identify the paths of unstable node—stable node type and out of these, we can single out the paths that contain boundary singular points localizing the distillation subregions.

The number of distillation regions is defined by the number of the shortest paths that link unstable and stable singular points of the system; such paths are eight, viz., 25-2; 25-5; 25-4; 25-3; 345-2; 345-5; 345-4; and 345-3. The number of maximum-length links defines the number of distillation subregions; all of the mixture components must be involved in the buildup of link chains (paths), and each link must be either conjugate, or semi-conjugate (that is, characteristic of an excess distillation subregion), which corresponds to Scheme (20). For example, a stable node cannot be linked through a conjugate or a semi-conjugate link to the second-order saddle point.

We now construct paths for the distillation regions as defined by the above end routes.

Route 25-2. The paths that define distillation subregions are: 25-2345-24-2 and 25-2345-32-2; the paths 25-24-2 and 25-23-2 are not boundary paths, since they do not contain all the mixture components.

Route 25-5. The paths that define a distillation subregion are 25-2345-35-5 and 25-2345-45-5; the paths 25-35-5 and 25-45-5 do not contain all the mixture components.

Route 25-4. The paths that define a distillation subregion are: 25-2345-24-4 and 25-2345-45-4; the paths 25-24-4 and 25-45-4 do not contain all the mixture components.

Route 25-3. The paths that define a distillation region are 25-2345-23-3 and 25-2345-35-3; the paths 25-32-3 and 25-35-3 do not contain all the mixture components.

Route 345-2. It has paths 345-2345-32-2, 345-2345-24-2, 345-43-32-2, and 345-43-24-2, each defining a distillation subregion.

Route 345-5. The paths that define a distillation subregion are 345-2345-54-5 and 345-2345-53-5; the paths 345-54-5 and 345-53-5 do not contain all the mixture components.

Route 345-4. The paths that define a distillation subregion are 345-2345-24-4, 345-2345-54-4, and 345-43-24-4; the paths 345-54-4 and 345-43-4 do not contain all the mixture components.

Route 345-3. The paths that define a distillation subregion are 345-2345-23-3, 345-2345-35-3, and 345-43-32-3; the paths 345-43-3 and 345-35-3 do not contain all the mixture components.

To summarize, the system of interest has a total of 20 distillation subregions defined by maximum-length paths.

5.4.2 Five-Component and n-Component Systems

As the dimension of a system becomes larger, the number of azeotropic types becomes greater and their behaviour becomes increasingly complex; nonetheless, the approach as outlined above allows to reveal specific structural features for a concentration simplex of any dimension.

In a five-component system, the major links can be defined taking into account the specificity for formation of singular points of different type in the pentatope as outlined by Zharov in Rev. [24].

For a five-component system, the types of ternary, binary, and unary singular points can be defined following the line of reasoning as previously applied to the four-component system; therefore, we have to define the type of quaternary and quinary singular points with respect to the five-component simplex and to identify possibly the major links.

1. If a five-component system is devoid of quinary azeotropes, then conjugate links are first defined between quaternary azeotropes exhibiting the

same type relative to the composition tetrahedron, that is, between two stable (unstable) nodes or between two saddle points of the same order. If the system lacks such two quaternary singular points, then conjugate links are established between quaternary and ternary singular points, or between quaternary and binary singular points and, subsequently, between quaternary and unary singular points; if the system lacks quaternary azeotropes, conjugate links are traced between ternary singular points, or between ternary and binary singular points.

Schematically, the conditions for conjugate linkage in five-component systems devoid of internal azeotropes may be represented as shown below:

$$C^2_{35}(U_{33}), C^2_{45}(S^1_{44}) \quad\quad S^1_{45}(S^1_{44}), S^1_{35}(S^1_{33}), S^1_{25}(N_{22})$$

$$\begin{vmatrix} - \\ + \\ + \\ - \end{vmatrix} \longrightarrow \begin{vmatrix} - \\ + \\ + \\ + \end{vmatrix}$$

$$C^2_{35}(N_{33}), C^2_{45}(S^2_{44}) \quad\quad S^3_{45}(S^2_{44}), S^3_{35}(S^1_{33}), S^3_{25}(U_{22})$$

$$\begin{vmatrix} + \\ - \\ - \\ + \end{vmatrix} \longleftarrow \begin{vmatrix} + \\ - \\ - \\ - \end{vmatrix}$$

$$C^1_{45}(U_{44}) \quad\quad U_{45}, U_{35}, U_{25}, U_{15}$$

$$\begin{vmatrix} + \\ + \\ + \\ - \end{vmatrix} \longrightarrow \begin{vmatrix} + \\ + \\ + \\ + \end{vmatrix}$$

$$C^3_{45}(N_{44}) \quad\quad N_{45}, N_{35}, N_{25}, N_{15}$$

$$\begin{vmatrix} - \\ - \\ - \\ + \end{vmatrix} \longleftarrow \begin{vmatrix} - \\ - \\ - \\ - \end{vmatrix}$$

(22)

Here, the parenthetic symbols specify the type of the singular point relative to the space within which the point is contained. The root signs in a column are read from top to bottom in the order: for a binary azeotrope, it is the upper sign; for a ternary, two uppermost signs, and for a quaternary azeotrope, three uppermost signs.

2. If a five-component system has a quinary maximum-pressure (minimum-boiling) azeotrope, this can serve as an origin for conjugate links pointed only to quaternary azeotropes which, in a four-component system, are maximum-

pressure (minimum-boiling) azeotropes. For a quinary minimum-pressure (maximum-boiling) azeotrope, the conjugate points can be only quaternary minimum-pressure (maximum-boiling) azeotropes.

In constructing the incidence matrix for a stable internal azeotrope exhibiting positive roots, the excess incoming links (denoted by 1) are enclosed in parentheses; by contrast, if the number of incoming links is less than four, an appropriate number of symbols 1 are added into one of the cells of the incidence matrix containing already 1 therein.

The first-order quinary saddle-point azeotrope possessing a separating hypersurface of stable type, in which the separatrices originate, has two conjugate links with stable nodes; also, this azeotrope is linked to the system's singular points which are, relative to the five-component system, second-order saddle points not included in the separating hypersurface.

If the number of incoming links within a stable-type nodal surface is greater than three, the excess 1's in the incidence matrix are enclosed in parentheses; if the number of incoming links is less than three, additional 1's are entered.

If in a system possessing a first-order quinary azeotrope, more than two stable nodes are available, then a semi-conjugate link arises between the quinary azeotrope and the first-order quaternary saddle-point azeotrope located within a separating hypersurface.

The first-order saddle-point azeotrope, possessing an unstable-type separating hypersurface with incoming separatrices, is linked to unstable nodes through two conjugate links and also to the singular points which are, with respect to the five-component system, second-order saddles not located in the separating hypersurface (stable-type surface).

If a system of this type has more than two unstable nodes, a semi-conjugate link arises between the quinary azeotrope and the second-order quaternary saddle-point azeotrope located in a separating surface.

If the system has a quinary azeotrope with no separating hypersurface (that is, possessing two two-dimensional hypersurfaces of stable or unstable type), then conjugate links can be established for the singular points that are located in a hypersurface of a stable or unstable type relative to the five-component simplex.

The number of incoming links for a quinary saddle-point azeotrope with no separating surface being more than two, in the incidence matrix the excess 1's are enclosed in parentheses, and an additional 1 is entered into the matrix for one of the incoming links.

The conjugation conditions for singular points in a five-component system with an internal azeotrope can be presented as shown in the scheme below:

$$\begin{array}{c}
U_{55} \quad\quad\quad C^1_{45}(U_{44}) \\
\begin{vmatrix}+\\+\\+\\+\end{vmatrix} \leftarrow \begin{vmatrix}+\\+\\+\\-\end{vmatrix} \\[1em]
N_{55} \quad\quad\quad C^3_{45}(N_{44}) \\
\begin{vmatrix}-\\-\\-\\-\end{vmatrix} \rightarrow \begin{vmatrix}-\\-\\-\\+\end{vmatrix}
\end{array}$$

$C^2_{35}(U_{33}), C^2_{45}(S^1_{44})\quad\quad S^1_{55} \quad\quad\quad U_{45}, U_{35}, U_{25}, U_{15}$

$\begin{vmatrix}+\\+\\+\\-\end{vmatrix} \rightarrow \begin{vmatrix}+\\+\\+\\-\end{vmatrix} \rightarrow \begin{vmatrix}+\\+\\+\\+\end{vmatrix} \nearrow$

$S^1_{45}(S^1_{44}), C^1_{35}(C^1_{33}), S^1_{25}(N_{22})$

$\begin{vmatrix}-\\+\\+\\+\end{vmatrix}$

(23)

$C^2_{35}(N_{33}), C^2_{45}(S^2_{44})\quad\quad S^2_{55} \quad\quad\quad N_{45}, N_{35}, N_{25}, N_{15}$

$\begin{vmatrix}-\\-\\+\\+\end{vmatrix} \leftarrow \begin{vmatrix}-\\-\\-\\+\end{vmatrix} \leftarrow \begin{vmatrix}-\\-\\-\\-\end{vmatrix} \nwarrow$

$S^3_{45}(S^2_{44}), S^3_{35}(C^1_{33}), S^3_{25}(U_{22})$

$\begin{vmatrix}-\\-\\-\\+\end{vmatrix}$

$S^3_{25}(U_{22}), S^3_{35}(S^1_{33}), S^3_{45}(S^2_{44})\quad\quad\quad\quad S^1_{45}(S^1_{44}), S^1_{35}(S^1_{33}), S^1_{25}(N_{22})$

$\begin{vmatrix}+\\-\\-\\-\end{vmatrix} \rightarrow \begin{vmatrix}+\\+\\-\\-\end{vmatrix}\; C^2_{55} \;\rightarrow \begin{vmatrix}+\\+\\+\\-\end{vmatrix}$

$C^2_{45}(S^1_{44}), C^2_{35}(U_{33}) \quad\quad C^2_{45}(S^2_{44}), C^2_{35}(N_{33})$

$\begin{vmatrix}+\\+\\+\\-\end{vmatrix} \nearrow \quad\quad \begin{vmatrix}+\\-\\-\\+\end{vmatrix} \nearrow$

The type of saddle-point azeotrope of a five-component system (with a separating hypersurface present or none) is defined by the azeotropy rule for quinary systems [24] after all the quaternary-azeotrope conjugate links have been identified (the link type relative to the five-component system cannot be defined starting only from the vapour-liquid equilibrium diagrams for four-component systems):

$$1 - \sum_{k=1}^{4} 2^{k-1}(N_k + C_4^2 - C_k^1 - C_k^3) = 16(C_5^2 - C_5^1 - C_5^3) \quad (24)$$

If the left-hand side of Eq. (24) is positive, the system of interest exhibits a quinary saddle-point azeotrope with no separating surface, and the missing conjugate links are added according to Scheme (23). In the case of a negative left-hand side, a saddle-point azeotrope with a separating surface occurs; its order is determined by the type of quaternary azeotropes relative to the four-component composition simplex (see Scheme (23)).

In a five-component system that lacks quaternary azeotropes, the type of the internal saddle-point azeotrope is defined by a simplified equation:

$$1 - \sum_{k=1}^{3} 2^{k-1}(N_k + C_k^2 - C_k^1 - C_k^3) = 16(C_5^2 - C_5^1 - C_5^3) \quad (25)$$

The lhs of Eq. (25) being positive, a quinary saddle-point azeotrope with no separating surface occurs. In the case of a negative lhs, a saddle-point azeotrope with a separating surface takes place, its order being defined by the type of ternary azeotropes relative to a three-component system (of stable or unstable type); these azeotropes in a five-component system are saddle-point azeotropes of type C_{35}^2 as defined at the earlier stage of four-component systems.

The missing conjugate links in the matrix are completed in accordance with Scheme (23).

If a five-component system has none of quinary, quaternary, or ternary azeotropes, one does not stand in need of defining conjugate links, since these have already been established at the stage of four-component system analysis.

As exemplified by a model five-component system with an internal saddle-point azeotrope (the adopted arbitrary pressures for components are: $P_1 = 1.0$; $P_2 = 1.0375$; $P_3 = 1.0485$; $P_4 = 1.0744$; $P_5 = 1.215$; $P_{12} = 1.0503$; $P_{13} = 1.361$; $P_{14} = 1.2667$; $P_{15} = 1.4823$; $P_{23} = 1.1905$; $P_{24} = 1.1132$; $P_{25} = 1.6422$; $P_{34} = 1.49$; $P_{35} = 1.283$; $P_{45} = 1.352$; $P_{134} = 1.5455$; $P_{135} = 1.5035$; $P_{145} = 1.5036$; $P_{345} = 1.492$; $P_{1345} = 1.5585$; $P_{2345} = 1.477$; $P_{12345} = 1.5348$), we specify now all the types for singular points of the composition simplex and define the paths that originate in the boundary singular points and delimit the distillation subregions.

The conjugate links are established sequentially at the stage of binary, ternary, and quaternary systems (the graphical method is illustrated by the scheme in Fig. 5.12, and the matrix method, by Table 5.2). In the columns of the incidence matrix, the incoming links are denoted by the code "1", and the exit, or missing, links, by the code "0"; the number and the sign of the roots of the characteristic equation determinative of the singular-point type relative to the multidimensional simplex are brought in correspondence with the number of incoming and exit links. The nonparenthetic symbols "1" in the matrix columns, Table 5.2 (the number of positive roots of the characteristic equation), allows an easy identification of the singular-point type relative to the composition simplex.

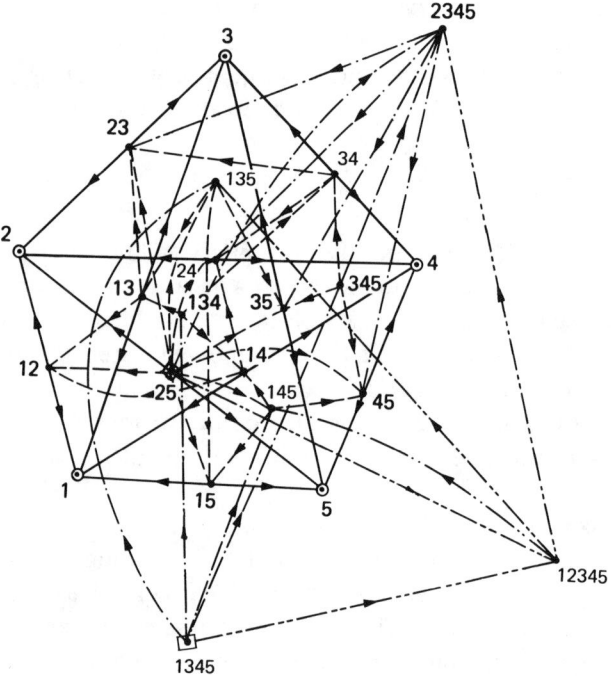

Fig. 5.12 Structural graph of a quinary system with an internal saddle-point azeotrope:

——— links defined at a binary system stage;
- - - - links defined at a ternary system stage;
—·—·— links defined at a quaternary system stage;
—··—·· links defined at a quinary system stage

Table 5.2 Incidence Matrix for the System Shown in Fig. 5.12
(SP signifies "singular point")

SP	[1]	[2]	[3]	[4]	[5]	12	13	14	15	23	24	{25}	34	35	45	134	135	145	345	{1345}	2345	12345
1	1	0	0	0	0	0	0	0	0	0	0	0	0	0	0	0	0	0	0	0	0	0
2	0	1	0	0	0	0	0	0	0	0	0	0	0	0	0	0	0	0	0	0	0	0
3	0	0	1	0	0	0	0	0	0	0	0	0	0	0	0	0	0	0	0	0	0	0
4	0	0	0	1	0	0	0	0	0	0	0	0	0	0	0	0	0	0	0	0	0	0
5	0	0	0	0	1	0	0	0	0	0	0	0	0	0	0	0	0	0	0	0	0	0
12	1	−1	0	0	0	1	0	0	0	0	0	0	0	0	0	0	0	0	0	0	0	0
13	1	0	−1	0	0	0	1	0	0	0	0	0	0	0	0	0	0	0	0	0	0	0
14	1	0	0	−1	0	0	0	1	0	0	0	0	0	0	0	0	0	0	0	0	0	0
15	1	0	0	0	−1	0	0	0	1	0	0	0	0	0	0	0	0	0	0	0	0	0
23	0	1	−1	0	0	0	0	0	0	1	0	0	0	0	0	0	0	0	0	0	0	0
24	0	1	0	−1	0	0	0	0	0	0	1	0	0	0	0	0	0	0	0	0	0	0
25	0	1	0	0	−1	0	0	0	0	0	0	1	0	0	0	0	0	0	0	0	0	0
34	0	0	1	−1	0	0	0	0	0	0	0	0	1	0	0	0	0	0	0	0	0	0
35	0	0	1	0	−1	0	0	0	0	0	0	0	0	1	0	0	0	0	0	0	0	0
45	0	0	0	1	−1	0	0	0	0	0	0	0	0	0	1	0	0	0	0	0	0	0
134	0	0	0	0	0	0	1	−1	0	0	0	0	1	0	0	1	0	0	0	0	0	0
135	0	0	0	0	0	0	1	0	−1	0	0	0	0	1	0	0	1	0	0	0	0	0
145	0	0	0	0	0	0	0	1	−1	0	0	0	0	0	1	0	0	1	0	0	0	0
345	0	0	0	0	0	0	0	0	0	0	0	0	1	−1	1	0	0	0	1	0	0	0
1345	0	0	0	0	0	0	0	0	0	(1)	0	0	0	(1)	(1)	1	(1)	(1)	1	(1)	0	0
2345	0	0	0	0	0	0	0	0	0	0	0	(1)	0	0	0	0	0	0	1	0	1	0
12345	0	0	0	0	0	0	0	0	0	0	0	0	0	0	0	0	0	0	0	(1)	0	1
SP type	U_{15}	U_{15}	U_{15}	U_{15}	U_{15}	S^1_{25}	C^2_{25}	C^2_{25}	S^1_{25}	S^1_{25}	S^1_{25}	N_{25}	C^2_{25}	S^1_{25}	S^1_{25}	C^3_{35}	C^2_{35}	C^2_{35}	C^3_{35}	N_{45}	C^2_{45}	S^3_5

As is apparent from examining the singular-point pressure set of the system, the internal azeotrope 12345 occurring in the five-component system of interest is a saddle point whose type can be specified after the conjugate links between this point and two quaternary azeotropes (1345 and 2345) have been defined. The azeotrope 12345 is a third-order saddle point (in the incidence matrix, it corresponds to the column with a single nonparenthetic "1").

The type of quaternary azeotropes relative to pentatope is also specified after the conjugate links have been defined: the azeotrope 1345 is an unstable node (the all-zero column in the incidence matrix), the azeotrope 2345 is a saddle point of C_{45}^2-type (its correspondent in the incidence matrix is the column with two nonparenthetic "1"s).

The type of internal azeotrope having been established, we can trace additional conjugate links (not interfering with the singular-point type defined earlier at the quaternary-simplex stage) toward the singular points as specified by Scheme (23). The link 25-12345 is a quinary-azeotrope separatrix, whereas the links 12345-135 and 12345-145 characterize, together with the link 12345-2345, an unstable nodal surface for the internal saddle-point azeotrope.

The system has five stable singular points which are pure components 1, 2, 3, 4, 5 (their matrical correspondents are the columns with four nonparenthetic codes "1") and two unstable singular points, viz., the binary azeotrope 25 and the quaternary azeotrope 1345.

By making use of the incidence matrix for singular points, we identify all the paths leading from an unstable to a stable node; of these, we single out the paths that contain the boundary singular points localizing the distillation subregions.

The number of the shortest pathways linking the unstable and stable singular points of the system determines the number of the distillation regions; the total of these is ten: 25-1, 25-2, 25-3, 25-4, 25-5, 1345-1, 1345-2, 1345-3, 1345-4, and 1345-5.

For each distillation region, the following maximum-length paths originating in the boundary singular points have been derived (the paths that do not contain all the mixture components have been eliminated from consideration):

Route 25-1: 25-12345-135-13-12-1; 25-12345-135-15-1; 25-12345-145-15-1; 25-12345-145-14-12-1.

Route 25-2: 25-12345-2345-24-2; 25-12345-2345-23-2; 25-12345-135-13-23-2; 25-12345-135-13-12-2; 25-12345-145-14-12-2; 25-12345-145-14-24-2.

Route 25-3: 25-12345-2345-35-3; 25-12345-2345-23-3; 25-12345-135-13-23-3; 25-12345-135-35-3.

Route 25-4: 25-12345-2345-54-4; 25-12345-2345-24-4; 25-12345-145-45-4; 25-12345-145-14-24-4.

Route 25-5: 25-12345-2345-54-5; 25-12345-2345-35-5; 25-12345-135-35-5; 25-12345-135-15-5; 25-12345-145-45-5; 25-12345-145-15-5.

Route 1345-1: 1345-12345-145-14-12-1; 1345-12345-145-15-1; 1345-12345-135-13-12-1; 1345-12345-135-15-1; 1345-134-14-12-1; 1345-134-13-12-1.

Route 1345-2: 1345-12345-2345-23-2; 1345-12345-2345-24-2; 1345-12345-145-14-12-2; 1345-12345-145-14-24-2; 1345-345-2345-24-2; 1345-345-2345-23-2; 1345-12345-135-13-12-2; 1345-12345-135-13-23-2; 1345-134-14-24-2; 1345-134-14-12-2; 1345-134-13-12-2; 1345-134-13-23-2; 1345-134-34-23-2; 1345-134-34-24-2; 1345-345-34-23-2; 1345-345-34-24-2.

Route 1345-3: 1345-12345-2345-23-3; 1345-12345-2345-35-3; 1345-345-2345-23-3; 1345-345-2345-35-3; 1345-12345-135-13-23-3; 1345-12345-135-35-3; 1345-134-13-23-3; 1345-134-34-23-3; 1345-345-34-23-3.

Route 1345-4: 1345-12345-2345-24-4; 1345-12345-2345-45-4; 1345-12345-145-45-4; 1345-12345-145-14-4; 1345-345-2345-45-4; 1345-345-2345-24-4; 1345-134-14-24-4; 1345-134-34-24-4; 1345-345-34-24-4.

Route 1345-5: 1345-12345-2345-54-5; 1345-12345-2345-35-5; 1345-12345-145-15-5; 1345-12345-145-54-5; 1345--345-2345-54-5; 1345-345-2345-35-5; 1345-12345-135-35-5; 1345-12345-135-15-5.

The derived distillation subregions total 72.

Extending the above approach to n-component systems, the condition for conjugation of such systems with no internal azeotropes can be represented by the scheme given on page 456.

For systems with internal azeotropes, the conjugation conditions can be represented as shown on page 457.

The method for structural analysis of multidimensional polyazeotropic compositional simplices that has been suggested in this paper makes use of the conjugative properties of the singular points of a system and is based, in its either graphical or incidence-matrix implementation, on the same principles of structural element construction. This method enables one to effectively solve problems concerned with a preliminary estimation of the possible sets of separation products characteristic of a given composition simplex.

5.4.3 Universality of the Azeotropy Rule

The concept of a contact-free manifold as suggested by Zharov in Rev. [25] constitutes the basis for deriving the index azeotropy rule for an

$$N_{n-1,n}, N_{n-2,n}, \ldots, N_{2,n}, N_{1,n} \to S_{n-1,n}^{n-2}(N_{n-1,n-1})$$

$$S_{n-1,n}^{n-2}(S_{n-1,n-1}^{n-3}), S_{n-2,n}^{n-3}(S_{n-2,n-2}^{n-3}), \ldots, S_{3,n}^{n-3}(S_{33}^1), S_{2,n}^{n-2}(U_{22}) \to C_{n-1,n}^{n-3}(S_{n-1,n-1}^{n-3}), C_{n-2,n}^{n-3}(N_{n-2,n-2})$$

$$C_{n-1,n}^{n-3}(C_{n-1,n-1}^{n-4}), C_{n-2,n}^{n-3}(C_{n-2,n-2}^{n-5}), \ldots, C_{4,n}^{n-3}(S_{44}^1), C_{3,n}^{n-3}(U_{33}) \to C_{n-1,n}^{n-4}(C_{n-1,n-1}^{n-4}), C_{n-2,n}^{n-4}(C_{n-2,n-2}^{n-4}), C_{n-3,n}^{n-4}(N_{n-3,n-3})$$

$$\ldots\ldots\ldots\ldots\ldots\ldots\ldots\ldots\ldots\ldots\ldots\ldots\ldots\ldots\ldots\ldots\ldots\ldots\ldots\ldots\ldots\ldots\ldots\ldots\ldots \quad (26)$$

$$C_{n-1,n}^3(C_{n-1,n-1}^2), C_{n-2,n}^3(S_{n-2,n-2}^1), C_{n-3}^3(U_{n-3,n-3}) \to C_{n-1,n}^2(C_{n-1,n-1}^2), C_{n-2,n}^2(C_{n-2,n-2}^2), \ldots, S_{4,n}^2(S_{44}^2), C_{3,n}^2(N_{33})$$

$$C_{n-1,n}^2(S_{n-1,n-1}^1), C_{n-2,n}^2(U_{n-2,n-2}) \to S_{n-1,n}^1(S_{n-1,n-1}^1), S_{n-2,n}^1(S_{n-2,n-2}^1), \ldots, S_{3,n}^1(S_{33}^1), S_{2,n}^1(N_{22})$$

$$S_{n-1,n}^1(U_{n-1,n-1}) \to U_{n-1,n}, U_{n-2,n}, \ldots, U_{2,n}, U_{1,n}$$

$$N_{n,n} \to C_{n-1,n}^{n-2}(N_{n-1,n-1})$$

$$N_{1,n}, N_{2,n}, \ldots, N_{n-2,n}, N_{n-1,n} \to S_{n,n}^{n-2} \to C_{n-1,n}^{n-3}(S_{n-1,n-1}^{n-3}), C_{n-2,n}^{n-3}(N_{n-2,n-2})$$

$$S_{n-2,n}^{n-2}(S_{n-2,n-2}^{n-4}), \ldots, S_{n-2,n}^{n-2}(S_{33}^1), S_{n-1,n}^{n-2}(S_{n-1,n-1}^{n-3})$$

$$S_{3,n}^{n-3}(U_{33}), C_{4,n}^{n-3}(C_{44}^1), \ldots, C_{n-2,n}^{n-3}(C_{n-2,n-2}^{n-5}), C_{n-1,n}^{n-3}(C_{n-1,n-1}^{n-4})$$

$$S_{2,n}^{n-2}(U_{22}), S_{3,n}^{n-2}(S_{33}^1), \ldots, S_{n-2,n}^{n-2}(S_{n-2,n-2}^{n-4}), S_{n-1,n}^{n-2}(S_{n-1,n-1}^{n-3}) \to C_{n,n}^{n-3} \to C_{n-1,n}^{n-4}(C_{n-1,n-1}^{n-4}), C_{n-2,n}^{n-4}(S_{n-2,n-2}^{n-4}), C_{n-3,n}^{n-4}(N_{n-3,n-3})$$

$$C_{n-1,n}^{n-3} \quad (S_{n-1,n-1}^{n-3}), \quad C_{n-2,n}^{n-3}(N_{n-2,n-2})$$

$$C_{3,n}^{n-3}(U_{33}), C_{4,n}^{n-3}(S_{44}^1), \ldots, C_{n-2,n}^{n-3}(C_{n-2,n-2}^{n-5}), C_{n-1,n}^{n-3}(C_{n-1,n-1}^{n-4})$$

$$C_{n-1,n} (N_{n-1,n-1})$$

$$C_{4,n}^{n-4}(U_{44}), C_{5,n}^{n-4}(S_{55}^1), \ldots, C_{n-2,n}^{n-4}(C_{n-2,n-2}^{n-6}), C_{n-1,n}^{n-4}(C_{n-1,n-1}^{n-5})$$

$$\cdots \cdots \cdots \cdots \cdots \cdots \cdots \cdots \cdots \cdots \cdots \cdots \cdots \cdots \cdots \cdots \quad (27)$$

$$C_{n-2,n}^2(U_{n-2,n-2}), C_{n-1,n}^2(S_{n-1,n-1}^1) \to C_{n,n}^2 \to S_{n-1,n}^1(S_{n-1,n-1}^1), S_{n-2,n}^1(S_{n-2,n-2}^1), \ldots, S_{3,n}^1(S_{33}^1), S_{2,n}^1(N_{22})$$

$$C_{n-1,n}^2(C_{n-1,n-1}^2) \quad C_{4,n}^2(S_{44}^2), C_{3,n}^2(N_{33})$$

$$C_{n-2,n}^2(U_{n-2,n-2}), C_{n-1,n}^3(C_{n-1,n-1}^2), C_{n-1,n}^2(C_{n-1,n-1}^1) \to C_{n-1,n}^3 \quad C_{5,n}^3(S_{55}^3), C_{4,n}^3(N_{44})$$

$$C_{n-1,n}^3(U_{n-2,n-2}), C_{n-1,n}^2(S_{n-1,n-1}^1) \to U_{n-1,n}, U_{n-2,n}, \ldots, U_{2,n}, U_{1,n}$$

$$S_{n-1,n}^1(S_{n-1,n-1}^1), S_{n-2,n}^1(S_{n-2,n-2}^1), \ldots, S_{3,n}^1(S_{33}^1), S_{2,n}^1(N_{22})$$

$$C_{n-1,n}^2(C_{n-1,n-1}^2), C_{n-2,n}^2(C_{n-2,n-2}^2), \ldots, C_{4,n}^2(S_{44}^2), C_{3,n}^2(N_{33})$$

$$C_{n-1,n}^1(U_{n-1,n-1})$$

$$U_{n,n} \quad \text{———} \quad C_{n-1,n}^1(U_{n-1,n-1})$$

n-component simplex. For first-order singular points, the contact-free manifold is an isothermal-isobaric surface, that is, a manifold to which no vector of the virtual vector field is tangent (in the vector field, the vector of a vapour-liquid node is equal to zero, and the temperature surface exhibits an extremum (minimax)).

It may be presumed, in applying this approach to a derivation of the azeotropy rule, that, if there exists a one-to-one correspondence between the degenerate second-order singular points and the contact-free manifold $dT(x) = $ const, then, by performing a similar analysis, we must obtain a completely analogous result, that is, the same expression for the azeotropy rule.

In order to verify this correspondence, we must prove that, if at the point Z taken for a singular point in the second-order temperature surface $T(x)$ (an elementary singular point in the surface $dT(x)$) the determinant of the matrix I_T is equal to zero,

$$I_T = \begin{vmatrix} \dfrac{\partial^2 T}{\partial x_1^2}, & \cdots & \dfrac{\partial^2 T}{\partial x_1 \partial x_n} \\ \cdots & \cdots & \cdots \\ \dfrac{\partial^2 T}{\partial x_n \partial x_1} & \cdots & \dfrac{\partial^2 T}{\partial x_n^2} \end{vmatrix} \tag{28}$$

the determinant of the matrix I_x is also equal to zero,

$$I_x = \begin{vmatrix} \dfrac{\partial f_1}{\partial x_1} & & \dfrac{\partial f_1}{\partial x_n} \\ \cdots & \cdots & \cdots \\ \dfrac{\partial f_n}{\partial x_1} & & \dfrac{\partial f_n}{\partial x_n} \end{vmatrix} \tag{29}$$

where $f(x) = x - y$.

By substituting the quantities

$$\frac{\partial f_i}{\partial x_j} = \frac{\partial x_i}{\partial x_j}(1 - K_i) - x_i \frac{\partial K_i}{\partial x_j}$$

$$\frac{\partial x_i}{\partial x_j} = \frac{x_i - y_i}{x_j - y_j}$$

$$\left.\frac{\partial K_i}{\partial x_j}\right|_z = 0 \quad \text{(the condition for extremality of } K_i \text{ at an inflexion point)}$$

into matrix (28), we obtain, after having performed an appropriate transposition and transformation, the desired result (see Ref. [7]).

5.4 Regime of Infinite Separation Efficiency

The boundaries for distillation regions in an $(n - 1)$-dimensional simplex are surfaces which pass through definite singular points of the simplex and which represent a family of possible limiting compositions of the end separation products for an infinite set of initial compositions belonging to a given distillation region.

The boundaries in the $(n - 1)$-dimensional composition simplex can be determined by making use of a special computer program for simulating distillation regimes at infinite reflux and at infinite number of plates, since the distillation line under such conditions passes in a close neighbourhood of all the singular points that belong to the given distillation region.

Using the Wilson-Renon equations for local compositions, one can calculate the constraints imposed on the separation products typical of a given initial mixture composition.

The procedure for defining the distillation subregion to which the initial mixture belongs enables: (1) defining the type and the coordinates of singular points that impose thermodynamic constraints on the composition of distillate products; (2) studying the effect of varying pressure or of a separating agent used as an additive to the mixture for improved separation into technologically pure individual components; (3) choosing an economically effective separation scheme.

A problem for defining the distillation region boundaries is thus seen to consist of the following stages:

1. Finding the coordinates and extrema in temperature (pressure) for the singular points of an $(n - 1)$-dimensional composition simplex, as well as the establishment of the type of these singular points.

2. Identification, in the distillation curves at infinite reflux and infinite number of equilibrium contacts, of the singular points that belong to the boundary of a given distillation region.

The former problem of finding the singular point coordinates reduces to a solution of the system of nonlinear equations

$$\begin{cases} y_i - x_i = 0 \text{ (with } i \text{ varied from 1 to } n) \\ \sum_{i=1}^{n} y_i - 1 = 0 \end{cases} \quad (30)$$

where

$$y_i = P_i^0 \gamma_i / P$$

$$\gamma_i = \frac{\exp\left\{1 - \sum_{k=1}^{n}\left[\dfrac{x_k \Lambda_{ki}}{\sum_{j=1}^{n} x_j \Lambda_{jk}}\right]\right\}}{\sum_{j=1}^{n} x_j \Lambda_{ij}}$$

The parameters Λ_{ij} can be obtained, to a first approximation, by computer-aided processing of the equilibrium data for binary systems constitutive of the given multicomponent mixture.

The type of singular points in an n-component system is specified by making use of the computed partition coefficients $K_i^{(j+1...n)}$ for j components at a given singular point of $(n - j)$ multiplicity (with i run 1 through j) and the type of the extremum containing $(n - j)$ components.

At a fixed feedstock composition and a varied amount of the distillate (within the range of $D/F \to 0$ and $D/F \to 1$) one can, as shown in Ref. [26], staying within the given distillation region, specify in succession all the singular points of a multidimensional space that belong to the given region. The pattern of concentration distribution through the column height varies continuously to reveal concentration extremes in the vicinity of singular saddle-type points.

For a graphical representation of distillation regions, only results for four component mixtures have been used in this paper. The number of types for concentration distribution through the distillation column height as a function of the D/F ratio at a constant feed composition for a four-component mixture can be equal to 7 or even greater and is determined by the distillation region pattern, that is, by the number of singular saddle points occurring in the distillation region of interest.

5 System Studies of Energy-Saving Schemes

For an ideal four-component mixtures, seven types of concentration distribution as a function of D/F ratio are possible at an infinite number of plates and at a fixed feed composition.

The distillation region for a mixture of this type is confined within the entire space of the composition tetrahedron. The distillation curve, at the separation regimes chosen, passes sequentially through all the tetrahedral vertices M_1, M_2, M_3, and M_4, in a close neighbourhood of the singular saddle points M_2 and M_3, and forms, in the intermediate section of the column, constant concentration zones in these components.

For a nonideal four-component mixture ($x_1^f = 0.315$; $x_2^f = 0.14$; $x_3^f = 0.245$; $x_4^f = 0.3$), characterized by the following binary interaction parameters Λ_{ij}:

$$\begin{vmatrix} 1.0 & 0.5 & 1.0 & 1.0 \\ 1.3 & 1.0 & 0.6 & 1.0 \\ 0.6 & 0.8 & 1.0 & 1.0 \\ 1.0 & 1.0 & 1.0 & 1.0 \end{vmatrix}, \Lambda$$

the types for all singular points have been identified and the component-to-component composition at the extremes determined. The computational results are given in Table 5.3.

Table 5.3 Computational Results for Singular Points of a Four-Component Nonideal Mixture as Obtained Using Binary Interaction Parameters (Λ)

Singular points	Component-to-component composition at singular points				Maximum pressure, P	Type of singular point in quaternary system
	x_1	x_2	x_3	x_4		
M_{1234}	—	—	—	—	—	—
M_{123}	0.420	0.415	0.165	—	1.432	Node
M_{124}	—	—	—	—	—	—
M_{134}	—	—	—	—	—	—
M_{234}	—	—	—	—	—	—
M_{12}	0.549	0.451	—	—	1.420	Saddle
M_{13}	0.802	—	0.198	—	1.386	Saddle
M_{14}	—	—	—	—	—	—
M_{23}	—	0.600	0.400	—	1.407	Saddle
M_{24}	—	—	—	—	—	—
M_1	1	—	—	—	1.360	Saddle
M_2	—	1	—	—	1.277	Saddle
M_3	—	—	1	—	1.101	Saddle
M_4	—	—	—	1	1.0	Node

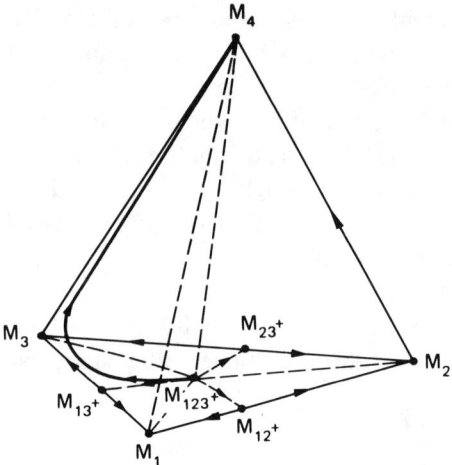

Fig. 5.13 Rectification subregions for a four-component system containing binary and ternary azeotropes

The distillation region for the initial mixture in question is confined within a space delimited by surfaces that pass through the singular points M_{13}, M_3, M_4, M_{123} (Fig. 5.13). This distillation region is characterized by seven types of concentration changes through the column height enabling one to withdraw a certain amount of a particular component (Fig. 5.14). The constant-composition zone in the intermediate section of the column is located at the singular saddle points M_3 and M_{13} which in fact correspond to a pure component 3 and a binary azeotrope 1.3. The concentration vs. column-height curves, shown in Fig. 5.14, allow suggesting a number of design variants for the component draw-off within the given distillation region at a constant pressure in the system. For the distillation region under consideration (Fig. 5.13), only two technologically pure components 3 and 4 can be drawn off. Five design variants for mixture separation can be envisaged (Fig. 5.15); for each distillation region, the simplex dimension in each case after the first column is reduced by one (in principle, the fractionation in separating the azeotropic mixtures in a column must not necessarily result in the reduction of simplex dimension as exemplified by the second column in the first scheme).

If a change in pressure causes a shift in the position of azeotropic points of an $(n - 1)$-dimensional composition simplex with the ensuing composition-

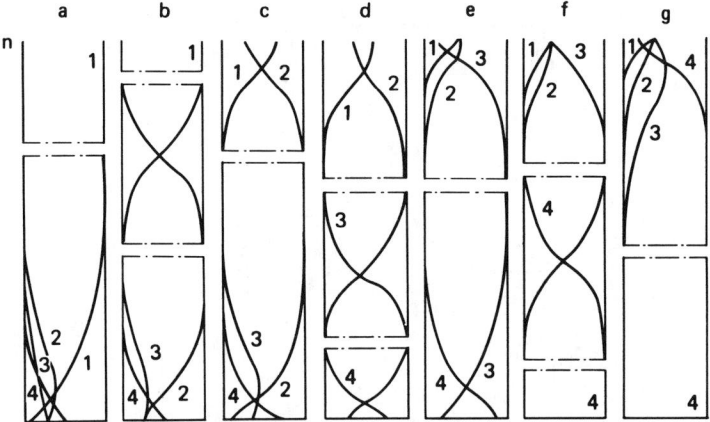

Fig. 5.14 Composition variation through the column height for a four-component mixture containing binary and ternary azeotropes at $n \to \infty$:

(a) $B/F < x_4^f$; (b) $B/F = x_4^f$;
(c) $(1 - x_{as123}^f - x_{as13}^f) > B/F > x_4^f$;
(d) $D/F = x_{as123}^f + x_{as13}^f$;
(e) $x_{as123}^f + x_{as13}^f > D/F > x_{as123}^f$;
(f) $D/F = x_{as123}^f$;
(g) $D/F < x_{as123}^f$

al redistribution between the regions of continuous distillation, the displacement of the initial mixture to another distillation region with a desired separation potential is possible. For a deeper study of this problem, additional data on vapour-liquid equilibria for binary systems at several pressures are needed.

The preferential choice of a separating agent is a computer-aided procedure which is carried out using the preliminary binary interaction parameters estimated for this agent. As the pressure of the system is made to change, or a separating component has been added to it, the computing procedure outlined above is repeated.

In such a manner, alternative engineering schemes for separation of complex azeotropic mixtures are designed. Further on, these schemes are analyzed and compared by applying the conventional methods for optimum design of distillation units. The presented approach enables one to carry out studies of the vapour-liquid diagram for any multicomponent mixture and to design an optimum engineering scheme for its separation.

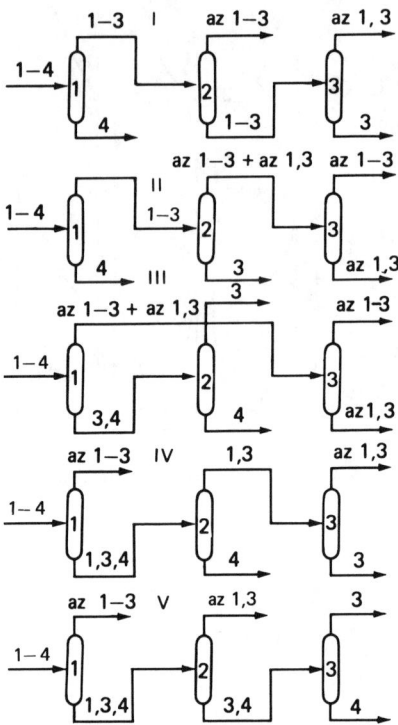

Fig. 5.15 Engineering schemes for separation of a four-component azeotropic mixture (the initial mixture composition corresponds to that of the distillation region shown in Fig. 5.13)

5.4.5 Minimum Reflux Theory

Exergy losses that may arise in mixing the column feed flows in supplying external energy to the column bear no relevance to the number of separation stages in the column and are rather associated with the extent of product draw-off and with the reflux-to-product ratio. Therefore, in designing an optimum flow sheet for separation of a specific mixture, it appears convenient to exploit one of the limiting distillation regimes where the number of vapour-reflux contacts regarded as a process variable is completely levelled off: this is the infinite-column minimum-reflux regime which provides for a desired separation of the mixture into key components, these being in this particular case the extremes in volatility. To make such a rigorous approach practicable, especially for mixtures whose separation admits of a vast variety of solutions (the case of pseudo-ideal mixtures), it was necessary to develop a mathematical description and a theory of minimum reflux regime in a functionally separate column section and in complex columns (engineering schemes).

Complex columns operated singly or as part of an engineering scheme—have been designed on the basis of studies aimed at reducing the extent of irreversibility of a distillation process and at saving thermodynamic losses. An increase, not necessarily large, in the total number of separation stages in complex-column schemes results in a 20-70% energy saving in comparison to conventional-column schemes. Serious limitations to a wide-scale application of such schemes in both design and engineering have been computational difficulties and uncertainty in the right guess of a concrete scheme, especially for multicomponent separation. Viewed from this standpoint, a study of the minimum reflux regime as effected in a column with an infinite number of separation stages in each section appears to be of great practical importance. First, the characteristics of this regime constitute a basis for process design; second, they provide a measure for estimating energy consumed in mixture separation (the major stage in distillation), which allows their use in the preferential choice of an optimum engineering scheme.

The following issues have been dealt with in developing a comprehensive theory of the minimum reflux regime in complex columns and in designing reliable computational methods:

(1) construction and study of a mathematical model for the counter-current mass-transfer section, which is the most common structural element of a column of any configuration [27, 28];

(2) analysis of all possible types of functional relation between the sections of a complex column [29, 30];

(3) description of a set of feasible regimes in a complex column [29, 30];

Let us consider a mathematical model for the countercurrent infinite column section (Fig. 5.16) consisting of a system of linear-fractional difference equations for separatory stage-to-stage transition, for example, for component-to-component reflux flows,

$$V_i(n) = L \frac{V_i(n-1)/\alpha_i}{\sum_{j=1}^{m} v_j(n-1)/\alpha_j} \tag{31}$$

$$q_i = V_i(n) - l_i(n), \quad i = 1, 2, \ldots, m$$

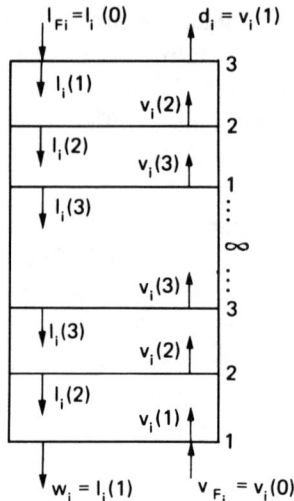

Fig. 5.16 Counter-current column section

A solution to Eq. (31) is the upper and lower concentration profiles through the section height which must satisfy the following conditions: (i) positivity of concentration profile, $V_i(n) \geq 0$, $l_i(n) \geq 0$; (ii) concentration profile matching at the section ends $l_i(0) = l_{Fi}$, $V_i(0) = v_{Fi}$; (iii) matching of the upper and lower concentration profiles at a limiting point (which is the constant composition zone).

It has been proved that if there exists a positive concentration profile which is the solution to Eq. (31), then at least one constant composition zone occurs in the column section. The constant composition zone may be related to a function

$$U(\lambda) = \sum_{i=1}^{m} \frac{\alpha_i q_i}{\alpha_i - \lambda} - V \tag{32}$$

which, in the general case, has a set of m roots $\Lambda = (\lambda_1, \ldots, \lambda_m)$, all of them being real; of these, only one is physically meaningful, $\lambda_I = \varkappa = L/V_{k_B}^{\infty}$, and can have a multiplicity of two. A solution to Eq. (31) could be obtained through the change to new variables $L(n) = (L_1(n), \ldots, L_m(n))$. The vectors $L(n)$ at two neighbouring separation stages are related to each other through

5 System Studies of Energy-Saving Schemes

a linear operator A, $L(n) = AL(n - 1)$. Two cases may be envisaged: (1) all the eigenvalues of the operator A are different; (2) among the eigenvalues, there is one of multiplicity two.

In the former case, the operator A can be diagonalized in the basis of its eigenvectors, and a solution to Eq. (31) takes the form:

$$V_i(n) = \alpha_i q_i \sum_{j=1}^{I} \frac{S_j(n)}{\alpha_i - \lambda_j}; \quad n \geqslant 0, \; i = 1, 2, \ldots, m \quad (33)$$

In the latter case, the operator A is reduced, using a basis of its $(m - 1)$ eigenvectors and one augmented vector, to a Jordan form, and the solution to Eq. (31) is represented as follows:

$$V_i(n) = \alpha_i q_i \left(\sum_{j=1}^{I} \frac{b_j(n)}{\alpha_i - \lambda_j} + \frac{b_{I+1}(n)}{(\alpha_i - \lambda_I)^2} \right);$$

$$n \geqslant 0, \; i = 1, 2, \ldots, m \quad (34)$$

Regimes in which the concentration profiles are described by formulas (33) have been called the rectification regimes, and those by formulas (34), the absorption regimes.

The upper feed may be related to a function $\Phi_1(\psi') = \sum_{i=1}^{m} \frac{l_{F_i} \alpha_i}{\alpha_i - \psi'}$ with a set of roots $\psi' = (\psi_1', \ldots, \psi_{m-1}')$, and the lower feed, to a function

$$\Phi_2(\psi'') = \sum_{i=1}^{m} \frac{V_{F_i} \alpha_i}{\alpha_i - \psi''} - V_F$$

with a set of roots $\psi'' = (\psi_0'', \ldots, \psi_{m-1}'')$ (see Fig. 5.17).

As has been proved in Refs. [28, 29], if there exist upper and lower concentration profiles, both being solutions to system (31) and satisfying the conditions of positivity and concentrational matching, then the roots of the constant-composition zone function (CCZ function) on the left (right) side of $\varkappa$ coincide with the respective roots of function $\Phi_1(\psi')$ (function $\Phi_2(\psi'')$); that is, $\lambda_j = \psi_j'$, if $\lambda_j < \varkappa$, and $\lambda_j = \psi_j''$, if $\lambda_j > \varkappa$ (Fig. 5.18).

Thus, specifying the upper and lower feed concentrations makes it possible to determine all the roots of the CCZ function, except one root $\varkappa$. Since the complete set of parameters Λ allows calculating all the design parameters by

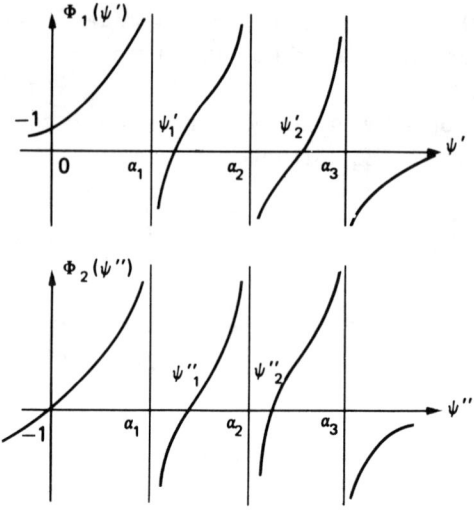

Fig. 5.17 Feed function diagrams

the formulas

$$q_i = \frac{V \prod_{j=1}^{m} (\alpha_i - \lambda_j)}{\alpha_i \prod_{j=1, i \neq j}^{m} (\alpha_i - \alpha_j)} \; ; \quad \frac{L}{V} = \prod_{j=1}^{m} \frac{\lambda_j}{\alpha_i} \tag{35}$$

the set of operating regimes of an infinite countercurrent section is one-dimensional and is completely defined through specifying the value of $\varkappa$.

As has been proved in the works [28, 29]: (1) a feasible domain for $\varkappa$ is the total set of real numbers, except intervals (ψ_i', ψ_i'') at $\psi_i' < \psi_i''$, that is, the concentration profile, which is a solution to (31) and matched to CCZ, is the solution only if $\varkappa$ falls within the feasible region, for those $\varkappa$'s that lie within interval (ψ_i', ψ_i''), ... the concentration profile necessarily transverses a negative domain; (2) the absorption regime is effected only at $\varkappa \in (\psi_i'', \psi_i')$, $\psi_i' > \psi_i''$ at any other feasible values of $\varkappa$, the rectification regime is effectuated (see Fig. 5.18).

The results obtained have been extended to a general case where both the upper and lower feeds to the section contain different or intersecting sets of components. For a multicomponent mixture, a variety of relations between

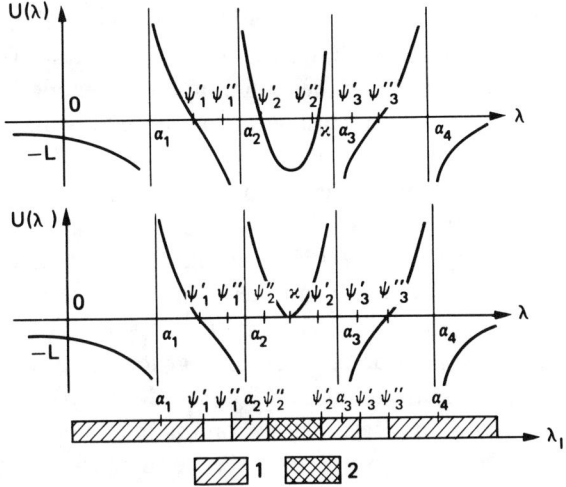

Fig. 5.18 Diagram for a constant-composition zone function:

1, rectification regime;
2, absorption regime

the upper and the lower feed compositions (and, consequently, between the ψ' and ψ'' elements) have been considered and their respective temperature profiles through the section height defined. The temperature profiles have been shown to be either monotonic, or exhibiting an extremum.

An analysis of the boundary regimes arising as $\varkappa$ assumes the values ψ_i', ψ_i'' or α_i has revealed that the boundary regimes in the rectifying section exhibit two constant-composition zones related to each other through a two-way compositional sequence; as distinct, only one constant-composition regime is typical for the boundary absorption regime.

The modelling experiments have laid the basis for developing methods and computational algorithms for operating regimes of the column section: (1) in design statement, when a minimum ratio L/V providing for a given concentration of one of the components is to be determined; (2) in control statement, when the composition of outlet streams is to be determined, given the L/V ratio; (3) in calculating separate stripping and rectifying sections in the two above statements; (4) in developing an algorithm that permits extracting information about the complete set of feasible operating regimes of the section.

In Refs. [29, 30] from an analysis of the counter-current section operation, all possible types of functional coupling (feed regime, side draw-off, and thermal coupling) have been studied, and the minimum-reflux regime in a complex column has been described. The operating regime of any distillation column is characterized by a set of process variables V^j, L^j, R, S, W^j, D^j, w_i^j, and d_i^j. A number of material balance relations and the conditions for concentration profiles impose constraints on these variables. If k variables from the above set are independent, then the k-dimensional set of regimes does not suffice to fill the space R^k completely and therefore occupies only certain domains in this space. These domains, defined by k independent variables, have a complex structure and curved boundaries, and, generally speaking, any attempt to define exactly the configuration and to provide a satisfactory description of these domains in a multisectional column is impracticable. However, the situation becomes more optimistic if we choose to characterize the minimum reflux regime in terms of variables $\Lambda^j = (\lambda_1^\lambda, \ldots, \varkappa^j, \ldots, \lambda_m^j)$, which are the roots to CCZ functions in the column section. A reverse transformation changing the variables Λ^j to process variable is accomplished by formulas (35). As has turned out, the variables Λ^j are the best choice for a structural description of the set of feasible operating regimes in a complex column; as independent variables, the quantities $\varkappa^j = L^j/V^j k_B^\infty$ should be recommended. A vast variety of regimes are amenable to a simple description in terms of these variables. Sufficient and necessary conditions for the operating regimes can easily be formulated; the regions of feasible regimes are formally reduced to readily tractable k-dimensional parallelepipeds. Relations between the independent variables for each column section type have been established.

The sections functionally interrelated through a feed system constitute an essential element in a column of any configuration (see Fig. 5.19). The relation between the CCZ functions of feed-coupled sections is expressed by material balance conditions,

$$u^j(\lambda) + \Phi^j(\lambda) = u^{j+1}(\lambda) \tag{36}$$

where

$$\Phi^j(\psi) = \sum \frac{f_i^j \alpha_i}{\alpha_i - \psi} - V_F^j$$

It has been shown that: (1) in the general case, the relation $\lambda^j > \varkappa^{j+1}$ must hold; (2) the positive concentration profile through the height of two sections is matched in the constant-composition zone and the feed zone only if the

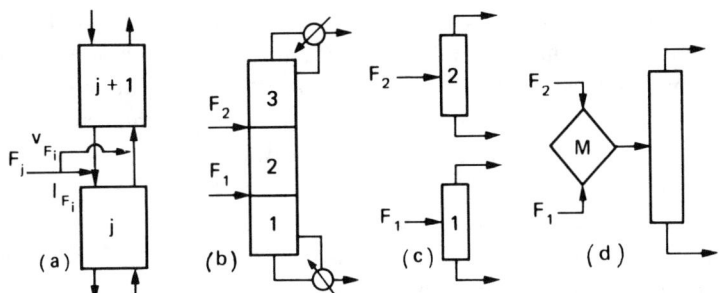

Fig. 5.19 Feed-coupled column sections (*a*); two-feed column (*b*); alternative schemes for separations (*c, d*)

roots to the CCZ functions of these sections, falling in the range between x^j and x^{j+1}, coincide.

In a two-feed column, in which the relation of type (36) is the only feasible one, the set of components supplied to the column is divided, by choosing appropriate values of x^1 and x^3, into three subsets representing heavy, distributed, and light components.

The minimum reflux regimes in the upper and lower sections do not differ essentially from the regimes in the respective sections of conventional (two-sectional) columns. In the intermediate section, depending on the ratio of the roots of functions $\Phi^1(\psi)$ and $\Phi^2(\psi)$, both rectification and absorption regimes are possible. As has been shown, at a fixed feed rate, the two-feed column, similar to a conventional column, has two degrees of freedom implying that only two of the three variables $\varkappa_1$, $\varkappa_2$, $\varkappa_3$ are independent. Systems of equations have been derived enabling one to compute the parameters of minimum reflux regime.

For two-feed columns, two additional and related to each other questions are to be answered: (1) which of the two feed streams is to be fed in at a higher level? (2) is the mixture separation in a two-feed column energetically always preferable?

To solve these problems, a boundary regime of complete separation has been considered. Here, of major importance are the relative positions of the roots ψ_T^1 and ψ_T^2 to the respective functions $\Phi^1(\psi)$ and $\Phi^2(\psi)$ within the intervals $(\alpha_T, \alpha_\Lambda)$. The derived formulas have been used to compute and compare the results for a number of variants of feed-stream separation (Fig. 5.19*b, c, d*): (1) the feed-stream separation is effected in a two-feed column; (2) each of the streams is split in a conventional column; (3) both streams are mixed and fed into a conventional column. The minimum vapour flow rate in the lower section has been used as a criterion to test the efficiency of separation. An analysis has shown that one can arrive at the right choice of the feed regime through comparing the values of ψ_T^1 and ψ_T^2; to the upper feed, a stream should be supplied such that $\psi_T^{upper} < \psi_T^{lower}$. The right choice of feed regime for a two-feed column will always result in an efficiency greater over that of a conventional column; otherwise, the former efficiency may happen to be even lower. It has been shown, for example, theat the temperature data alone are not sufficient for a right choice among the available variants, that is, a situation can be envisaged where $t_b^{upper} < t_b^{lower}$ and $\psi_T^{upper} > \psi_T^{lower}$. The analysis as outlined above can be extended to columns with an arbitrary number of feeds.

Columns with side-cut streams allow ration of the initial mixture into several streams of different composition and drawing off moderately volatile components. In such columns, a new type of functional coupling associated with side-cut streams arises between the sections (Fig. 5.20). In the general case, when the side-cut streams are withdrawn from the liquid and vapour phases, a relation between the CCZ functions of two such sections has been established,

$$u^j(\lambda) = u^{j+1}(\lambda) + f^s(\lambda), \tag{37}$$

where

$$f_s(\lambda) = \sum_{i=1}^{m} \frac{(V_i^s + l_i^s)\alpha_i}{\alpha_i - \lambda} - V^s$$

It has been proved that, in order to comply with the material balance conditions and the concentration profile matching in the sections coupled through side-cut streams, the following constraints should be obeyed: (1) $x^{j+1} \geq x^j$; (2) the roots to the CCZ functions $u^j(\lambda)$ and $u^{j+1}(\lambda)$, both on the left side of x^j and on the right side of x^{j+1}, must coincide, that is, if $\lambda_i^j < x^j$ and $\lambda_i^{j+1} < x^{j+1}$, then $\lambda_i^j = \lambda_i^{j+1}$; if $\lambda_i^j > x^{j+1}$ and $\lambda_i^{j+1} > x$, then $\lambda_i^j = \lambda_i^{j+1}$.

An analysis of the set of feasible regimes has shown that each minimum-reflux regime in a two-sectional column is brought in correspondence to a class of regimes in a side-stripper column. For all regimes of this class, the respective process parameters in the feed zone and its two neighbouring constant-composition zones are coincident. However, a multisectional side-

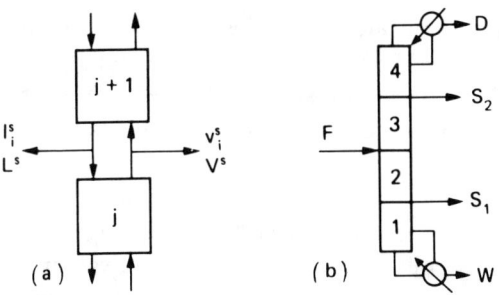

Fig. 5.20 Side-stream-coupled sections (*a*); side-stripper column (*b*)

stripper column provides a larger possibility for process optimization, since the regimes in other sections can be varied to a better advantage. All the necessary control systems have been defined and a computational algorithm has been developed.

In distillation systems with partly or completely coupled thermal streams (Fig. 5.21), one stream between two neighbouring sections is split and fed into a third (side-cut) section, and one of the outlet streams of this section is returned into the column. Depending on the phase state of the inlet and outlet streams, two variants—stripping and rectifying—are distinguished for the side-stream sections. A relation between CCZ functions for the stripping sections (Fig. 5.21a) has been established:

$$u^{s+1}(\lambda) = u^s(\lambda) + u^{s-1}(\lambda) \tag{38}$$

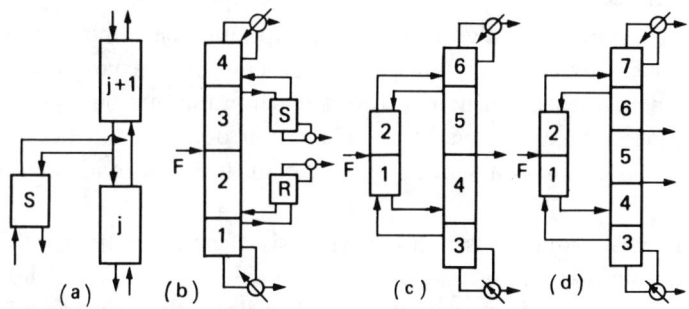

Fig. 5.21 Thermal coupling between sections (*a*); thermally coupled columns (*b*); partial coupling (*c*); complete coupling (*d*)

With the stripping section located above the feed inlet, the following conditions are fulfilled: (1) $x^s < x^{s+1} < x^{s-1}$; (2) the roots of CCZ functions for each of the three sections, which are smaller than x^s and fall between x^{s+1} and x^{s-1}, are coincident. Analogous results have been obtained for the rectifying section.

In Refs. [29, 30], formulas and non-iterative algorithms for simulating the performance of side-stripper columns have been obtained. The problem resides in determining the overall flow rates for liquid and vapour in all sections in the course of separating the initial mixture into pure components. It has been presumed, in developing the main algorithm, that the relative fugacities remain invariant over the entire temperature range inside the column. The modified algorithm allows calculating the separation parameters for a large class of zeotropic nonideal mixtures whose components have fugacities largely varied through the column height; however, the relation $\alpha_1 < \alpha_2 \ldots < \alpha_m$ is kept over the entire temperature range inside the column.

In columns with completely coupled thermal streams, both types of coupling, associated with the stripping and rectifying sections, are operative. The minimum-reflux regimes for several types of such schemes have been considered (Fig. 5.21). In the first columns (sections 1-2) of these schemes, a regime with one distributed component is effected. There exists a one-dimensional set of such regimes corresponding to different streams in the column. The least flow-rate streams correspond to a boundary regime at $x^2 = \psi_1$, $x^1 = \psi_2$. It is only in one pair of sections 3-4 or 5-6 (6-7 for the case in Fig. 5.21c), that the boundary regime of complete separation is accomplished; in the other pair, the reflux flow rates are greater over the boundary flow rates. For each scheme, formulas and computational algorithms have been derived.

The methods and algorithms for computing the minimum-reflux regime have been applied to designing a process for separation of 1, 2, 4, 5-tetramethylbenzene (durene) from a reaction mixture obtained in alkylation of pseudocumene with methanol. The major objective has been an optimization of the distillation scheme and its main technical and economic characteristics.

Durene is primarily used in the synthesis of pyromellitic dianhydride which is a starting material for aromatic polyimides. The whole-sale users of polyimides are space-aircraft, electric-power and electronic industries. The commercial interest in polyimides is also borne out by their valuable physico-chemical and electrical properties resistent to wide temperature variation, which ranks

5 System Studies of Energy-Saving Schemes

them among the most stable and durable polymer materials. A handicap to a wider commercial market of these engineering thermoplastics is their high cost. Therefore, the design of economic methods for production of durene—the starting material for these polymers—is actually an urgent problem.

On the basis of a preliminary analysis, a scheme for product separation has been proposed. The synthetic products from a reactor are delivered to a system of condensers for separating the unreacted gases, and the liquid phase is then directed to a florence flask for demixing. The lower water-methanol layer is fed into a column for separating water and methanol, and the upper layer, to a distillation column for separating the C_5-C_7 fraction from aromatic hydrocarbons (the bottoms). The bottoms are then delivered to a unit for separating aromatic hydrocarbons; the bottoms composition is given in Table 5.4.

Table 5.4 Composition of Aromatic Hydrocarbons (Bottoms Product)

Component	Mass, %	Molar fraction	Molecular mass	Boiling point, °C
Prehnitene	0.847	0.00783	134.2	205
Durene	28.249	0.26094	134.3	196
Pseudocumene	66.667	0.68753	120.2	169
Mesitylene	4.237	0.04370	120.2	165

For an optimal choice of an energy-saving scheme for aromatic hydrocarbon separation, initially a number of flow charts made up of conventional and complex columns operating in a minimum reflux regime have been calculated and compared. The mixture of polymethylbenzenes (homologous and isomeric) is practically an ideal one; the variation of relative fugacities within the separation temperature range is rather small, and the presumption of α_i = const appears to be quite acceptable.

The net vapour flow rate from all the boilers and the net liquid flow rate from all the reflux condensers were the criterion for the preferential choice of a scheme. These values were proportional, respectively, to the amounts of heat and cold consumed in mixture separation.

Five schemes made up of three conventional columns and eight schemes of complex columns have been compared (Fig. 5.22).

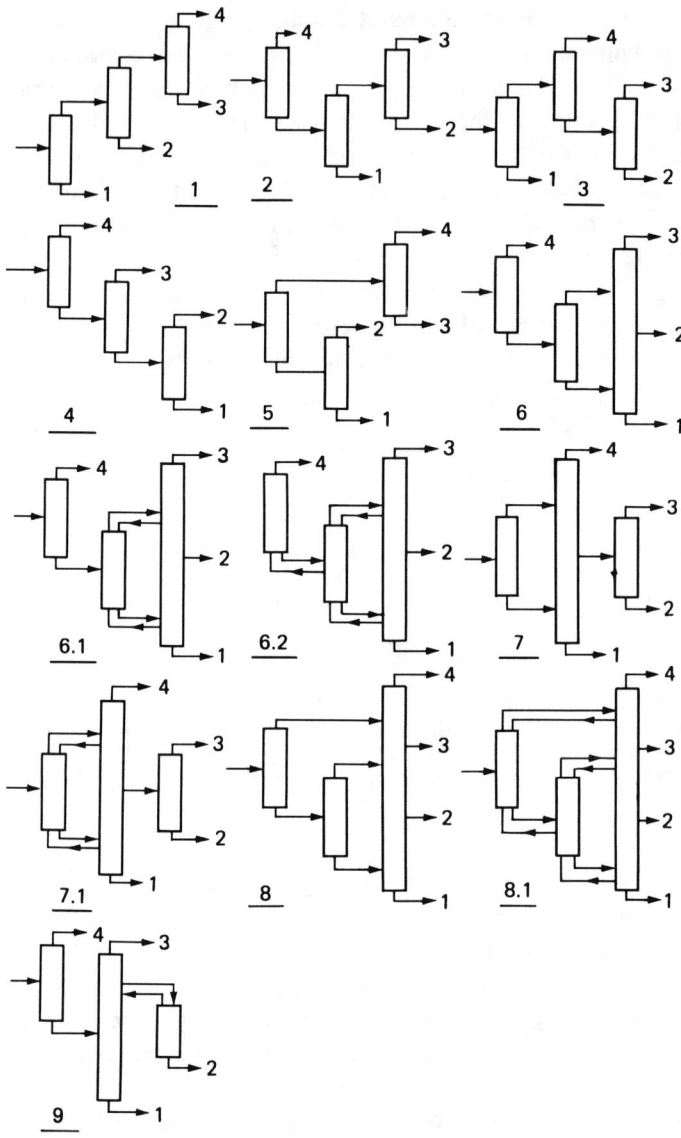

Fig. 5.22 Column configurations considered in designing a scheme for durene separation

Among the conventional column schemes, energetically advantageous is Scheme 5 which, however, has a technological shortcoming. The point is that the end product durene (component 2) obtained as a distillate is difficult to handle, since, at a high vapour concentration, it is prone to undergo an easy desublimation on cooling, and deposits crystals which clog the "rear" reflux condenser and the air ducts. A number of special technical measures must be implemented to prevent the system from clogging.

In view of this drawback, the complex column schemes were only those in which the product durene was obtained either as the bottoms, or as a liquid draw-off. The computed results are compared to the preferred conventional column scheme in Table 5.5.

Table 5.5 Comparative Results for Separation Scheme Parameters

Scheme No.	Overall vapour flow rate from the stills, kmol/h	Overall liquid flow rate from reflux condensers, kmol/h	Heating cost saving (in % against Scheme 5)	Cooling cost saving (in % against Scheme 5)
8.1	548	544	33	23
8	601	521	26	27
7.1	682	609	16	14
7	695	557	15	22
6.2	790	726	3	-2
6.1	800	727	2	-2
6	813	661	0	7
9	839	766	-3	-8
5	814	711	0	0

The best of the Schemes considered are 8.1 and 8. A slight energetic advantage of Scheme 8.1 cannot apparently offset a more complicated process control system incurring an additional capital cost. For this reason, Scheme 8 has been recommended for engineering implementation.

On the basis of the derived minimum reflux ratios, plate-to-plate calculations for the columns in Schemes 8 and 5 have been performed, with the reflux-to-product ratio varying within 1.1-1.5. The calculations have shown that at an equal number of reflux condensers and boilers, Scheme 8 has a 6% larger number of theoretical plates; however, the overall vapour load is 35%, and the liquid load 36% less than those in Scheme 5. A large number of theoretical plates (70) in one of the Scheme 8 columns is not seen as a handicap to engineering implementation, since this column may easily be adapted to this mode of operation.

List of Symbols

x	liquid mole fraction of a component
y	vapour mole fraction of a component
L	liquid flow rate in the column, mol/h
V	vapour flow rate in the column, mol/h
F	feed flow rate, mol/h
D	distillate (amount of)
B	bottoms (amount of)
T	temperature
P	total pressure
P_i^0	vapour pressure of a pure component
R	gas constant
r_{12}, g_{12}	parameters of Renon equation
K	vapour-liquid partition coefficient of a component
$U_{kr}(N_{kr})$	k-nary singular nodal point, stable (unstable) relative to the r-nary system
$S_{kr}^N(C_{kr}^N)$	k-nary singular point which in the r-nary system is a saddle point of order q (the number of negative roots of the characteristic equation) and generates (not generates) a separating manifold
α	line connecting all points of equal relative volatility,

$$\alpha_{ij} = P_i^0 \gamma_i / P_j^0 \gamma_j = 1$$

γ_i	activity coefficient
δ_{ij}	Kronecker delta
λ	eigenvalue (root of characteristic equation)
Λ_{ij}	parameters in Wilson equation

Superscripts

f	the feed plate number
$'$	(prime) related to the rectifying section of a column
0	pure component
O	singular point

Subscripts

as	related to an azeotropic point
i	the component number
p	the least volatile component

References

1. T. W. Mix, J. S. Dweck, M. Weinberg, and R. C. Armstrong, *AIChE Symp. Ser.*, **76**, No. 192: 15 (1980).
2. V. M. Platonov and I. B. Zhvanetsky, *Teor. Osnovy Khim. Tekhnol.*, **14**, No. 1: 3 (1980).
3. V. M. Platonov and B. G. Bergo, *Razdelenie mnogokomponentnykh smesei* (Separation of Multicomponent Mixtures), Khimiya, Moscow, 1965 (in Russian).
4. J. Grunberg, *Advances in Cryogenic Engineering*, Vol. 2, Timmerhaus, New York, 1960.
5. V. M. Platonov and F. B. Petlyuk, *Khim. Prom.*, No. 8: 488 (1982).
6. W. J. Stupin and F. J. Lockhart, *Chem. Eng. Progr.*, **68**, No. 10: 71 (1972).
7. F. B. Petlyuk, V. M. Platonov, and V. S. Avet'yan, *Khim. Prom.*, No. 11: 865 (1966).
8. G. M. Wilson, *J. Amer. Chem. Soc.*, **86**, No. 2: 127-133, (1964).
9. V. M. Platonov, F. B. Petlyuk and V. S. Avet'yan, *Teor. Osnovy Khim. Tekhnol.*, **5**, No. 1: 122-127 (1971).
10. M. S. Vrevsky, *Raboty po teorii rastvorov* (Works on Theory of Solutions), Akad. Nauk SSSR, Moscow, Leningrad, p. 335, 1953 (in Russian).
11. V. Yu. Aristovich, A. T. Polozov, A. K. Terpugov, and I. I. Sabylin, *Zhurn. Prikl. Khim.*, **42**, No. 7: 1531-1540 (1969).
12. N. Silverman, and D. Tassios, *Ind. Engng Chem. Process Design. Dev.*, **16**, No. 1: 13-20 (1977).
13. L. V. Morozova, V. M. Platonov, and E. V. Orlova, *Teor. Osnovy Khim. Tekhnol.*, **14**, No. 2: 206-213 (1980).
14. L. V. Morozova, V. M. Platonov, and E. V. Orlova, *Teor. Osnovy Khim. Tekhnol.*, **15**, No. 2: 180-186 (1981).
15. H. Renon and J. M. Prausnitz, Am. Inst. Chem. Engng Journal, **14**, No. 1, 135-144 (1968).
16. D. P. Tassios, *Ind. Engng Chem. Process Design Dev.* **18**, No. 1: 182-186 (1979).
17. J. M. Marina and D. P. Tassios, *Ind. Engng Chem. Process. Design. Dev.*, **12**, No. 1: 67-71 (1973).
18. L. V. Morozova and V. M. Platonov, *Teor. Osnovy Khim. Tekhnol.*, **16**, No. 5: 585-591 (1982).
19. L. V. Baburina and V. M. Platonov, Teor. Osnovy Khim. Tekhnol., **19**, No. 3: 317-321 (1985).
20. L. V. Baburina and V. M. Platonov, *Khim. Prom.*, No. 3: 178-180 (1984).
21. V. T. Zharov and L. A. Serafimov, *Fiziko-khimicheskie osnovy distillyatsii i rektifikatsii* (Physico-Chemical Fundamentals of Distillation and Rectification), Khimiya, Leningrad, p. 240, 1975 (in Russian).
22. F. B. Petlyuk and L. A. Serafimov, *Mnogokomponentnaya rektifikatsiya. Teoriya i raschet* (Multicomponent Distillation. Theory and Calculations), Khimia, Moscow, p. 304, 1983, (in Russian).
23. L. A. Serafimov, F. B. Petlyuk, and I. B. Aleksandrov, *Teor. Osnovy Khim. Tekhnol.*, **8**, No. 6: 911-914 (1974).
24. *Voprosy termodinamiki geterogennykh sistem i teorii poverkhnostnykh yavlenii* (Problems in Thermodynamics of Heterogeneous Systems and Theory of Surface Phenomena), Vol. 1, Izd. LGU, Leningrad, pp. 70-124, 1971 (in Russian).
25. E. S. Nikolaev, V. M. Platonov, and M. G. Slin'ko, *Dokl. Akad. Nauk SSSR*, **266**, No. 1: 187 (1982).
26. L. V. Morozova and V. M. Platonov, *Teor. Osnovy Khim. Tekhnol.*, **8**, No. 4: 483-488 (1974).
27. A. G. Kolokol'nikov, G. A. Meskhi, and V. M. Platonov, *Teor. Osnovy Khim. Tekhnol.*, **20**, No. 2: 136-149 (1986).
28. A. G. Kolokol'nikov, G. A. Meskhi, and V. M. Platonov, *Teor. Osnovy Khim. Tekhnol.*, **20**, No. 6: 723-732 (1986).
29. G. A. Meskhi, A. G. Kolokol'nikov, and V. M. Platonov, *Zh. Prikl. Khimii*, **59**, No. 9: 2127-2135 (1986).
30. G. A. Meskhi, A. G. Kolokol'nikov, and V. M. Platonov, *Khim. Prom.*, No. 8: 480-482 (1988).

6 Optimum Design of Chemical Plants

G. M. Ostrovsky, Yu. M. Volin, and T. A. Berezhinsky

Karpov Institute of Physical Chemistry, Obukha 10, Moscow 103064, USSR

6.1 Introduction

This paper is concerned with two classes of optimization problems encountered in chemical engineering. The first class includes problems associated with the optimization of process flow sheet (henceforth abbreviated PFS) with a fixed structure. We consider the "classical" problem of PFS optimization and its more sophisticated variants—discrete optimization, PFS optimization with uncertain simulation parameters, and multiobjective optimization. Problems of this kind arise in both the design of innovatory chemical engineering processes and the operational intensification of the existing ones, including computer-assisted CFS design. The second class comprises problems mainly concerned with the choice of optimal PFS structures (PFS synthesis). Despite the relatively recent interest in PFS synthesis, the development of methods for solution of PFS synthesis problems has become a major issue for the effective use of computing machinery in chemical engineering.

6.2 Traditional Problem of PFS Optimization

The problem of steady-state PFS regimes is defined in the following manner. Suppose there are N blocks, each of these described by a vector equation

$$y^{(k)} = f^{(k)}(x^{(k)}, u^{(k)}, \theta^{(k)}) \tag{1}$$

where $y^{(k)}$, $x^{(k)}$, and $u^{(k)}$ are, respectively, the vectors of output, input, and control variables for the kth block; $\theta^{(k)}$ is the model coefficient vector. We presume, unless specified otherwise, that the vector $\theta^{(k)}$ is known.

For simplicity, each block is presumed to have one input stream and one output stream. The bounding relations between the blocks take the form

$$x^{(i)} = \sum_{j=1}^{N} \alpha_{ij} y^{(j)} \tag{2}$$

where the structural parameters α_{ij} are defined by the conditions

$$\alpha_{ij} = \begin{cases} 1, & \text{if the output stream of the } j\text{th block is} \\ & \text{the input stream for the } i\text{th block} \\ 0, & \text{if otherwise} \end{cases} \quad (3)$$

All the structural parameters are presumed to be known. Also, the output stream from a block can serve as the input stream only for one of the blocks, that is, the relation

$$\varphi_j = \sum_{i=1}^{N} \alpha_{ij} - 1 = 0 \quad j = \overline{1, N} \qquad (4)$$

holds.

The optimization criterion is defined as

$$F = \sum_{k=1}^{N} F^{(k)}(x^{(k)}, y^{(k)}, u^{(k)})$$

where $F^{(k)}$ is the partial test related to the kth block.

The constraint imposed on the control variables is

$$u \in U \qquad (5)$$

where u is the vector whose components are the components of $u^{(k)}$ vectors and U is the specified region in the space of variables $u^{(k)}$ ($k = \overline{1, N}$).

The PFS optimization problem is thus formulated as: state and control variables $u^{(k)}$, $x^{(k)}$ are to be determined satisfying the conditions (4) and (5) where the optimization test F, with allowance made for (1) and (2), takes a minimum value. In formal terms, the optimization problem may be written as shown below:

$$\min_{x, u} \tilde{F}(x, u) \qquad (6)$$

$$\varphi(x, u, \theta) = 0 \quad (\dim \varphi = \dim x) \qquad (7)$$

$$\tilde{\psi}(x, u, \theta) \leqslant 0 \qquad (8)$$

$$a \leqslant u \leqslant b \qquad (9)$$

where x is the state variable vector; u is the control variable vector; and relations (7) define the material and heat balance equations for PFS and are actually equivalent to Eqs. (1) and (2). A common procedure in optimization is the so-called simulation approach, when the search for solution is carried out in the u variable, rather than in the (x, u) variable, space. In point of fact, this signifies that the x variables are eliminated from the (6)-(9) problem. Indeed, system (7) defines the x variables as implicit functions of u variables:

$$\varphi(x, u, \theta) = 0 \Rightarrow x = x(u, \theta)$$

Substituting the expression for x into (6), (8), we come to a problem

$$\min_{u} F(u, \theta) \tag{10}$$

$$\psi(u, \theta) \leqslant 0 \tag{11}$$

$$a \leqslant u \leqslant b \tag{12}$$

where $F = \tilde{F}(u, x(u, \theta))$, $\psi = \tilde{\psi}(u, x(u, \theta), \theta)$. For simplicity, the argument θ in functions F and ψ is occasionally omitted. As a rule, the explicit form of the function $x(u, \theta)$ is virtually unknown; therefore, for each set of variables u, the variables x are defined by solving a system of nonlinear equations (7); to solve this system is equivalent to computing a steady-state PFS. Thus, the algorithm for PFS optimization is a multilevel procedure whose lower level is represented by an algorithm for calculating a steady-state PFS, and the upper level, by an optimization algorithm. Let us consider the two algorithms in some detail.

First, we turn our attention to the lower level algorithm. As has already been noted the procedure for determining a steady-state PFS reduces to the solution of a system of nonlinear equations

$$f(x) = 0 \tag{13}$$

A common approach to the solution of (13)-type problems is presented by the so-called quasi-Newton methods [1]. With an appropriately chosen initial approximation, they provide a superlinear convergence of the iterative procedure. However, with a poorly chosen initial approximation, difficulties may arise as, for example, specified by a situation below. In the iterative solution of system (13), the search may lead to a region in which the mathematical models for operation regimes of separate PFS units have not been defined. A major reason for such a procedural misguidance is that the design of a mathematical model for a separate PFS apparatus is based on a limited range of input variables, and any departure from the preset conditions causes an untimely stoppage of the iterative procedure. In addition, the computational procedure for system (13) may lead to physically meaningless results, for example, negative concentrations.

6 Optimum Design of Chemical Plants

These circumstances often necessitate the imposition of constraints on the region within which the solution is sought. We presume that constraints take the form

$$a_j \leqslant x_j \leqslant b_j \tag{14}$$

where x_j is the jth component of vector x.

Denote a region to which constraint (14) applies as S:

$$S = \{x: a_j \leqslant x_j \leqslant b_j\}$$

Let $S(l_1, l_2, \ldots, l_k)$ denote a boundary region within the region S:

$$S(l_1, l_2, \ldots, l_k) = \{x: x \in S, x_{l_1} = c_{l_1}, \ldots, x_{l_k} = c_{l_k}\}$$

where c_{l_i} can take values either a_{l_i} or b_{l_i}. Westerberg and Benjamin [2] suggested to reduce the solution of problem (13), (14) to the minimization of a function

$$F = \frac{1}{2} \sum_{j=1}^{n} f_j^2$$

with the constraint (14) imposed, by making use of the well-developed methods of conditional optimization [3]. However, this route has a number of limitations:

1. The Hessian G of function F in the vicinity of a minimum is approximately expressed [4] as

$$G = J^T J \tag{15}$$

where $J = (\partial f_i/\partial x_j)$ is the Jacobian matrix of system (1). As is known [5], the condition number of matrix $J^T J$ is equal to the square of the condition number of matrix J. This means that the matrix $J^T J$ can become ill-conditioned, and difficulties may arise in minimizing function F.

2. The use of quasi-Newton optimization methods requires the computation of partial derivatives of function F at each iteration step.

Let us consider another route to a solution of problems (13), (14). In constructing the algorithm, we make use of the strategy of active constraints which is instrumental as a common method in solving a large class of linearly constrained optimization problems [3]. The iterative procedure for the problem's solution is represented as

$$x_{i+1} = x_i + \alpha_i p_i$$

where x_i is the value of vector x at the ith point; α_i is a scalar; and p_i is the direction of search. The jth component of vector x_i is denoted x_{ij}. Constraint (14) with the index number j at the ith point is named active if x_{ij} is equal to either b_j or a_j. A certain set of active constraints

$$x_{l_1} = c_{l_1}, \ldots, x_{l_{k_i}} = c_{l_{k_i}}$$

where $c_{l_r} = a_{l_r}$ or $c_{l_r} = b_{l_r}$ is brought in correspondence to each point x_i. In a particular case, this set can be empty. We denote part of region S, defined by active constraints at the ith point, as $S_i(l_1, \ldots, l_{k_i})$. The direction of search at the jth point will be always confined to region S_i.

The iterative procedure includes the following algorithms.

1. The algorithm for search in the interior of region S, that is, a case of $a_j < x_j < b_j$.

2. The algorithm for search at a boundary of region S.

3. The algorithm to increase the number of active constraints, that is, an algorithm for transition from a certain region $S_i(l_1, \ldots, l_k)$ to a region $S_i(l_1, \ldots, l_k, l_{k+1})$, in a particular case, the transition from the interior of region S onto its boundary.

4. The algorithm for reduction in the number of active constraints, that is, the transition from a certain region $S_i(l_1, \ldots, l_{r-1}, l_r, l_{r+1}, \ldots, l_k)$ to a region $S_i(l_1, \ldots, l_{r-1}, l_{r+1}, \ldots, l_k)$, in a particular case, the transition from the boundary of region S into its interior. We now consider the above algorithms in greater detail.

Algorithm for Search in the Interior of Region S. In the sequel, we shall need the expression for the gradient of function F,

$$\text{grad } F = J^T(x) f(x) \qquad (16)$$

For search in the interior of region S, we make use of the hybrid Powell method [6]. In this case,

$$p_i = \gamma_i p_i^{(1)} + (1 - \gamma_i) p_i^{(2)}$$

where γ_i is a scalar; $p_i^{(1)} = B_i^{-1} f(x)$ is the quasi-Newton direction; and $p_i^{(2)}$ is the steepest descent direction for function $F(x)$ defined by formula (16) in which the exact Jacobian matrix is replaced by its approximation,

$$p_i^{(2)} = -B_i^T f(x)$$

Matrix B_i is defined at each point by the Broyden formula [1]

$$B_{i+1} = B_i + \frac{(y_i - B_i s_i) s_i^T}{s_i^T s_i}$$

where $s_i = x_{i+1} - x_i$; $y_i = f_{i+1} - f_i$; f_i is the value of vector function f at the ith point.

Algorithm for Search Along Region S Boundary. Let the search be effected in the $S(l_1, \ldots, l_k)$ subspace. In moving along the region boundary, the number of search variables is always smaller than the number of equations. Therefore, the boundary search reduces to the minimization of function F,

$$\min F(x)$$
$$x \in S(l_1, \ldots, l_k)$$

By analogy with the movement in the interior of region S, the direction of search is defined as a combination of the gradient and quasi-Newton directions:

$$p_i = -[\gamma_i, E_{n-k} + (1 - \gamma_i)\tilde{G}_i^{-1}(l_1, \ldots, l_k)]B_i^T(l_1, \ldots, l_k)f(x)$$

where E_{n-k} is the identity matrix of size $(n - k) \times (n - k)$; $B_i(l_1, \ldots, l_k)$ is the approximation of the Jacobian matrix of function system $f(x)$ with respect to the variables x_j ($j = \overline{1, n}$; $j \neq l_1, \ldots, j \neq l_k$);

$$B_i(l_1, \ldots, l_k) = (b_{rj}) \quad r = \overline{1, n}$$
$$j = \overline{1, n}; \quad j \neq l_1, \ldots, j \neq l_k$$
$$\dim B_i(l_1, \ldots, l_k) = n \times (n - k)$$

and $\tilde{G}_i(l_1, \ldots, l_k)$ is the approximation of the Hessian of function F defined as

$$\tilde{G}_i = B_i(l_1, \ldots, l_k)^T B_i(l_1, \ldots, l_k)$$
$$\dim \tilde{G}_i(l_1, \ldots, l_k) = (n - k) \times (n - k)$$

which is derived from Eqs. (15) through replacing the Jacobian matrix by its approximation B_i.

In using formula (15), an approximation of the rectangular Jacobian matrix is to be sought for. The procedure is as follows. The $n \times n$ matrix B_i is again calculated by making use of the Broyden formula, this time, however, taking into account that, in search within region $S(l_1, \ldots, l_k)$,

$$s_{ij} = 0, \quad j = l_1, \ldots, j = l_k$$

It is easily seen that the matrix B_i columns with index numbers $l_1, \ldots, l_k$ remain invariant, and the elements of these columns do not affect the rest of the matrix B_i elements. It is to be inferred, therefore, that matrix $B_i(l_1, \ldots, l_k)$ is obtained by merely deleting the columns with index numbers $l_1, \ldots, l_k$.

To stop the search within region $S(l_1, \ldots, l_k)$, the following test can be made use of. The search is terminated as a minimum at x^* for function F within region $S(l_1, \ldots, l_k)$ has been found. The next step is to decide which of the active constraints at point x^* must be rejected.

Algorithm to Reduce the Number of Active Constraints. Suppose the search is being made within a subspace $S(l_1, \ldots, l_k)$. The active constraint with the index number l at point x_i is rejected if the function F gradient is pointed inside the feasible region, that is, if the condition $(-\operatorname{grad} F, n^{(l)}) \leqslant 0$ is fulfilled, where $n^{(l)}$ is the normal to the lth hyperplane $(n^{(l)})^T x - d_l = 0$, $d_l = a_l$ or $d_l = b_l$; $n^{(l)} = (0, 0, \ldots, n_l, 0, \ldots, 0)$

$$n_l = \begin{cases} +1, & \text{if } c_l = b_l \\ -1, & \text{if } c_l = a_l \end{cases}$$

This condition can be rewritten in the form

$$n_l \frac{\partial F}{\partial x_l} \geqslant 0$$

It is clear that not all of the gradient components should be necessarily calculated; it suffices to determine only the components numbered $l_1, \ldots, l_k$.

Suppose, at a point x_i, the constraint $x_{l_r} = a_{l_r}$ (here the subscript denotes the index number of the component) must not be active any longer, that is, the subsequent search is to be carried out within the region $S(l_1, \ldots, l_{r-1}, l_{r+1}, \ldots, l_k)$. As has been mentioned above, at each search point with the constraint $x_{l_r} = a_{l_r}$ remaining active, the l_r-th column of matrix B_i remains invariant. It is to be inferred, therefore, that if the movement along the active constraint $x_{l_r} = a_{l_r}$ is extended enough in time, the l_r-th column of matrix B_i will provide little information about the l_r-th column

$$\left(\frac{\partial f_1}{\partial x_{l_r}}, \frac{\partial f_2}{\partial x_{l_r}}, \ldots, \frac{\partial f_n}{\partial x_{l_r}} \right)^T$$

of the Jacobian matrix of the system of functions $f(x)$. Therefore it appears expedient, at the point x_i, to replace the l_r-th column of matrix B_i by the l_r-th column of Jacobian matrix calculated by the difference method.

Algorithm to Increase the Number of Active Constraints. Let us consider an ith iteration point for which known are both the active constraint region $S_i(l_1, \ldots, l_k)$ and matrix B_i. Suppose, in the search along the p_i-direction, a number of inactive constraints become violated. Then the step in the p_i-direction is equal to the distance from point x_i to the nearest inactive constraint. Let it be a $x_r \leq a_r$ constraint. Then the intersection point at which the straight line in the p_i-direction meets the $x_r = a_r$ hyperplane is chosen for point x_{i+1}. At point x_{i+1}, an iterative step is attempted once again in the $S(l_1, \ldots, l_k)$ subspace. If the step in the p_{i+1} direction does not violate the $x_r \leq a_r$ constraint, the search is further continued within the $S(l_1, \ldots, l_k)$ subspace; otherwise, this constraint is classified as an active one, and the search is switched to the S_{i+1} subspace,

$$S_{i+1} = \{x: x \in S_i, x_r = a_r\} = S_{i+1}(l_1, \ldots, l_k, r)$$

At point x_{i+1}, matrix B_{i+1} is derived in a usual manner.

Now, let us turn our attention to two computational aspects in the use of the suggested algorithm, namely, the calculation of the partial derivatives of a system of functions $f(x)$, and the inverse matrix.

As with any quasi-Newton method, when using this approach, we must calculate the partial derivatives of the system of functions $f(x)$ at the starting point. In addition, the partial derivatives of $f(x)$ are to be determined only at certain points of the S region boundary.

As has already been mentioned, at each search point, an inverse matrix — either B_i^{-1}, or G_i^{-1}—is to be determined. In determining these matrices, we have made use of the QR-factorization of matrix B_i:

$$B_i = Q_i R_i$$

where Q_i is an orthogonal matrix, and R_i is an upper triangular matrix. A complete QR-factorization of matrix B_i is performed only at the first point; at other points, corrections are applied to matrices Q_i and R_i in accordance with the Broyden formulas [4].

To test the efficiency of the described method, a steady-state of Williams-Otto process [7] has been calculated (see Fig. 6.1).

The mathematical models for separate blocks of this process are written as shown below:

Reactor:
$$F_{RA} = F_A + R_B - k_1 F_{RA} F_{RB} V \varrho / F_R^2$$
$$F_{RB} = F_B + R_B + (-k_1 F_{RA} F_{RB} - k_2 F_{RB} F_{RC}) V \varrho / F_R^2$$
$$F_{RC} = R_C + (2k_1 F_{RA} F_{RB} - 2k_2 F_{RB} F_{RC} - k_3 F_{RC} F_{RP}) / F_R^2$$
$$F_{RE} = R_E + 2k_2 F_{RB} F_{RC} V \varrho / F_R^2$$
$$F_{RP} = R_P + (k_2 F_{RB} F_{RC} - 0.5 k_3 F_{RC} F_{RP}) V \varrho / F_R^2$$
$$F_R = F_A + F_B + R_A + R_B + R_C + R_E + R_P$$

where $k_i = A_i \exp(-B_i/T)$.

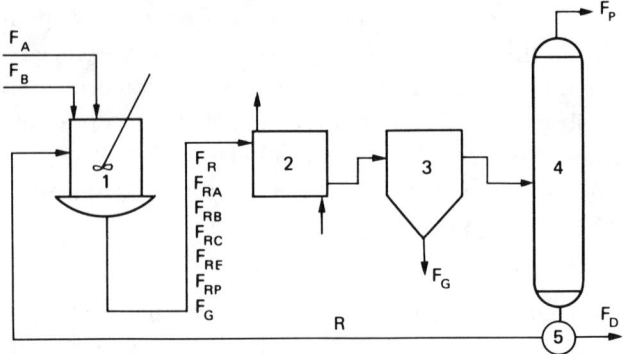

Fig. 6.1 From chart of Williams-Otto process:

(1) stirred-tank reactor, (2) heat exchanger, (3) filter, (4) distillation column, (5) stream splitter

Distillation column:

$$F_{Si} = F_{Ri}, \quad i = A, B, C, E; \quad F_{SP} = 0.1 F_{RE}$$

Stream splitter:

$$R_i = \alpha F_{Si}, \quad i = A, B, C, E, P$$

The values of A_i and B_i have been borrowed from [7]. The input variables and parameters chosen for computation are:

$$F_A = 10\ 000; \quad F_B = 40\ 000; \quad \varrho = 50; \quad T = 610 \text{ K};$$
$$\alpha = 0.74; \quad V = 60$$

The iteration variables are F_{RA}, F_{RB}, F_{RC}, F_{RE}, and F_{RP}. The additional constraints introduced into the problem are:

$$F_{RA} \geqslant 50; \quad F_{RC} \geqslant 1000; \quad F_{RE} \geqslant 50$$

The computed results are given below:
Suggested algorithm: $K_f = 73$; $K_c = 50$.
Reduction to constrained minimization problem: $K_f = 259$.

Here K_f denotes the total number of computations performed on the left-hand side of system (13);

K_c is the number of computations performed on the left-hand side of system (13) when moving along the region boundary.

6 Optimum Design of Chemical Plants

For comparison, the same problem has been solved via its reduction to a constrained minimization problem treated by the method of successive quadratic programming (SQP) [8] and via suggested algorithm.

A comparison of the results has shown the adopted approach to be more efficient over the constrained minimization methods as applied to problems (12), (13).

We now consider the higher-level algorithms. Earlier, of common use were the penalty method and the augmented Lagrangian method [3]. Recently, a novel and very efficient SQP method for solving constrained nonlinear programming problems has been developed [8, 9]. Biegler and coauthors [10-12] have noticeably contributed to further development of that method as applied to solving CFS optimization problems. The acquired computational experience has provided evidence that the efficiency of the latter method is an order of magnitude higher over the augmented Lagrangian method, let alone the penalty method. In a number of cases, a successive linearly constrained optimization method also leads to good results. To this effect, the use of a generalized adjoint-direction method in solving linearly constrained problems has been reported [13].

It should be noted, however, that the use of efficient algorithms at each level does not by any means warrant an effective functioning of the entire multilevel optimization procedure. Moreover, the autonomous operation of both levels can altogether result either in a failure, or in a very long iterative procedure. Indeed, the control variables u, liable to variation at the upper level, can enter a region where either the PFS steady-state equation (see Eq. (7)) has no solution, or the lower-level algorithm convergence under the specified initial conditions becomes extremely slow. These facts attest to that, firstly, the operative interaction between the different level algorithms must be taken into account; in particular, the operation of the lower-level algorithm must exert influence on the operation of the upper-level algorithm. Secondly, a fact should also be taken into consideration that the steady-state equations (7) at the lower level are multiply solved for a varied set of control variable values. In this connection, a comprehensive system of programs for multilevel optimization of process flow sheets has been developed at the L. Ya. Karpov Physico-Chemical Institute [14]. This system makes use of the augmented Lagrangian method [15] at the upper level, and the Broyden method [1], at the lower. We have performed the optimization of the Otto-Williams process [7] using the two techniques for a set of starting points. In the first case, the optimization has been carried out by means of a multilevel procedure with an autonomous operation of the upper- and lower-level algorithms. In this optimization, about 10% of the total set of starting points gave no solution, since the lower-level algorithm diverged at certain intermediate points. By contrast, the use of the interactive multilevel system has provided for the solution from all the starting points.

Let us consider the problem concerned with the computation of derivatives. Quasi-Newton methods, which are those most commonly used and which are the most efficient for the solution of nonlinear equation systems and for the minimization of functions of several variables, require the computation of derivatives. However, the classical methods of higher mathematics as applied to programming and computation of analytical derivatives require a large amount of preparatory work. For this reason, in most applied problems, the derivatives are computed by the difference method. However, this procedure is very laborious, since at each iteration point it requires $(n + 1)$ computations

$$\frac{\partial f}{\partial x_i} = \frac{f(x_1, \ldots, x_i + \Delta x_i, \ldots, x_n) - f(x_1, \ldots, x_i, \ldots, x_n)}{\Delta x_i}$$

This circumstance has stimulated the emergence of a novel field in applied mathematics concerned with the development of methods of analytical differentiation which enable performing the differentiation of arbitrary functions. It is noteworthy that the REDUCE-2 program complex [16] allows computing the first derivatives of functions constructed by superposition of a set of elementary functions.

Volin and Ostrovsky [17, 18] suggested a method (the so-called "adjoint process") for computing the optimization test derivatives of an arbitrary PFS with respect to input and control variables. In principle, this method can be applied to partial differentiation of arbitrary analytical functions [19]. In [20, 21], an approach has been proposed for computing partial derivatives, which is a combination of the "adjoint process" method and variational method. Regrettably, the well-known simulation systems of FLOWTRAN and FLOW-PACK type are lacking in software facilities enabling the computation of analytical derivatives. In the ROSS simulation system that has been designed at the L. Ya. Karpov Physico-Chemical Institute [22] the "adjoint process" method constitutes the basis for a computer-aided differentiation of the optimization test.

In summary, one may say that the problem of search for the local minimum of optimization problems for small- and medium-size PFS can be handled with well-designed computational methods. However, much more complex problems are quite common in practice; these will be dealt with in the subsequent sections.

6.3 Optimization of Chemical Processes with Uncertain Parameters

The problems of process flow sheet optimization with uncertainty are recently receiving much attention of researchers engaged in the field [23-26]. Commonly, in flow-sheet design many of the mathematical model parameters remain largely uncertain, and this fact is to be taken into account in solving the problems of PFS optimization. As has been shown in the foregoing Sections, the problem of PFS optimization is defined in terms of (10)-(12) formulations. In contrast to the former statements, the parameters θ are now presumed to be unknown. The only condition is that these parameters must lay within a certain region

$$T = \{\theta: \theta_i^L \leqslant \theta_i \leqslant \theta_i^U\}$$

We denote the set of vertices of this polyhedron by T_0.

The traditional approach where the vector θ is presumed equal to a certain average θ^0 ($\theta^0 \in T$), and then the problem, as defined by (10)-(12), is solved by conventional nonlinear programming methods suffers from a number of limitations. Since the actual value of θ is different from the nominal, it follows that (i) the computed regime will not be an optimal one; (ii) constraints (11) may be violated. For these reasons, the problem in question requires special methods to be developed for its solution. The definition of PFS optimization in the form of (10)-(12) is not complete, since the values of θ that are used in the optimization test (10) and constraints (11) remain largely unknown. In fact, two approaches to specifying the optimization test parameter θ are known; these are the probabilistic approach and the minimax strategy [27]. We shall use, in what follows, the probabilistic approach. A specific feature of PFS optimization problems is that constraints (11) are expected to hold true at any θ contained in the T region. This having been taken into account, the PFS optimization problem may be defined in the following manner:

$$\min_{u} E\{F(u, c)\}$$
$$\psi(u, \theta) \leqslant 0 \qquad \forall \theta \in T$$
$$a \leqslant u \leqslant b$$

where $E\{F\}$ is the expected value of F.

Commonly, one-stage and two-stage PFS optimization strategies are distinguished. The one-stage optimization problem emerges as an attempt to improve the actual process under a deficient knowledge of certain model parameters of separate PFS blocks. The two-stage optimization problem arises in designing novel chemical processes [24, 25]. To start with, we consider the one-stage optimization.

6.3.1 One-Stage Problem

In this case, the variables u are control variables. Let us simplify our problem. To this effect, we replace the expected values of F by a sum [24]:

$$\min_u F^* \tag{17}$$

$$\psi(u, \theta) \leq 0, \qquad \forall \theta \in T \tag{18}$$

$$a \leq u \leq b \tag{19}$$

where

$$F^* = \sum_{i \in I_1} w_i F(u, \theta^i) \tag{20}$$

w_i is the weight coefficients and I_1 is a set of the point numbers. We name a set of points θ^j, $j \in I_1$, the approximation set and denote it S_1,

$$S_1 = \{\theta^{(j)} : j \in I_1\} \tag{21}$$

Since the number of points contained in the region T is infinite, problem (17)-(19) is a nonlinear programming problem with infinite constraints. Two approaches to a solution of this problem are possible. Let us consider the first route. We introduce a set of critical points S_2:

$$S_2 = \{\theta^{(j)} : j \in I_2\} \tag{22}$$

where I_2 is a set of the point numbers. The set is presumed to be dense enough to cover the whole region T. Then problem (17)-(19) can be rewritten in the form:

$$\min_u F^* \tag{23}$$

$$\psi(u, \theta^{(j)}) \leq 0, \quad j \in I_2 \tag{24}$$

$$a \leq u \leq b \tag{25}$$

A limitation of this approach is that the number of constraints may be large, which makes the nonlinear programming problem an exceptionally laborious procedure. Another drawback is that the solution of problem (23)-(25) does not guarantee a complete fulfilment of constraints (18), since the number of the S_2 set points is finite. Therefore, another approach may be envisaged. The problem (17)-(19) can be rewritten as [28]

$$\min \sum_{j \in I_2} w_j F(u, \theta^{(j)}) \tag{26}$$

$$\max_{\theta \in T} \psi_i(u, \theta) \leqslant 0, \quad i = \overline{1, m} \tag{27}$$

$$a \leqslant u \leqslant b$$

A straightforward solution of problem (26), (27) leads to a two-level procedure functionally composed of two cycles: the outer cycle is concerned with the minimization of function F^*, whereas the inner cycle is assigned m problems for optimization of functions ψ_i (see (27)) which are to be solved at each step of the outer cycle. This having been taken into account, we consider the approach which is a combination of the above approaches. To this effect, we make the critical point set a variable set, with the number of its points increasing at each step of iteration procedure. Let $S_2^{(k)}$ be the set of critical points at the kth step of iteration procedure,

$$S_2^{(k)} = \{\theta^{(j)}: j \in I_2^{(k)}\} \tag{28}$$

where $I_2^{(k)}$ is the set of point numbers.

The following algorithms for solution of problem (17)-(19) can be suggested.

Algorithm 1.

Step 1. Set $k = 0$. Choose a set S_1 of approximation points and a starting set $S_2^{(0)}$ of critical points.

Step 2. Solve the problem

$$\min_{u} \sum_{j \in I_1} F(u, \theta^{(j)}) \tag{29}$$

$$\psi_i(u, \theta^{(j)}) \leqslant 0, \quad i = \overline{1, m}; \; j \in I_2^{(k)} \tag{30}$$

$$a \leqslant u \leqslant b \tag{31}$$

The $u^{(k)}$ is thus derived.

Step 3. Solve the problems, in number m

$$\max_{\theta \in T} \psi_j(u^{(k)}, \theta), \quad j = \overline{1, m} \tag{32}$$

The m points $\theta^{(j)*}$ ($j = \overline{1, m}$) for solutions of these problems are obtained.

Step 4. Construct a set $R^{(k)}$

$$R^{(k)} = \{\theta^{(j)*}: \psi_j(u^{(k)}, \theta^{(j)*}) > 0, j = \overline{1, m}\} \tag{33}$$

If this set is empty, the solution of the problem is thus terminated. Otherwise, proceed to the next step 5.

Step 5. Define $S_2^{(k+1)} = S_2^{(k)} \cup R^{(k)}$. Set $k = k + 1$ and return to Step 2.

Thus, at the kth iteration, one has to solve one problem (29)-(31) and m problems (32).

We now consider operation (32) in greater detail. Generally speaking, this must be an operation of search for global maximum. Such operations are known to be very tedious. We shall confine ourselves to considering this operation as applied to certain specific functions ψ_j.

1. Let function ψ_j be monotonic with respect to variables θ. Then the solution of problem (32) is located at a vertex θ^* of T polyhedron, with its coordinates defined as shown below:

$$\theta_i^* = [b_1 \alpha_{1i} + (1 - \alpha_{1i})a_1, \ldots, b_r \alpha_{ri} + (1 - \alpha_{ri})a_r] \tag{34}$$

$$\alpha_{ji} = \begin{cases} 1, & \text{if } \left.\dfrac{\partial \psi_i}{\partial \theta_j}\right|_{u=u^{(k)}} > 0 \\ 0, & \text{if } \left.\dfrac{\partial \psi_i}{\partial \theta_j}\right|_{u=u^{(k)}} < 0 \end{cases}$$

If all the partial derivatives $\partial \psi_i / \partial \theta_j$ retain their respective sign at all feasible values of variables θ and u, then point θ^*, once located in iteration, will remain invariant in the sequel. It is clear that in this particular case one has to do with a single-step procedure only.

2. The function ψ_j is concave with respect to variables θ_i.

In this case, the function ψ_j has one maximum, and the known methods of nonlinear programming can be effectively applied.

3. The function ψ_j is convex. In this case, the solution of problem (32) is located at a vertex of T polyhedron [25]. To this effect, it appears expedient to use a modification of Algorithm 1. The starting set of critical points $S_2^{(0)}$ is augmented by a certain number of corner points of the T polyhedron, and at Step 3, the functions $\psi_j(u^{(k)}, \theta)$ are computed for all corner points of polyhedron T that are not contained in the set $S_2^{(k)}$. Among these, m points

6 Optimum Design of Chemical Plants   495

are selected at which the functions $\psi_j(u^{(k)}, \theta)$ attain their maximum. Thus, Step 3 can be specified in the following fashion.

Step 3. Define points $\theta^{(j)*}$ ($j = \overline{1, m}$), in number m, that satisfy the condition

$$\psi_j(u^{(k)}, \theta^{(j)*}) = \max \psi_j(u^{(k)}, \theta^{(i)}), \tag{35}$$
$$\theta^{(i)} \in T^0/(S_1 \cup S_2^{(k)})$$

where $T^0/(S_1 \cup S_2^{(k)})$ is the difference between the set T^0 and the union of sets S_1 and $S_2^{(k)}$. All other steps of Algorithm 1 remain the same. We name Algorithm 1' a modified Algorithm 1 with operation (35) performed at step 3. One can easily see that the derived algorithm is an analogue of Algorithm II suggested in [25] for solving a two-phase program problem for the case where the left-hand sides of inequality constraints are convex functions with respect to the control functions and parameters θ.

4. The behaviour of function ψ_j is unknown. This appears to be the most common case, since both the convexity and concavity of a function ψ_j is, as a rule, difficult to prove. In this case, an approach as outlined below may be, for example, recommended. Presuming that the functions ψ_j are convex, one can attempt to solve the problem by using Algorithm 1'. Suppose, a certain set S_2^* of critical points and the respective value u^* have been obtained. The value u^* cannot be regarded as a solution because of the lack of a rigorous proof of the function ψ_j convexity. Therefore, Algorithm 1' can be further applied by making use of the former set S_2^* for $S_2^{(0)}$. Algorithm 1 can also be used; however, at step 3, a local maximum must be initially sought for. This having been performed, if the set $R^{(s)}$, at a certain sth iteration, has become empty, one may proceed to a global minimization at step 3.

6.3.2 Two-Stage Problem

Such problems emerge in the design of innovatory flow sheets.
Let us attend to the basic aspect of this problem. In the first place, the search variable vector must be divided into two subsets of variables: vector s of the design variables d, and vector q of the control variables z. Thus, two stages may be distinguished in this problem: the design stage of the process flow sheet and the operation stage of the process. Let us see how variables d and z and the parameters θ are related to one another at these two stages. As regards parameters θ, at the design stage they are known to belong to the T region. At the same time, at the operation stage of the process, their role

may be reassigned in more precise terms. Indeed, we can, on the basis of experimental evidence by making use of identification methods, define more accurately the coefficients in the mathematical models. Obviously, in this case there remain parametric uncertainties associated with measurement errors; however, the degree of uncertainty here tends to decrease, so that at the second stage, the parameters θ, to a good approximation, may be taken as known. The roles of the design variables d and control variables z at the two stages are markedly distinct. The variables d, specified at the design stage, remain fixed within the whole extent of process operation. By contrast, the control variables z can be optimally readjusted depending on the actual values of parameters θ. In fact, the problem thus reduces to an optimal choice of safety factors for the design variables that could ensure the fulfilments of constraints (11) at any θ satisfying the condition $\theta \in T$. Undoubtedly, in this case one may consider the variables d and z to be on equal grounds and, by adopting the one-stage strategy, attempt to solve the problem.

However, this approach can hardly be recommended, since in this case we admit implicitly that the control variables, once defined at the design stage, remain invariant in the sequel irrespective of the actual values the parameters θ may adopt. This may lead to unnecessarily large safety factors. By contrast, the possibility of manipulating the variables z at the operational stage of the process provides "favourable conditions" for the variables d to satisfy constraints (11), which, in turn, enables one to reduce the safety factors to a reasonable level.

In the light of the aforesaid, it can be shown that in this particular case the optimization problem is defined as follows [25]:

$$\min_d E\{\min_z F(d, z, \theta) / \psi_j(d, z, \theta) \leq 0, j \in J\} \tag{36}$$

$$\forall \theta \in T\{\exists z(\forall_j \in J[\psi_j(d, z, \theta) \leq 0])\} \tag{37}$$

$$J = 1, 2, \ldots, m$$

Verbally, condition (37) may be formulated as: "For each θ contained in T, there exists a vector z such that for any j, the constraint $\psi_j(d, z, \theta) \leq 0$ is fulfilled".

Using the discretization technique and replacing logical constraint (37) by an equivalent analytical formulation, we rewrite problem (36), (37) in the form [25]:

$$\min \sum_{i \in I_1} w_i F(d, z^{(i)}, \theta^{(i)}) \tag{38}$$

6 Optimum Design of Chemical Plants

$$\psi_j(d, z^{(i)}, \theta^{(i)}) \leq 0; \quad j = \overline{1, m}; \ i \in I_1 \qquad (39)$$
$$d \in D$$

$$\max_{\theta \in T} \min_{z \in Z} \max_{j \in J} \psi_j(d, z, \theta) \leq 0 \qquad (40)$$

where I_1 is the set of the approximation point numbers (see Eq. (21)).

As shown in [25], if the functions ψ_j are convex with respect to variables z, θ, then the solution of the problem

$$\max_{\theta \in T} \min_{z \in Z} \max_{j \in J} \psi_j(d, z, \theta) \qquad (41)$$

is located at a vertex of the multidimensional polyhedron (16). Problem (41) may be represented in the form

$$\max_{\theta \in T} f(d, \theta) \qquad (42)$$

where

$$f(d, \theta) = \min_{z \in Z} \max_{j \in J} \psi_j(d, z, \theta) \qquad (43)$$

Halemane and Grossmann [25] proposed the following algorithm for solution of problem (42) for the case of the functions ψ_j being convex with respect to variables z and θ.

Algorithm 2.

Step 1. Set $k = 0$. Assuming that the functions ψ_j are monotonic, select m approximation points $\theta^{(i)}$ at the vertices of the multidimensional polyhedron in accordance with formula (34). The derivatives $\partial \psi_i / \partial \theta_j$ in Eq. (34) are determined for the starting values of vectors z. Let $I_1^{(k)}$ be a set of the point numbers (34) that form the initial approximation set.

Step 2. Solve the problem

$$\min_{d, z^{(i)}} \sum_{i \in I_1^{(k)}} w_i F(d, z^{(i)}, \theta^{(i)}) \qquad (44)$$

$$\psi_j(d, z^{(i)}, \theta^{(i)}) \leq 0, \quad j = \overline{1, m}; \ i \in I_1^{(k)} \qquad (45)$$

and determine $d^{(k)}$.

Step 3. Solve problem (43) at each vertex $\beta^{(j)}$ ($\theta^{(j)} \in T^0/S_1^{(k)}$) of polyhedron T not contained in the set $S_1^{(k)}$. Assume

$$f(d^{(k)}, \theta^{(j)}) = \min_{z} \max_{j \in J} \psi_j(d^{(k)}, z, \theta^{(j)})$$

Next, the vertex is to be determined at which the value of $f(d^{(k)}, \theta^{(j)})$ attains its maximum. Let i_k is the number of this vertex. If $f(d^{(k)}, \theta_k^{(i)}) \leqslant 0$, then the solution of the problem is terminated; otherwise, go to Step 4.

Step 4. Set $S_1^{(k+1)} = S_1^{(k)} \cup \{\theta^{(i_k)}\}$ and go back to Step 2.

It is to be noted that in this particular case only one set of approximation points is used, which serves simultaneously as a set of critical points. Concerning the use of this algorithm, it should be kept in mind that:

1. The use of this algorithm is strictly justified only in the case where the functions $\psi_j (j = \overline{1, m})$ exhibit the properties of both convexity and monotonicity.

2. The rigorous proof of these properties is an arduous task. Obviously, one may resort to this algorithm assuming that the system in question is in possession of both convexity and monotonicity; however, one can never be sure that the solution found is a feasible one.

In this connection, we consider here novel approaches to the solution of problem (38)-(40). In what follows, we will need a solution of the problem as defined below:

$$\min_{d,\ z^{(k)}} \sum_{j \in I_1} w_j F(d, z^{(j)}, \theta^{(j)}), \quad k \in I \tag{46}$$

$$\psi(d, z^{(j)}, \theta^{(j)}) \leqslant 0, \qquad j \in I_1 \tag{47}$$

$$\psi(d, z^{(i)}, \theta^{(i)}) \leqslant 0, \qquad i \in I_2 \tag{48}$$

$$I = I_1 \cup I_2^{(k)}$$

where $I_2^{(k)}$ is the set of the critical point numbers (see Eq. (28)).

The proposed algorithm is as follows:

Algorithm 3.

Step 1. Set $k = 0$. Select a set of approximation points S_1 and a starting set of critical points $S_2^{(0)}$.

Step 2. Solve problem (46)-(48). The values of $d^{(k)}$ and $z^{(j)k}$ are thus obtained.

Step 3. Determine a supplementary set of critical points $R^{(k)}$.

Step 4. If the set $R^{(k)}$ is empty, stop the algorithm performance. Otherwise, proceed to Step 5.

Step 5. Determine $S_2^{(k+1)} = S_2^{(k)} \cup R^{(k)}$. Set $k = k + 1$ and go back to Step 2.

We now consider procedures for determination of the set $R^{(k)}$ presuming that both $d^{(k)}$ and $z^{(j)k}$ have been obtained by performance of Step 2 in Algorithm 3.

6 Optimum Design of Chemical Plants

Procedure 1. Apparently, a straightforward route is the solution of problem (41) at $d = d^{(k)}$. If

$$\Phi(d^{(k)}) = \max_{\theta} \min_{z} \max_{i} \psi_i(d^{(k)}, z, \theta) \leq 0$$

then the guessed regime is feasible, and problem (38)-(40) is thus solved. In this case, we assume the set $R^{(k)}$ to be empty and, in accordance with Step 4, the solution of the problem is terminated. On the contrary, if

$$\Phi(d^{(k)}) > 0$$

then the operative regime is infeasible. In this case, the set $R^{(k)}$ is represented by a single point $\theta^{(k)*}$, which is the solution of problem (41). Halemane and Grossmann [25] have shown that the function $f(d, \theta)$ is an unsmooth and multiextremal one. Hence, the solution of problem (41) reduces to the maximization of a multiextremal unsmooth function. However, it is a well-known fact that the search for an extremum of an unsmooth multiextremal function is an exceptionally difficult task. In [29], problem (41) has been reduced to the solution of a mixed-integer nonlinear programming problem. But the solution of such a problem is likewise a sufficiently laborious procedure. It has been also noted in the cited paper that in a special case of functions ψ_j monotonic with respect to variables z, this problem reduces to the enumeration of all possible sets of active constraints and to the solution of an appropriate nonlinear programming problem for each set. However, in certain situations, the number of possible sets of active constraints may be large enough.

Procedure 2. To start with, let us prove a statement. The regime $d = d^{(k)}$ is feasible, if there is at least one point $z^{(i)k}$, $i \in I_2^{(k)}$, at which the conditions

$$\max_{\theta} \psi_j(d^{(k)}, z^{(i)k}, \theta) \leq 0, \quad j = \overline{1, m} \tag{49}$$

are fulfilled. The validity of this statement is evident. Hence, the following technique for constructing $R^{(k)}$ may be suggested. Let us perform the following operation as defined below for all $i \in I_2^{(k)}$:

$$\max_{\theta} \psi_j(d^{(k)}, z^{(i)k}, \theta), \quad j = \overline{1, m} \tag{50}$$

Let $\theta^{(ji)*}$ be a solution of this problem for all $i \in I_2^{(k)}$, $j = \overline{1, m}$. We now determine points $\theta^{(j)*}$, in number m, satisfying the condition

$$\psi_j(d^{(k)}, z^{(i)k}, \theta^{(j)*}) = \max_i \psi_j(d^{(k)}, z^{(i)k}, \theta^{(i)*})$$

Then the set $R^{(k)}$ is constructed in the following manner:

$$R^{(k)} = \{\theta^{(j)*}: \psi_j(d^{(k)}, z^{(i)k}, \theta^{(j)*}) > 0\} \quad (j = \overline{1, m}) \tag{51}$$

Let, in Algorithm 4, the set $R^{(k)}$ have been constructed in accordance with Eq. (51). It is clear that, if at a certain step the set $R^{(k)}$ proves to be empty, problem (38)-(40) may be regarded as solved. It should be noted, however, that there may be envisaged a situation where this algorithm fails to yield a solution. Indeed, let at the kth and $(k + 1)$th iterations, the relations

$$d^{(k)} = d^{(k+1)}, \quad S_2^{(k)} = S_2^{(k+1)} \tag{52}$$

hold true, and at least one $\psi_j(d^{(k)}, z^{(i)k}, \theta^{(j)*}) > 0$, $\theta^{(j)*} \in S_2^{(k)}$; otherwise stated, the conditions

$$\exists j \theta^{(j)*} \in S_2^{(k)}, \quad (\psi_j(d^{(k)}, z^{(i)k}, \theta^{(j)*}) > 0) \tag{53}$$

are fulfilled.

It is clear that if $\psi_j(d^{(k)}, z^{(i)k}, \theta^{(j)*}) < 0$ for all j's, then problem (38)-(40) has a solution. It is easy to see that if relations (52)-(53) are fulfilled, then the values of $d^{(j)}$, $z^{(j)}$ ($j \geqslant k + 1$) remain invariant at each subsequent iteration, and the actual operative regime is infeasible because of relation (53). One would hesitate to predict with certainty how frequently such a situation may occur; nonetheless, one will require a mechanism to circumvent the hindrance. To that effect, other procedures for constructing the set $R^{(k)}$ can be used.

Procedure 3. First, we formulate a lemma.

Lemma 1. The regime $d = d^{(k)}$ is infeasible, if at least for one i ($i = \overline{1, m}$) there holds the condition

$$\max_{\theta} \min_{z} \psi_i(d^{(k)}, z, \theta) > 0 \tag{54}$$

Indeed, condition (54) signifies that, at a given $d^{(k)}$, there is such $\theta^{(i)*}$ that condition (40) is never fulfilled whichever be z. The converse is not, however, true. In fact, let the condition

$$\max_{\theta \in T} \min_{z \in Z} \psi_i(d, z, \theta) \leqslant 0, \quad i = \overline{1, m} \tag{55}$$

be fulfilled for any i. Let us denote the solutions for the left-hand side of inequalities (55) by $z^{(i)*}$. Generally speaking, these solutions are not necessarily the same. However, for the regime to be feasible, there must exist a single vector z^* that should provide for the fulfilment of all inequalities (46). Thus, condition (54) is a sufficient condition for the regime infeasibility. Consequently, operation (54) can be used for constructing the set $R^{(k)}$.

Procedure 4. First, we formulate a lemma.

Lemma 2. The regime $d = d^{(k)}$ is feasible if there exists at least one i ($i = \overline{1,}$

m) for which the relation

$$\max_{\theta \in T} \min_{z \in \overline{Z}} \psi_i(d^{(k)}, z, \theta) \leqslant 0 \tag{56}$$

holds. Here

$$\overline{Z_i} = \{z : z \in Z, \ \psi_i(d, z, \theta) \geqslant \psi_j(d, z, \theta) \ j \neq i\} \tag{57}$$

Indeed, we have, for any i,

$$\min_{z \in Z} \max_l \psi_l(d, z, \theta) \leqslant \min_{z \in \overline{Z}} \psi_i(d, z, \theta) \tag{58}$$

Let inequality (56) be fulfilled for a certain i. Then for all $\theta \in T$ we have

$$\min_{z \in Z} \max_l \psi_l(d^{(k)}, z, \theta) \leqslant \min_{z \in \overline{Z_i}} \psi_i(d^{(k)}, z, \theta) \tag{59}$$

which implies that the regime $d = d^{(k)}$ is feasible.

Let us consider the use of Procedures 1, 2, 3, and 4 in Algorithm 3.

Suppose, for example, that Procedure 2 is used in Algorithm 3 and that, at the kth and $(k + 1)$th steps, relations (52), (53) are fulfilled; then, at the $(k + 1)$th step, a set $R^{(k+1)}$ is constructed in the following fashion:

$$R^{(k+1)} = \{\theta^{(i)} : \psi_j(d^{(k)}, z^{(i)k}, \theta^{(i)*}\} \tag{60}$$

where $\theta^{(i)*}$ is a solution either for the problem in the left-hand side of inequality (54), or for the problem in the left-hand side of inequality (56). Procedure 2 is, in all evidence, essentially more simple than Procedure 1 (see (40)). However, the former has a limitation. In Procedure 1, the variable z plays a constraining role (at each θ, it is selected in such a manner as to reduce the maximal of functions ψ_j). On the contrary, operation (50) is performed at a fixed z. This may lead to a redundancy in the number of critical points. Therefore, in forming the critical point set at the initial steps, it may appear expedient to use Procedure 3 or 4. However, the final choice can be made only on the basis of a large computational experiment.

We now touch briefly upon operation (50). All that has been said above about operation (32) for a one-stage problem applies with equal right to operation (50). Apparently, if the function ψ_j is definitely convex with respect to variables θ and z, Algorithm 2 must be used. If a rigorous proof of the function ψ_j convexity with respect to variables z and θ is lacking, the obtained solution cannot be regarded with all certainty as a true solution of problem (35), (36). Therefore, it is recommended to resort to Algorithm 3 by making use of the specified set of critical points.

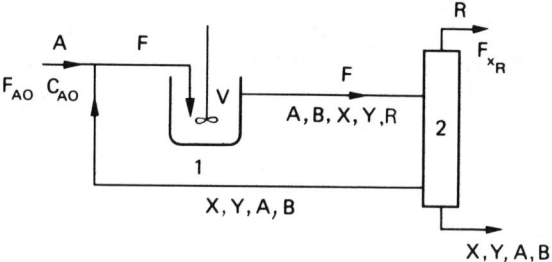

Fig. 6.2 Process flow chart:

(1) stirred-tank reactor, (2) separator

Let us consider, by way of example, the optimization problem for a flow sheet composed of a stirred-tank reactor, a separator, and a recycle (Fig. 6.2), after [23]. In the reactor of volume V, the process is carried out in accordance with Denbig reaction (Fig. 6.3), where k_i ($i = $ B, X, Y, R) are rate constants. The flow F at the reactor outlet is a mixture of five components with concentrations x_A, x_B, x_R, x_X, and x_Y. The separator is presumed to perform a perfect separation. The desired product R is drawn off at the column top; the outlet flow at the column bottom is a mixture of A, B, X, and Y components, of which X and Y are undesirable by-products. Components A, B, X, and Y are partially recycled. The recycle ratio for components A and B is α, and that for components X and Y, β. The recycle flow is returned to the reactor inlet and is mixed with the input flow which is actually a single component A at concentration c_{A0}. The input flow rate is F_{A0}. The stationary isothermal regime is described by a system of equations:

$$F_{A0} - x_A F(1 - \alpha) - V(k_B + k_X)c_{A0}x_A = 0$$
$$-Fx_B(1 - \alpha) + Vc_{A0}[k_B x_A - (k_R + k_Y)x_B] = 0$$
$$-Fx_X(1 - \beta) + Vc_{A0}k_X x_A = 0$$
$$-Fx_Y(1 - \beta) + Vc_{A0}k_Y x_B = 0$$
$$-Fx_R + Vc_{A0}k_Y x_B = 0$$
$$x_A + x_B + x_R + x_X + x_Y = 1$$

The optimization test takes the form

$$F = d_1 V + d_2 F[\alpha(x_A + x_B) + \beta(x_X + x_Y)]$$

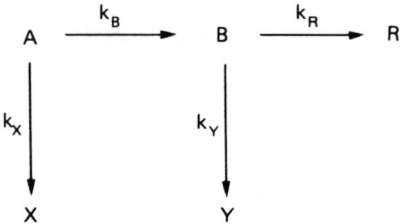

Fig. 6.3 A scheme of Denbig reaction

where d_1 is the cost per unit reactor volume and d_2 is the product transport cost per unit recycle flow volume. The condition for problem solution is that the end product flow rate must be greater than a certain prefixed value F_R,

$$F_R - F x_R \leqslant 0$$

The search variables are the reactor volume V and the recycle ratios α and β. The rate constants k_B, k_R, k_X, and k_Y are partially indeterminate. They may fall in the ranges

$$0.32 \leqslant k_B \leqslant 0.48, \qquad 0.016 \leqslant k_X \leqslant 0.024,$$
$$0.008 \leqslant k_Y \leqslant 0.012, \qquad 0.08 \leqslant k_R \leqslant 0.12.$$

The other parameters are:

$$F_{A0} = 100 \text{ mol/l}; \qquad c_{A0} = 100 \text{ mol/m}^3;$$
$$F_R = 70 \text{ mol/g}; \qquad d_1 = 10 \text{ \$/m}^3; \qquad d_2 = 0.125 \text{ \$/mol/h}.$$

The optimization problem has been considered as a one-stage one. The set of approximation points $S_1(\theta^{(1)}, \theta^{(2)}, \theta^{(3)})$ is

$$\theta^{(1)} = (0.4, 0.02, 0.01, 0.1);$$
$$\theta^{(2)} = (0.32, 0.024, 0.012, 0.08)$$
$$\theta^{(3)} = (0.48, 0.016, 0.008, 0.12)$$

The chosen weight factors are $w_1 = w_2 = w_3 = 1/3$. The starting set of critical points is $S_2^{(0)} = \theta^{(1)}$. For solving the problem, Algorithm 1 has been used. The solution has required two iterations. At the second step of the first iteration (see (32)), a point $\theta^{(2)}$ has been obtained. Therefore, the set of critical points for the second iteration contained two points, $\theta^{(1)}$ and $\theta^{(2)}$. The results for solution of this problem at nominal constants equal to the coordinates of the point $\theta^{(1)}$ and with allowance made for uncertainty are summarized in Table 6.1.

Table 6.1 Results of Optimization Problem Solutions for Reactor Flow Sheet with Uncertain Kinetic Parameters

Computation	F	V	α	β
Nominal	151.9	1.23×10^4	0.925	4.25×10^{-4}
With allowance for uncertainty	173.9	1.43×10^4	0.963	2.49×10^{-5}

As is seen, the allowance for uncertainty has resulted in a near 16% reactor volume increase.

6.4 Discrete Optimization

For a long period of time, the major issue in chemical process optimization was concerned with continuous optimization problems, that is, those in which the search variables could take any value within the given region of real space. However, a common case in design optimization is when part of the variables to be optimized can assume discrete values only. In view of this, we will consider PFS optimization problems using two groups of search variables: continuous variables x_i ($i = \overline{1, n}$) and discrete variables y_j ($j = \overline{1, m}$). Each of the y_j variables is contained within an interval

$$\bar{c}_j \leqslant y_j \leqslant \bar{\bar{c}}_j \tag{61}$$

and can assume discrete values $c_{j1}, c_{j2}, \ldots, c_{jk_j}$ such that

$$\bar{c}_j \leqslant c_{j1} < c_{j2} < \ldots < c_{jk_j} \leqslant \bar{\bar{c}}_j \tag{62}$$

For each discrete variable y_j we define a quantity

$$\gamma_j = \frac{1}{k_j} \tag{63}$$

and name it the discreteness degree. In terms of this definition, three classes of discrete optimization problems may be distinguished. The first class is associated with sufficiently small values of γ_j (the number of points k_j is sufficiently large). In this case the discreteness of variables y_j can be neglected, and we can use nonlinear programming methods taking the y_j's for continuous variables. For example, in solving integer linear programming problems with $\gamma_j \leqslant 0.05$, the variable y_j can preferably be treated as a continuous one.

The second class includes problems with $\gamma_j = 1/2$, that is, with $k_j = 2$, and the variables y_j take values extreme for the given interval. In a particular case, these are 0 and 1. This case in fact typifies the problem of an optimal PFS synthesis. Finally, problems assigned to the third class are those with the discreteness degree γ_j taking a medium value,

$$0 \leqslant \gamma_j \leqslant 0.5 \tag{64}$$

Here, two groups of problems are to be distinguished. The first group includes problems in which the discrete variables y_j can, in principle, take any value within the interval (61); these discrete values must comply with the size specifications for a given apparatus (in accordance with the adopted All-Union Standard (GOST) regulations). Such variables may be exemplified by the design variables of a shell-and-tube heat exchanger, e.g., the length and diameter of the tubes, the shell diameter. In designing a heat exchanger, these variables can take any values within a certain range; however, according to the GOST regulations, each variable can take only one value out of a specified set of discrete values. The second group comprises problems in which the variables y_j in principle cannot be continuous and can take only discrete values. Such variables may be exemplified by the total number of trays and by the ordinal number of the feed tray of a distillation column, by the total number of reactors in a cascade, and so forth.

The methods for solving problems of these two types are essentially distinct. We will confine ourselves to problems of the first type; by convention, we name them the design optimization problems.

We introduce a vector $z(z_i, \ldots, z_p)$, $(p = n + m)$, whose components are defined as $z_i = x_i$, $i = \overline{1, n}$; $z_{j+n} = y_j$, $j = \overline{1, m}$. Then the problem of mixed discrete-continuous optimization can be written in the form

$$\min_{z \in Z_1} f(z) \tag{65}$$

where $Z_1 = \{z: \psi(z) \leqslant 0, a \leqslant z \leqslant b\}$.

In discrete optimization problems, of frequent use is the branch-and-bound (BB) method [30-33]; therefore, we now briefly outline the procedure of this method as applied to our problems. The BB method is a multistep procedure, in which the ith step is brought into correspondence with a graph tree (Fig. 6.4). The total set Z_1 corresponds to the node 1 of the graph, and the set Z_j ($Z_j \subset Z_1$), to the node j. Let $R^i = (k_1, \ldots, k_s)$ denote the set of index numbers of the graph's leaf nodes at the ith step. We introduce now some

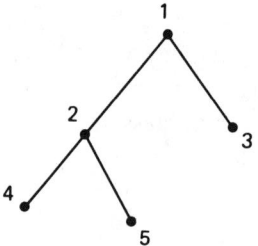

Fig. 6.4 Graph corresponding to the 3rd procedural step of branch-and-bound method

definitions. Let

$$f_j^* = \min_{z \in Z_j} f(z) \qquad (66)$$

$$z^{j*} = \arg \min_{z \in Z_j} f(z) \qquad (67)$$

The quantity f_j^* is named the lower bound of the set Z_j (or of the node j); f_1^* is the value of optimization test at the optimal point of problem 1. The lower-bound estimate of the set Z_i is called a number exhibiting the property

$$\varrho_i \leqslant f_i^* \qquad (68)$$

Commonly, the quantity ϱ_i is determined by solving an auxiliary optimization problem. If the pth node is a descendant of the node j, then, as a rule, the condition

$$\varrho_p \geqslant \varrho_j$$

holds. The set R^i is divided into three types of nodes: the quasi-optimum node, perspective nodes, and nonperspective nodes. The quasi-optimum node is a node, numbered k_i, for which the lower-bound estimate is the least one among the estimates for all the nodes of the set R^i, that is,

$$\varrho_{k_i} = \min_{j \in R^i} \varrho_j \qquad (69)$$

To be noted, if condition (68) was an exact equality, then, in accordance with Eq. (69), the optimal solution would necessarily be contained in the set Z_{k_i}. Be this the case, only the set Z_{k_i} would be liable to further analysis, omitting all other nodes $j \in R^i$ as not relevant to the optimum solution. If condition (68) is an inequality, one cannot exclude the possibility that the optimum solution is not contained in the set Z_{k_i} (however, the closer is ϱ_i to f_i^*, the less probable is such a situation). The remaining nodes are classified into perspective and nonperspective ones. The node j is called nonperspective if the set Z_i with certainty contains no optimum solution. The rest of the nodes in the set R^i are perspective ones; these, generally speaking, can contain an optimum solution.

At the ith step of the BB procedure the following operations are performed:
1. The lower-bound estimates ϱ_i for all the leaf nodes are determined.
2. The condition for procedure termination is checked. If the condition is fulfilled, the optimum solution is regarded as found, and the search is terminated. Otherwise, the search proceeds to the next operation.
3. The least lower-bound estimate is determined and its respective quasi-optimum point (quasi-optimum set) is located.
4. A graph is constructed, which corresponds to the $(i + 1)$th step. To this effect, two new nodes are constructed; they are descendants of the quasi-optimum node, and are assigned the numbers $s_i + 1$ and $s_i + 2$. These nodes are associated with the respective sets Z_{s_i+1} and Z_{s_i+2} that satisfy the condition

$$Z_{k_i} = Z_{s_i+1} \cup Z_{s_i+2}; \quad Z_{s_i+1} \cap Z_{s_i+2} = 0.$$

Thus, the node k_i is subject to branching. We assign the numbers $s_i + 1$ and $s_i + 2$ to the descendants of the k_ith node that are linked to this node through the left-hand and the right-hand edges (Fig. 6.5). Accordingly, the $(s_i + 1)$th node is henceforth called the left descendant, and the $(s_i + 2)$th node, the right descendant. The set R^i now takes the form

$$R^{i+1} = \{j: j \in R^i, s_i + 1, s_i + 2\}$$

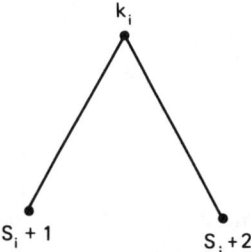

Fig. 6.5 Branching of a quasi-optimum node

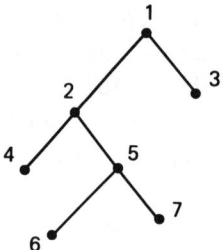

Fig. 6.6 Graph corresponding to the 4th procedural step of branch-and-bound method

The perspective nodes at the ith step remain invariant, but at subsequent steps some of them may be subject to branching. The nonperspective nodes are excluded from the graph. Shown in Fig. 6.4 is the graph pattern at the third iteration; here $R^3 = (4, 5, 3)$. Let the fifth node be quasi-optimum, the third node be perspective, and the fourth, nonperspective. Then at the fourth step, the graph takes the form as shown in Fig. 6.6, the set R^4 being (6, 7, 3).

We now consider a general approach to the construction of algorithms for the major operations of the branch-and-bound method.

Determination of Lower-Bound Estimate. Suppose there is a problem,

$$\min f(x), \quad x \in G, \quad f_i^* = \min f(x), \quad x \in G \tag{70}$$

The lower-bound estimate for the function $f(x)$ on the set (region) G is to be found. We consider three routes to the lower-bound estimate determination.

The first route is the transition to a larger region. Let $\overline{G}$ be a set such that $G \subset \overline{G}$; it is clear then that the following relation is true:

$$\min_{x \in \overline{G}} f(x) \leqslant \min_{x \in G} f(x), \quad G \subset \overline{G} \tag{71}$$

It follows therefore that the value ϱ,

$$\varrho = \min_{x \in \overline{G}} f(x)$$

can be taken for the lower-bound estimate of function $f(x)$ on the set G. The region $\overline{G}$ should be chosen in such a manner as to provide, in a sense, a more facile solution of problem (71) in comparison to the solution of problem (70). For example, if G is a nonconvex region, then problem (70) may have more than one minimum even if the function $f(x)$ itself has a single extremum and its solution requires the use of complex global search methods. If a convex envelope about the region G is taken for $\overline{G}$, then the well-developed local descent methods can be used for solving problem (71).

The second route is a transition from finite-dimensional to infinite-dimensional problems. The known problems that were solved for optimal temperature profile, pressure, and reactor load (Fan, 1966) give, in essence, the lower-bound (upper-bound) estimate for all possible variations of the said parameters as a function of the reactor length.

The third route. Let the function $f(x)$ be a nonconvex one; then problem (70) is, generally speaking, multiextremal and requires the use of global search methods. We introduce a convex function $\overline{f}(x)$ which satisfies the condition

$$\overline{f}(x) \leqslant f(x), \quad x \in G$$

Then

$$\varrho = \min_{x \in G} \overline{f}(x) \tag{72}$$

gives a lower-bound estimate for the function $f(x)$ on the set G. Usually, the function $f(x)$ satisfies the condition

$$\exists x(\overline{f}(x) = f(x))$$

We denote the region within which this condition holds by $\overline{\overline{G}}$. As is clearly seen, if a solution x^* for problem (72) is contained in the region $\overline{\overline{G}}$ ($x^* \in \overline{\overline{G}}$), then this solution gives the exact lower-bound estimate of function $f(x)$ in the region G. It should be emphasized that the lower-bound estimation for the sets of graph nodes is a central and quite complicated problem in the BB method. Regrettably, it is this part of the procedure that is nonstandard and should be worked out separately for each particular problem. The efficiency of this procedure predetermines, in many respects, a successful application of the BB method. In developing this algorithm, one must compromise with conflicting conditions: on the one hand, the closer is ϱ_i to the lower bound f_i^*, the more probable is the optimum solution to locate at quasi-optimum node and the lower is the probability for false branches to occur. On the other hand, an accurate algorithm is, as a rule, a more complex one and requires more time for determining ϱ_i.

Branching of Quasi-Optimum Node. The branching algorithm is fully defined by the concrete problem to be solved. Therefore, we content ourselves with a general remark only. Division of a quasi-optimum set k_i should preferably be carried out in such a way as to make the lower-bound estimates ϱ_{s_i+1}, ϱ_{s_i+2} for the quasi-optimum node descendants differ as much as possible. This is essential, since the greater the difference

$$\varrho_{s_i+1} - \varrho_{s_i+2} \tag{73}$$

the more probable is that the optimum solution will be an element of the set with the least estimate.

Determination of Nonperspective Nodes. For each node j, an element $\bar{z}^j$ is chosen. Let be

$$\bar{f}_j = f(\bar{z}^j) \tag{74}$$

It is clear that the condition

$$\bar{f}_j \geqslant f_j^* \geqslant \varrho_j \tag{75}$$

is valid. For this reason, $\bar{f}_j$ will be called the upper-bound estimate for the node k (the set j). Commonly, an element that can be regarded as the candidate for optimum solution is chosen as the element $\bar{z}^j$. The actual algorithm for choosing this element is defined by the specificity of the problem to be solved; however, a common feature of all potential algorithms is that in locating the point $\bar{z}^j$, they use the information derived from the optimization problem solution for the lower-bound estimate ϱ_j.

We now introduce the quantity

$$r^i = \min_{j \in R^i} \bar{f}_j \tag{76}$$

It is clear that the inequality

$$r^i > f_1^*$$

holds. For this reason, the quantity r^i is the upper-bound estimate for the solution of problem (65). We denote by x^{i**}, y^{i**} the point at which the optimization test takes the value of r^i. To be noted, the quantity $\bar{f}_j$ and, consequently, the quantity r^i define the optimization test at certain points of a feasible set. Let us assume that for the jth leaf node the inequality

$$\varrho_j > r^i \tag{77}$$

at the ith step is valid. This means, in accordance with (68), that $f_j^* > r^i$, and the optimum solution, contained in the set z_j, is clearly worse than the formerly derived solution z^{i**}. For this reason, the jth leaf node is nonperspective and can be omitted from consideration. To summarize, the set Q^i for the index numbers of nonperspective nodes at the ith step is defined by the following

6 Optimum Design of Chemical Plants

condition:

$$Q^i = \{j: \varrho_j > r^i\}$$

Termination Test. The BB procedure terminates at the ith step providing that the condition

$$|r^i - \varrho_{k_i}| \leqslant \delta \tag{78}$$

is fulfilled, where δ is a prefixed small number and ϱ_{k_i} is the lower-bound estimate at the quasi-optimum point.

Let us estimate the BB steps for a particular case. Suppose the lower-bound estimate is coincident with the lower bound: this implies that inequality (68) becomes an equality. Then at each step all the nodes, except the quasi-optimum one, are nonperspective. Let a set Z be composed of N elements. We assume that the set correspondent to the quasi-optimum node is each time divided by branching into two equivalent subsets. It is clear that in this particular case the branching continues until the leaf node contains no more than one node. The first branching having been accomplished, each of the node 1 descendants contains $N/2$ elements. The second branching yields $N/2^2$ elements for each descendent and, the process continued, the mth branching yields $N/2^m$ elements. The branching terminates as

$$\frac{N}{2^m} = 1$$

or

$$m = \frac{\ln N}{\ln 2}$$

The quantity m gives the lower estimate for the number of branchings. In reality, the number of branchings is most commonly higher than m, since in relation (68), the inequality remains valid.

6.4.1 Design Optimization Problem

We denote by I_1 a set of points in the space of variables $y_1, \ldots, y_m$, each point has coordinates as described by discrete quantities from (62). Then problem (65) may be rewritten as

$$\min_{x, y} f(x, y) \tag{79}$$

$$x, y \in Q(x, y) \tag{80}$$

$$y \in I_1 \tag{81}$$

where $Q(x, y) = \{x_i, y_j: \overline{\overline{c}} \leqslant y \leqslant c; \psi_j(x, y) \leqslant 0, j = \overline{p + 1, q}\}$; x and y are the vectors with components x_i and y_j, respectively; $\overline{c}$, $\overline{\overline{c}}$ are the vectors with components $\overline{c}_i$, $\overline{\overline{c}}_i$ ($i = 1, m$), respectively. The set Z is defined as

$$Z = Q \cup I_1$$

Two distinct approaches to the solution of this problem may be envisaged.

We now consider the first approach. As has already been noted, a typical feature of problems such as (79)-(81) is that the optimization test $f(x, y)$ can be calculated not only at discrete points of the set I_1, but also for all variables y_j ($j = \overline{1, m}$) that satisfy constraints (61). This provides the possibility of replacing mixed-integer nonlinear programming problem MI NLP (79)-(81) by a corresponding problem with continuous variables:

$$\min_{x, y} f(x, y) \tag{82}$$

$$x, y \in Q(x, y) \tag{83}$$

$$\overline{c} \leqslant y \leqslant \overline{\overline{c}} \tag{84}$$

This problem can be solved by the well-known nonlinear programming methods. Let x^*, y^* be a solution of this problem. Then the value of $\bar{y}_i$, obtained from y_i^* by rounding off to its nearest discrete value, is taken as the "optimum" variable y_i^* ($i = \overline{1, m}$). One must be aware, however, that such a route, at large discreteness of variables y_i, can lead to large errors. The second approach reduces to the solution of N problems

$$\min_x f(x, y^{(j)}) \quad j = \overline{1, N}$$

$$x \in Q(x, y^j)$$

where y^j is a point of the set I_1; $N = i_1, i_2, \ldots, i_m$ at all the points of the set I_1. If the problem is large in dimension, this procedure may require much computer time. We now consider a route which is based on the BB procedure using elements of the two above approaches, namely, (i) a purposeful enumeration of the points on the set I_1 and (ii) the transition to a continuous problem for deriving the lower-bound estimates by BB method [34]. The major operations of BB method as applied to this particular case are outlined below.

Division of Quasi-Optimum Set into Two Subsets (Node Branching). Initially, a variable y_j is chosen, and the set Z_{k_i} (k_i being the number of a quasi-optimum node at the ith step) is divided into two subsets $Z_{s_i + 1}$, $Z_{s_i + 2}$ in the following manner:

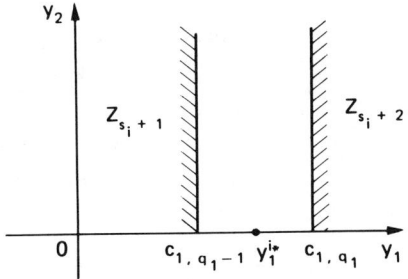

Fig. 6.7 Division of the set, corresponding to a quasi-optimum node, into two subsets Z_{s_i+1} and Z_{s_i+2}

$$Z_{s_i+1} = \{y \in Z_{k_i}, \quad y_j \leqslant c_{j,q_j-1}\} \tag{85}$$

$$Z_{s_i+2} = \{y \in Z_{k_i}, \quad y_j \geqslant c_{j,q_j}\} \tag{86}$$

Thus, in the set Z_{s_i+1}, all the points must lie on the left side of the plane $y_j = c_{j,q_j-1}$, and in the set Z_{s_i+2}, on the right side of the plane $y_j = c_{j,q_j}$, where $c_{j,q-1}$, c_{j,q_j} are two neighbouring discrete values of the variable y_j (see Fig. 6.7). The variable y_j will be named the branching variable. The determination of both the branching variable and the number q will be discussed below. Since any set Z_i is produced by a number of branchings, it may be represented in the following manner:

$$Z_i = \{y \in I_1, \quad y \in J_i\} \tag{87}$$

where

$$J_i = \{y: y_j \leqslant c_{j,q_j-1}, \quad j \in P_i; \quad y_j \geqslant c_{j,r_j}, \quad j \in T_i\}$$

We now describe how the sets P_i and T_i can be constructed. To be noted, in accordance with Eq. (85), the constraint $y_j \leqslant c_{j,q_j-1}$ corresponds to the left descendent of quasi-optimum node, and the constraint $y_j \geqslant c_{j,q_j}$, to the right descendent. For this reason, the number of the branching variable y_j belongs to the set P_i, if the quasi-optimum node (with which this variable is associated) and its left descendent both belong to the path that links the first and the ith nodes. Analogously, the number of the branching variable y_j belongs to the set T_i if the quasi-optimum node to which it corresponds and its right descendent both belong to the path that links the first and the ith nodes. Let the branching variables y_2, y_1 and y_5 correspond to the graph nodes 1, 2, 5 in Fig. 6.6; then

$$P_7 = \{2\}; \quad T_7 = \{1, 5\}$$

Lower-Bound Estimation of a Graph Node. We shall make use of the first route to the estimation which in this particular case is accomplished as the conversion from discrete to continuous variables. In what follows, we need the solutions of two optimization problems that correspond to the ith leaf node:

Problem 1

$$\min f(x, y)$$
$$x, y \in Q, \quad y \in I_1, \quad y \in J_i, \quad \bar{c} \leqslant y \leqslant \bar{\bar{c}}$$

Problem 2

$$\min f(x, y)$$
$$x, y \in Q, \quad y \in J_i, \quad \bar{c} \leqslant y \leqslant \bar{\bar{c}}$$

We denote the values of vectors x, y at the extremal point of problem 2 by x^{i*} and y^{i*}. We denote the optimum test values in problems 1 and 2 by f_i and $\bar{f}_i$. Since problem 1 differs from problem 2 only by the presence of the constraint $y \in I_1$ (this is a requirement for the variables y_j to take discrete values), the inequality

$$\bar{f}_i \leqslant f_i$$

holds. Consequently, the value of $\bar{f}_i$ may be taken as a lower-bound estimate ϱ_i,

$$\varrho_i = \bar{f}_i = \min f(x, y)$$
$$x, y \in Q, \quad y \in J_i, \quad \bar{c} \leqslant y \leqslant \bar{\bar{c}}$$

Termination Test. Let us consider the way to choose the point $\bar{y}^i$ which is a candidate for optimum solution. To this end, we take a point with its coordinates obtained by rounding off the corresponding components of vector y^{i*}:

$$\bar{y}_j^i = \begin{cases} c_{j, q_j}, & \text{if } y_j^{i*} > \dfrac{c_{j, q_j-1} + c_{j, q_j}}{2} \\[1ex] c_{j, q_j-1}, & \text{if } y_j^{i*} \leqslant \dfrac{c_{j, q_j-1} + c_{j, q_j}}{2} \end{cases}$$

where y_j^{i*}, $\bar{y}_j^i$ are the jth components of vectors y^{i*} and $\bar{y}^i$, respectively; c_{j, q_j-1} is the largest of the values of $c_{j, l}$ ($l = \overline{1, i_j}$) which are smaller than y_j^{i*} (Fig. 6.7),

6 Optimum Design of Chemical Plants

$$c_{j,q_j-1} = \max_l c_{j,l} \tag{88}$$

$$c_{j,l} \leqslant y_j^{i*}$$

c_{j,q_j} is the smallest of the values of $c_{j,l}$ ($l = \overline{1, i_j}$), which are larger than y_j^{i*} (Fig. 6.7),

$$c_{j,q_j} = \min_l c_{j,l} \tag{89}$$

$$c_{j,l} > y_j^{i*}$$

Let us consider another problem.

Problem 3

$$\min f(x, \bar{y}^i)$$

$$x \in Q(x, \bar{y}^i)$$

Since $x, \bar{y}^i \in Z_i$, relation (75) is valid. The BB procedure terminates as condition (78) is fulfilled. We now can consider the choice of the branching variables y_j and index numbers q_j in Eqs. (85) and (86), provided that problem 2 has been solved and the minimum point x^{i*}, y^{i*} identified. We focus on two approaches to the choice of this variable. The first approach makes use of a general statement, formerly expounded in describing BB method, that the quantity (73) must take its maximum value. It appears reasonable to take for a branching variable the variable in which the sensitivity of the objective function is maximal in absolute value. Commonly, to characterize the sensitivity, first derivatives

$$\frac{\partial f}{\partial y_i}$$

are used. However, in this case all the derivatives are equal to zero, since the point x^{i*}, y^{i*} is a minimum point (it is presumed, for simplicity, that only constraints (61) are imposed on variables y_j). Therefore, to characterize the sensitivity at the point x^{i*}, y^{i*} in variables y_j, it is necessary to use the second derivatives and to choose, for a branching variable, the variable whose second partial derivative

$$\frac{\partial^2 f}{\partial y_j^2}$$

is maximal in absolute value. Then the quantities c_{j,q_j-1} and c_{j,q_j} are chosen, taking into account conditions (88), (89).

In the second approach, for a branching variable the variable with the largest relative round-off error in determining $\bar{y}^i$ is chosen, that is, the one, for which the value of

$$\left| \frac{y_j^{i*} - \bar{y}_j^i}{y_j^{i*}} \right| \tag{89a}$$

is maximal.

Let us consider some of the computational aspects of the method. At each step, both problems 2 and 3 must be solved for two descendents of the quasi-optimum node. Since the condition of discreteness is not obligatory for these problems, they can be solved by the well-known nonlinear programming methods. The solution x^{i*}, y^{i*} of problem 2 for quasi-optimum node being known, this point can conveniently be chosen as initial approximation in solving problem 2 for two descendents of the quasi-optimum node. Obviously, the lower the degree of discreteness of the branching variable, the better the initial approximation. In addition, if a solution of problem 2 for one of the descendents has been found, this solution is suitable for use as an initial approximation in solving problem 3 for this descendent. Therefore, it is necessary to provide for storing the vectors x^{i*}, y^{i*} for all the leaf nodes. Thus, the solution of problem 2 is rather laborious only for the node 1; for the rest of the nodes, one is in possession of good initial approximations for solving problems 2 and 3 (in the case of a low degree of discreteness of variables).

6.4.2 Synthesis of Optimum Process Flow-Sheet Structure

In designing a novel chemical engineering process, the major concern of the designer always was the construction of efficient and economical schemes. The choice of the best version out of a host of alternatives was commonly confided to the creative care and intuition of the designer. Nowadays, a challenging problem is to hand over this job or part of it to a computer. In other words, the problem is stated as the creation of a formal theory for construction (synthesis) of chemical flow-sheet structures (CFS). A number of routes to its implementation may be envisaged. A first route is the so-called heuristic route which consists in a formalization of the mode of thinking specific of the creative activity of the human designer, in a formalization of the extant heuristic rules and creation of novel ones, in the development of a new methodology of using these rules, in a judicial choice of their priority, and so forth [35].

A second route is the use of thermodynamic approaches [36]. And, finally, a third route is a completely algorithmized approach consisting in a mathematical formulation of the process flow-sheet synthesis problem and in the elaboration of an appropriate mathematical apparatus for its solution. The two former approaches do not by any means guarantee the finding of an optimum flow-sheet structure, whereas the third one is commonly conducive to tedious mathematical problems. The works concerned with all of the three trends has been considered in [37]. Here we confine ourselves to the algorithmic methods of PFS synthesis. In the first place, the structural parameter (SP) method should be mentioned [38]. Recently, Grossmann and coworkers have published a number of papers concerned with the algorithmic methods of PFS synthesis [39, 40].

In this paper we wish to propose a generalization of the structural parameter method. By analogy with the PFS optimization problem, the mathematical formulation of PFS synthesis is given by Eqs. (1)-(5). However, in this case the binary variables are unknown and are included among the search variables, whereas Eqs. (4) are treated as constraints of equality type. Thus, in this special case, the binary variables α_{ij} and control variables $u^{(k)}$ must be determined to satisfy conditions (3), (4), (5) under which the optimization test F, with allowance made for (1), (2), takes a minimum value. Otherwise stated, the PFS synthesis problem reduces to a problem of mixed-integer nonlinear programming (MINLP). In the SP method, conditions (3) in problem (1)-(5) are replaced by conditions

$$0 \leqslant \alpha_{ij} \leqslant 1 \tag{90}$$

that is, the binary variables are replaced by continuous variables within the interval [0, 1]. A flow-sheet interpretation of SP method is known [38]. More exactly, the SP method is equivalent to the optimization of a certain global flow sheet constructed in the following manner. At the inlet and the outlet of each main block, a flow mixer and a stream splitter, respectively, are inserted. The blocks are interconnected through flow lines in such a manner that Eqs. (2) become material balance equations for the flow mixer at the ith block inlet, and Eqs. (4) become material balance equations for the stream splitter at the jth block outlet. The global flow sheet includes all alternative variants, from among which an optimum scheme is to be chosen. A global flow sheet for the problem of optimal flow-sheet structure design in which a stirred-tank reactor, or a plug-flow reactor, and a separator can be used is shown in Fig. 6.8 [41]. Commonly, two essential limitations of this approach are commented upon. The first shortcoming is that the structural parameters derived in the course of problem optimization can assume noninteger values. The roundoff

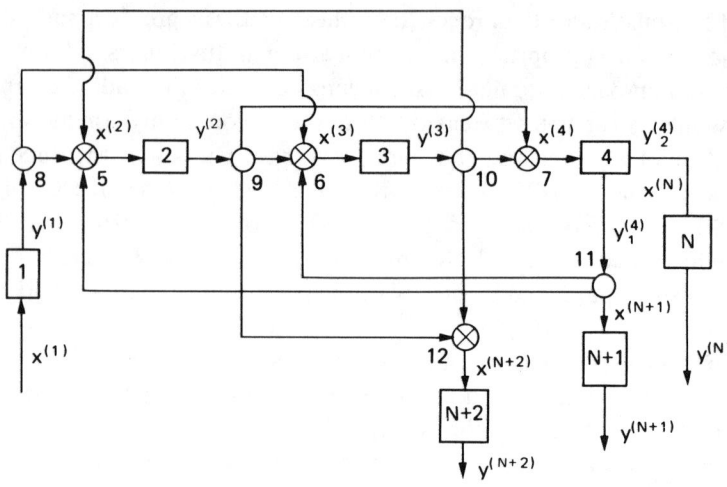

Fig. 6.8 Global reactor-separator flow sheet

of a noninteger value to the nearest integer can lead to a point distant from the optimum. Let us consider this issue in greater detail. In principle, the integrity of structural parameters α_{ij} is not a necessary condition. Suppose, noninteger values of α_{kj} and α_{ik}^* have been obtained. We can accept the solution obtained, although in this case we must provide for a flow mixer at the inlet and for a stream splitter at the outlet of the kth block. Simultaneously, there may happen situations when the structural parameters can take values either 0, or 1. In this case, a flow-sheet design optimization of the first type, based on the BB method [42, 43], can be used. Here, the structural parameters α_{ij} can, in search performance, take noninteger values (they must be integers only in the final representation of problem solution). Therefore, we can use the PFS design optimization method as it has been previously outlined, with no modification made. The only essential point is to specify a rule for choosing the branching variable. One must be aware that the rule according to which the necessary condition for choosing the branching variable is a maximum of quantity (89a) may happen to be ineffective. Suppose, for example, the parameter α_{ik} has been chosen as a branching variable. If the stream at the kth block outlet is close to zero, then two descendents of the quasi-optimum node as estimated by optimum tests will be close in value to each other.

The second essential limitation of the structural parameter method is that the objective function can have several local minima [44], which thus suggests the preferential use of global optimization methods; however, these are not as yet sufficiently elaborate to guarantee a good solution.

In this connection we consider now a problem of search for a global extremum by the structural parameter method [51].

First let us take a closer look at the objective function (5). Suppose, the profit

$$Q = \sum_{k=1}^{N} Q^{(k)}(x^{(k)}, y^{(k)}, u^{(k)})$$

is to be maximized, where $Q^{(k)}$ is part of profit from the kth block. We divide the function $Q^{(k)}$ into two parts:

$$Q^{(k)} = Q_1^{(k)} - Q_2^{(k)}$$

where $Q_2^{(k)}$ is the investment cost in the kth block, and $Q_1^{(k)}$ is the remainder of function $Q^{(k)}$. We now consider three types of function $Q_2^{(k)}$. The first type takes the form [44]

$$Q_2^{(k)} = \varphi_k(V^{(k)}) = b_k(V^{(k)})^{a_k}, \quad b_k > 0, \quad 0 \leqslant a_k \leqslant 1 \tag{90a}$$

where $V^{(k)}$ is a variable characteristic of the apparatus size. This form of function $Q_2^{(k)}$ is typical of a heat exchanger ($V^{(k)}$ being the heat transfer surface), or of a reactor ($V^{(k)}$ being the reactor volume). It should be emphasized that in this case $Q_2^{(k)}$ is the function of a single variable that can be chosen as an independent variable.

The second type of functions $Q_2^{(k)}$ can be written in the form

$$Q_2^{(k)} = \varphi_k(V_1^{(k)})\chi_k(V_2^{(k)}) \tag{91}$$

where

$$\varphi_k(V_1^{(k)}) = c_1^{(k)}(V_1^{(k)})^{a_k}; \quad \chi_k(V_2^{(k)}) = c_2^{(k)}(V_2^{(k)})^{b_k};$$
$$c_1^{(k)} > 0; \quad c_2^{(k)} > 0; \quad 0 \leqslant a_k \leqslant 1; \quad b_k \geqslant 1$$

where $V_1^{(k)}$, $V_2^{(k)}$ are variables characteristic of the apparatus size. For example, for a distillation column, $V_1^{(k)}$ may be the column weight, and $V_2^{(k)}$, the column diameter [45].

We now consider the third type of the investment cost function. In a simple modelling of distillation columns, the investment cost function is often brought into the form of (90a), but here the variable $V_1^{(k)}$ serves as a quantitative characteristic of the inlet flow. A basic distinction of this type of investment cost function from the first type is that here the quantity $V^{(k)}$ is a dependent variable and is, generally speaking, a nonlinear function of independent search variables.

Since the synthesis problem is posed as a minimization problem, then

$$F^{(k)} = -Q^{(k)} = -Q_1^{(k)} + Q_2^{(k)}$$

To start with, we consider the synthesis problem for a case where the first type of investment cost function is used for all flow-sheet blocks. As has already been mentioned, the arguments of functions $Q_2(V^{(k)})$ in this case are independent variables.

We denote by z a vector of state variables whose components are both the components of the vector u and variables α_{ij} ($i = \overline{1, N}$; $j = \overline{1, N}$).

Suppose, for simplicity, the variables $V^{(k)}$ are the first N variables z_i ($z_i = V^{(i)}$), ($i = \overline{1, N}$).

By deleting the state variables x in a manner as this has been done in handling problems (10)-(12), it is easy to show that the PFS synthesis problem can be written as

$$\min_z f(z) \tag{92}$$

$$\psi_i(z) \leq 0, \quad i = \overline{1, p} \tag{93}$$

$$0 \leq z_i \leq \overline{z}_i \tag{94}$$

$$f(z) = \theta(z) + \sum_{i=1}^{N} \varphi_i(z)$$

where function $\theta(z)$ is

$$\theta(z) = -\sum_{k=1}^{N} Q_1^{(k)} \tag{95}$$

As is easily seen, the functions $\varphi_i(z)$ are concave (see Fig. 6.9). We denote the feasible region for this problem (92)-(94) by Z_1:

$$Z_1 = \{z: \psi_i(z) \leq 0, \; i = 1, p; \; 0 \leq z \leq \overline{z}\}$$

Generally speaking, the problem in question is a nonconvex nonlinear programming problem and may be a multiextremal one.

The concavity of functions $\varphi_i(z_i)$ [44] is a source of the problem's global nonconvexity. Let, for example, the function $\theta(z)$ be a convex and one-extremal function. Since the relation

$$\frac{\partial \varphi_i}{\partial z_i} \to \infty \text{ at } z_i \to 0$$

holds, the function $f(z)$, as a function of one variable z_i ($i \leqslant N$), has two minima: at $z_i = 0$ and at a certain $z_i = z_i^*$. This is the case for all z_i ($i = \overline{1, N}$). It is clear that the coordinates of function $f(z)$ have nonzero values only at one local minimum whereas at the other local minima, one or several coordinates z_i ($i \leqslant N$) are zero. It is easily seen that the case $z_i = 0$ corresponds to the nonexistence of the ith apparatus. Thus, the local minimum of function $f(z)$ corresponds to a flow sheet with one or several blocks missing, that is, to an alternative of the flow sheet. To briefly summarize, the structural parameter method actually reduces the problem of enumeration of the alternative flow-sheet variants to the enumeration of local minima for function $f(z)$.

As stated by Grossmann [44], the concavity of function $\varphi_i(z_i)$ is a major source of nonconvexity for a CFS synthesis problem. Accordingly, we presume that the function $\theta(z)$ is convex in a feasible region Z_1, and the feasible region Z_1 is also convex. If the functions $\varphi_i(z_i)$ are linear, the problem is a one-extremal one. Let now the functions $\varphi_i(z_i)$ be concave ($b_i < 1$).

We consider here a method of search for a global extremum in problem (92)-(94) based on the BB method. In this approach, the original problem reduces to a sequence of subproblems in whose solving the functions $\varphi_i(z_i)$ are approximated by linear functions.

We have shown earlier that the BB method reduces to a multistep procedure in which the ith step is brought in correspondence to a binary tree. Each node of this tree corresponds to a set of points Z_i in the space of variables z,

$$Z_i = \{z: z \in Z_1, \overline{c}_j^{(i)} \leqslant z_j \leqslant \overline{\overline{c}}_j^{(i)}, j \in P_i\}$$

where

$$P_i = (i_1, i_2, \ldots, i_{r_i})$$

is a set of the variable indices. At the node 1, the set P_i is empty. The way to forming the sets P_i will be dealt with below.

The set Z_i is distinguished from the set Z_1 by the occurrence of additional constraints

$$\overline{c}_j^{(i)} \leqslant z_j \leqslant \overline{\overline{c}}_j^{(i)} \qquad (96)$$

imposed, however, only on some of the variables on which the functions φ_i ($i = \overline{1, N}$) are dependent. The choice of constants $\overline{c}_j^{(i)}$, $\overline{\overline{c}}_j^{(i)}$ will also be dealt with below. Let us consider the major operations of the BB method as applied to the case in question.

Lower-Bound Estimation. We consider two optimization problems associated with the ith node.

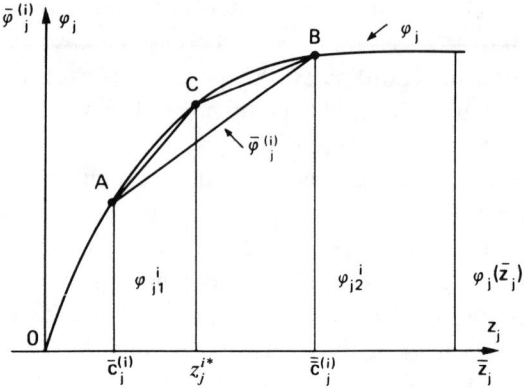

Fig. 6.9. Graphical $\bar{\varphi}_j^{(i)}$, φ_j versus z_j relationships

Problem 1

$$\min f(z)$$
$$z \in Z_i$$

Problem 2

$$\min \bar{f}^{(i)}(z)$$
$$z \in Z_i$$

where

$$\bar{f}^{(i)}(z) = \theta(z) + \sum_{j=1}^{N} \bar{\varphi}_j^{(i)}(z_j)$$

$$\bar{\varphi}_j^{(i)}(z_j) = \varphi_{j1}^i + \gamma_j^i(z_j - \bar{c}_j^{(i)}) \qquad (97)$$

$$\gamma_j^i = \frac{\varphi_{j2}^{(i)} - \varphi_{j1}^{(i)}}{\bar{\bar{c}}_j^{(i)} - \bar{c}_j^{(i)}}; \quad \varphi_{j1}^i = \varphi_j(\bar{c}_j^{(i)}); \quad \varphi_{j2}^i = \varphi_j(\bar{\bar{c}}_j^{(i)})$$

Relation (97) is the equation of a straight line passing through points A and B (see Fig. 6.9). It is clear that the inequality below is always valid:

$$\bar{f}^{(i)}(z) \leqslant f(z), \; z \in Z_i \qquad (98)$$

We denote the value of vector z at the extremal point of problem 2 by z^{i*}. It can easily be shown, with reference to relation (98), that the quantity

$$\varrho_i = \min \bar{f}^{(i)}(z) \qquad (99)$$
$$z \in Z_i$$

can be taken as the lower-bound estimate.

Node Branching. We adopt the following procedure to divide the set Z_{k_i}, corresponding to the quasi-optimum node, into two subsets Z_{s_i+1} and Z_{s_i+2}. A variable z_j, henceforth named the branching variable, is chosen. Then the sets Z_{s_i+1} and Z_{s_i+2} are constructed in the following manner:

$$Z_{s_i+1} = \{z \in Z_{k_i}, z_j \leq z_j^{i*}\}$$
$$Z_{s_i+2} = \{z \in Z_{k_i}, z_j > z_j^{i*}\}$$

where z_j^{i*} is the jth component of vector z^{i*}. For the nodes s_{i+1} and s_{i+2} we have

$$\underline{c}_l^{(s_i+1)} = \underline{c}_l^{(k_i)} \ (l \in P_{k_i}); \quad \overline{c}_l^{(s_i+2)} = \overline{c}_l^{(k_i)} \ (l \in P_{k_i})$$

$$\overline{c}_l^{(s_i+1)} = \begin{cases} \overline{c}_l^{(k_i)} & \text{if } l \neq j; \\ z_j^{i*} & \text{if } l = j; \end{cases} \quad \underline{c}_l^{(s_i+2)} = \begin{cases} \underline{c}_l^{(k_i)} & \text{if } l \neq j \\ z_j^{i*} & \text{if } l = j \end{cases}$$

The sets P_{s_i+1} and P_{s_i+2} are:

$$P_{s_i+1} = \{P_{k_i}, j\}; \quad P_{s_i+2} = \{P_{k_i}, j\}$$

In Fig. 6.9, the straight line AC corresponds to the function $\overline{\varphi}_j^{(s_i+1)}$ and the straight line CB, to the function $\overline{\varphi}_j^{(s_i+2)}$.

We now consider the choice of branching variable z_j. For simplicity, it is presumed that constraints (94) are imposed on variables z_j ($j = \overline{1, N}$). We seek the derivatives of functions $f^{(k_i)}(z)$, $f^{(s_i+1)}(z)$, $f^{(s_i+2)}(z)$ with respect to variable z_j:

$$\frac{\partial f^{(k_i)}}{\partial z_j} = \frac{\partial \theta}{\partial z_j} + \gamma_j^{k_i}; \quad \frac{\partial f^{(s_i+1)}}{\partial z_j} = \frac{\partial \theta}{\partial z_j} + \gamma_j^{s_i+1};$$

$$\frac{\partial f^{(s_i+2)}}{\partial z_j} = \frac{\partial \theta}{\partial z_j} + \gamma_j^{s_i+2} \qquad (100)$$

Let the value of z_j^{i*} be found within an interval $\underline{c}_j^{(i)}$, $\overline{c}_j^{(i)}$; then at an extremal point,

$$\frac{\partial f^{(k_i)}}{\partial z_j} = \frac{\partial \theta}{\partial z_j} + \gamma_j^{k_i} = 0 \qquad (101)$$

Since the slope of the straight line CB is less steep than that of AB, $\gamma_j^{s_i+2} < \gamma_j^{k_i}$, and we obtain from (100), (101):

$$\left.\frac{\partial f^{(s_i+2)}}{\partial z_j}\right|_{z_j=z_j^{i*}} < 0 \qquad (102)$$

In a similar way, we obtain

$$\left.\frac{\partial f^{(s_i+1)}}{\partial z_j}\right|_{z_j=z_j^{i*}} > 0 \tag{103}$$

Let us form a quantity

$$\beta_j = \frac{\partial f^{(s_i+1)}}{\partial z_j} - \frac{\partial f^{(s_i+2)}}{\partial z_j} = \gamma_j^{(s_i+1)} + \gamma_j^{(s_i+2)} \tag{104}$$

From now on we presume, in seeking a solution for nodes $s_i + 1$ and $s_i + 2$, that the point z^{k_i*} (which is the solution of problem 2 for the node k_i) is taken as a starting point. In solving problem 2 for the node $s_i + 2$, the variable z_j must first diminish in accordance with inequality (102). By analogy, in solving problem 2 for the nodes $s_i + 1$, the variable z_j must increase in accordance with inequality (103). In this connection, the value of β_j in Eq. (104) characterizes the rate of divergence of the coordinate z_j in problem 2 for the nodes $s_i + 1$ and $s_i + 2$. As we have already noted, the usual procedure in the BB method is to perform branching at the k_i node in such a way as to make the lower-bound estimates of its two descendents differ as much as possible. The value of β_j is recommended to be taken as a characteristic measure of this difference. Then the variable with the number l_i is chosen for a branching variable such that the quantity β_{l_i} attains a maximum value,

$$l_i = \arg\max_j \beta_j \; (j = \overline{1, N})$$

However, one must be aware that the maximum of β_j does not provide a complete characterization of the difference between the lower-bound estimates for nodes $s_i + 1$ and $s_i + 2$.

Upper-Bound Estimation. Two procedures can be suggested for estimating the upper bound. We consider the first procedure. We evaluate the function f at the point z^{i*}. Let $\delta_i^{(1)} = f(z^{i*})$. It is clear that

$$\delta_i^{(1)} \geq \min_{z \in Z_i} f(z) \tag{105}$$

The value of $\delta_i^{(1)}$ is thus the upper-bound estimate for the set Z_i. Let us turn now to the second procedure. We make a search for a local minimum of function $f(z)$ in the region Z_i, with z^{i*} chosen for the starting point. Let

$$\delta_i^{(2)} = \min_{z \in Z_i} f(z) \tag{106}$$

To be noted, the right-hand side of inequality (105) specifies the operation of search for global minimum, whereas the right-hand side of Eq. (106) defines the operation of search for local minimum. It is clear that $\delta_i^{(2)} \leqslant \delta_i^{(1)}$, that is, the second procedure gives a more exact upper-bound estimate. It must be admitted, however, that this procedure is much more laborious, since it requires local minimization. We introduce a quantity

$$r^i = \min_{j \in S_i} \delta_j$$

where S_i is the set of all graph nodes, and δ_i may be either $\delta_i^{(1)}$ or $\delta_i^{(2)}$. The value of r^i is the upper-bound estimate of the optimization test. The point at which the function $f(z)$ attains the value of r^i is denoted by z^{i**}.

Determination of Nonperspective Nodes. Let, at the ith step for the jth leaf node, the following inequality hold

$$\varrho_j = \min_{z \in Z_j} \bar{f}^{(j)}(z) > r^{(i)},$$

Hence, the optimum solution contained in the set Z_j is clearly worse than the solution at the point z^{i**}. Thus, the set of numbers A^i for nonperspective nodes is defined in the following manner:

$$A^i = \{j: \varrho_j > r^i\}$$

Termination Test. The iteration procedure terminates if, at the quasi-optimum node, the relation $|\varrho_i - r^i| \leqslant \varepsilon$ is fulfilled, where ε is a small number.

Let us consider the computational aspect of the method. At each step, we must solve problem 2 for the two descendents of the quasi-optimum node. In addition, in the second procedure of upper-bound estimation, two nonlinear programming problems are to be solved. Since the solution z^{i*} of problem 2 for the quasi-optimum node is known, the point z^{i*} is a good initial approximation to the solution of problem 2 for its two descendents. Therefore, it appears expedient to store the vectors z^{i*} for all leaf nodes. Let us now take a closer look at problem 2 for node 1. In this case valid are the relations

$$\bar{c}_j^{(1)} = 0; \quad \bar{\bar{c}}_j^{(1)} = \bar{z}_j; \quad \varphi_{j1}^1 = 0; \quad \varphi_{j2}^1 = \varphi(\bar{z}_j)$$

and the function $\bar{\varphi}_j^{(1)}$ is a straight line passing through the points 0 and B (Fig. 6.10). Let, in solving problem 2, a solution z^{1*} has been found. Suppose, the variable z_j is a branching variable. Then, for one of the node 1 descendents, problem 2 is to be solved, with z_j varied within the range of $0 \leqslant z_j \leqslant z_j^{1*}$, and with the other variables varied within the former intervals $(0, \bar{z}_j)$; the function $\bar{\varphi}_j^{(2)}$ is represented by a straight line passing through the points 0 and

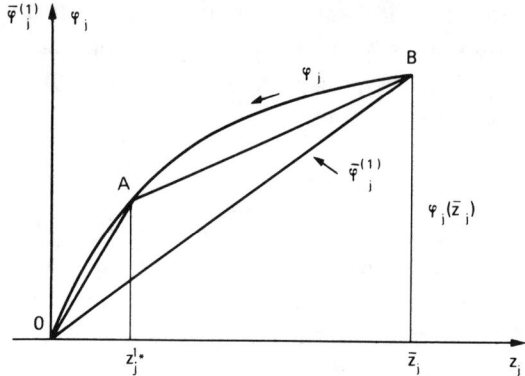

Fig. 6.10 Graphical $\bar{\varphi}_j^{(1)}$, φ_j versus z_j relationships

A in Fig. 6.10. For the other descendent in problem 2, the variable is allowed to vary within $(z_j^{1*} \leqslant z_j \leqslant \bar{z}_j)$, whereas the function $\bar{\varphi}_j^{(3)}$ is represented by the straight line AB, and so forth.

Thus, the sets Z_2, Z_3 and P_2, P_3 for the nodes 2 and 3 take the form

$$Z_2 = \{z \colon z \in Z_1, \ \bar{c}_j^{(2)} \leqslant z_j \leqslant \bar{\bar{c}}_j^{(2)}\}$$
$$Z_3 = \{z \colon z \in Z_1, \ \bar{c}_j^{(3)} \leqslant z_j \leqslant \bar{\bar{c}}_j^{(3)}\}$$

where

$$\bar{c}_j^{(2)} = 0; \ \bar{\bar{c}}_j^{(2)} = z_j^{1*}; \ \bar{c}_j^{(3)} = z_j^{1*}; \ \bar{\bar{c}}_j^{(3)} = \bar{z}_j$$

$$\bar{c}_l^{(2)} = 0; \ \bar{\bar{c}}_l^{(2)} = \bar{z}_l; \ \bar{c}_l^{(3)} = 0; \ \bar{\bar{c}}_l^{(3)} = \bar{z}_l \ (l \neq j)$$

We now turn to problem 2. Since the region Z_1 has been assumed to be convex, all the regions Z_i are also convex because they are derived from Z_1 through imposing simple constraints (96). Since the function $\bar{f}^{(i)}(z)$ in problem 2 is the sum of a convex function and linear functions, it is a convex function and has a single minimum in the convex region Z_i. For this reason, the solution of problem 2 is a simpler task in comparison to problem 1. The known non-linear programming methods can be applied to its solution.

Consider a case where the function $Q_2^{(k)}$ takes the form (91). Let, for the first N_1 blocks, the function $Q_2^{(k)}$ take form (90a), and for the other blocks, form (91). The problem of PFS synthesis may be formulated in the following

manner:
$$\min f(z)$$
$$\varphi_i(z) \leqslant 0 \quad i = \overline{1, p}$$
$$0 \leqslant z \leqslant \overline{z}_i$$
$$f(z) = \theta(z) + \sum_{i=1}^{N_1} \varphi_i(z_i) + \sum_{k=N_1+1}^{N} Q_2^{(k)}(z_k, z_{jk})$$

where $Q_2^{(k)} = \varphi_k(z_k) \cdot \chi_k(z_{jk})$, and z_{jk} is an independent variable. The functions φ_k are concave, and the functions $\chi_k(z_{jk})$ are convex. In this particular case the procedure for solving the PFS synthesis problem remains basically the same as that described for the investment cost function of the first type. However, it has some specific features:

1. In problem 2, the function $\overline{f}^{(i)}(z)$ is written as

$$\overline{f}^{(i)}(z) = \theta(z) + \sum_{j=1}^{N} \overline{\varphi}^{(i)}(z_i) + \sum_{k=N_1+1}^{N} \overline{Q}_2^{(k)}(z_k, z_{jk})$$

where $\overline{Q}_2^{(k)} = \overline{\varphi}_k^{(i)}(z_k) \chi_k(z_{jk})$; $\overline{\varphi}_k^{(i)}$, as formerly, is defined by expression (97).

2. The quantity β_j takes the form

$$\beta_j = \chi(z_{jk}^{i*})(\gamma_j^{s_i+1} + \gamma_j^{s_i+2})$$

We consider problem 2 as applied to this particular case. The objective function $\overline{f}^{(i)}$ is derived through replacing all concave functions $\varphi_i(z_i)$ by linear functions (97). The function $\overline{f}^{(i)}$ is the sum of convex functions $\theta(z)$ and $\overline{\varphi}^{(i)}(z)$ and also of function $\overline{Q}_2^{(k)}$. The function $\overline{Q}_2^{(k)}$, generally speaking, is not a convex one despite its being the produce of the convex function $\chi_k(z_{jk})$ and the convex (linear) function $\overline{\varphi}_k^{(i)}(z_k)$. In this particular case we cannot with certainty assert that the function $f^{(i)}(x)$ is convex. However, any function $\overline{Q}_2^{(k)}$ has the property of being convex at any section $Z_k = \text{const}$ or $z_{jk} = \text{const}$ ($k = \overline{N_1 + 1, N}$). Therefore, there is good reason to believe that the function $\overline{f}^{(i)}(z)$ is one-extremal, which, of course, is not a rigorous statement.

Finally, we consider a case where the flow sheet has blocks specified by the third type of investment cost function. As has already been noted, the function $Q_2^{(k)}$ takes the form (90a); however, $V^{(k)}$ is an implicit and, generally speaking, nonlinear function of independent variables,

$$V^{(k)} = \psi^{(k)}(z_{i_1}, \ldots, z_{i_p})$$

Let the investment cost function of the first type be used for the first N_1 blocks, and the function of the third type, for the rest of the blocks. The

PFS synthesis problem is defined in this case as shown below:

$$\min f(z)$$
$$\psi_i(z) \leqslant 0$$
$$0 \leqslant z_i \leqslant \bar{\bar{z_i}}$$

$$f(z) = \theta(z) + \sum_{i=1}^{N_1} \varphi_i(z_i) + \sum_{i=N_1+1}^{N} \varphi_i(V^{(i)})$$

where $V^{(i)}$ are functions of independent variables. The procedure for solving the PFS synthesis problem in this case is basically the same as in treating the investment cost functions of the first type. In a similar manner, the function $\varphi_i(z_i)$ is subject to linearization both for $i \leqslant N_1$ and $i \geqslant N_1$; either the independent variable z_i ($i \leqslant N_1$), or the dependent variable $V^{(i)}$ ($N_1 < i \leqslant N$) can be chosen for a branching variable. However, the principle underlying the choice of branching variable must be appropriately modified, since in this case the variables $V^{(i)}$ are no longer independent.

Nonetheless, there is a basic distinction from the first case. In the first case, the additional constraints (96) that emerge as a result of branching process are simple constraints which set a limit to the greatest and the least values of independent variables. Actually, they make the feasible region convex.

In this particular case, if the variable $V^{(i)}$ ($i > N_1$) is a branching variable, then the nonlinear constraints imposed on the quasi-optimum node descendents are

$$\chi^{(i)}(z_{i1}, \ldots, z_{ik}) \leqslant V^{(i)*} \qquad (107)$$
$$\chi^{(i)}(z_{i1}, \ldots, z_{ik}) \geqslant V^{(i)*} \qquad (108)$$

where $V^{(i)*}$ is the value of dependent variable $V^{(i)}$ at the extremal point of problem 2. In this case, one cannot assert that the regions Z_i are convex, since they are produced by nonlinear constraints of type (107), (108). Let, for example, the function $\chi^{(i)}$ be a convex one. Then region (107) is convex; however, region (108) is not convex. Therefore, in this particular case one cannot be confident of problem 2 having only one minimum.

We now wish to show the application of our method to the synthesis of a flow-sheet system composed of a stirred tank reactor (STR), a plug-flow reactor (PFR) and a separator [41]. In the reactors, an irreversible isothermal reaction A → B is carried out, where A is the initial reactant and B is the desired product. The separator is presumed to be perfect in performance, that is, capable of separating completely a mixture of A + B into pure components

A and B. A global flow sheet for this problem is shown in Fig. 6.8. Henceforth, the species A and B in each apparatus are designed by the respective subscripts 1 and 2.

We first consider mathematical models for separate blocks and their optimization tests.

Stirred-tank reactor. The mathematical model for STR is defined as

$$y_1^{(i)} = \frac{x_1^{(i)}}{1 + k\tau^{(i)}}; \quad y_2^{(i)} = \frac{k\tau^{(i)} x_1^{(i)}}{1 + k\tau^{(i)}} + x_2^{(i)} \tag{109}$$

where $y_1^{(i)}$ and $x_1^{(i)}$ are the flow rates of species A at the inlet and the outlet of the kth reactor, respectively; $y_2^{(i)}$ and $x_2^{(i)}$ are the flow rates of species B at the inlet and the outlet of the ith reactor, respectively; k is the reaction rate constant; and $\tau^{(i)}$ is the holdup time.

Plug-flow reactor. The mathematical model for PFR takes the form

$$y_1^{(i)} = x_1^{(i)} e^{-k\tau^{(i)}}; \quad y_2^{(i)} = x_1^{(i)}(1 - e^{-k\tau^{(i)}}) + x_2^{(i)} \tag{110}$$

The optimization test for both STR and PFR takes the form

$$F^{(i)} = a^{(i)} (v^{(i)})^{0.6}$$

where $v^{(i)} = \tau^{(i)}(x_1^{(i)} + x_2^{(i)})$; $a^{(i)}$ is a coefficient.

Separator. The mathematical model for the separator takes the form

$$y_1^{(i)} = x_1^{(i)}; \quad y_2^{(i)} = x_2^{(i)}$$

The optimization test for the separator is

$$F^{(i)} = b(x_1^{(i)} + x_2^{(i)})$$

where b is a coefficient.

Imaginary blocks. The global flow sheet has one imaginary input block 1 and three imaginary output blocks N, $N + 1$, $N + 2$. Their respective mathematical models take the form

$$y_j^{(i)} = x_j^{(i)}, \quad j = 1, 2; \quad k = 1, N, N + 1, N + 2$$

The optimization tests for imaginary blocks are:

$$F^{(i)} = -c_2 y_2^{(i)} + c_1 y_1^{(i)}; \quad i = N, N + 1, N + 2$$
$$F^{(i)} = 0$$

In addition, mathematical models (2) have been defined for mixers 5, 6, 7, and 12. The search variables are variables $\tau^{(2)}$, $\tau^{(3)}$, and α_{ij}. The constraints take the forms (4) and (90). The following numerical parameters have been

used in the problem:
$$x_1^{(1)} = 1, \ x_2^{(1)} = 0, \ c_2 = 1, \ c_1 = +1, \ k = 10, \ b = 0.04,$$
$$a^{(2)} = a^{(3)} = 1.0$$

To start with, we have considered a case where the block investment cost functions are linear functions, with $b_2 = b_3 = 1$. In this case, problem (92)-(94) has one local minimum.

The test values at the minimum have been found to be -0.782. The substructure corresponding to the minimum is lacking in the second block (stirred-tank reactor).

Next, we have explored problem (92)-(94) for a case of the functions $\varphi_i(z_i)$ being concave functions (see (90a)), $b_2 = b_3 = 0.6$. Problem (92)-(94) has been found to have two minima.

First local minimum:

$$V^{(2)} = 0; \ V^{(3)} = 0.138;$$
$$\alpha_{21} = \alpha_{24} = \alpha_{23} = \alpha_{32} = \alpha_{N+2,3} = \alpha_{N+2,4} \quad (111)$$
$$= \alpha_{42} = \alpha_{N+1,N} = 0; \ \alpha_{31} = \alpha_{34} = 1; \ F = -0.61$$

Second local minimum:

$$V^{(2)} = 0.163; \ V^{(3)} = 0.0; \ \alpha_{23} = \alpha_{31} = \alpha_{32}$$
$$\alpha_{34} = \alpha_{N+2,3} = \alpha_{N+2,2} = \alpha_{43} = \alpha_{N+1,4} = 0;$$
$$= \alpha_{21} = \alpha_{24} = 1.0; \ F = -0.58$$

The first minimum is a global one. Its structural parameters are coincident with those at the minimum point for the case of linear functions $\varphi_i(z_i)$. By making use of the branch-and-bound procedure, we have identified a global minimum. In this search, five graph nodes have been constructed, 154 computations performed on an objective function and 124 computations, on a gradient function. A scheme corresponding to the global minimum is shown in Fig. 6.11.

We now consider a somewhat more complex problem of flow-sheet synthesis which is a further development of the former one. Suppose, blocks *2*, *3*, and *4* of the scheme in Fig. 6.8 are superblocks, each composed of several minor blocks. Thus, superblock *2* is functionally brought in correspondence to stream-splitter *12*, mixer *15*, and two stirred-tank reactors *21* and *22*, operating in parallel; of these reactors, one has a higher degree of conversion (the holdup time for both reactors being the same) and, simultaneously, has a higher operating cost (Fig. 6.12). Analogously, superblock *3* is functionally equivalent to

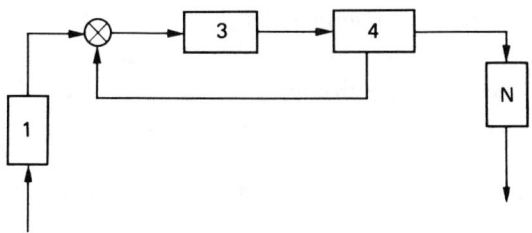

Fig. 6.11 Optimum reactor-separator flow sheet

stream-splitter *13*, mixer *16*, and two operating in parallel plug-flow reactors *31* and *32*, of which one has a higher degree of conversion and is also more costly (Fig. 6.12). Finally, superblock *4* is composed of stream-splitter *14* which distributes the input flow among separator *41*, distillation column *42*, and two mixers *17* and *18*. The distillation column is here a more efficient apparatus for mixture separation, and simultaneously it is more costly in operation. By deleting the imaginary blocks $N + 1$ and $N + 2$ and all incoming streams therein, we obtain the global flow sheet as shown in Fig. 6.12.

If in the previous flow-sheet synthesis problem it was necessary to choose the type of reactor and to specify the structure of flow-sheet streams, in the given problem an additional task is that a decision must be made, reactors of which type (those more efficient and more costly, or those less efficient and less costly) are to be preferred.

Analogously, a decision is to be made, what type of separating apparatus (if any) is to be employed.

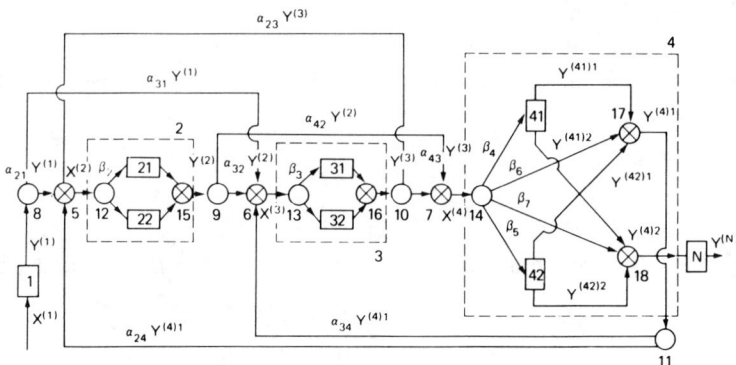

Fig. 6.12 Global flow sheet of a sophisticated reactor-separator process

We focus now on mathematical models for separate blocks and their optimization tests. We denote the structural parameters that characterize the streams between the superblocks, as previously, by α_{ij}, whereas the structural parameters that characterize the streams within a superblock are denoted by β_i (see Fig. 6.12). The mathematical models for stirred-tank reactors *21* and *22* and for plug-flow reactors *31* and *32* as well as their optimization tests are the same as those in the previous problem.

The mathematical model for separator *41* takes the form [46]:

$$y_1^{(41)1} = 0.85 x_1^{(41)}, \quad y_1^{(41)2} = 0.15 x_1^{(41)}$$
$$y_2^{(41)1} = 0.2 x_2^{(41)}, \quad y_2^{(41)2} = 0.80 x_2^{(41)} \tag{112}$$

The mathematical model for distillation column *42* takes the form [46]:

$$y_1^{(42)1} = 0.975 x_1^{(42)}, \quad y_1^{(42)2} = 0.025 x_1^{(42)}$$
$$y_2^{(42)1} = 0.050 x_2^{(42)}, \quad y_2^{(42)2} = 0.950 x_2^{(42)}$$

The optimization tests for separator and distillation column are those as specified above. The mathematical models for mixers *5*, *6*, and *7* take the form (2).

The mathematical models for stream-splitters *12* and *13* are:

$$x^{(21)} = \beta_2 x^{(2)}, \quad x^{(22)} = (1 - \beta_2) x^{(2)}$$
$$x^{(31)} = \beta_3 x^{(3)}, \quad x^{(32)} = (1 - \beta_3) x^{(3)}$$

The mathematical models for mixers *15* and *16* take the form:

$$y^{(2)} = y^{(21)} + y^{(22)}, \quad y^{(3)} = y^{(31)} + y^{(32)}$$

The mathematical models for stream-splitter *14* and mixers *17*, *18* are written in a similar manner.

As in the previous case, the profit with a negative sign which takes into account both the equipment cost and the recycle cost is used as an optimization test:

$$F = -c_2 y_2^{(N)} + c_1 y_1^{(N)} + \sum_{k \in I} F^{(k)} + \gamma(y_1^{(41)1} + y_2^{(42)2})$$

where
$$I = (21, 22, 31, 32, 41, 42)$$

The following constants have been used for the models and optimization tests:

$$X_1^{(1)} = 1; \; X_2^{(1)} = 0, \; C_2 = 1, \; C_1 = 1; \; k_{21} = k_{31} = 10,$$
$$k_{22} = k_{32} = 5, \; d_{41} = 0.02, \; d_{42} = 0.08,$$
$$a_{21} = a_{31} = 10, \; a_{22} = a_{32} = 0.5, \; \gamma = 0.1$$

6 Optimum Design of Chemical Plants

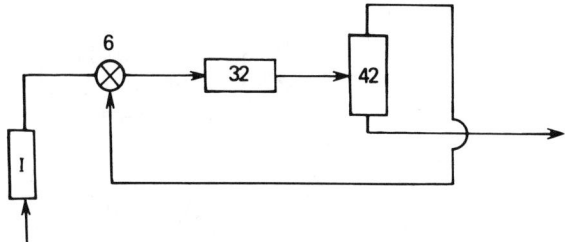

Fig. 6.13 Optimum flow sheet

The search variables are the structural parameters α_{ij}, β_i, and the volumes v_{21}, v_{22}, v_{31}, v_{32} of stirred-tank and plug-flow reactors.

The constraints imposed on the purity of end product are:

$$y_2^{(4)1} \geq 5 y_1^{(4)2}$$

To start with, we have explored a case of the investment cost functions being linear functions: $b_{21} = b_{22} = b_{31} = b_{32} = 1.0$. Here the problem has been shown to have one local (global) minimum. The optimization test at this point gives a value

$$F = -0.6466$$

The optimum structure is represented in Fig. 6.13. The preferable variant for the flow-sheet system is a "cheap" plug-flow reactor of volume $V_{32} = 0.401$.

Next, a case has been considered where the investment cost functions are concave, with $b_{21} = b_{22} = b_{31} = b_{32} = 0.6$. Thus, the problem to be solved is a multiextremal one. Using the above method, we have found a global minimum. In the search procedure, a 13-node graph has been constructed (Fig. 6.14).

In Fig. 6.14, each quasi-optimum node is supplemented with the lower-bound (ϱ_i) and upper-bound ($\delta_i^{(2)}$) estimates and with the branch variables. The other nodes are supplemented only with the lower-bound estimates. The search as defined by the graph tree in Fig. 6.14 terminated, since all the leaf nodes were found to be nonperspective.

The optimum flow sheet is identical to that shown in Fig. 6.13. The optimization test and the reactor volume have been found to be

$$F = -0.549; \quad V_{32} = 0.417$$

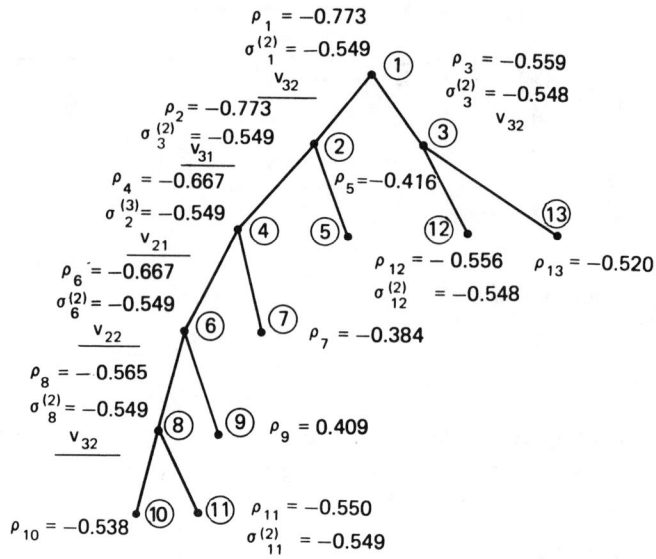

Fig. 6.14. The tree corresponding to branch-and-bound procedure in search for an optimum reactor-separator flow sheet

We will consider now another approach to the global minimum search in problem (92)-(94) presuming, as previously, that the function $\theta(z)$ is convex in the region $Z_1(z)$. At first, we reduce our problem to a mixed-integer nonlinear programming problem. As has been shown in [47], the functions $\varphi_i(z_i)$ can be represented in the form (see Fig. 6.15):

$$z_i = a_1^{(i)}\lambda_1^{(i)} + a_2^{(i)}\lambda_2^{(i)} + \ldots + a_{n_i}^{(i)}\lambda_{n_i}^{(i)}, \quad (a_1^{(0)} = 0; \, a_{n_i}^{(i)} = \bar{z}_i) \quad (113)$$

$$\varphi(z_i) = b_1^{(i)}\lambda_1^{(i)} + b_2^{(i)}\lambda_2^{(i)} + \ldots + b_{n_i}^{(i)}\lambda_{n_i}^{(i)}, \quad b_j^{(i)} = \varphi(a_j^{(i)}) \quad (114)$$

where

$$\sum_{j=1}^{n_i} \lambda_j^{(i)} = 1 \quad (115)$$

$$\lambda_j^{(i)} \geq 0, \, j = \overline{1, n_i} \quad (116)$$

$$\lambda_k^{(i)}\lambda_j^{(i)} = 0, \, j > k+1, \, k = 1, 2, \ldots, n_i - 1 \quad (117)$$

We introduce binary variables $z_j^{(i)}$. One will have seen then that condition (117) can be rewritten in the form

$$\lambda_j^{(i)} + \lambda_{j+1}^{(i)} = \beta_j^{(i)}, \quad j = \overline{1, n_i - 1} \quad (118)$$

$$\beta_j^{(i)} = 0 \text{ or } 1 \quad (119)$$

6 Optimum Design of Chemical Plants

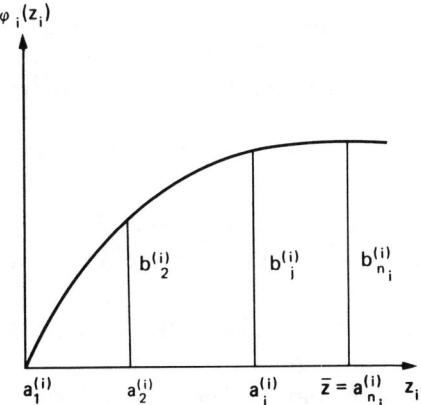

Fig. 6.15 Graphical φ_i versus z_i relationship

We substitute z_j (113) and $\varphi_j(z_j)$ ($j = \overline{1, N}$) (114) into (92)-(94) to define a mixed integer nonlinear programming problem; here, the search variables are continuous variables λ and binary variables $\beta_j^{(i)}$. Similar to the method [39], the solving procedure consists of two stages. At stage I, a solution of problem (92)-(94), (113)-(116), (118), (119), which is a common nonlinear programming problem, is sought, at fixed binary variables $\beta_j^{(i)}$.

Since the function $\varphi_i(z)$ has been replaced by a linear function (113), (114), the formulated problem is thus one-extremal. Let the solution of the problem is z^*, λ^*. At the point z^*, λ^*, the functions $\psi_i(z)$, $\theta(z)$ are replaced by their linear approximations, and we have to solve the following mixed-integer linear programming problem:

$$\min_{\lambda, \beta} \Delta f$$

$$\psi_i(z^*) + \sum_{j=1}^{p} \frac{\partial \psi_i}{\partial z_j}(z_j - z_j^*) \leq 0$$

$$0 \leq z_j \leq \bar{z}_j$$

where

$$\Delta f = \sum_{j=1}^{p} \frac{\partial \theta}{\partial z_j}(z_j - z_j^*) + \sum_{i=1}^{N} \varphi_i(z_i) \qquad (120)$$

In (120) the quantities $\varphi_i(z_i)$ and z_i have been replaced by their expansions (113) and (114). Using the derived values of λ and β, problem (92)-(94), (113), (114), (118), (119) is solved once again, and this procedure is repeated until a final solution is obtained.

Let us consider another approach to the solution of PFS synthesis problem, also based on the BB procedure. As distinct from the foregoing case, we presume that the search variables include, along with the variables α_{ij} ($i, j = \overline{1, N}$) and the control variables $u^{(k)}$ ($k = \overline{1, N}$), also the input variables $x^{(k)}$ ($k = \overline{1, N}$) of all PFS blocks. We denote the vector of all search variables by z. Equations (2), with the quantities $y^{(k)}$ replaced by their expressions (1), become now equality constraints. We presume that the first type of investment cost functions (90) applies to all the blocks. The PFS synthesis problem can be written in the form

$$\min f(z) \tag{121}$$
$$\chi^{(i)}(z) = 0, \quad i = \overline{1, N} \tag{122}$$
$$\psi^i(z) \leq 0, \quad i = \overline{1, N} \tag{123}$$
$$0 \leq z \leq \bar{z}; \tag{124}$$
$$W(Z) \equiv \sum_{i=1}^{N} \alpha_{ij} - 1 = 0; \quad j = \overline{1, N} \tag{125}$$

where

$$f(z) = \theta(z) + \sum_{i=1}^{N} \varphi_i(z);$$

$$\theta(z) = \sum_{i=1}^{N} F(x^{(k)}, f^{(k)}(x^{(k)}, u^{(k)}), u^{(k)}) = \sum_{i=1}^{N} \overline{F}(x^{(k)}, y^{(k)}); \tag{126}$$

$$\chi^{(i)}(z) = x^{(i)} - \sum_{j=1}^{N} \alpha_{ij} y^{(j)} \equiv x^{(i)} - \sum_{j=1}^{N} \alpha_{ij} f^{(j)}(x^{(j)}, u^{(j)})$$

$$\psi^{(i)}(z) = \psi_i(u^{(i)})$$

Problem (121)-(125) can be written in a more concise form:

$$\min_{z \in \tilde{Z}_1} f(z) \tag{127}$$

where

$$\tilde{Z}_1 = \{z: 0 \leq z \leq \bar{z}; \psi^{(i)}(z) \leq 0; \chi^{(i)}(z) = 0; \sum_{j=1}^{N} \alpha_{ji} = 1; i = \overline{1, N}\}$$

6 Optimum Design of Chemical Plants

Let us consider the following optimization problem:

$$\min_{z \in \tilde{Z}_1} L(z, \lambda) \tag{128}$$

where L is the Lagrangian function,

$$L = f + \sum_{k=1}^{N} \lambda^{(k)T} \chi^{(k)} \tag{129}$$

$$\tilde{Z}_1 = \{z: 0 \leq z \leq \bar{z};\ \psi^{(i)}(z) \leq 0;\ \sum_{j=1}^{N} \alpha_{ji} = 1;\ i = \overline{1, N}\}$$

The solution of this problem is not necessarily existent; for this reason, we impose additional constraints on the variables $x^{(k)}$, $(k = \overline{1, N})$:

$$0 \leq x^{(k)} \leq b^{(k)}, \quad (k = \overline{1, N}) \tag{130}$$

The constant $b^{(k)}$ should be chosen such as to fulfil the constraint

$$0 \leq x^{(k)*} \leq b^{(k)} \tag{131}$$

where $x^{(k)*}$ is the value of vector $x^{(k)}$ at the optimum point of problem (127). Taking into account (131), we rewrite problem (128) as shown below:

$$\min_{z \in Z_1} L(z, \lambda) \tag{132}$$

where $Z_1 = \{z: z \in \tilde{Z}_1;\ 0 \leq x^{(k)} \leq b^{(k)}\ (k = \overline{1, N})\}$ Let

$$h^{(1)}(\lambda) = \min_{z \in Z_1} L(z, \lambda) \tag{133}$$

The function $h^{(1)}(\lambda)$ is called the dual function [47]. We presume that the dual function exists within a certain region M of definite values of λ: $M = \{\lambda$: there exists $h^{(1)}(\lambda)\}$. As has been shown in [48], an equality holds:

$$h^{(1)}(\lambda) \leq \min_{z \in Z_1} f(z), \quad \forall \lambda \in M$$

$$Z_1 = \{z: z \in \tilde{Z}_1;\ 0 \leq x^{(k)} \leq b^{(k)}\}$$

Since condition (131) is valid, the following relation holds:

$$\min_{z \in Z_1} f(z) = \min_{z \in \tilde{Z}_1} f(z)$$

Hence, we have

$$h^{(1)}(\lambda) \leq f(z^*) \tag{134}$$

where z^* is the solution of problem (127). Thus, the value of $h^{(1)}(\lambda)$ gives the lower-bound estimate of function $f(z)$ in problem (127). By substituting $\chi^i(z)$ as defined by (126) into the function L, we obtain, after simple transformations,

$$L = \sum_{k=1}^{N} L^{(k)}(x^{(k)}, u^{(k)}, \alpha_{jk}) \tag{135}$$

where

$$L^{(k)}(z^{(k)}) = F^{(k)} + (\lambda^{(k)})^T x^{(k)} - \left[\sum_{j=1}^{N} (\lambda^j)^T \alpha_{jk}\right] y^{(k)}(x^{(k)}, u^{(k)})$$

the vector of variables $x^{(k)}$, $u^{(k)}$, α_{jk} $(j = \overline{1,N})$ is denoted by $z^{(k)}$. It is to be noted that $L^{(k)}(z^{(k)})$ is dependent on the variables of the kth block only. Therefore, problem (132) may be rewritten in the form [48]:

$$h^{(1)}(\lambda) = \min_{z \in Z_1} \sum_{k=1}^{N} L^{(k)}(z^{(k)}) = \sum_{k=1}^{N} \min_{z^{(k)}} L^{(k)}(z^{(k)}) \tag{136}$$

Thus, the problem of determining $h^{(1)}(\lambda)$ reduces to the solution of N minimization problems for each block,

$$\min_{z^{(k)}} L^{(k)}(x^{(k)}, u^{(k)}, \alpha_{jk}) \tag{137}$$

$$\sum_{j=1}^{N} \alpha_{jk} = 1 \tag{138}$$

$$\psi^{(k)}(u^{(k)}) \leq 0$$
$$0 \leq x^{(k)} \leq b^{(k)} \tag{139}$$

As we have already mentioned, the BB method reduces to a multistep procedure. The flow sheet which is derived from the global flow sheet by deleting p_i blocks (with the respective numbers $j = l_1^{(i)}, \ldots, l_{p_i}^{(i)}$) and all their links to other blocks will correspond to the node with the number i. We denote the flow sheet that corresponds to the ith node by $S^{(i)}$. The global flow sheet made up of N blocks corresponds to the first node with $p_1 = 0$. By definition, the ith node can be brought in correspondence to the following set of points $\tilde{Z}_i$ within the space of variables z:

$$\tilde{Z}_i = \{z\colon z \in Z_1;\ u^{(s)} = 0;\ x^{(s)} = 0;\ \chi^{(i)}(z) = 0;$$

$$\sum_{j=1}^{N} \alpha_{jt} = 1;\ t \neq l_1^{(i)}, \ldots, l_{p_i}^{(i)};\ \alpha_{js} = 0;$$

$$\alpha_{sj} = 0;\ j = \overline{1,N};\ s = l_1^{(i)}, \ldots, l_{p_i}^{(i)}\}$$

6 Optimum Design of Chemical Plants

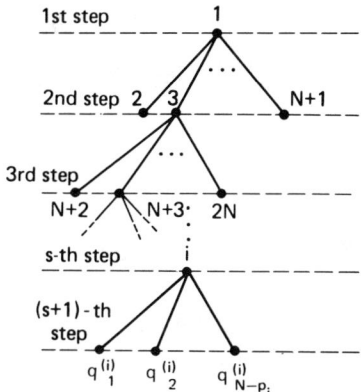

Fig. 6.16 The tree corresponding to a general procedure in branch-and-bound method

The set $\tilde{Z}_1$ corresponds to the first node (see Eq. (127)). The set $\tilde{Z}_i$ ($\tilde{Z}_i \subset \tilde{Z}_1$) is distinct from the set $\tilde{Z}_1$ in that the structural parameters of input variables and control variables of the blocks lacking in the PFS $S^{(i)}$ are set equal to zero. As distinct from the former version of the BB method, in this case any quasi-optimum node (suppose, with the number i) has $(N - p_i)$, rather than two, descendents (Fig. 6.16), in accordance with the number of blocks available in $S^{(i)}$. For example, node 1 has N descendents. Let us consider major operations of the BB method as applied to this particular case.

Node Branching. Suppose, the flow sheet corresponding to the ith node has blocks with the numbers $s_1^{(i)}$, $s_2^{(i)}$, ..., $s_{N-p_i}^{(i)}$. As has already been pointed out, the ith node has $(N - p_i)$ descendents; suppose, their numbers are $q_1^{(i)}$, $q_2^{(i)}$, ..., $q_{N-p_i}^{(i)}$. To each descendent of the ith node, a flow sheet corresponds derived from the flow sheet $S^{(i)}$ by deleting one of its blocks. For example, a flow sheet consisting of $N - 1$ blocks corresponds to any of the descendents $2, \ldots, N + 1$ of the node 1. Suppose, the flow sheet, correspondent to the descendent q_j ($1 \leq j \leq N - p_i$), is lacking in the block numbered s_j. It is clear that the set $\tilde{Z}_{q_j}$ will take the form:

$$\tilde{Z}_{q_j} = \{z: z \in \tilde{Z}_i; \; x^{(l_j)} = 0; \; u^{(l_j)} = 0; \; \alpha_{k s_j} = 0;$$
$$\alpha_{s_j k} = 0; \; k = 1, N; \; j = 1, \ldots, N - p_i\} \tag{140}$$

Lower-Bound Estimation. We now consider two optimization problems associated with the ith node.

Problem 3

$$\min_{z \in Z_i} f(z)$$

Problem 4

$$\min_{z \in Z_i} L_i(z, \lambda^{(i)})$$

$$L_i(z, \lambda^{(i)}) = \sum_{k=s_1}^{S_{N-p_i}} L_i^{(k)}(z^{(k)}, \lambda^{(i)})$$

$$L_i^{(k)} = F^{(k)} + (\lambda^{(k)i})^T x^{(k)} - (a_k^{(i)})^T f^{(k)}(x^{(k)}, u^{(k)}) \tag{141}$$

where

$$a_k^{(i)} = \sum_{j=s_1}^{S_{N-p_i}} (\lambda^{(k)i}) \alpha_{jk}$$

$\lambda^{(k)i}$ is the vector of Lagrangian multipliers $\lambda^{(k)}$ at the ith node. In both problems 3 and 4, a global minimum is sought for. Problem 3 represents a problem for optimization of a flow sheet corresponding to the ith block. We denote the value of function $L^{(i)}$ at the optimum point of problem 4 by $h^{(i)}(\lambda)$. It is clear that there exists a relation

$$h^{(i)}(\lambda) \leq f(z^{(i)*}), \quad \forall \lambda \in M^{(i)} \tag{142}$$

where $z^{(i)*}$ is the value of the vector z at the optimum point of problem 3; $M^{(i)} = \{\lambda^{(i)}: \text{there exists } h^{(i)}(\lambda)\}$. Thus, the value of $h^{(i)}(\lambda)$ is the lower-bound estimate for the node i. By analogy with (136), we obtain

$$h^{(i)}(\lambda) = \sum_{k=s_1}^{S_{N-p_i}} \min_{z^{(k)}} L_i^{(k)}(z^{(k)}, \lambda^{(i)}) \tag{143}$$

Thus, the problem for determination of $h^{(i)}(\lambda)$ reduces to the solution of $(N - p_i)$ problems

$$\min_{z^{(k)} \in Z_i^{(k)}} L_i^{(k)}(z^{(k)}, \lambda^{(i)}), \quad k = s_1^{(i)}, \ldots, s_{N-p_i}^{(i)} \tag{144}$$

where

$$Z_i^{(k)} = \{z^{(k)}: \psi^{(k)} \leqslant 0;\ 0 \leqslant x^{(k)} \leqslant b^{(k)};\ \sum_{j=s_1^{(i)}}^{S_{N-p_i}^{(i)}} \alpha_{jk} = 1\}$$

Upper-Bound Estimation. We take the value of $f_{\mathrm{loc}}^{(i)}$ of function f at one of the local minima of problem 3 for an upper-bound estimate of this problem corresponding to the node i. This minimum can be located using the known nonlinear programming methods.

Similar to the foregoing case, we define the upper-bound estimate $r^{(i)}$ for the ith step as

$$r^{(i)} = \min_{j \in Q_i} f_{\mathrm{loc}}^{(i)} \tag{145}$$

where Q_i is the set of all examined quasi-optimum nodes.

The determination of nonperspective nodes is carried out in the same manner as in the foregoing case. Similarly, we use the condition

$$|h^{(i)}(\lambda) - r^{(i)}| \leqslant \varepsilon \tag{146}$$

for the termination test of iteration procedure, ε being a small number. One must be, however, aware of the fact that in this case, as distinct from the foregoing, condition (146) may fail to provide for the iteration procedure termination, which will eventually necessitate the use of heuristic rules for the termination of an iterative process.

Let us turn our attention to computational aspects of the method. First, we consider the lower-bound estimation. Let a quasi-optimum node with the number i be known. We assume that we have identified the local minimum needed for upper-bound estimation in problem 3 corresponding to the node i. At this point, all the Lagrangian multipliers $\lambda^{(i)}$ are known. We construct, by making use of these factors, a Lagrangian function $L^{(i)}(z, \lambda^{(i)})$. As has already been noted, in both problems 3 and 4 a global maximum is sought for. However, problem 4, appears to be a simpler one, since the problem of search for global minimum in the flow sheet $S^{(i)}$ reduces to $(N - p_i)$ problems of search for global minimum for $(N - p_i)$ separate blocks of this flow sheet (see (144)). It can easily be seen that the dimension of problem 3 is equal to $(N - p_i)(n + r_i + N)$, where, in the right-hand parenthesis, r_i is the dimension of the control variable vector; n is the dimension of the input variable vector; and N is the number of the PFS blocks. This problem reduces to $(N - p_i)$ problems of dimension $(n + r_i + N)$. The reduction of dimension always facilitates essentially the search for a global minimum. To be recalled, we are considering the global minimum search in the case where there are two sources of problem nonconvexity—a bilinear form of Eqs. (2) [46] and a convexity of investment cost function (90). In this connection, let us focus on problem (144). It is clear that the first cause of nonconvexity—the form of Eqs. (2)—can be dismissed from consideration, since these equations are not present in problem (144). Therefore, the second source of nonconvexity is the only feasible one. We can immediately identify a local minimum. Indeed, let us set $V^{(k)} = 0$; consequently, $\varphi^{(k)}(V^{(k)}) = 0$. In addition, since $V^{(k)}$ characterizes the size of an apparatus, at its zero value (for example, the apparatus length is zero), the inlet flow is not affected by the apparatus and, consequently, $y^{(k)} = x^{(k)}$, and $u^{(k)} = 0$, $F^{(k)} = 0$. Therefore, problem (144) in this particular case takes the form

$$\left.\begin{array}{l} \min_{z^k} \bar{L}_i^{(k)}(z^{(k)}, \lambda^{(k)i}) \\[6pt] 0 \leqslant x^{(k)} \leqslant b^{(k)}; \quad \sum_{j=s_1}^{S_N - p_i} \alpha_{jk} = 1 \\[6pt] \bar{L}_i^{(k)} = (x^{(k)})^T \left(\lambda^{(k)i} - \sum_{j=s_1}^{S_N - p_i} \lambda^{(k)i} \alpha_{jk} \right) \end{array}\right\} \quad (147)$$

We denote the solution of this problem by $\bar{x}^{(k)}$. Then the point $\bar{z}^{(k)} = \{V^{(k)} = 0, x^{(k)} = \bar{x}^{(k)}, u^{(k)}\}$ is a local minimum for problem (144). Since $\bar{x}^{(k)}$, is a solution of problem (147), the derivative with respect to any direction $l^{(j)}$ defined within the subspace $R^{(k)} = \{z: z^{(k)} \in Z_i^{(k)}, V^{(k)} = 0\}$ satisfies the condition

$$\frac{\partial \bar{L}_i^{(k)}}{\partial l^{(j)}} \geqslant 0; \quad \forall l^{(j)} \in R^{(k)}$$

But $\bar{L}_i^{(k)} \equiv L_i^{(k)}$, subject to $z^{(k)} \in R^{(k)}$; hence

$$\frac{\partial L_i}{\partial l^{(j)}} \geqslant 0, \quad \forall l^{(j)} \in R^{(k)}$$

We have shown earlier that

$$\left.\frac{\partial F^{(k)}}{\partial V^{(k)}}\right|_{R^{(k)} = 0} = \infty$$

One can show that

$$\frac{\partial L_i^{(k)}}{\partial l^{(j)}} \geqslant 0; \quad \forall l^{(j)} \in Z_i^{(k)}$$

Thus, the point $\bar{z}^{(k)}$ is a local minimum of problem (144). Let $L_i^{(k)*}$ be the value of function $L_i^{(k)}$ at the point $\bar{z}^{(k)}$. Then, considering that $L_i^{(k)} = 0$ at $V^{(k)} = 0$; $x^{(k)} = 0$, a relation

$$L_i^{(k)}(\bar{z}^{(k)}) \leqslant 0$$

holds. Since the only source of the function $L_i^{(k)}$ nonconvexity is concavity of function $F^{(k)}$, the function

$$\tilde{L}_i^{(k)} = (\lambda^{(k)i})^T x^{(k)} - \sum_{j=s_1}^{s_{N-p_i}} (\lambda^{(k)i})^T \alpha_{jk} y^{(k)} \ (x^{(k)}, u^{(k)})$$

is convex. Hence, $L_i^{(k)}$ plotted as a function of variable $V^{(k)}$, will always take a form as shown in Fig. 6.17, that is $L_i^{(k)}$ as a function of variable $V^{(k)}$ has two local minima. Since the function $\tilde{L}_i^{(k)}$ is convex, the function $L_i^{(k)}$ will exhibit, in all likelihood, two local minima, of which one, $\bar{z}^{(k)}$, is known. To identify the other local minimum, one must provide for the search procedure not to fall into the attraction region due to minimum $\bar{z}^{(k)}$. To this effect, an additional constraint

$$z_k \geqslant c^{(k)} \tag{148}$$

must be introduced into problem (144); the constant $c^{(k)}$ can easily be determined. Indeed, if, with a certain $c^{(k)0}$ chosen, the search has led to the point $\bar{z}^{(k)}$, then the constant $c^{(k)0}$ must be raised in value, and the search repeated once again, etc. It is clear that the relation

$$L_i^{(k)}(\bar{\bar{z}}^{(k)}) \leqslant 0$$

is valid, $\bar{\bar{z}}^{(k)}$ being the second local minimum of the problem solution.

Thus, the solution of problem (144) reduces to a search for two local minima, one of which is obtained by solving problem (147), and the other one, by solving problem (144), with additional constraint (148) imposed. It is to be noted, however, that the above reasoning is not quite rigorous.

The lower-bound estimate for the ith quasi-optimum node having been determined, we proceed to the lower-bound estimation for all of its $(N - p_i)$ descendents. To this end, by making use of the Lagrangian multipliers $\lambda^{(k)i}$, we construct $(N - p_i)$ functions L_j $(j = l_1^{(i)}, \ldots, l_{N-p_i}^{(i)})$ for all the descendents of the ith node; for each descendent, two problems (147) and (144), (148) are solved and the lower-bound estimates are thus determined for all the descendents. Suppose, the new quasi-optimum node has a number $l_p^{(i)}$. Then

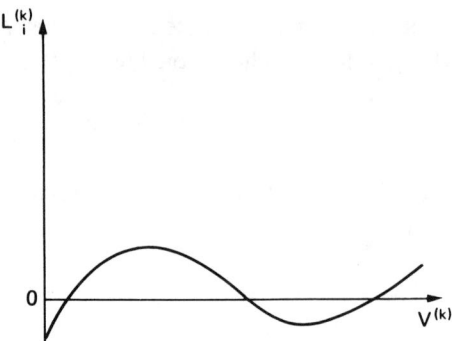

Fig. 6.17 Graphical L_i^k versus $V^{(k)}$ relationship

for this node problem 3 is solved once again (a local minimum is being sought for) and the derived Lagrangian multipliers are used for constructing a new Lagrangian function, for the $l_p^{(i)}$-th node as well as for all of its descendents, and so forth. It is easy to show that, in order to obtain the lower-bound estimate for the ith node and all its descendents, problem (144) must be solved $(N - p_i)$ times for each block. These problems will differ only in the values of vector components $a_k^{(i)}$. Therefore, the solution of one of problems (144) may happen to be a good initial approximation for another problem. Since the flow sheet $S^{(i)}$ contains $(N - p_i)$ blocks, the total number of problems (144) to be solved is $(N - p_i)^2$. Thus, at the ith step, it is necessary to solve only one problem 3 to determine the upper-bound estimate (local minimum search) and $(N - p_i)^2$ problems (144).

In using the above procedure, of particular importance is the right selection of constants $b^{(k)}$ in the constraint (148). If the $b^{(k)}$ constants have been chosen large, then the lower-bound estimates $h^i(\lambda)$ in (143) may prove to be too rough; conversely, if the $b^{(k)}$ have been chosen small, inequalities (131) may be violated, and relation (133) will not yield a good lower-bound estimate.

We now consider a method for upper-bound estimation. As has already been mentioned, to this effect we must find a local minimum of problem (121)-(125). To solve this problem, it is expedient to make use of the sequential quadratic programming (SQP) method [8] which is currently believed to be one of the most effective methods for solving nonlinear programming problems of small and medium dimension. However, the straightforward application of SQP method to solving a synthesis problem of type (121)-(125), usually of large dimension, may run into great difficulties [49].

For this reason, we consider here a modified version of SQP method. Our approach is in fact a further development of the method for optimization of large systems described in [43]. Let $z^{(k)}$ be an $n^{(k)}$-dimensional vector whose component vectors are $x^{(k)}$, $u^{(k)}$ and structural parameters are α_{jk} ($j = 1, N$). Commonly, $n^{(k)} \ll n$. The SQP iteration procedure may be written in the following manner:

$$z_{i+1} = z_i + \alpha p_i$$

where z_i is the value of vector z at the ith point; α is a scalar; p_i is the direction of search. At each iteration, the vector p_i is determined through solving the following quadratic programming problem [8]:

$$\min_{p} (p^T B_i p + g_i^T p)$$
$$J_1 p + \chi(z_i) = 0$$
$$J_2 p + \psi(z_i) \leqslant 0$$
$$J_3 p + W(z_i) = 0$$
$$0 \leqslant z_i + p \leqslant \bar{z}$$

where J_1, J_2, J_3 are the Jacobian matrices of the left-hand sides of relations (122), (123), (125), respectively. B_i is the approximation of the Hessian of Lagrangian function L in problem (121)-(125) at the point i; $g = \text{grad } L$. In this particular case

$$L = f + \sum_{k=1}^{N} (\lambda^{(k)})^T \chi^{(k)} + \sum_{k=1}^{N} (\mu_1^{(k)})^T \psi^{(k)}$$

$$+ \sum_{k=1}^{N} \mu_2^{(k)} \left(\sum_{k=1}^{N} \alpha_{jk} - 1 \right) + \sum_{k=1}^{N} (\mu_3^{(k)})^T (-z^{(k)})$$

$$+ \sum_{k=1}^{N} (\mu_4^{(k)})^T (z^{(k)} - \bar{z}^{(k)})$$

where $\lambda^{(k)}$, $\mu_1^{(k)}$, $\mu_3^{(k)}$, $\mu_4^{(k)}$ are the vectorial Lagrangian multipliers corresponding to the kth block constraints: $\mu_2^{(k)}$ is the scalar Lagrangian multiplier.

At each iteration, matrix B_i is calculated my means of a modified BFGS formula:

$$B_{i+1} = B_i + \frac{y_i V_i^T}{V_i^T s_i} - \frac{B_i s_i s_i^T B_i}{s_i^T B_i s_i} \qquad (149)$$

where

$$V_i = \theta y_i + (1 - \theta) B_i s_i$$

$$\theta = \begin{cases} 1 & \text{if } s_i^T y_i > 0.2 s_i^T B_i s_i \\ \dfrac{0.8 s_i^T B_i s_i}{s_i^T B_i s_i - s_i^T y_i} & \text{if } s_i^T y_i \leqslant 0.25 s_i^T B_i s_i \end{cases}$$

$$y_i = g_{i+1} - g_i; \quad s_i = z_{i+1} - z_i$$
$$g = \operatorname{grad} L$$

As is known, transformation (149) provides for positive definiteness of matrix B_{i+1}, if matrix B_i is positively definite.

In PFS synthesis problems, a large number of structural parameters α_{ij} are commonly equal to zero. Therefore, the Hessian matrix of Lagrangian function L is commonly a sparse matrix. However, the Hessian matrix, when subjected to transformation, usually does not retain its original structure, and the resultant matrix B_i is a full matrix. In this connection, we consider an alternative procedure for constructing matrix B_i.

In a manner much similar to that previously applied to deriving relation (135), one easily obtains

$$L = \sum_{k=1}^{N} L^{(k)}(x^{(k)}, u^{(k)}, \alpha_{jk})$$

where

$$L^{(k)} = F^{(k)} + (\lambda^{(k)})^T x^{(k)} - \sum_{j=1}^{N} (\lambda^{(j)})^T \alpha_{jk} y^{(k)} \quad (x^{(k)}, u^{(k)})$$
$$+ (\mu_1^{(k)})^T \psi^{(k)} + \mu_2^{(k)} \left(\sum_{j=1}^{N} \alpha_{ij} - 1 \right)$$
$$- (\mu_3^{(k)})^T z^{(k)} + (\mu_4^{(k)})^T (z^{(k)} - \bar{z}^{(k)})$$

From the above relation an equality

$$G = \sum_{k=1}^{N} G^{(k)} \tag{150}$$

ensues, where G and $G^{(k)}$ are the Hessians of functions L and $L^{(k)}$, respectively.

We denote by $B_i^{(k)}$ the Hessian approximation of function $L^{(k)}$ at the ith iteration. A reasonable condition is to require that the Hessian approximation of Lagrangian function L and the Hessian approximations of functions $L^{(k)}$ be related as

$$B_i = \sum_{k=1}^{N} B_i^{(k)} \tag{151}$$

similar to (150). Matrix B_i is determined in the following manner. Initially, matrices $B_i^{(k)}$ are determined at the ith iteration; then matrix B is determined using formula (151).

Let us take a closer look an the mode the matrices $B_i^{(k)}$ are determined. First, we consider matrix $G^{(k)}$. This matrix has zero elements in its rows and columns corresponding to the components $z^{(j)}$ ($j = \overline{1, N}$; $j \neq k$), that is, the number of zeros in it is quite large. The dimension of this matrix is $n \times n$. We denote by $\overline{G}^{(k)}$ an $n^{(k)} \times n^{(k)}$ matrix which is derived from matrix $G^{(k)}$ by deleting the rows and columns corresponding to variables $z^{(j)}$ ($j = \overline{1, N}$; $j \neq k$), that is, by deleting all its zero rows and columns. If matrix $G^{(k)}$ is known, then matrix $\overline{G}^{(k)}$ is easily found, and vice versa. Obviously, matrix $G^{(k)}$ is a full matrix. We impose a constraint requiring the matrix $B^{(k)}$ to have the same structure as that of matrix $G^{(k)}$. We denote by $\overline{B}^{(k)}$ an $n^{(k)} \times n^{(k)}$ matrix which approximates matrix $\overline{G}^{(k)}$. Matrix $B_i^{(k)}$ is determined by the modified BFGS formula (see (149)). If matrix $\overline{B}_i^{(k)}$ is known, determination of matrix $B_i^{(k)}$ becomes an easy procedure.

Certainly, other known methods of matrix transformation can be used. However, in their use may cause difficulties with providing for the positive definiteness of matrix $B_i^{(k)}$ and, consequently, of matrix B_i.

Let us briefly comment on the advantage of such a mode for constructing matrix B. For one thing, matrix B has a structure identical to that of matrix G, that is, the zero elements of matrix B correspond to those of matrix G. Therefore, in solving quadratic programming problems, we can use methods which take into account the sparsity of matrix B. This can substantially save time in solving quadratic programming problems; in addition, the overall number of iterations can be reduced. Not pretending to a rigorous argumentation, we wish to put forward reasons in favour of this supposition. Commonly, in quasi-Newton methods, a good approximation by matrix B to the Hessian is obtained only after n steps, n being the dimension of a problem. Now, if f is a quadratic function, then at the nth step we obtain an exact value of Hessian. However, at $i < n$ (i is the step number), matrix B provides a poor approximation to Hessian. In our particular case, the dimension n may happen to be very large. If we use the conventional SQP, then a good approximation of Hessian by matrix B is to be expected only after the accomplishment of n steps. However, using the newly suggested procedure, we merely need $n^{(k)}$ ($n^{(k)} \ll n$) steps to arrive at a good approximation to the Hessians of functions $L^{(k)}$ and, consequently, to the complete Hessian of function L. Since the required number of iteration steps is significantly smaller, the iteration procedure is much less time-consuming.

The above SQP approach may also be used, as an independent technique, for optimization of large systems.

6.5 Multiobjective Optimization

Quite often, the use of only one optimization test fails to characterize the operation of a chemical process. In such a case, one is encountered with the necessity of solving a multiobjective optimization problem,

$$\min_{x \in Q} [f_1(x), f_2(x), \ldots, f_m(x)], \quad m \geq 2$$

where x is an n-dimensional vector; $f_1(x), \ldots, f_m(x)$ is a set of optimization tests; and $Q \in E^n$ is a certain domain within the space X. The solution of a multiobjective problem is commonly carried out in two stages: (i) construction of a noninferior set (NIS) of solutions (or the Pareto set); (ii) choice of a preferable solution out of the NIS elements by making use of a certain global criterion. Here we are concerned only with the NIS construction. Let f_i^* be a solution of the following problem,

$$f_i^* = \min_{x \in Q(x)} f_i(x)$$

Let

$$x^{i*} = \arg \min_{x \in Q} f_i(x)$$

We consider an m-dimensional criterion space. The values of variables $f_1, \ldots, f_m$ are plotted on the coordinate axes of this space. We now consider in what a manner the set Q can be mapped onto the criterion space. We denote the image of the set Q by Λ, and the boundary surface by L. A point $C(f_1^*, f_2^*, \ldots, f_m^*)$ will be called the ideal point. It is clear that this point, generally speaking, is inaccessible at any x. It can be rigorously proved that the NIS coincides with a portion of the Λ region boundary. As shown in Fig. 6.18, NIS lies between two points A and B. At the point $A(f_1(x^{2*}), f_2^*)$, a straight line passing through the point C parallel to the f_1-axis is tangent to the Λ

6 Optimum Design of Chemical Plants   549

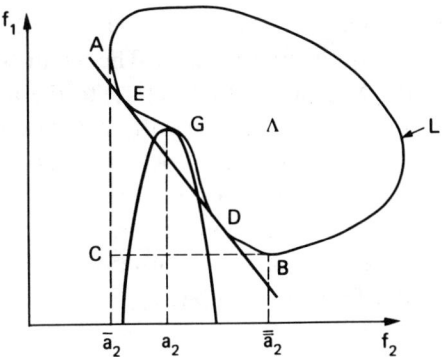

Fig. 6.18 Two-criterion optimization space

region boundary; at the point $B(f_1^*, f_2^*(x^{1*}))$, a straight line passing through the point C parallel to the f_2 axis is tangent to the Λ region boundary. In constructing the NIS, of frequent use is the so-called parametric method [50]. The basic idea of the method is as follows. Let us consider a problem

$$\min F_1(f_1, \ldots, f_m, \omega_1, \ldots, \omega_m) \tag{152}$$

where

$$F_1 = \sum_{i=1}^{m} \omega_i f_i, \quad \omega_i \in \Omega \tag{153}$$

$$\Omega = \left\{ \omega: \omega_i \geq 0, \sum_{i=1}^{m} \omega_1 = 1 \right\} \tag{154}$$

Let $\bar{x} = \arg\min F_1 \mid x \in Q$. The point $(\bar{f}_1, \ldots, \bar{f}_m)$, where $\bar{f}_i = f_i(\bar{x})$, is known to belong to the NIS [50]. For this reason, the NIS points are obtained by solving problem (152) for various ω_i that satisfy condition (153). However, this approach has an essential limitation in that serves to obtain all of the NIS points only if the region Λ is convex. The region Λ being nonconvex, this method enables deriving only part of NIS. Suppose, problem (152) has been solved for a certain set ω. Then, in the criterion space, the hyperplane

$$\sum_{\omega=1}^{m} \omega_i f_i = F_1^*$$

where $F_1^* = \min_{x} F_1 \mid x \in Q$ is a supporting plane with respect to the region Λ. Shown in Fig. 6.18 is a supporting line ED which identifies the NIS points that cannot be obtained by the parametric method (that is, all the points in the curve L that are confined within the interval ED). The parametric method is effective only in obtaining the NIS points that belong to the arcs AE and BD. In view of that, another approach was proposed (see, for example, [50]). This approach consists in that an optimization problem is defined in which one of the functions $f_1, \ldots, f_m$ is taken for an optimization test, the other functions serving as a basis for constructing constraints. For definiteness, we assume that the function f_1 has been taken as optimization test. Then, a NIS point is obtained by solving the following problem:

$$\min_{x \in Q} f_1(x)$$

$$f_i(x) \leq a_i, \quad i = \overline{2, m} \tag{155}$$

where a_i are prefixed constants. By making the parameters a_i ($i = \overline{2, m}$) vary within certain limits ($\bar{a}_i \leq a_i \leq \bar{\bar{a}}_i$), one can obtain various NIS points. Let us consider the definition of limits $\bar{a}_i$ and $\bar{\bar{a}}_i$. First, let us turn our attention to the case of $m = 2$ (see Fig. 6.18). As is seen in Fig. 6.18, if $\bar{a}_2 < f_2^*$ is chosen, problem (154), (155) has no solution. If, however, $\bar{\bar{a}}_2 > f_2(x^{1*})$ is taken, the solution of problem (154), (155) always yields the point B. For this reason, the values of $\bar{a}_2$ and $\bar{\bar{a}}_2$ must be

$$\bar{a}_2 = f_2^*; \quad \bar{\bar{a}}_2 = f_2(x^{1*})$$

Let now be $m = 3$. It is a natural way to construct the NIS in the form of a section for

$$f_3 = a_3^{(l)}, \quad l = 1, 2, \ldots \tag{156}$$

Since the NIS contains no points for which $f_3 < f_3^*$, it appears reasonable to take the values of $a_3^{(l)}$ in the form

$$a_3^{(l)} = f_3^* + l\delta, \quad l = 1, 2, \ldots$$

where δ is the chosen iteration step. In the criterion space within the plane (156), the NIS section is represented by a curve $f_1 = F(f_2)$ whose points are defined by the solution of the problem

$$\min_{x \in Q} f_1$$

$$f_3(x) = a_3^{(l)}$$

$$f_2(x) \leq a_2$$

6 Optimum Design of Chemical Plants

for a number of values of a_2

$$\bar{a}_2^{(l)} \leqslant a_2 \leqslant \bar{\bar{a}}_2^{(l)}$$

Following the same line of reasoning for the case $m = 2$, it can easily be shown that the values of $\bar{a}_2^{(l)}$, $\bar{\bar{a}}_2^{(l)}$ must be defined as follows:

$$\bar{a}_2^{(l)} = \min_{x \in \Omega} f_2(x)$$

$$f_3(x) = a_3^{(l)}$$

$$\bar{\bar{a}}_2^{(l)} = f_2(x^{1,3*})$$

$$x^{1,3*} = \arg \min f_1(x) \,|\, x \in Q, \, f_3(x) = a_3^{(l)}$$

It is clear that in this particular case the construction of NIS is a more laborious procedure, since in problem (154), (155), in comparison to problem (152)-(153), additional $(m - 1)$ nonlinear constraints (155) should be taken into account. Now, if the region Q is defined by linear constraints, for its solution the use of linearly constrained optimization methods is required. As to problem (154), (155), its solution, strictly speaking, requires the use of more elaborate methods of nonlinearly constrained optimization.

We wish to show, however, that the derivation of a NIS point in the case of a nonconvex region Λ can be carried out by a procedure not much more complex than the parametric method. Consider the case $m = 2$. Let the penalty function be

$$F_2 = f_1 + K(f_2 - a_2)^2 \tag{157}$$

where K is a constant penalty coefficient. Let

$$F_2^* = \min_{x \in Q} F_2(x) \tag{158}$$

In the criterion space, the curve

$$f_1 + K(f_2 - a_2)^2 = F_2^* \tag{159}$$

is a parabola with the axis $f_2 = a_2$. This parabola is tangent to the curve L at a certain point G [43]. Therefore, with a_2 chosen within a range of $\bar{a}_2 \leqslant a_2 \leqslant \bar{\bar{a}}_2$, all of the NIS points can be derived. To be noted, as distinct from the former penalty method, in this case the penalty coefficient K remains always constant. However, the greater is K, the more "narrow" is parabola (159). As is seen in Fig. 6.18, parabola (159) must be sufficiently "narrow" (that is, coefficient K must be sufficiently large) so that a sufficiently large number of points could be obtained in that part of the NIS where the region

Λ is nonconvex. As is seen in Fig. 6.18, the values of $\bar{a}_2$ and $\bar{\bar{a}}_2$ can be selected in the following manner:

$$\bar{a}_2 < f_2^*, \quad \bar{\bar{a}}_2 = f_2(x^{1*})$$

The choice of the variation interval $(\bar{a}_i, \bar{\bar{a}}_i)$ for the case $m \geqslant 3$ aapears to be somewhat more complicated. In this instance, another route may be suggested. We construct a penalty function

$$F_2 = f_1 + K \sum_{i=2}^{m} (f_i - a_i)^2$$

By having solved problem (158), we obtain a point $(\bar{f}_1, \ldots, \bar{f}_m)$ which may belong to the surface L but, on the other hand, may not belong to the NIS. Therefore, we must test this point, by making use of the definition of NIS, for its belonging, or not, to the NIS. According to the test, the given point is either included in, or excluded from, the NIS.

References

1. C. G. Broyden, *Mathematics of Computation*, **19**: 577-593 (1965).
2. A. W. Westerberg and Benjamin D. R., *Comput. Chem. Engng*, **9**, No. 5: 517 (1985).
3. P. E. Gill, W. Murray, and M. H. Wright, *Practical Optimization*, Academic Press, London, 1981.
4. J. E. Dennis and K. B. Schnabel, *Numerical Methods for Unconstrained and Nonlinear Equations*, Prentice-Hall, Englewood, Cliffs, New Jersey, 1983.
5. G. Strang, *Linear Algebra and Its Applications*, Academic Press, New York, 1976.
6. M. J. D. Powell, *A Hybrid Method for Nonlinear Equations*, in: Numerical Methods (Ed. P. Rabinowitz), Gordon and Breach, London, 1970.
7. H. W. Ray and J. Szekely, *Process Optimization*, John Wiley and Sons, London, 1973.
8. K. Schnitkowski, *Annals of Operational Research*, **5**: 485 (1985/6).
9. M. J. D. Powell, *A Fast Algorithm for Nonlinearly Constrained Optimization Calculations*, Lecture Notes in Mathematics, 630, Springer Verlag, p. 144, 1978.
10. L. T. Biegler and R. H. Hughes, *Comput. Chem. Engng*, **9**, No. 4; 379-394 (1985).
11. L. T. Biegler, *Comput. Chem. Engng*, **9**, No. 3: 245-265 (1985).
12. Y.-D. Lang and L. T. Biegler, *Comput. Chem. Engng*, **11**, No. 2: 143-158 (1987).
13. G. M. Ostrovsky, M. G. Ostrovsky, and T. A. Berezhinsky, *Comput. Chem. Engng*, **12**, No. 4: 289-296 (1988).
14. G. M. Ostrovsky et al., *Byulleten' po Khimicheskoi Promyshlennosti*, No. 2: 62-64 (1986) (in Russian).
15. D. A. Pierre and M. J. Lowe, *Mathematical Programming via Augmented Lagrangians*, Addison-Wesley Publishing Company, London, 1975.
16. A. C. Heurn, *REDUCE-2 User's Manual*, University of Utah, UCP-19, 1973.
17. Yu. M. Volin and G. M. Ostrovsky, *Automated Remote Control*, No. 12: 29-36 (1966).
18. Yu. M. Volin and G. M. Ostrovsky, *Comput. Chem. Engng*, **5**: 21-30, 31-40 (1981).
19. G. M. Ostrovsky, Yu. M. Volin, and W. W. Borisow, *Wiss. Z. Techn. Hochsch. Chem. Lenna-Merseburg*, **13**, No. 4: 382-384 (1971).

20. Yu. M. Volin and G. M. Ostrovsky, *Comp. and Maths with Appl.*, **11**, No. 11: 1099-1144 (1985).
21. Yu. M. Volin, A. R. Belyaeva and G. M. Ostrovsky, *Programmirovanie*, No. 2: 77-87 (1989).
22. S. P. Pevzner, Yu. M. Volin, and G. M. Ostrovsky, *Program System ROSS-1980 for Simulating Chemical Engineering Processes, Informatsionn. Byulleten' po Khimicheskoi Promyshlennosti*, No. 5: 21-23 (1982) (in Russian).
23. T. Takamatsu, I. Hashimoto, and S. Shioya, *J. Chem. Eng. Japan*, **6**: 453 (1973).
24. I. E. Grossman and R. W. H. Sargent, *AIChE Journal*, **24**, No. 6: 1021-1028 (1978).
25. K. P. Halemane and I. E. Grossmann, *AIChE Journal*, **29**, No. 3: 425-433 (1983).
26. A. Palazoglu and Y. Arkun, *Comput. Chem. Engng*, **11**, No. 3: 205-216 (1987).
27. N. Nishida, A. Ichikawa, and E. Tazaki, *Ind. Eng. Chem. Process Des. Develop.*, **13**, No. 3: 209-214 (1974).
28. B. M. Kwak and E. J. Haug, *J. Opt. Theory Appl.*, **19**: 527 (1976).
29. I. E. Grossmann and C. A. Floudas, *Comput. Chem. Engng*, **11**, No. 6: 675-693 (1987).
30. C. H. Papadimitriou, *Combinatorial Optimization: Algorithms and Complexity*, Prentice-Hall, New Jersey, 1982.
31. M. A. Gomez and J. D. Seader, *AIChE Journal*, **22**: 970 (1976).
32. A. W. Westerberg and G. Stephanopoulos, *Chem. Eng. Sci.*, **10**: 963 (1975).
33. L. Pibouleau and S. Domenech, *Comput. Chem. Engng*, **10**, No. 5: 479 (1986).
34. G. M. Ostrovsky, M. G. Ostrovsky, and G. V. Mikhailov, *Comput. Chem. Engng*, **14**, No. 1: 111-117 (1990).
35. D. F. Rudd, G. J. Powell, and J. J. Siirola, *Process Synthesis*, Prentice-Hall, Englewood Cliffs, New York, 1973.
36. B. Linnhoff, in: *Foundations of Computer-Aided Chemical Process Design* (Eds R. S. H. Mah and W. D. Seader), Engineering Foundation, New York, Vol. II: pp. 537-572 (1981).
37. A. W. Westerberg, G. Stephanopoulos, and N. Nishida, *AIChE Journal*, **27**, No. 3: 321-351 (1981).
38. T. A. Umeda, A. Hirai, and A. Ichikawa, *Chem. Eng. Sci.*, **27**: 795 (1972).
39. M. A. Duran and I. E. Grossmann, *AIChE Journal*, **32**, No. 4: 592 (1986).
40. G. R. Kocis and I. E. Grossmann, *Ind. Eng. Chem. Res.*, **27**: 1407-1421 (1988).
41. K. Osakada and L. T. Fan, *Can. J. Chem. Eng.*, **1**: 94-101 (1973).
42. G. M. Ostrovsky and A. L. Shevchenko, *Chem. Eng. Sci.*, **34**: 1243-1246 (1979).
43. G. M. Ostrovsky and T. A. Berezhinsky, *Optimizatsiya khimiko-tekhnologicheskikh protsessov: teoriya i praktika* (Optimization of Chemical Engineering Processes: Theory and Practice), Khimiya, Moscow, 1984 (in Russian).
44. I. E. Grossmann, *Comput. Chem. Engng*, 9, No. 5: 463 (1985).
45. D. R. Hoerner and O. K. Crosser, *Ind. Eng. Chem. Process Design and Development*, **20**, No. 2: 210-218 (1981).
46. G. R. Kocis and I. E. Grossmann, *Comput. Chem. Engng*, **13**, No. 7: 797-819 (1989).
47. M. S. Bazaraa and C. M. Shetty, *Nonlinear Programming, Theory and Algorithms*, John Wiley and Sons, New York, 1979.
48. L. S. Lasdon, *Optimization Theory for Large Systems*, McMillan Company, London, 1970.
49. S. Vasantharayan, L. T. Biegler, *Comput. Chem. Eng.*, **12**, No. 11: 1089-1092 (1988).
50. A. Sophos, E. Rotstein, and G. Stephanopoulos, *Chem. Eng. Sci.*, **35**, No. 12-D: 2415-2426 (1980).
51. G. M. Ostrovsky and T. A. Berezhinsky, *Computers and Chem. Eng.*, **15** (1991) (in press).